Polymer Science and Technology

nova
science publishers
www.novapublishers.com

Polymer Science and Technology

Far Infrared and Terahertz Spectroscopy of Polymers
Valery A. Ryzhov (Author)
2022. ISBN: 978-1-68507-662-7 (Softcover)
2022. ISBN: 978-1-68507-830-0 (eBook)

A Study of Polymer Dynamics by Solid-State NMR
John Nobleman (Author)
2021. ISBN: 978-1-53619-715-0 (Hardcover)
2021. ISBN: 978-1-53619-803-4 (eBook)

What to Know about Lignin
María González Alriols, PhD (Editor)
Jalel Labidi, PhD (Editor)
M. Özgür Seydibeyoğlu, PhD (Editor)
2021. ISBN: 978-1-53619-152-3 (Hardcover)
2021. ISBN: 978-1-53619-222-3 (eBook)

Polydimethylsiloxane: Structure and Applications
Philip N. Carlsen (Editor)
2020. ISBN: 978-1-53617-590-5 (Hardcover)
2020. ISBN: 978-1-53617-611-7 (eBook)

An Introduction to Molecular Dynamics Simulation of Polymer Composites
Sumit Sharma
2020. ISBN: 978-1-53617-408-3 (Hardcover)
2020. ISBN: 978-1-53617-409-0 (eBook)

More information about this series can be found at https://novapublishers.com/product-category/series/polymer-science-and-technology-series/

Mohammad Sirousazar
Editor

Nanocomposite Hydrogels and Their Emerging Applications

nova
science publishers
www.novapublishers.com

DOI: https://doi.org/10.52305/TQLX5032.

NOTICE TO THE READER

The Publisher has taken reasonable care in the preparation of this book but makes no expressed or implied warranty of any kind and assumes no responsibility for any errors or omissions. No liability is assumed for incidental or consequential damages in connection with or arising out of information contained in this book. The Publisher shall not be liable for any special, consequential, or exemplary damages resulting, in whole or in part, from the readers' use of, or reliance upon, this material. Any parts of this book based on government reports are so indicated and copyright is claimed for those parts to the extent applicable to compilations of such works.

Independent verification should be sought for any data, advice or recommendations contained in this book. In addition, no responsibility is assumed by the Publisher for any injury and/or damage to persons or property arising from any methods, products, instructions, ideas or otherwise contained in this publication.

This publication is designed to provide accurate and authoritative information with regards to the subject matter covered herein. It is sold with the clear understanding that the Publisher is not engaged in rendering legal or any other professional services. If legal or any other expert assistance is required, the services of a competent person should be sought. FROM A DECLARATION OF PARTICIPANTS JOINTLY ADOPTED BY A COMMITTEE OF THE AMERICAN BAR ASSOCIATION AND A COMMITTEE OF PUBLISHERS.

Library of Congress Cataloging-in-Publication Data

Names: Sirousazar, Mohammad, editor.
Title: Nanocomposite hydrogels and their emerging applications / Mohammad Sirousazar (editor)
Description: New York : Nova Science Publishers, [2023] | Series: Polymer science and technology | Includes bibliographical references and index. |
Identifiers: LCCN 2023021938 (print) | LCCN 2023021939 (ebook) | ISBN 9798886976755 (hardcover) | ISBN 9798886978957 (adobe pdf)
Subjects: LCSH: Colloids--Industrial applications. | Colloids in medicine. | Nanocomposites (Materials)
Classification: LCC QD549 .N256 2023 (print) | LCC QD549 (ebook) | DDC 660/.2945--dc23/eng20230708
LC record available at https://lccn.loc.gov/2023021938
LC ebook record available at https://lccn.loc.gov/2023021939

Published by Nova Science Publishers, Inc. † New York

Contents

Preface

With the beginning of the era of modern technology in the middle of the 20th century, the design, manufacture, and development of advanced engineered products expanded and as a result, a drastic mutation occurred in various industrial fields. It was not possible to produce such products using ordinary materials, but this required the design and manufacture of advanced multi-functional materials. Six decades ago, at the same time as the production of advanced materials flourished, a new category of polymeric materials, called "polymeric hydrogels", was produced for the first time and used in the preparation of modern products. Although the first application of polymeric hydrogels was in the field of medicine, their use in the following years was not limited to this field and they were then successfully utilized in numerous practical applications.

Polymeric hydrogels are solid-liquid porous materials that are produced by crosslinking the hydrophilic polymer chains for the creation of three-dimensional crosslinked networks. They are hydrophilic water-swollen polymeric networks, with a huge tendency to absorb significant amounts of water, while maintaining their structure and being insoluble in aqueous solutions. They are also inherently soft and elastic due to their thermodynamic compatibility with aqueous solutions. As biomimetic materials, the main application of hydrogels due to their similarity to the tissues of the human body is in biomedicine. However, the unique characteristics of hydrogels have caused them to be utilized in many other fields, such as separation, molecular filtration, imaging technologies, food and agricultural industries, energy conversion, and biosensing.

In practical applications, high mechanical tenacity is always desired for hydrogels in addition to the softness and high water content. However, in most cases, hydrogels do not meet the mechanical requirements and are seldom used as mechanical devices because of their weak mechanical properties, for instance, low strength, poor stability, and low fracture toughness, especially in the swollen state. These disadvantages are mainly attributed to the constrained molecular motion of the polymer chains due to the large number of randomly arranged crosslinks in the hydrogel network. In order to overcome this limitation, several efforts have been made to reinforce the structure of hydrogels using different aspects of nanotechnology by incorporating nanoparticles or nanostructures to prepare a new class of hydrogels, the so-called "nanocomposite hydrogels", with improved mechanical, physical, and chemical properties.

For the first time, Haraguchi and Takehisa reported the creation of a novel nanocomposite hydrogel with a unique organic-inorganic network structure in 2002, which was fabricated by in situ polymerization of N-isopropyl acrylamide in aqueous suspension of laponite nanoclay. In general, nanocomposite hydrogel is defined as a reinforced hydrophilic three-dimensional network, consisting of at least a crosslinked polymeric matrix and a nanoscale (1-100 nm) reinforcing agent. Nanocomposite hydrogels have unique superior properties and tailored functionalities, such as enhanced mechanical properties, swelling/deswelling rates, stimuli-responsive behavior, biological interactions, thermal and optical properties, and the ability of remote actuation. Various nanoparticles are used in the production of nanocomposite hydrogels, including ceramic nanoparticles (e.g., silica and titanium oxide), metal or metal-oxide nanoparticles (e.g., gold, silver, and iron oxide), carbon-based nanoparticles (e.g., carbon nanotubes and graphene), polymeric nanoparticles (e.g., micelles, dendrimers, and nanogels) and nanoclays (e.g., montmorillonite, laponite, and kaolinite). In recent years, nanocomposite hydrogels, as newly born multi-functional materials in the 21st century, have attracted much scientific attention and they are believed to be a revolutionary type of material. Recently, they have found numerous applications in different areas, such as tissue engineering, drug delivery, wound dressing, food industry, agricultural irrigation systems, adsorption and separation processes, oil and gas industry, and sensor-actuator systems.

The objective of this book is to introduce the fundamentals of nanocomposite hydrogels and review their emerging application in different practical fields. For this purpose, in the couple of first chapters of the book, the basic concepts of nanocomposite hydrogels are introduced, including the preparation methods, characterizations, and properties, along with their self-healing mechanisms The next two chapters deal with the general applications of nanocomposite hydrogels in tissue engineering. The application of nanocomposite hydrogels in cancer therapy is discussed in Chapters 5 and 6. In Chapters 7 and 8, the recent researches regarding the application of nanocomposite hydrogels as novel wound dressing materials for the treatment of wounds and also their applications in drug delivery are reviewed. The emerging applications of nanocomposite hydrogels in orthopedics are introduced in Chapter 9. Chapter 10 deals with the performance of nanocomposite hydrogels as novel packaging materials in the food industry and the recent research in this field are reviewed. Chapters 11-13 focus on the application of nanocomposite hydrogels as novel adsorbents in the treatment of wastewater for the adsorptive removal of contaminants from water by focusing on iron- and polysaccharide-based nanocomposite hydrogels. Finally, a more recent application of nanocomposite hydrogels in the production of novel pipes for molecular irrigating in the agricultural sector is introduced in the last chapter.

Chapter 1

Nanocomposite Hydrogels: Basic Principles

Ali Samadi[1,*]
and Payam Zarrintaj[1]
[1] Polymer Engineering Department, Faculty of Engineering,
Urmia University, Urmia, Iran
[2] Department of Biomedical and Pharmaceutical Sciences,
University of Montana, Missoula, Montana, USA

Abstract

Hydrogels are three-dimensional water-absorbent polymeric networks used in various fields such as food, packaging, pharmaceutical, agricultural, electrical, environmental, and biomedical applications. Those 3-dimensional structures are formed by chemical or physical crosslinking techniques in water. Physical hydrogels are interesting due to their relative convenience and the lack of chemicals in their synthesis, while chemical hydrogels are due to their proper mechanical strengths. The mechanical, physical, chemical, optical, and biological properties can be enhanced by the incorporation of nanomaterials into the network of the hydrogel. Therefore, nanocomposite hydrogels have been considered in recent advanced research. Various NMs and natural or synthetic polymers have been used to acquire novel enhanced hydrogels with tailored applications. This chapter presents the polymer-nanomaterials micro-interactions and their effect on hydrogel macro-properties, the classification of different types of nanocomposite hydrogels, preparation and characterization methods, and some of their essential properties and applications in various fields.

Keywords: nanocomposite, hydrogel, nanomaterials, polymer

1.1. Introduction

Hydrogels are 3D crosslinked hydrophilic polymeric networks that absorb and retain a substantial amount of water or aqueous solutions without dissolving (Varaprasad et al., 2017). Hydrogels are soft natural or synthetic polymers with simultaneous solid (defined shapes) and liquid (containing free water) properties. Hydrogels have various advantages such as

---

[*] Corresponding Author's Email: al.samadi@urmia.ac.ir.

In: Nanocomposite Hydrogels and their Emerging Applications
Editor: Mohammad Sirousazar
ISBN: 979-8-88697-675-5

biocompatibility, biodegradability, functionalizability, and processability which case applications such as tissue engineering, drug delivery, contact lenses, super-absorbents, and biomedical applications (Chacko et al., 2012; Haider et al., 2004; Huglin, 2019; Jeong and Gutowska, 2002; Langer and Tirrell, 2004; Merino et al., 2015; Stuart et al., 2010). Engineering swelling/deswelling properties of hydrogels in the variation of pH, light, temperature, or specific biomarkers have provided the biomedical applications of hydrogels as stimuli-sensitive materials. However, the mechanical properties of hydrogels such as mechanical strength and fracture toughness are weak. Therefore, a lot of strategies have been done to improve the mechanical properties of hydrogel networks. These strategies include synthesizing topological hydrogels (Slide-ring gel) by figure-of-eight crosslinks (freely movable crosslinks) (Bin Imran et al., 2014; Ito, 2007; Karino et al., 2006; Okumura and Ito, 2001), synthesizing double-network hydrogels (Gong et al., 2003), and incorporation of nanoparticles (NPs) to polymer, called nanocomposite (NC) hydrogels (Haraguchi and Takehisa, 2002).

The hydrodynamic effect and dissipative mechanisms between NPs and polymer chains enhance the mechanical properties of NC hydrogels (Creton, 2017). Also, various NPs including carbon allotropes, organic, inorganic, metallic, or polymeric nanomaterials are incorporated within hydrogels to enhance desired properties of the hydrogel network, such as optical, mechanical, magnetic, electrical, and biomedical properties. The polymer-NP interactions can physically or covalently provide adjustable functionalities to the NC hydrogels. For instance, the incorporation of magnetic NPs makes the magnetic hydrogel (Zhu et al., 2012). The incorporation of drugs containing NPs in stimuli-sensitive hydrogels is applicable in drug delivery (Gaharwar et al., 2014). Hydrogel biosensors can be made by incorporating special NPs in hydrogels, which makes the swelling properties dependent on the stimulus (Jung et al., 2016; Yetisen et al., 2014). The addition of conductive NPs within polymeric hydrogel provides electrically conductive hydrogels for engineering muscle tissues or cardiac patches. The number of publications in the NC hydrogels field has increased in recent years, which demonstrates the rising interest in NC hydrogels and their applications (Figure 1.1). This chapter briefly presents the basic principles and classification of NC hydrogels, their preparation methods, and some of their essential properties and applications in different fields.

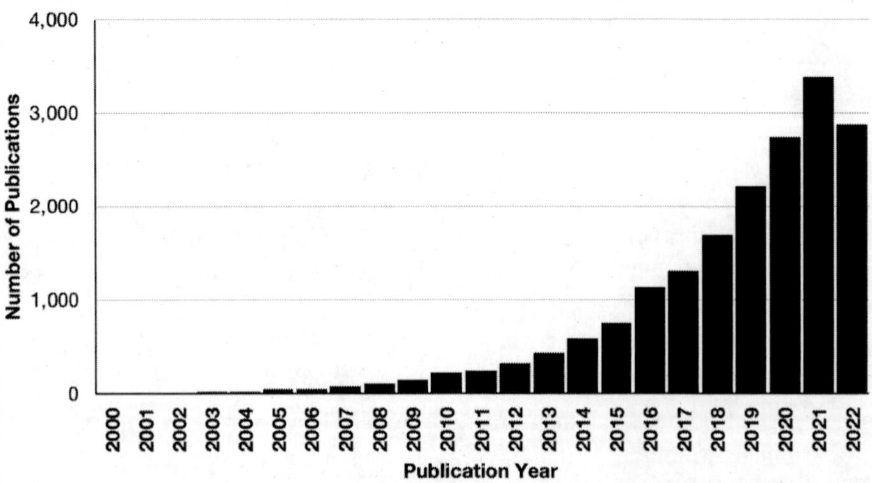

Figure 1.1. The number of publications related to NC hydrogels (until September 2022).

1.2. Molecular and Thermodynamic Concepts of Hydrogels

Hydrogels could be natural or synthetic. Natural hydrogels are natural macromolecules that have a 3D network structure, but synthetic hydrogels are composed of multiples of simpler units called monomers. Various monomers are used for the synthesis of hydrogels, the most important of them are listed in Table 1.1.

Table 1.1. The most important monomers used in the synthesis of synthetic hydrogels (Sirousazar et al., 2014)

Name of Monomer	Structure of Monomer
Acrylic acid	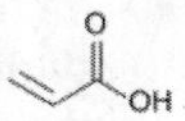
Vinyl acetate	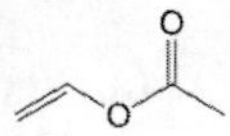
Hydroxyethyl methacrylate	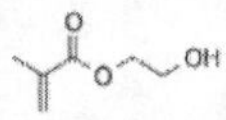
Methacrylic acid	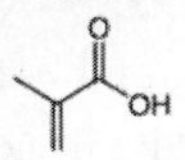
Ethylene glycol	
Ethylene glycol methacrylate	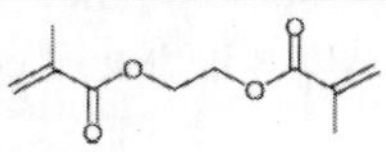
Ethylene glycol diacrylate	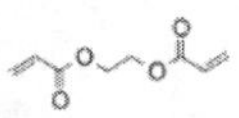
Ethylene glycol dimethacrylate	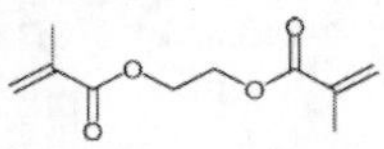
Methoxyethyl methacrylate	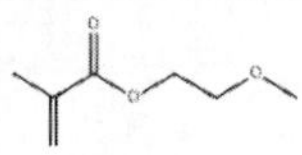
Glyceryl methacrylate	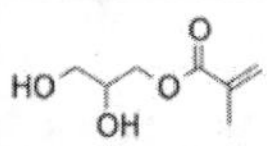
Methyl methacrylate	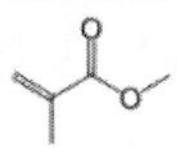
Methoxyethoxyethyl methacrylate	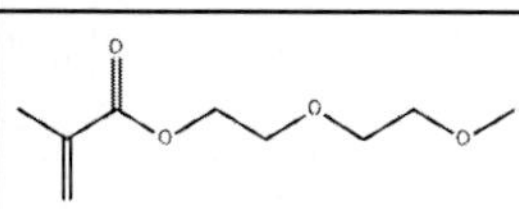
Hydroxyethoxyethyl methacrylate	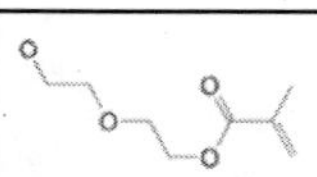
N-(2-hydroxypropyl) methacrylamide	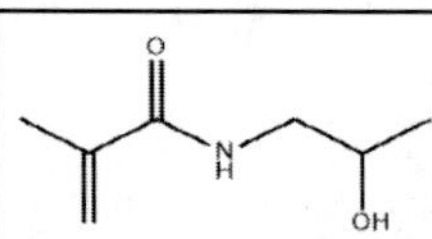

Table 1.1. (Continued)

Name of Monomer	Structure of Monomer
N-vinyl-2-pyrrolidone	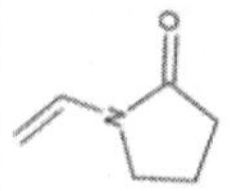
Dimethylacrylamide	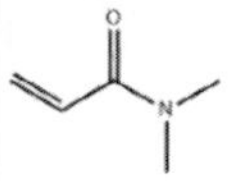

There are various methods for crosslinking hydrogels including physical and chemical methods. Physical methods include ionic interactions (Bajpai et al., 2008; Magnin et al., 2004; Zhao et al., 2009), hydrogen bonds (Eagland et al., 1994; Gayet and Fortier, 1996; Pongjanyakul and Puttipipatkhachorn, 2007), thermal (cooling/heating) (Bajpai et al., 2008), crystallization (Giannouli and Morris, 2003; Gupta et al., 2011; Hassan and Peppas, 2000), and maturation (Aoki et al., 2007). Chemical crosslinking methods include crosslinking of the pendant hydroxyl groups of polymer chains by crosslinking agents (Chang and Zhang, 2011), grafting (Huglin and Zakaria, 1986; Jabbari and Nozari, 2000; Makuuchi, 2010), and crosslinking by enzymes (is used in polyethylene glycol) (Sperinde and Griffith, 2000; Wang et al., 2012).

According to the Flory-Rehner (FR) theory (Flory and Rehner, 1943a, 1943b; Hill, 1960), the free energy of hydrogels (ΔF_{gel}) is described as Equation (1-1):

$$\Delta F_{gel} = \Delta F_{el} + \Delta F_{mix} + \Delta F_{ion} \quad (1\text{-}1)$$

where ΔF_{el} is the free energy of elastic deformation, ΔF_{mix} is the free energy of the polymer-solvent mixing, and ΔF_{ion} is the electrostatic effects of ion-solvent. ΔF_{el} *is* calculated by Equation (1-2):

$$\Delta F_{el} = \frac{v k_B T}{2}(\lambda_1^2 + \lambda_2^2 + \lambda_3^2 - 3) \quad (1\text{-}2)$$

where v is the density of elastic strands, k_B is the Boltzmann constant, T is the absolute temperature, and λ_1^2, λ_2^2, and λ_3^2 are three deformation ratios. Then, if λ is the stretch in uniaxial tension direction, for incompressible materials $\lambda_1\lambda_2\lambda_3 = 1$, therefor $\lambda_1 = \lambda_2 = \lambda_3^{-1/2}$. Then, the normal engineering stress for small and moderate strains in unentangled crosslinked hydrogels can be calculated from Equation (1-3):

$$\sigma_N = v k_B T(\lambda - \lambda^{-2}) \quad (1\text{-}3)$$

Young's modulus (E) is defined as the derivative of the stress to strain:

$$E = \frac{d\sigma_N}{d\lambda} = 3 v k_B T \quad (1\text{-}4)$$

The ΔF_{mix} is given by the Flory-Huggins theory as below (Equation (1-5)):

$$\Delta F_{mix} = N_s K_B T[\ln(1-\phi) + \chi\phi] \quad (1\text{-}5)$$

Here, N_s is the number of solvent molecules, χ is the interaction parameter of Flory-Huggins theory, and ϕ is the volume fraction of the gel (Creton, 2017; Quesada-Pérez et al., 2011). The driving force of swelling/deswelling is the gradient of osmotic pressure (or ion concentration in ionic hydrogels) between the polymer network and solvent resulting for the non-ionic hydrogels according to Equation (1-6) (Gunasekaran et al., 2006):

$$P = \pi - p \quad (1\text{-}6)$$

Here, P is the swelling/deswelling pressure of hydrogels, π is the osmotic pressure, and p is the elastic pressure of the polymer network. At the equilibrium, there is no swelling/deswelling pressure, then $\pi = p$. According to the FR theory (Flory and Rehner, 1943a, 1943b; Hill, 1960), in thermodynamic equilibrium, the inside and outside chemical potential of the solvent is equal (Equation (1-7)):

$$\mu_s^{in} = \mu_s^{out} \quad (1\text{-}7)$$

In the term of osmotic pressure (π) the Equation (1-8) is resulted:

$$\pi = -\frac{\mu_s^{in} - \mu_s^{out}}{v_s} \quad (1\text{-}8)$$

where v_s is the molar volume of the solvent. Calculation of the osmotic pressure from free energy is as Equation (1-9):

$$\pi = -\frac{N_A}{v_s}\frac{\partial \Delta F}{\partial N_s} \quad (1\text{-}9)$$

where N_A is the Avogadro number and N_s is the number of solvent molecules. The same as equation (1-1), the osmotic pressure can also be written as Equation (1-10):

$$\pi = \pi_{el} + \pi_{mix} + \pi_{ion} \quad (1\text{-}10)$$

Obviously, osmotic pressure is a sum of deformation, mixing, and ion effects that can affect the equilibrium state of hydrogels (Peppas et al., 2006). Therefore, controlling parameters like network elasticity, mesh size, temperature, pKa, pKb, charge density, solvent composition, hydrogen bonding, etc. can control the swelling/deswelling of hydrogels (Bhardwaj et al., 2003; Díez-Peña et al., 2002; Peppas and Sahlin, 1989; Ruan et al., 2003; Wu et al., 2004). The mesh size is the space between the crosslinked points of the network, and by increasing the mesh size, the water absorption capacity of hydrogels increases (Omidian et al., 2005). Then, the crosslinking nature and density and the arrangement of solvent within the hydrogel network (degree of swelling) affect the equilibrium state of hydrogel (Kizilay and Okay, 2003). By increasing the charge density, the swelling of the hydrogel at higher pH increases (De et al., 2002; Kiser et al., 2000). Ionization of the pendent groups such as sulfonic or carboxylic acid occurs at pH above the pKa value, and the ionization of amine groups occur at pH below the pKb value in the hydrogels, which increases the swelling capacity of the hydrogel network due to the increasing electrostatic repulsions capacity (Gupta et al., 2002; Jianqi and Lixia, 2002).

1.3. NP-Hydrogel Composites

NP-hydrogel composites are classified into two categories depending on the polymer-NPs ratio and dimensions, including core-shell NP-nanogel complexes (nanogels) and macroscopic NC hydrogels.

1.3.1. Core-Shell NP-Nanogel Complexes

Nanogels are composed of coated hydrogel on the NP surface, giving a core-shell structure. Nanogels have a wide application, especially in drug delivery systems (Figure 1.2). Shell polymer layer can be composed of various hydrogels like protein, fibrin, collagen, chitosan, DNA, alginate, hyaluronic acid (HA), etc. (Ahmed, 2015; Hoare and Kohane, 2008). Coating of the NPs prevents them from aggregation and increases their efficiency and protects coated molecules from elimination or degradation resulting in higher circulation time in drug delivery applications (Kabanov and Vinogradov, 2009; Lin et al., 2017). Coating stimuli-responsive gels on NPs can be used to make drug carriers with release control capability by stimulating its swelling/deswelling properties (Barbucci, 2010). For instance, pH-sensitive nanogels can be used in chemotherapy by absorption and preservation of nuclear drugs in normal cells (the lysosomal lumen pH = 5-6) and released in tumor cells (the lysosomal lumen pH = 4.5-5.5) (Oishi et al., 2007). The application of magnetic-responsive, temperature-responsive, and light-responsive nanogels has been investigated as sensors, robots, and carriers (Chen et al., 2018). Magnetic-responsive nanogels have the potential to be used in diagnostics and imaging applications (Ma et al., 2017).

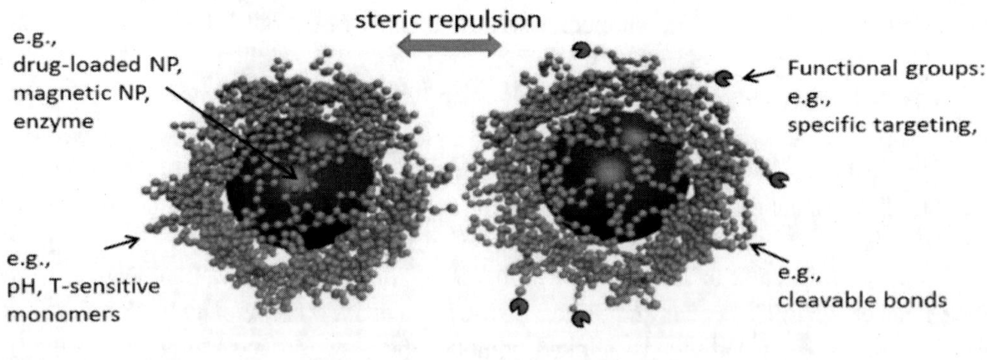

Figure 1.2. Schematic example of core-shell NP-nanogel complexes used for drug delivery application (Dannert et al., 2019).

1.3.2. Macroscopic NC Hydrogels

Macroscopic NC hydrogels have much higher hydrogel content than nanogels. Various NPs are loaded in hydrogels in order to achieve the desired properties. NP-hydrogel interactions could be physical, chemical, or topological. There are three approaches to making NC hydrogels including the addition of NPs (or NP precursors) to preformed hydrogels, the addition of NPs (or NP precursors) to the polymer solution, and the addition of NPs to monomers (Dannert et al., 2019).

1.3.2.1. Addition of NPs (or NP Precursors) to Preformed Hydrogels

The addition of NPs to preformed hydrogels is classified into: (i) incorporation of NPs into hydrogels, and (ii) synthesizing NPs in a polymer network (*in situ* synthesizing). The incorporation of NPs into hydrogels is done by the procedure named the "breathing" process. The breathing process is the penetration of NPs from aqueous suspension into the hydrogel network by the cyclic deswelling/swelling process of polymer hydrogel in poor/good solvents. The breathing process is suggested when the formation of a polymer network could not be done in presence of NPs, due to its aggregation. During breathing cycles, the NPs move from water into the polymer network (diffusion process) and remain inside due to the physical entanglements and polymer-NP hydrogen bonding. The application of the breathing process has been used for the incorporation of Au NPs in polyacrylamide (PAM) hydrogel (Pardo-Yissar et al., 2001). However, it has not been proved the stability of NPs in hydrogel due to the possibility of leaching NPs with time. Another procedure for the incorporation of NPs into hydrogels is the mixing of colloidal NPs with microgels, which has been reported for making Au-poly(N-isopropyl acrylamide) hydrogel (Jones and Lyon, 2003).

In situ synthesizing of NPs in a polymer network has been practically used for making metal NP microgels, which was first reported by Antonietti for synthesizing gold and iron oxide in polystyrene sulfonate. Metal NP microgels use in making biosensors (Kim et al., 2018), optical devices (Hsu et al., 2018), catalysts (Palioura et al., 2007), and biomedical and antimicrobial materials (Strong and West, 2011; Wahid et al., 2017). The final shape and aggregation size of NPs controlled by the density of crosslinking in hydrogels. Higher crosslinking density makes spherical and smaller NPs structures. Incorporating chemical nucleation groups on the hydrogel chains and controlling their concentration provide the controllability of the morphology, size, and stability of NPs in the hydrogel network (Wang et al., 2004).

1.3.2.2. Addition of NPs (or NP Precursors) to Polymer Solution

In this method, the NPs blend with polymer chains and then crosslink the polymer chains, or NP precursors blend with polymer chains and then simultaneously synthesize NPs and network of polymer chains. In the first method, NPs could make the polymer network and instantaneous gelation, as reported in the formation of polyvinyl alcohol (PVA)/chitosan/Fe_3O_4 hydrogel (Zhu et al., 2012). This method is applicable in making economically recyclable hydrogels. For instance, applying a magnetic field separates the magnetic NPs from hydrogels, then the network structure of the hydrogel collapse. The formation of the polymer network could be done due to the polymer-NP electrostatic interactions, for instance, polyhedral oligomeric silsesquioxane (POSS) NPs-carboxymethyl cellulose (CMC) hydrogel. But in these NC hydrogels, changing pH led to the precipitation of NPs (Stornes et al., 2018). Other than the electrostatic interactions, the formation of hydrogen bonds between NPs and polymers can lead to the formation of a polymer network, for instance, poly(sodium acrylate) (PANa) incorporated with SiO_2 NPs hydrogels.

Chemical crosslinking of polymer chains in presence of NPs is a practical method for preparing stronger NC hydrogels. This strategy can be done by chemical functionalization of the NPs. In this method, NPs as crosslinkers lead to mechanically stronger hydrogel without disadvantages like leaching NPs from hydrogel compared to only physically interacted NPs (Pasqui et al., 2011). This strategy has been done in Amin-grafted-TiO_2/CMC hydrogels (Pasqui et al., 2011), pectin/hydroxyethyl methacrylate/TiO_2 (da Silva et al., 2016), gelatin/Ag (García-Astrain et al., 2015), and chitosan/Au (García-Astrain et al., 2016) hydrogels. Besides the advantages of covalent bonding of NP to polymer chains, this restricts the potential of self-healing and extensibility in hydrogels compared with non-covalent bonded NPs. The best method for preparing NC hydrogels is mixing NP precursors with a polymer solution and simultaneous formation of NPs (from NP precursors) and crosslinking the polymer chains. This strategy leads to higher NP dispersion and stability. For instance, it has been used for making Fe_3O_4/PVA/poly(N-isopropylacrylamide)(PNIPAm) hydrogels obtained by the addition of mixed Fe_3O_4 salt with the PVA/PNIPAm into ammonia (Zhou et al., 2012).

1.3.2.3. Addition of NPs to Monomers

The most common procedure to make NC hydrogel is the polymerization of polymer in presence of NPs which was firstly down by Haraguchi (Haraguchi, 2002; Haraguchi and Takehisa, 2002; Haraguchi et al., 2007). He polymerized N, N-dimethyl acrylamide (DMAm) (or NIPAm) in the presence of clay NPs. The polymerization started on clay surfaces and finally, clay played the role of physical crosslinker in the polymer network (Miyazaki et al., 2006). Therefore, the physical gel network was formed due to the proper NP-polymer interaction without further chemical crosslinker agents. This approach can be used in the *in situ* polymerization of acrylamide monomers (and derivatives) and other NC hydrogels containing clay, metal, graphitic carbon nitride (g-CN), titanium oxide, or metal oxide (Tan et al., 2017). Although this method leads to the formation of physical NC hydrogels, bud by adding chemical crosslinkers, the mechanical strength of the polymer gel network significantly increases (Sun et al., 2017).

1.4. Molecular Interactions in Hydrogels

Understanding interactions between NPs and polymers in NC hydrogels is fundamental for choosing the right NC to achieve desired properties in the hydrogel. When the NP acts as a physical crosslinker, several polymer chains are attached to the NP surface so that the polymer chains act as a bridge between NPs. Therefore, the formation and strength of the hydrogel network depend on the strength of polymer-NP attachment depending on their affinity. The adsorption energy of the polymer on the NP surface should be at least equal to the conformational entropy reduction caused by the kinetic restriction of binding to the NP surface. Polymer chains could adapt to the different conformations and form umpteen links, which lead to providing enough interaction strength to swelling without dissolving or gluing gels by NPs (Rose et al., 2014). Understanding this mechanism is needed to know the solubility parameters that have developed by Hansen in 1967 (Wiśniewski et al., 1995). According to Hansen solubility parameters, each molecule has three Hansen parameters, including: (i) dispersion forces (δ_d), (ii) dipolar intermolecular force (δ_p), and (iii) hydrogen bond forces (δ_h). These parameters could be considered as three dimensions for a point known as Hansen space. Therefore, for two neighbor molecules to interact strongly with each other, the radius of the sphere (R_0) and the center of the sphere must be closer to each other. The distance between Hansen parameters (Ra) is calculated by Equation (1-11) as follows:

$$R_a^2 = 4(\delta_{d2} - \delta_{d1})^2 + (\delta_{p2} - \delta_{p1})^2 + (\delta_{h2} - \delta_{h1})^2 \qquad (1\text{-}11)$$

The relative energy difference of the system (RED) is defined as Equation (1-12):

$$RED = \frac{R_a}{R_0} \qquad (1\text{-}12)$$

When RED > 1, the system is not miscible and there is no interaction between the two phases. When RED < 1, the system is miscible and there is a strong interaction between the two phases. When RED = 1, the system is partially miscible. Higher system interaction in NC hydrogels results in higher filler dispersion, stronger filler-polymer linkage, and higher network strength and stability. Hence, the chemistry of polymer chains and the NP surface are important parameters in the interactions between the NPs and the polymer, and higher interaction between polymer chains and NP surfaces results in stronger NP-polymer adhesion. For instance, polymers having amino end groups dendrites (positively charged) have strong interaction with negatively charged NPs (like inorganic nano clays or silica that have silanol groups on the surface) due to hydrogen bond forces (Wang et al., 2010).

Similarly, strong polymer-clay interaction forms in block polymers having a positive charge end block in clay hydrogels (Tamesue et al., 2013). As well as, polymers having either oxygen or carboxyl groups will make hydrogen bonds with hydroxyl groups at mineral NP surfaces, and their strength of interactions depends on the group's chemistry. It should be considered that the interaction between polymer and NPs is in competition with NP-solvent interactions and polymer-solvent interactions. An instance is the interaction between graphene oxide (GO) and PVA. The physical crosslinking effect of GO was proved due to the hydrogen bonding between hydroxyl, carboxyl, epoxy, and groups of the GO surface with the hydroxyl-riches PVA chains. The weak interaction of GO nanosheets with a pure solution led to the domination of the interaction between PVA and GO in the network structure of the hydrogel (Bai et al., 2010; Liu et al., 2005).

In addition, the physical linkage between polymer chains and filler surface may break down in tension conditions but that is replaced with neighbor chain strands and as a result, the failure is retarded (Montarnal et al., 2011). That mechanism supports the capability of self-healing that appears in numerous NC hydrogels. The chemistry of the NP surface and the composition of the hydrogel are important parameters in the adhesion of NPs to the polymer network and can be explored to increase the strength of the interaction and thus adhesion. Another application of NP-polymer interactions in hydrogels is the engineering of the variation of hydrogel network strength by controlling electrostatic interactions via changing pH in ionic hydrogels (Abe et al., 2014; Nejadnik et al., 2014).

1.5. Properties of NC Hydrogels

The incorporation of NPs into hydrogels influences various properties of hydrogels that need to know and care about based on the required properties and intended applications.

1.5.1. Mechanical Properties

The incorporation of NPs in hydrogels enhances the mechanical properties due to the physical linkage between NPs and polymer chains. The mechanical properties of NC hydrogels have a relation to the NP concentration, but this effect is different in physically or chemically crosslinked hydrogels (Carlsson et al., 2010). Generally, the chain length between the chemical crosslinking or physical joints has an important role in the mechanical properties of hydrogels. The more flexible and longer the chain between crosslinks, the more flexibility of the NC hydrogel is. In contrast to chemical crosslinked hydrogels, in physical NC hydrogels, mechanical properties like elongation, tearing, compression, and bending usually improved in presence of NPs (Haraguchi et al., 2011; Kostina et al., 2013; Shao et al., 2017).

The increase of NP content in physical NC hydrogels increases Young's modulus (E), shear storage modulus (G') (Boonmahitthisud et al., 2017; Haraguchi et al., 2002; Su and Okay, 2018), and network stiffness, and decreases the maximum elongation (Carlsson et al., 2010; Pasqui et al., 2011). But, above a certain concentration, the storage modulus decreased rapidly and hydrogel collapses and turns into a quasi-liquid state. In one case, the elastic modulus of silica NP-filled HA hydrogels decreases above the concentration of 2%, which that do not occur in chemically crosslinked HA hydrogels (Vallés-Lluch et al., 2013). The elongation at break in physical NC hydrogels is higher than in the chemical crosslinked hydrogel having an equal crosslinking density (Boonmahitthisud et al., 2017; Du et al., 2015; Haraguchi et al., 2002), which shows the role of NC-polymer interaction in the improvements of the mechanical properties. Some NC hydrogels can elongate until form a knot without breaking down in the elongation (Haraguchi et al., 2002) and 100% shape recovery was observed (Haraguchi et al., 2002; Kostina et al., 2013; Shao et al., 2017).

The interaction between NPs and hydrogel chains also affects the stress relation of NC hydrogels. Stress-relaxation in physical NC hydrogels is under the influence of breaking and reforming physical links between NPs and polymer chains (Zhao et al., 2010), while in chemically crosslinked hydrogels it is mainly under the influence of water migration (Boonmahitthisud et al., 2017).

1.5.2. Optical Properties

Hydrogels are usually transparent at low crosslinking degrees and become opaquer with increasing crosslink density (Mallam et al., 1989) which is related to the increases in system inhomogeneity (Okajima et al., 2000). In most cases, physical NC hydrogels have higher transparency than chemically crosslinked hydrogels due to heterogeneity caused by chemical crosslinks (Shibayama et al., 2005). Generally, any factor that increases structural heterogeneity or diffusive scattering causes a decrease in the transparency of hydrogel

(Boonmahitthisud et al., 2017). Accordingly, the optical behavior of hydrogels is dependent on the concentration and cluster size of NPs in the hydrogel network. Therefore, a bigger aggregation size of NPs increases the heterogeneity and diffusive scattering and as a result decreases the transparency of hydrogel (Boonmahitthisud et al., 2017; Du et al., 2015). The transparency of hydrogels is decreased above the critical concentration of NPs. The critical concentration of NPs depends on the NPs properties, NPs-polymer interactions, and efficiency of the mixing method on the NP dispersion. Also, any factors that affect the NPs aggregations may affect the hydrogel transparency, for instance, variation of temperature, pH, deswelling/swelling, etc. For instance, silicates-in-silica NC hydrogel in presence of Ludox silica becomes opaque with decreasing pH from 7 to 5, but in presence of sodium silicate NPs, on the opposite, it becomes opaque with increasing pH from 5 to 6 (Ahmed et al., 2017).

1.5.3. Self-Healing Properties

Self-healing hydrogels can regenerate their cracks or damages automatically or in the presence of NPs as physical crosslinkers. NC hydrogels consisting of inorganic NPs and long polymer chains may possibly behave as self-healing. Physical links between the NP surface and polymer chain in NC hydrogels can easily break and form new bonds with another NP surface, this ability is necessary for the self-healing phenomenon. The self-healing phenomenon occurs due to the diffusion of polymer chains from one side of the cut surface into the polymer network of another surface and makes bonds with their NPs contents. The temperature affects the rate of self-healing of hydrogels (Haraguchi et al., 2011; Jiang et al., 2017; Shao et al., 2017; Wang et al., 2014; Wang et al., 2017; Wu et al., 2018; Yang et al., 2016; Zhong et al., 2015). For achieving the fastest self-healing, the length of polymer chains between two NPs must not be too long or too short. If the NP content in the hydrogel is high, the polymer chain between two NPs is too short and led to difficulty in self-healing (Haraguchi et al., 2011). If the polymer chain between two NPs is too long, long polymer chains need a longer time for diffusion, finding, and attaching to NPs (Wang et al., 2017). As well as, when the polymer chains are short, only fresh-cut surfaces are able to heal due to the binding of the chains on the first surface (Wang et al., 2010). Interestingly, the healing phenomenon can be applied to the combination of cut surfaces of two different hydrogels. In the latter case study, the treated surface was not detached even with applying periodic swelling/deswelling (Haraguchi et al., 2011).

1.5.4. Adhesion Properties

There are reports on the application of NC hydrogels for adhesion to glass, plastic, and tissue (Peak et al., 2013). This kind of adhesion is reversible due to the ability of polymer chains to reform new physical linkage after delink (Peak et al., 2013). Research has shown that the adhesion property is higher at lower NP loading and is lower at higher crosslink density. Higher adhesion was due to the higher flexibility of the NC hydrogel chains leading to higher stress losses (Wu et al., 2013). There are reports on the application of NP solutions as adhesives for hydrogels and biological tissues due to the connection of polymer chains to the NPs named nanobridging process (Cao and Dobrynin, 2016; Meddahi-Pellé et al., 2014; Rose et al., 2014). When NPs have comparative diameters to the mesh size of the gel network, multiple polymer chains can adsorb to an NP and act as bridges, making a nano adhesive junction layer. In the tension condition, some polymer chains detached from the NP's surface, but they can re-attach or other polymer chains can attach, resulting in high tensile strength in the adhesive layer. In other words, the adhesive layer is stronger than the side hydrogel networks, so a fracture of the adhesive layer does not occur (Rose et al., 2014).

1.5.5. Stimuli-Responsive Properties

Stimuli-responsive hydrogels react spontaneously in response to physical changes (such as light, pressure, temperature, electrical field, magnetic field, and mechanical stress), or chemical changes (such as pH, ionic groups, and other chemicals). These types of polymers are also known by other names such as smart, intelligent, or environmentally-sensitive polymers (Kikuchi and Okano, 2002; Stuart et al., 2010). The performance of smart materials has three main parts: receiving changes in environmental conditions, processing the received information, and finally reacting to the stimulus (Filipcsei et al., 2000). Some NC hydrogels have the ability to react to several types of stimuli, which makes them more efficient in different applications in different environmental conditions (Gil and Hudson, 2004; Kikuchi and Okano, 2002). Some of the most widely used types of stimuli-responsive hydrogels are temperature, pH, Electro, and magnetic field responsive NC hydrogels.

1.5.5.1. pH-Sensitivity

pH-responsive NC Hydrogels, at the appropriate pH, the degree of ionization (pKa or pKb) changes significantly. This change in the net density of the ionized chains causes a sudden change in volume by creating electrostatic repulsion forces between the ionized groups. As a result, it causes an increase in osmotic pressure (Ullah et al., 2015). These hydrogels contain acidic (poly-ionic) or alkaline (poly-cationic) polyelectrolyte chains that can receive or release protons in an aqueous solution in response to pH changes. In ionic hydrogels, swelling increases when the external pH is higher than pKa. On the other hand, when the external pH is lower than pKb, swelling increases due to electrostatic repulsion (Li et al., 2009; Sirousazar and Kokabi, 2012).

1.5.5.2. Temperature Sensitivity

The most basic response to temperature sensitivity is the swelling or destruction behavior that occurs by reducing or increasing the temperature of the swelling environment to a certain temperature under the name of phase transition temperature (or critical temperature) (Peppas et al., 2006). In general, temperature-sensitive NC hydrogels are divided into positive-temperature and negative-temperature sensitive hydrogels. Positive-temperature-sensitive NC hydrogel has upper critical solution temperature (UCST). Below the UCST the hydrogel chains become hydrophobic and above they become hydrophilic. As a result, in positive-temperature-sensitive NC hydrogels, the degree of swelling increases as the temperature increases above UCST.

Negative-temperature-sensitive NC hydrogel has the opposite behavior, and its critical temperature is known as the lower critical solution temperature (LCST). Above LCST, the hydrophilic nature of hydrogel polymer chains becomes hydrophobic. As a result, lowering the temperature below the LCST increases swelling (Sirousazar and Kokabi, 2012). For example, thermally responsive NC hydrogels composed of poly(ethylene glycol)-poly(L-alanine) (PEG-L-PA) and reduced GO (rGO) as 2D/3D hybrid systems, with increasing temperature from 4 to 37°C show the sol-to-gel transition (Park et al., 2015).

1.5.5.3. Electro-Sensitivity

Electrically-responsive NC hydrogels can be prepared by using conductive polymers or adding conductive NPs to the polymer. Changes in the electrical field cause electrical changes in the properties of NC hydrogels, such as swelling/de-swelling or bending of the NC hydrogel. Some of the most important conductive NPs for making electrically responsive hydrogels are gold, carbon nanotubes, and graphene. For example, pristine multi-walled carbon nanotubes (pMWNTs) have been used to make NC hydrogels based on electro-responsive poly(methacrylic acid) (PMAA) in drug delivery applications (Servant et al., 2013). Although the presence of pMWNTs in the NC hydrogel network had no significant effect on the degree of swelling of NC hydrogel, but the degree of deswelling increased. In addition, applying an electrical stimulation caused the NC hydrogel to release heat due to its high impedance properties. Therefore, the electrical response of the NC PMAA/pMWNT hydrogel resulted in controlled drug release through the application of an external electrical field. The PMAA/graphene NC hydrogels compared to the PMAA/pMWNT NC hydrogels (in the same amount of crosslinking agent) and due to better distribution of graphene in the hydrogel network, the PMAA/graphene NC hydrogels showed a better degree of swelling/deswelling properties, which led to improved mechanical and electrical properties. In general, PMAA/graphene NC hydrogel shows lower bulk strength than PMAA/pMWNT NC hydrogel (Servant et al., 2014).

In smart GO-filled poly (acrylic acid) (GO-PAA) NC, the presence of GO increased the compressive strength and water absorption capacity of the NC hydrogel (Tai et al., 2013). Compared to PAA hydrogel, GO-PAA hydrogel has better mechanical strength and swelling properties and showed higher electrical response under stronger electrical fields. Both PAA hydrogel and GO-PAA NC hydrogel showed similar behavior under DC electrical fields, but the response rate of GO-PAA NC hydrogel under the same DC electrical field was higher than pure hydrogel.

1.5.5.4. Magnetic Field Sensitivity

The magnetic field-responsive NC hydrogels are usually prepared by incorporating magnetic NPs into the hydrogel network. The mechanical and morphological properties of magnetic field-responsive NC hydrogels such as elasticity, degree of swelling, or deformation can be reversibly controlled by an external magnetic field. For example, NC hydrogels made of NPs of poly (2-(2-methoxyethoxy) ethyl methacrylate-co-oligo (ethylene glycol) methacrylate-co-acrylic acid) (PMOA) and magnetic attapulgite/iron oxide (AT-Fe_3O_4) NPs showed better super magnetic properties than AT-Fe_3O_4 (Wang et al., 2012). Applying an external magnetic field changes the arrangement of AT-Fe_3O_4 in the hydrogel network. The swelling capacity of this hydrogel increased in the presence of an alternating magnetic field (AMF). The PMOA/AT-Fe_3O_4 NC hydrogel was also sensitive to pH and temperature stimuli. At room temperature, changing the pH to lower and higher than 4.5 caused deswelling and swelling of the NC hydrogel, respectively. The mechanical properties of the PMOA/AT-Fe_3O_4 NC hydrogel depend on the concentration of AT-Fe_3O_4. The PMOA network forms a physical linkage on the

AT-Fe_3O_4 surface and with the increase of AT-Fe_3O_4 content, the tensile strength and intensity of the magnetic response increase.

Another example of magnetic field-responsive NC hydrogel is hemicellulose hydrogels containing magnetic Fe_3O_4 NPs, which are made through the *in situ* synthesizing of Fe_3O_4 NPs simultaneously with covalent crosslinking of hemicellulose. The hemicellulose/Fe_3O_4 hydrogel showed a magnetic and super-paramagnetic response, which increases with the increase of Fe_3O_4 NPs content. This magnetic field-responsive hemicellulose hydrogel was used to investigate the ability of controlling the absorption and release of drugs (Zhao et al., 2015).

1.5.6. Thermal Properties

The behavior of the material against temperature changes is an important component in determining the performance conditions of the material. In fact, thermal stability represents the ability of matter to preserve its physical properties and structural strength at high temperatures. The results of the studies in this regard indicate the improvement of the thermal stability of the materials in the presence of NPs (Haraguchi and Li, 2006). The noteworthy point is the noticeable effect of the small number of NPs in the hydrogel structure. So that by adding only 0.5 wt.% of silica NPs, the energy required for thermal degradation will increase from 69 to 99 kJ/mol. Such studies showed that adding 0.8 wt.% of silica increases the temperature degradation of NC hydrogel up to 18°C (Sengupta et al., 2005). In general, compared to pure polymer, the coefficient of thermal expansion of NC hydrogel is reduced. The reduction of the thermal expansion coefficient occurs due to the creation of a link between NPs and the polymer network resulting in an increase in the thermal stability of the hydrogel (Hsiao et al., 2001).

1.5.7. Biocompatibility

Enhancing the biochemical properties of hydrogels by adding NPs, has developed broad biomedical applicability including use as drug delivery agents, adhesives, wound dressings, scaffolds, bio-printing inks, cell culture systems, and flexible biosensors (He et al., 2020; Lai et al., 2020; Mendes et al., 2019; Wang et al., 2019; Wei et al., 2020). Investigations have demonstrated that the incorporation of NPs in hydrogels can provide and enhance the anti-pathogenic (Agnihotri et al., 2012; Dai et al., 2018; Farag et al., 2019; Jayaramudu et al., 2013; Wang et al., 2016), angiogenic, and adhesive properties (Arno et al., 2020; Baït et al., 2011; Haraguchi et al., 2018; Liu et al., 2013; Rajabi et al., 2020; Wu et al., 2013) in the application as wound dressings (Basha et al., 2020; Mai et al., 2020; Zhu et al., 2018).

Adding a type of clay NP, to PNIPA hydrogels improved their suitability for tissue engineering and cell culture and applications (Haraguchi and Takada, 2010; Haraguchi et al., 2006). NPs are added to hydrogels such as gelatin methacryloyl (GelMA) in the application of cell culture including 3D cell culture (Wang et al., 2018). The incorporation of silica NPs into GelMA enabled it to guide the differentiation of bone marrow stem cells. Furthermore, it is reported that the NPs help the loading and releasing of pinacidil to improve cell viability (Wang et al., 2018). Incorporating DNA hydrogel with silicate nanosheets absorbs stem cells and improves the healing time of wounds (Basu et al., 2019). There are reports on the application of NPs in hydrogel as scaffolds for artificial bone grafts (Parisi-Amon et al., 2017), cancer treatment (Wang et al., 2020), and neural regeneration (Cheng et al., 2019). HA NPs are incorporated in hydrogel scaffolds in order to enhance self-healing properties by the recombination of proteins to add osteoconductivity as a bone scaffold implant (Parisi-Amon et al., 2017). Cellulose nanofibers (CNF) increase the self-healing rate, extend the degradation time, and enhance neural stem cell viability and differentiation of chitosan hydrogels as neural regenerators (Cheng et al., 2019). The incorporation of NPs in hydrogels can control the degree of porosity and crosslinking density and improve their injectability and application as drug delivery systems. For another instance, the incorporation of GO-HA NPs into PVA hydrogel enhanced its printability and printing accuracy as hydrogel bio-ink in printing artificial cartilage due to the improving the biocompatibility and mechanical properties (viscosity, shear thinning, compressive, and tribological properties) (Meng et al., 2020).

1.6. Types of NPs in NC Hydrogels

The most crucial parameters in the application of various NPs in NC hydrogels are the shape, size, and surface properties, in order to obtain unique properties. In general, nanomaterials can be in the shape of spheres, rods, plates, or tubes. The size of NPs determines their effective surface (surface-to-volume ratio) and affects their dispersion ability. The surface properties of NPs determine their Hansen solubility parameters and molecular interaction with polymer chains. The NPs used in the production of NC hydrogels are divided into metals and metal oxide, carbon-based, inorganic, and organic NPs.

1.6.1. Metals and Metal Oxid NPs

Metal NPs include gold (Au), silver (Ag), platinum (Pt), and other noble elements, while metal oxide NPs include iron oxide (Fe_2O_3 and Fe_3O_4), copper oxide (CuO), zinc oxide (ZnO), alumina (Al_2O_3), and titania (TiO_2). These NPs interact with polymer chains in the network of NC hydrogels through physical or covalent bonds. Metal and metal oxide NPs create or enhance unique properties such as magnetic properties, electrical conductivity, stimulus-response, and mechanical strength. NC hydrogels with these properties can be used in the fabrication of biosensors, drug delivery systems, imaging materials, and responsive scaffolds.

1.6.1.1. Gold NPs

Gold NPs have been used in making different types of hydrogels. Au NPs exhibit localized surface plasmon resonance (LSPR) that convert light energy into thermal energy. This makes the Au-incorporated NC hydrogels a candidate for photo-thermal and photodynamic therapy, imaging materials, and stimuli-responsive drug carriers. NC hydrogels containing nanospheres or nanorods of Au have been used in photo-thermal cancer immunotherapy as DNA hydrogel materials. The Au NPs were modified by photosensitive material like oligodeoxynucleotide and hexapod-like structured DNA that has an affinity to DNA. This structure releases the DNA, in exposure to near-infrared (NIR) laser irradiation, following that releasing inflammatory cytokines in immune cells in order to kill cancer cells. The incorporation of Au nanowires in alginate hydrogel scaffolds improved its electrical conductivity and mechanical properties resulting in better seeding of cardiac cells due to the prevention of cell aggregation and higher alignment of cells (Khampieng et al., 2018).

1.6.1.2. Silver NPs

Ag NPs are widely used due to their antimicrobial properties. The antimicrobial effect of Ag-loaded PVA hydrogels against various bacteria has been introduced as much as ampicillin. Other applications of Ag NPs are the treatment of burn injuries and wounds. The incorporation of Ag NPs into alginate/chitosan/polyvinyl pyrrolidone (PVP) hydrogel (gamma radiation crosslinked) showed non-toxicity, anti-bacterial, antimicrobial, and high cellular viability properties in treating pressure ulcers (Khampieng et al., 2018).

1.6.1.3. Iron Oxide NPs

Many forms of iron oxides exist in nature. The most common types are magnetite (Fe_3O_4), hematite (α-Fe_2O_3), and maghemite (γ-Fe_2O_3). The iron atom has four unpaired electrons in 3d orbitals resulting in an intense magnetic moment. Among transition metal oxides, magnetite has the highest magnetism (Teja and Koh, 2009). Other than magnetic properties, iron oxide-filled hydrogels have shown low cytotoxicity and appropriate cell viability which indicates the potential of iron oxide in making magnetic-stimuli hydrogels for biomedical applications (Meenach et al., 2009). Magnetic NPs have the potential for aggregation that can be overcome by surface modification, resulting in higher dispersion and biocompatibility. There are various functional groups on the iron oxide surface, such as aldehyde (-CHO), amine ($-NH_2$), sulfhydryl (-SH), and carboxyl (-COOH) which can bond covalently to other molecules. Iron oxide NPs have been used in various biomedical applications because of their antiplasmodial, anticancer, antiviral, and antimicrobial properties. Diagnostic selection and separation, magneto-relaxometry, magnetic hyperthermia, magnetic resonance imaging, and drug delivery could be the principal biomedical application of iron oxide (Bustamante-Torres et al., 2022). For instance, magnetic poly (*N*-isopropyl acrylamide) hydrogels incorporated into Fe_3O_4 have been proposed for hyperthermia applications in cancer therapy (Ang et al., 2007).

1.6.1.4. Alumina and Titania NPs

Alumina and titania NPs have been incorporated into the hydrogel matrix to enhance the bioactivity of NC by increasing cellular adhesion and migration. Hydrogel composites of PVA/PVP incorporated with TiO_2 nanosticks have been used for cartilage tissue engineering applications. The TiO_2 nanosticks enhanced adhesion proliferation and migration of cells other than the mechanical and tribological properties suitable for cartilage tissue engineering (Cao et al., 2018).

1.6.1.5. Zinc Oxide and Copper Oxide NPs

Improvement of antimicrobial properties can be achieved by the incorporation of metal or metal oxide NPs in hydrogels. Recently, it was confirmed the excellent skin compatibility, hemostatic, ultraviolet (UV) blocking, antioxidant, and antibacterial effects of ZnO in silk fibroin-tannin multifunctional hydrogel sunscreen (Yang et al., 2022). The incorporation of ZnO NPs enhanced the blood clotting, antioxidant, and antimicrobial activity, of the carrageenan hydrogels, showing promising biological applications (Baghaie et al., 2017). Prepared carrageenan-based hydrogels with ZnO, CuO, and their combination NPs showed strong antibacterial effects against pathogenic bacteria such as *Listeria monocytogenes* and *Escherichia coli*. Additionally, ZnO-filled hydrogels exhibited higher thermal, mechanical, and antimicrobial properties than CuO-filled hydrogels in the application of wound dressing (Oun and Rhim, 2017).

1.6.2. Carbon-Based NPs

Carbon-based nanomaterials (carbon allotropes) including carbon nanotubes, graphene, fullerene, and nanodiamonds are being investigated for making NC hydrogels. Carbon-based NC hydrogels have shown high mechanical properties and electrical conductivity, which have made them suitable candidates in biomedicine, tissue engineering, drug delivery, biosensing, environmental applications, etc. (Kuilla et al., 2010).

1.6.2.1. Carbon Nanotubes

Carbon nanotubes (CNT) which were first introduced in 1991, are one of the widely used NPs in the production of NC hydrogels that have appropriate electrical, thermal, and mechanical properties. CNTs are classified into single-wall carbon nanotubes (SWCNT) and multi-walled carbon nanotubes (MWCNT). In order to increase the dispersion of CNTs in the NC hydrogels and improve polymer-NP interactions, some modifications are applied to their surfaces including the development of hydroxyl (OH), amine (NH_2), and carboxyl (COOH) groups. The incorporation of CNTs in NC hydrogels enhances the thermal, mechanical, biological, and electrical properties due to the physical and chemical bridging of CNTs to the polymer matrix. Grafting of the polymer chains on the CNT surface is another method for modification of the CNT surface resulting in surface-grafted CNTs (Liu et al., 2010; Ma et al., 2010).

1.6.2.2. Nanodiamonds and Carbon Dots

Nanodiamonds (ND) consist of tetrahedral sp_3 carbon atoms, but carbon dots (CD) contain various ratios of sp_3 and sp_2 carbon atoms. The application of ND in medicine is due to its low cost, low toxicity, small size, scalable production, surface chemical groups, fluorescence, and photobleaching resistance which make ND a suitable candidate in biomedical applications like drug delivery and theranostic nanomedicine. ND has other applications in NC hydrogels to produce tissue engineering scaffolds, biodegradable bone surgery devices, killing drug-resistant microbes, fighting viruses, and delivering genes into cells (Turcheniuk and Mochalin, 2017). The NC hydrogels containing CDs have benefits line low cost, simple synthesizing method, and low toxicity (Sharma et al., 2017). The chitosan/CD NC hydrogel shows blue color fluorescence. The chitosan/CD NC, which is made by a solvent casting method for wound healing applications, shows antibacterial and pH-sensitive behaviors, proving the non-toxicity and biocompatibility of the chitosan/CD NC (Omidi et al., 2017).

1.6.2.3. Graphene and GO

Graphene is a two-dimensional monolayer of carbon allotropes. The basic structure of carbon allotropes is a graphene single-layer structure. Graphene has high electrical conductivity and electrochemical and mechanical properties (Xu et al., 2010). Therefore, graphene-based NC hydrogels exhibit high strength and stimuli-responsive properties. Graphene surfaces can link physically or chemically to the hydrogel polymer chains. Graphene NC hydrogels showed pH sol-gel alteration sensitivity that can be used for selective drug release (Bai et al., 2010). Graphene NC hydrogels are applicable in biomedical imaging and magnetic resonance imaging (MRI) due to their light transmittance and photoluminescence properties. The GO protein-based hydrogels are suitable for tissue engineering applications due to their electrical and mechanical properties (Biru et al., 2022). The graphene/chitosan/PVA nanofibrous hydrogel has been investigated for wound-healing applications (Lu et al., 2012).

1.6.3. Inorganic NPs

Many inorganic NPs are used to prepare NC hydrogels. Nanoclays, hydroxyapatite, silica, and glass NPs are the most important inorganic NPs (Conte et al., 2019; Goetz et al., 2010).

1.6.3.1. Nanoclays

Nanoclays are a kind of layered mineral silicate NPs. There are different types of nanoclays including montmorillonite (MMT), hectorite, kaolinite, bentonite, and halloysite, which differ in morphology and chemical composition. One of the most important natural layered nanosilicates is MMT clay, which is widely used in making NC hydrogels. The structure of MMT is such that cations like Na^{+}, Mg^{2+}, Ca^{2+}, etc. are located in the spaces between the layers. In general, the existence of cations makes it possible to carry out the cation exchange process. In this order, the cations are replaced by water molecules and cause the creation of a single-layered or multilayered structure, and as a result, make MMT hydrophilic. This process of cation exchange for MMT can be done more easily compared to other natural layered silicates. Due to the hydrophilic nature of MMT, this NP is very compatible with water-soluble polymers such as PAM (Zareie et al., 2019).

Studies have shown that adding nanoclay to the hydrogel matrix leads to improved tensile strength (305 kPa), elongation at break (1000 times), elastic modulus (2780 Pa) (Jeon and Baek, 2010), polarization resistance (from 10 to 1340) (Yu et al., 2004) and to some extent it also becomes electric conduction (Olad and Rashidzadeh, 2008). The use of sodium-MMT nanoclay in the gel structure leads to an increase in the strength of polymer gels and a decrease in their sensitivity to ions existing in the solvent (Jeon and Baek, 2010; Loste et al., 2019). Also, the results of the tests conducted on the superabsorbent composite made of bentonite and poly (acrylic acid-acrylamide) show that this composite has a high swelling ability in aqueous solutions containing salt, and for this reason, it can be used in petroleum processes. NC hydrogels have less syneresis in the water of petroleum reservoirs compared to conventional PAM hydrogels (Zareie et al., 2019). Novel pipes of PVA/MMT NC hydrogel were prepared via a cyclic freezing-thawing method designed for subsurface irrigation in order to reduce water consumption. MMT layers were exfoliated and enhanced gel content, thermal, and mechanical properties, but decreased swelling and drying ability of PVA. A significant decrease in consumed water was observed in NC hydrogel compared to the pure PVA hydrogel pipe (Rezazadeh et al., 2020).

1.6.3.2. Hydroxyapatite NPs

Hydroxyapatite is a member of the calcium phosphate family with the general formula of $Ca_{10}(OH)_2(PO_4)_6$. Hydroxyapatite is chemically similar to teeth and natural bone, making it an attractive option for bone scaffolding applications. Uniform interconnected porous structures with high mechanical strengths and cell adhesion properties were observed in hydroxyapatite-incorporated PAM hydrogels (Li et al., 2014). Also, the incorporation of the hydroxyapatite enhanced the tensile and compression properties of the PAM hydrogel due to the hydrogen and cheating bonding between polymer chains and hydroxyapatite, acting as the crosslinking agent. Higher crosslink density in presence of hydroxyapatite results in a lower equilibrium swelling ratio of PAM hydrogel. Moreover, the differentiation of mesenchymal stem cells to osteoblast cells and adhesion to the hydrogel matrix, indicate the potential application of the NC in tissue engineering. Nano-hydroxyapatite incorporated GelMA was investigated as an elastomeric scaffold for bone tissue engineering (Thakur et al., 2016). The incorporation of nano-hydroxyapatite enhanced the elasticity, bioactivity, stiffness, and physiological stability of the hydrogel network. Furthermore, high functionality, cell viability, and the osteogenic differentiation of seeded pre-osteoblast cells were observed in presence of nano-hydroxyapatite.

1.6.3.3. Bioactive Glass NPs

Bioactive glass NPs have a particle size of less than 100 nm, and unique merits like physicochemical properties, rapid bioactive response, high specific surface area, biodegradability, high release rate of bioactive ions, injectability, excellent bioactivity, spherical morphology, and intracellular uptake. Bioactive glass-filled hydrogels have potential applications in bone and dental tissue engineering, skin regeneration, and wound healing (El-Fiqi, 2022).

1.6.4. Organic NPs

Organic NPs are natural or synthesized NPs prepared by "top-down" or "bottom-up" approaches. Top-down techniques including mechanical milling, microfluidics, and lithography, and bottom-up techniques are synthetic NPs from molecules or atoms by self-organization chemistry to generate various organic NPs such as micelles, capsules, liposomes and vesicles, polymersomes, polymer conjugates, dendrimers, and polymeric NPs. Most organic NPs involve several polymeric or organic molecules. Also, organic NPs, like vesicles or micelles, can change their shape and size through fusion or self-assembly due to their dynamic character. Various successful methods have been applied to load molecules like conjugation or physical encapsulation on the organic NPs, which make them the right choice in biomedical and drug delivery applications (Romero and Moya, 2012).

1.6.4.1. Dendrimers and Hyper-Branched NPs

Dendritic NPs have been interested due to their interconnected 3D hyper-branched, globular architecture, and highly porous structures that can affiliate to both hydrophobic and hydrophilic moieties, enhancing the mechanical and other desired properties of the NC hydrogels. These unique properties enable spatiotemporal control of drug delivery in order to sustain dosage resulting in improved therapeutic efficacy. The application of polypropylene imine dendrimer (PPI-G2) in poly (L-lactic acid) (PLLA) has resulted in the great promotion of hydrophilicity, cell viability, and adhesion of hydrogel as a biodegradable biomaterial as biomedical scaffolds (Ganjalinia et al., 2017). Similar results have been reported for the incorporation of pH-responsive poly(amidoamine) (PAMAM) dendrimers in hydrogel for drug delivery in cancer therapy that increased mitigated cell viability, induced apoptosis, cellular uptake, and sustained drug release, without obvious systemic toxicity (Hanurry et al., 2022).

1.6.4.2. Micelles and Liposomes

Liposomes are spherical artificial vesicles that have a minimum of one lipid bilayer. liposomes are hydrophobic tail-hydrophilic head NP vehicles that act as lipid carriers and are applicable for drugs, genes, and/or cosmetics delivery (Akbarzadeh et al., 2013). Micelles are block amphiphilic copolymers having high drug loading capacity in drug delivery due to hydrophobic and hydrophilic parts in their structure. The increasing transdermal hydrocortisone (HC) transportation in HC-loaded micelles in carbomer hydrogel was investigated, to improve topical therapy of atopic dermatitis (AD). The NC hydrogel increased the rate of skin permeation and the cumulative amount of transferred HC and reduced skin irritation and effectively ameliorates inflammatory symptoms, compared with the HC cream (Yuan et al., 2020).

1.7. Characterization of NC Hydrogels

The gel fraction of NC hydrogels can be calculated by the Khoylou method using Equation (1-13) (Azady et al., 2021):

$$\text{Gel farction}(\%) = \frac{W_d}{W_i} \times 100 \tag{1-13}$$

where W_d is the dry weight of crosslinked hydrogel after drying and W_i is the initial weight of the hydrogel. The Sol fraction of NC hydrogels is calculated using the following equation:

$$\text{Sol fraction}(\%) = \frac{W_d - Wi}{W_i} \times 100 \tag{1-14}$$

The swelling ratio (swelling degree) of the NC hydrogels can be calculated as Equation (1-15):

$$\text{Swelling Ratio}(\%) = \frac{W_t - W_d}{W_d} \times 100 \tag{1-15}$$

where W_t is the weight of swollen gel at time t. The relative weight of the hydrogel (W_R), which represents the rate of swelling/deswelling, is shown in Equation (1-16) (Sirousazar et al., 2014):

$$W_R = \frac{W_{24h}}{W_i} \tag{1-16}$$

Here, W_{24h} is the hydrogel weight after 24 hours. The maximum amount of water adsorbing by NC hydrogel is known as the equilibrium swelling ratio (ESR) which can be calculated by Equation (1-17) (Azady et al., 2021):

$$ESR = \frac{W_s - W_d}{W_d} \tag{1-17}$$

where Ws is the weight of swollen hydrogel at equilibrium.

The surface morphology of NC hydrogels could be studied by scanning electron microscopy (SEM). X-ray diffraction (XRD) is a useful investigation method to understand the dispersion of NPs in NC hydrogels. Using mechanical tests, the effect of the presence of NPs on mechanical properties (such as strength, elongation at break, and modulus) can be investigated in tension, compression, and shear states. The effect of NPs on the thermal properties of hydrogels can be studied by thermogravimetric analysis (TGA) which determines the mass change of the sample over increasing temperature. The effect of NPs on the adsorption capacity of NC hydrogel can be investigated using an adsorption investigation (Azady et al., 2021). Infrared spectroscopy (FTIR) and nuclear magnetic resonance (NMR) techniques are common techniques to investigate the chemical structure and NPs-polymer interactions. Atomic force microscopy (AFM) is a non-destructive technique that by drawing a topographic image of a hydrogel surface provides useful information about the surface properties of the hydrogel. Differential scanning calorimetry (DSC) is another technics widely used to investigate the crystalline behavior of NC hydrogel by calculation of heat capacity as a function of temperature. Transmission electron microscopy (TEM) investigation is a practical imaging method to inform about the size, shape, and dispersion of NPs in NC hydrogels (Sirousazar et al., 2014).

1.8. Applications of NC Hydrogels

NC hydrogels are used in various applications. The most important applications of NC hydrogels are in the fields of biomedicine, sensors, and the environment.

1.8.1. Biomedicine

Biomedical applications of NC hydrogels are in soft robots, drug delivery, wound care, tissue engineering, and bio-printing. Some stimuli-responsive NC hydrogels that can transit their shape under external stimuli, including light, temperature, humidity, volume, and phase change are the ideal candidates for making artificial soft robots. Artificial soft robots can perform programmed movements by applying macroscopic actuation as a command (Huang et al., 2018; Tan et al., 2018; Wirthl et al., 2017). In the application of softer robots, the structure of NC hydrogels must be designed as anisotropic, such as gradient structures, asymmetric structures, alignment structures, etc. (Lin et al., 2016).

The incorporation of NPs in hydrogels enables the scheduled release of the loaded drug at the desired target *in vivo* (Vermonden et al., 2012). For instance, neutral and cationic drugs due to the interaction with negatively charged or active sites of NPs surfaces, such as negatively charged MMT, bond to the NC hydrogel network. Breaking the interaction by hydrogel dissolution makes the release of the drugs. Although MMT enhances the dissolution resistance of hydrogel, the application of stimuli-responsive hydrogels enables the target releasing of the drug in response to the applied external stimuli. Another, NC gels can be used as programmed tools for delivering the drug to the target and releasing it. For instance, an external magnetic field can guide magnetic-responsive NC hydrogels toward a target (Breger et al., 2015).

The ideal wound dressing should be biocompatible, antibacterial, water absorber, water retainer, non-cytotoxic, moisture permeability, oxygen permeability, biodegradable, and act as physical barriers against microorganisms (Li et al., 2017; Zhao et al., 2018). The role of NPs in wound hydrogel includes improving mechanical performance, antibacterial effect, sustained drug release, and increasing adhesion (Liu et al., 2022).

Tissue engineering materials must be able to prepare and regenerate tissues that can make by the combination of biologically active factors, stem cells, and biomaterial scaffolds (Annabi et al., 2013). NC hydrogels are an ideal choice in tissue engineering due to their biocompatibility, mechanical properties, and biomimetic structures (Khan et al., 2018; Mehrali et al., 2017). Various NP types, such as CNTs, hydroxyapatite, and nanoclays have been successfully incorporated into hydrogels to fabricate NC hydrogel scaffolds. Some NC hydrogels have been proven for tissue engineering applications, like photo-crosslinkable gelatin/clay hydrogel due to the proliferation and differentiation of stem cells to repair the desired tissue (Paul et al., 2016).

Hydrogels are one suitable biomaterial for 3D and 4D bio-printing due to their high viscosity and shape fixing after gelation. 4D printing products are 3D printed materials whose functionality, shape, and property are self-transformed by exposure to the stimulus (Dong et al., 2020). The viscosity of synthetic hydrogels can be designed to be suitable for printing, but they need the development of cell adhesion sites. Some NC hydrogels have suitable bioactivity, but they need the enhancement of mechanical properties (Mendes et al., 2019; Nadernezhad et al., 2019). The incorporation of NPs can improve both mechanical and cell adhesion properties in bioprinting applications (Betsch et al., 2018; McCracken et al., 2019; Mendes et al., 2019; Nadernezhad et al., 2019; Shin et al., 2018). For instance, the incorporation of GO-hydroxyapatite NPs in PVA hydrogels enhanced its mechanical properties such as shear thinning, viscosity, compressive, and tribological properties, resulting in higher printability and printing accuracy as hydrogel bio-ink for 3D printing artificial cartilage (Meng et al., 2020).

1.8.2. Sensors

Sensors are devices that can convert receiving stimuli like light or strain to other signals. For instance, strain sensors can convert mechanical energy into optical signals (Yue et al., 2014) or electrical energy (Lei et al., 2017; Qiu et al., 2016). PAA/alginate/calcium carbonate NC hydrogel was investigated as a capacitive ionic skin sensor that could detect subtle stress of daily human activities resulting in the change of the capacitance (Lei et al., 2017). NC hydrogels as strain sensors could be applied in wearable devices, biosensors, artificial intelligence, and healthcare gadgets.

1.8.3. Environmental Applications

Environmental applications of NC hydrogels include irrigation, photocatalysis, and the adsorption of heavy metals from wastewater. Adsorption of heavy metals is made possible by the presence of ionic groups in NC hydrogels. Various NPs such as activated carbon and natural and modified clays have been used for the removal of organic dyes by the adsorption method (Ju et al., 2009; Kadirvelu et al., 2000; Orthman et al., 2003). Different NC hydrogels like poly(N-isopropyl acrylamide)/MMT, poly(sodium acrylate)/bentonite, PAA/mica, PAA/MMT, etc. have been used as water absorbent NC hydrogels (Santiago et al., 2007; Sharma et al., 2018).

Another method for removing organic dyes is the photocatalytic process. Most catalysts used are CdS, TiO_2, CuS, ZnO, ZnS, and Fe_3O_4 (Kumar et al., 2017). The polymer network provides a platform for catalytic NPs to achieve solar energy. Also, the presence of polar groups (epoxide, ether, hydroxyl, and carboxylate) in hydrogels enhances photocatalytic activity. For instance, PAA/PVA/GO/TiO_2 NC hydrogels have been applied as photocatalytic for the decomposition of dyes by UV radiations (Moon et al., 2013).

The addition of NC hydrogels as super-absorbents in the soil increases porosity and water retention and decreases the bulk density of the soil and water consumption in agriculture. NC hydrogels can also be used for continuous irrigation. For instance, PVA/MMT NC hydrogel has been proposed as a novel irrigation pipe that makes continued transportation of water from the pipe to the soil during the irrigation process (Rezazadeh et al., 2020).

1.8.4. Oil and Gas Industry

NC Hydrogels have various applications in the oil and gas industry as gas leakage stoppers (Sirousazar and Ghasemi, 2018), water shutoff gel (Fathima et al., 2018), oil-water separator (Su et al., 2019), and oil recovery devices (Long et al., 2019; Tongwa and Bai, 2014; Wang et al., 2018). The PVA/MMT NC hydrogel has been proposed as a novel stopper for preventing the leakage of natural gas during welding operations. The incorporation of MMT increased the crosslink density of PVA hydrogel, applicable as a perfect sealing and preventing gas leakage in natural gas pipelines (Sirousazar and Ghasemi, 2018). Another application of NC hydrogels is preventing water production from oil reservoirs. NPs can control the sensitivity of hydrogels to the downhole environment. Metal oxide NPs and graphene are two well-known nanomaterials for robust water shutoff gels, due to increased thermal stability, salt tolerance, and mechanical properties (Fathima et al., 2018). In addition, NC hydrogels have been investigated as superoleophobic non-swellable materials and showed long-term anti-crude-oil-adhesion properties under seawater (Li et al., 2019). In a study, the NC hydrogels showed higher elasticity and extensibility at low crosslinker density compared with the similar unfilled hydrogel, which makes the NC hydrogels suitable for in-depth profile modifiers and oil displacing agents in the heterogeneous oil reservoir in chemical-enhanced oil recovery (Zolfaghari et al., 2006).

Conclusion

Hydrogels have various applications that can be designed or enhanced required properties of them by the incorporation of appropriate NPs. This chapter briefly presented the basic principles of NC hydrogels including molecular and thermodynamic concepts, types of NC hydrogels and production methods, polymer-NP molecular interactions, properties, different NP types applied in hydrogels, characterization methods, and the important applications of NC hydrogels. In general, it can be said that the use of NC hydrogels is increasing day by day, and different types of NC hydrogels are made in order to improve the properties required in common applications or use in novel applications. Therefore, understanding the micro molecular concepts and relation to macro properties and applications of NC hydrogels is necessary to develop engineering knowledge for use in specific applications which were discussed in this chapter.

References

Abe, H., Hara, Y., Maeda, S., and Hashimoto, S. (2014). Adhesion of gels by silica particle. *The Journal of Physical Chemistry B*, 118(9), 2518-2522.

Agnihotri, S., Mukherji, S., and Mukherji, S. (2012). Antimicrobial chitosan-PVA hydrogel as a nanoreactor and immobilizing matrix for silver nanoparticles. *Applied Nanoscience*, 2(3), 179-188.

Ahmed, E. M. (2015). Hydrogel: preparation, characterization, and applications: a review. *Journal of Advanced Research*, 6(2), 105-121.

Ahmed, N. B., Ronsin, O., Mouton, L., Sicard, C., Yéprémian, C., Baumberger, T., Brayner, R., and Coradin, T. (2017). The physics and chemistry of silica-in-silicates nanocomposite hydrogels and their phycocompatibility. *Journal of Materials Chemistry B*, 5(16), 2931-2940.

Akbarzadeh, A., Rezaei-Sadabady, R., Davaran, S., Joo, S. W., Zarghami, N., Hanifehpour, Y., Samiei, M., Kouhi, M., and Nejati-Koshki, K. (2013). Liposome: classification, preparation, and applications. *Nanoscale Research Letters*, 8, 102.

Ang, K. L., Venkatraman, S., and Ramanujan, R. V. (2007). Magnetic PNIPA hydrogels for hyperthermia applications in cancer therapy. *Materials Science and Engineering: C*, 27(3), 347-351.

Annabi, N., Tsang, K., Mithieux, S. M., Nikkhah, M., Ameri, A., Khademhosseini, A., and Weiss, A. S. (2013). Highly elastic micropatterned hydrogel for engineering functional cardiac tissue. *Advanced Functional Materials*, 23(39), 4950-4959.

Aoki, H., Al-Assaf, S., Katayama, T., and Phillips, G. O. (2007). Characterization and properties of Acacia *senegal* (L.) Willd. var. *senegal* with enhanced properties (Acacia (*sen*) SUPER GUM™): Part 2-Mechanism of the maturation process. *Food Hydrocolloids*, 21(3), 329-337.

Arno, M. C., Inam, M., Weems, A. C., Li, Z., Binch, A. L. A., Platt, C. I., Richardson, S. M., Hoyland, J. A., Dove, A. P., and O'Reilly, R. K. (2020). Exploiting the role of nanoparticle shape in enhancing hydrogel adhesive and mechanical properties. *Nature Communications*, 11(1), 1420.

Azady, M. A. R., Ahmed, S., and Islam, M. S. (2021). A review on polymer nanocomposite hydrogel preparation, characterization, and applications. *European Journal of Chemistry*, 12(3), 11.

Baghaie, S., Khorasani, M. T., Zarrabi, A., and Moshtaghian, J. (2017). Wound healing properties of PVA/starch/chitosan hydrogel membranes with nano zinc oxide as antibacterial wound dressing material. *Journal of Biomaterials Science, Polymer Edition*, 28(18), 2220-2241.

Bai, H., Li, C., Wang, X., and Shi, G. (2010). A pH-sensitive graphene oxide composite hydrogel. *Chemical Communications*, 46(14), 2376-2378.

Baït, N., Grassl, B., Derail, C., and Benaboura, A. (2011). Hydrogel nanocomposites as pressure-sensitive adhesives for skin-contact applications. *Soft Matter*, 7(5), 2025-2032.

Bajpai, A. K., Shukla, S. K., Bhanu, S., and Kankane, S. (2008). Responsive polymers in controlled drug delivery. *Progress in Polymer Science*, 33(11), 1088-1118.

Barbucci, R. (2010). *Hydrogels Biological Properties and Applications*. Springer.

Basha, S. I., Ghosh, S., Vinothkumar, K., Ramesh, B., kumari, P. H. p., Mohan, K. V. M., and Sukumar, E. (2020). Fumaric acid incorporated Ag/agar-agar hybrid hydrogel: a multifunctional avenue to tackle wound healing. *Materials Science and Engineering: C*, 111, 110743.

Basu, S., Alkiswani, A. R., Pacelli, S., and Paul, A. (2019). Nucleic acid-based dual cross-linked hydrogels for *in situ* tissue repair via directional stem cell migration. *ACS Applied Materials & Interfaces*, 11(38), 34621-34633.

Betsch, M., Cristian, C., Lin, Y. Y., Blaeser, A., Schöneberg, J., Vogt, M., Buhl, E. M., Fischer, H., and Duarte Campos, D. F. (2018). Incorporating 4D into bioprinting: real-time magnetically directed collagen fiber alignment for generating complex multilayered tissues. *Advanced Healthcare Materials*, 7(21), 1800894.

Bhardwaj, Y. K., Kumar, V., and Sabharwal, S. (2003). Swelling behavior of radiation-polymerized polyampholytic two-component gels: dynamic and equilibrium swelling kinetics. *Journal of Applied Polymer Science*, 88(3), 730-742.

Bin Imran, A., Esaki, K., Gotoh, H., Seki, T., Ito, K., Sakai, Y., and Takeoka, Y. (2014). Extremely stretchable thermosensitive hydrogels by introducing slide-ring polyrotaxane cross-linkers and ionic groups into the polymer network. *Nature Communications*, 5(1), 5124.

Biru, E. I., Necolau, M. I., Zainea, A., and Iovu, H. (2022). Graphene oxide-protein-based scaffolds for tissue engineering: recent advances and applications. *Polymers*, 14(5), 1032.

Boonmahitthisud, A., Nakajima, L., Nguyen, K. D., and Kobayashi, T. (2017). Composite effect of silica nanoparticle on the mechanical properties of cellulose-based hydrogels derived from cottonseed hulls. *Journal of Applied Polymer Science*, 134(10), 44557.

Breger, J. C., Yoon, C., Xiao, R., Kwag, H. R., Wang, M. O., Fisher, J. P., Nguyen, T. D., and Gracias, D. H. (2015). Self-folding thermo-magnetically responsive soft microgrippers. *ACS Applied Materials & Interfaces*, 7(5), 3398-3405.

Bustamante-Torres, M., Romero-Fierro, D., Estrella-Nuñez, J., Arcentales-Vera, B., Chichande-Proaño, E., and Bucio, E. (2022). Polymeric composite of magnetite iron oxide nanoparticles and their application in biomedicine: a review. *Polymers*, 14(4), 752.

Cao, L., Wu, X., Wang, Q., and Wang, J. (2018). Biocompatible nanocomposite of TiO_2 incorporated bi-polymer for articular cartilage tissue regeneration: a facile material. *Journal of Photochemistry and Photobiology B: Biology*, 178, 440-446.

Cao, Z., and Dobrynin, A. V. (2016). Nanoparticles as adhesives for soft polymeric materials. *Macromolecules*, 49(9), 3586-3592.

Carlsson, L., Rose, S., Hourdet, D., and Marcellan, A. (2010). Nano-hybrid self-crosslinked PDMA/silica hydrogels. *Soft Matter*, 6(15), 3619-3631.

Chacko, R. T., Ventura, J., Zhuang, J., and Thayumanavan, S. (2012). Polymer nanogels: a versatile nanoscopic drug delivery platform. *Advanced Drug Delivery Reviews*, 64(9), 836-851.

Chang, C., and Zhang, L. (2011). Cellulose-based hydrogels: present status and application prospects. *Carbohydrate Polymers*, 84(1), 40-53.

Chen, T., Hou, K., Ren, Q., Chen, G., Wei, P., and Zhu, M. (2018). Nanoparticle-polymer synergies in nanocomposite hydrogels: from design to application. *Macromolecular Rapid Communications*, 39(21), 1800337.

Cheng, K. C., Huang, C. F., Wei, Y., and Hsu, S. H. (2019). Novel chitosan-cellulose nanofiber self-healing hydrogels to correlate self-healing properties of hydrogels with neural regeneration effects. *NPG Asia Materials*, 11(1), 25.

Conte, R., De Luise, A., Valentino, A., Di Cristo, F., Petillo, O., Riccitiello, F., Di Salle, A., Calarco, A., and Peluso, G. (2019). Hydrogel nanocomposite systems: characterization and application in drug-delivery systems, *Nanocarriers for Drug Delivery*, 319-349.

Creton, C. (2017). 50th anniversary perspective: networks and gels: soft but dynamic and tough. *Macromolecules*, 50(21), 8297-8316.

da Silva, E. P., Guilherme, M. R., Garcia, F. P., Nakamura, C. V., Cardozo-Filho, L., Alonso, C. G., Rubira, A. F., and Kunita, M. H. (2016). Drug release profile and reduction in the *in vitro* burst release from pectin/HEMA hydrogel nanocomposites crosslinked with titania. *RSC Advances*, 6(23), 19060-19068.

Dai, T., Wang, C., Wang, Y., Xu, W., Hu, J., and Cheng, Y. (2018). A nanocomposite hydrogel with potent and broad-spectrum antibacterial activity. *ACS Applied Materials & Interfaces*, 10(17), 15163-15173.

Dannert, C., Stokke, B. T., and Dias, R. S. (2019). Nanoparticle-hydrogel composites: from molecular interactions to macroscopic behavior. *Polymers*, 11(2), 275.

De, S. K., Aluru, N. R., Johnson, B., Crone, W. C., Beebe, D. J., and Moore, J. (2002). Equilibrium swelling and kinetics of pH-responsive hydrogels: models, experiments, and simulations. *Journal of Microelectromechanical Systems*, 11(5), 544-555.

Díez-Peña, E., Quijada-Garrido, I., and Barrales-Rienda, J. M. (2002). Hydrogen-bonding Effects on the dynamic swelling of P(N-iPAAm-co-MAA) copolymers. A case of autocatalytic swelling kinetics. *Macromolecules*, 35(23), 8882-8888.

Dong, Y., Wang, S., Ke, Y., Ding, L., Zeng, X., Magdassi, S., and Long, Y. (2020). 4D printed hydrogels: fabrication, materials, and applications. *Advanced Materials Technologies*, 5(6), 2000034.

Du, J., Wu, R., Liu, H., Nie, X., Li, H., Xu, S., and Wang, J. (2015). Synthesis of amphoteric nanocomposite hydrogels with ultrahigh tensibility. *Polymer Composites*, 36(3), 538-544.

Eagland, D., Crowther, N. J., and Butler, C. J. (1994). Complexation between polyoxyethylene and polymethacrylic acid-the importance of the molar mass of polyoxyethylene. *European Polymer Journal*, 30(7), 767-773.

El-Fiqi, A. (2022). Nano-bioactive glass: advances and applications, *Bioactive Glasses and Glass-Ceramics*, 173-201.

Farag, R. K., Labena, A., Fakhry, S. H., Safwat, G., Diab, A., and Atta, A. M. (2019). Antimicrobial activity of hybrids terpolymers based on magnetite hydrogel nanocomposites. *Materials*, 12(21), 3604.

Fathima, A., Almohsin, A., Michael, F. M., Bataweel, M., and Alsharaeh, E. H. (2018). Polymer nanocomposites for water shutoff application- a review. *Materials Research Express*, 6(3), 032001.

Filipcsei, G., Fehér, J., and Zrínyi, M. (2000). Electric field sensitive neutral polymer gels. *Journal of Molecular Structure*, 554(1), 248-255.

Flory, P. J., and Rehner Jr, J. (1943a). Statistical mechanics of cross-linked polymer networks I. Rubberlike elasticity. *The Journal of Chemical Physics*, 11(11), 512-520.

Flory, P. J., and Rehner Jr, J. (1943b). Statistical mechanics of cross-linked polymer networks II. Swelling. *The Journal of Chemical Physics*, 11(11), 521-526.

Gaharwar, A. K., Peppas, N. A., and Khademhosseini, A. (2014). Nanocomposite hydrogels for biomedical applications. *Biotechnology and Bioengineering*, 111(3), 441-453.

Ganjalinia, A., Akbari, S., and Solouk, A. (2017). PLLA scaffolds surface-engineered via poly (propylene imine) dendrimers for improvement on its biocompatibility/controlled pH biodegradability. *Applied Surface Science*, 394, 446-456.

García-Astrain, C., Chen, C., Burón, M., Palomares, T., Eceiza, A., Fruk, L., Corcuera, M. Á., and Gabilondo, N. (2015). Biocompatible hydrogel nanocomposite with covalently embedded silver nanoparticles. *Biomacromolecules*, 16(4), 1301-1310.

García-Astrain, C., Hernández, R., Guaresti, O., Fruk, L., Mijangos, C., Eceiza, A., and Gabilondo, N. (2016). Click crosslinked chitosan/gold nanocomposite hydrogels. *Macromolecular Materials and Engineering*, 301(11), 1295-1300.

Gayet, J. C., and Fortier, G. (1996). High water content BSA-PEG hydrogel for controlled release device: evaluation of the drug release properties. *Journal of Controlled Release*, 38(2), 177-184.

Giannouli, P., and Morris, E. R. (2003). Cryogelation of xanthan. *Food Hydrocolloids*, 17(4), 495-501.

Gil, E. S., and Hudson, S. M. (2004). Stimuli-reponsive polymers and their bioconjugates. *Progress in Polymer Science*, 29(12), 1173-1222.

Goetz, L., Foston, M., Mathew, A. P., Oksman, K., and Ragauskas, A. J. (2010). Poly(methyl vinyl ether-co-maleic acid)-polyethylene glycol nanocomposites cross-linked *in situ* with cellulose nanowhiskers. *Biomacromolecules*, 11(10), 2660-2666.

Gong, J. P., Katsuyama, Y., Kurokawa, T., and Osada, Y. (2003). Double-network hydrogels with extremely high mechanical strength. *Advanced Materials*, 15(14), 1155-1158.

Gunasekaran, S., Wang, T., and Chai, C. (2006). Swelling of pH-sensitive chitosan-poly(vinyl alcohol) hydrogels. *Journal of Applied Polymer Science*, 102(5), 4665-4671.

Gupta, P., Vermani, K., and Garg, S. (2002). Hydrogels: from controlled release to pH-responsive drug delivery. *Drug Discovery Today*, 7(10), 569-79.

Gupta, S., Webster, T. J., and Sinha, A. (2011). Evolution of PVA gels prepared without crosslinking agents as a cell adhesive surface. *Journal of Materials Science: Materials in Medicine*, 22(7), 1763-72.

Haider, M., Megeed, Z., and Ghandehari, H. (2004). Genetically engineered polymers: status and prospects for controlled release. *Journal of Controlled Release*, 95(1), 1-26.

Hanurry, E. Y., Birhan, Y. S., Darge, H. F., Mekonnen, T. W., Arunagiri, V., Chou, H. Y., Cheng, C. C., Lai, J. Y., and Tsai, H. C. (2022). PAMAM dendritic nanoparticle-incorporated hydrogel to enhance the immunogenic cell death and immune response of immunochemotherapy. *ACS Biomaterials Science & Engineering*, 8(6), 2403-2418.

Haraguchi, K. (2007). Nanocomposite hydrogels. *Current Opinion in Solid State and Materials Science*, 11(3), 47-54.

Haraguchi, K., and Li, H. J. (2006). Mechanical properties and structure of polymer-clay nanocomposite gels with high clay content. *Macromolecules*, 39(5), 1898-1905.

Haraguchi, K., Shimizu, S., and Tanaka, S. (2018). Instant strong adhesive behavior of nanocomposite gels toward hydrophilic porous materials. *Langmuir*, 34(29), 8480-8488.

Haraguchi, K., and Takada, T. (2010). Synthesis and characteristics of nanocomposite gels prepared by *in situ* photopolymerization in an aqueous system. *Macromolecules*, 43(9), 4294-4299.

Haraguchi, K., and Takehisa, T. (2002). Nanocomposite hydrogels: a unique organic-inorganic network structure with extraordinary mechanical, optical, and swelling/de-swelling properties. *Advanced Materials*, 14(16), 1120-1124.

Haraguchi, K., Takehisa, T., and Ebato, M. (2006). Control of cell cultivation and cell sheet detachment on the surface of polymer/clay nanocomposite hydrogels. *Biomacromolecules*, 7(11), 3267-3275.

Haraguchi, K., Takehisa, T., and Fan, S. (2002). Effects of clay content on the properties of nanocomposite hydrogels composed of poly(N-isopropylacrylamide) and clay. *Macromolecules*, 35(27), 10162-10171.

Haraguchi, K., Uyama, K., and Tanimoto, H. (2011). Self-healing in nanocomposite hydrogels. *Macromolecular Rapid Communications*, 32(16), 1253-1258.

Hassan, C. M., and Peppas, N. A. (2000). Structure and morphology of freeze/thawed PVA hydrogels. *Macromolecules*, 33(7), 2472-2479.

He, J., Shi, M., Liang, Y., and Guo, B. (2020). Conductive adhesive self-healing nanocomposite hydrogel wound dressing for photothermal therapy of infected full-thickness skin wounds. *Chemical Engineering Journal*, 394, 124888.

Hill, T. L. (1960). An introduction to statistical thermodynamics. Dover Publications.

Hoare, T. R., and Kohane, D. S. (2008). Hydrogels in drug delivery: progress and challenges. *Polymer*, 49(8), 1993-2007.

Hsiao, S. H., Liou, G. S., and Chang, L. M. (2001). Synthesis and properties of organosoluble polyimide/clay hybrids. *Journal of Applied Polymer Science*, 80(11), 2067-2072.

Hsu, S. W., Rodarte, A. L., Som, M., Arya, G., and Tao, A. R. (2018). Colloidal plasmonic nanocomposites: from fabrication to optical function. *Chemical Reviews*, 118(6), 3100-3120.

Huang, J., Liao, J., Wang, T., Sun, W., and Tong, Z. (2018). Super strong dopamine hydrogels with shape memory and bioinspired actuating behaviours modulated by solvent exchange. *Soft Matter*, 14(13), 2500-2507.

Huglin, M. B., and Zakaria, M. B. (1986). Swelling properties of copolymeric hydrogels prepared by gamma irradiation. *Journal of Applied Polymer Science*, 31(2), 457-475.

Huglin, M. R. (2019). Hydrogels in medicine and pharmacy: fundamentals. CRC Press.

Ito, K. (2007). Novel cross-linking concept of polymer network: synthesis, structure, and properties of slide-ring gels with freely movable junctions. *Polymer Journal*, 39(6), 489-499.

Jabbari, E., and Nozari, S. (2000). Swelling behavior of acrylic acid hydrogels prepared by γ-radiation crosslinking of polyacrylic acid in aqueous solution. *European Polymer Journal*, 36(12), 2685-2692.

Jayaramudu, T., Raghavendra, G. M., Varaprasad, K., Sadiku, R., Ramam, K., and Raju, K. M. (2013). Iota-Carrageenan-based biodegradable Ag0 nanocomposite hydrogels for the inactivation of bacteria. *Carbohydrate Polymers*, 95(1), 188-194.

Jeon, I. Y., and Baek, J. B. (2010). Nanocomposites derived from polymers and inorganic nanoparticles. *Materials*, 3(6), 3654-3674.

Jeong, B., and Gutowska, A. (2002). Lessons from nature: stimuli-responsive polymers and their biomedical applications. *Trends in Biotechnology*, 20(7), 305-311.

Jiang, H., Zhang, G., Feng, X., Liu, H., Li, F., Wang, M., and Li, H. (2017). Room-temperature self-healing tough nanocomposite hydrogel crosslinked by zirconium hydroxide nanoparticles. *Composites Science and Technology*, 140, 54-62.

Jianqi, F., and Lixia, G. (2002). PVA/PAA thermo-crosslinking hydrogel fiber: preparation and pH-sensitive properties in electrolyte solution. *European Polymer Journal*, 38(8), 1653-1658.

Jones, C. D., and Lyon, L. A. (2003). Photothermal patterning of microgel/gold nanoparticle composite colloidal crystals. *Journal of the American Chemical Society*, 125(2), 460-465.

Ju, X. J., Zhang, S. B., Zhou, M. Y., Xie, R., Yang, L., and Chu, L. Y. (2009). Novel heavy-metal adsorption material: ion-recognition P(NIPAM-co-BCAm) hydrogels for removal of lead(II) ions. *Journal of Hazardous Materials*, 167(1), 114-118.

Jung, S., Kaar, J. L., and Stoykovich, M. P. (2016). Design and functionalization of responsive hydrogels for photonic crystal biosensors. *Molecular Systems Design & Engineering*, 1(3), 225-241.

Kabanov, A. V., and Vinogradov, S. V. (2009). Nanogels as pharmaceutical carriers: finite networks of infinite capabilities. *Angewandte Chemie International Edition*, 48(30), 5418-5429.

Kadirvelu, K., Palanival, M., Kalpana, R., and Rajeswari, S. (2000). Activated carbon from an agricultural by-product, for the treatment of dyeing industry wastewater. *Bioresource Technology*, 74(3), 263-265.

Karino, T., Shibayama, M., and Ito, K. (2006). Slide-ring gel: topological gel with freely movable cross-links. *Physica B: Condensed Matter*, 385(1), 692-696.

Khampieng, T., Wongkittithavorn, S., Chaiarwut, S., Ekabutr, P., Pavasant, P., and Supaphol, P. (2018). Silver nanoparticles-based hydrogel: characterization of material parameters for pressure ulcer dressing applications. *Journal of Drug Delivery Science and Technology*, 44, 91-100.

Khan, M., Koivisto, J. T., Hukka, T. I., Hokka, M., and Kellomäki, M. (2018). Composite hydrogels using bioinspired approach with *in Situ* fast gelation and self-healing ability as future injectable biomaterial. *ACS Applied Materials & Interfaces*, 10(14), 11950-11960.

Kikuchi, A., and Okano, T. (2002). Intelligent thermoresponsive polymeric stationary phases for aqueous chromatography of biological compounds. *Progress in Polymer Science*, 27(6), 1165-1193.

Kim, M., Kwon, J. E., Lee, K., and Koh, W. G. (2018). Signal-amplifying nanoparticle/hydrogel hybrid microarray biosensor for metal-enhanced fluorescence detection of organophosphorus compounds. *Biofabrication*, 10(3), 035002.

Kiser, P. F., Wilson, G., and Needham, D. (2000). Lipid-coated microgels for the triggered release of doxorubicin. *Journal of Controlled Release*, 68(1), 9-22.

Kizilay, M. Y., and Okay, O. (2003). Effect of initial monomer concentration on spatial inhomogeneity in poly(acrylamide) gels. *Macromolecules*, 36(18), 6856-6862.

Kostina, N. Y., Sharifi, S., de los Santos Pereira, A., Michálek, J., Grijpma, D. W., and Rodriguez-Emmenegger, C. (2013). Novel antifouling self-healing poly(carboxybetaine methacrylamide-co-HEMA) nanocomposite hydrogels with superior mechanical properties. *Journal of Materials Chemistry B*, 1(41), 5644-5650.

Kuilla, T., Bhadra, S., Yao, D., Kim, N. H., Bose, S., and Lee, J. H. (2010). Recent advances in graphene based polymer composites. *Progress in Polymer Science*, 35(11), 1350-1375.

Kumar, A., Shalini, Sharma, G., Naushad, M., Kumar, A., Kalia, S., Guo, C., and Mola, G. T. (2017). Facile hetero-assembly of superparamagnetic Fe3O4/BiVO4 stacked on biochar for solar photo-degradation of methyl paraben and pesticide removal from soil. *Journal of Photochemistry and Photobiology A: Chemistry*, 337, 118-131.

Lai, F., Fang, Z., Cao, L., Li, W., Lin, Z., and Zhang, P. (2020). Self-healing flexible and strong hydrogel nanocomposites based on polyaniline for supercapacitors. *Ionics*, 26(6), 3015-3025.

Langer, R., and Tirrell, D. A. (2004). Designing materials for biology and medicine. *Nature*, 428(6982), 487-492.

Lei, Z., Wang, Q., Sun, S., Zhu, W., and Wu, P. (2017). A bioinspired mineral hydrogel as a self-healable, mechanically adaptable ionic skin for highly sensitive pressure sensing. *Advanced Materials*, 29(22), 1700321.

Li, F., Zhang, G., Wang, Z., Jiang, H., Feng, X., Yan, S., Zhang, L., Li, H., Zhao, T., and Liu, M. (2019). Bioinspired nonswellable ultrastrong nanocomposite hydrogels with long-term underwater superoleophobic behavior. *Chemical Engineering Journal*, 375, 122047.

Li, P., Kim, N. H., Siddaramaiah and Lee, J. H. (2009). Swelling behavior of polyacrylamide/laponite clay nanocomposite hydrogels: pH-sensitive property. *Composites Part B: Engineering*, 40(4), 275-283.

Li, Q., Lu, F., Zhou, G., Yu, K., Lu, B., Xiao, Y., Dai, F., Wu, D., and Lan, G. (2017). Silver inlaid with gold nanoparticle/chitosan wound dressing enhances antibacterial activity and porosity, and promotes wound healing. *Biomacromolecules*, 18(11), 3766-3775.

Li, Z., Mi, W., Wang, H., Su, Y., and He, C. (2014). Nano-hydroxyapatite/polyacrylamide composite hydrogels with high mechanical strengths and cell adhesion properties. *Colloids and Surfaces B: Biointerfaces*, 123, 959-964.

Lin, C., Sheng, D., Liu, X., Xu, S., Ji, F., Dong, L., Zhou, Y., and Yang, Y. (2017). A self-healable nanocomposite based on dual-crosslinked Graphene Oxide/Polyurethane. *Polymer*, 127, 241-250.

Lin, S., Yuk, H., Zhang, T., Parada, G. A., Koo, H., Yu, C., and Zhao, X. (2016). Stretchable hydrogel electronics and devices. *Advanced Materials*, 28(22), 4497-505.

Liu, H., Liu, M., Zhang, L., Ma, L., Chen, J., and Wang, Y. (2010). Dual-stimuli sensitive composites based on multi-walled carbon nanotubes and poly(N,N-diethylacrylamide-co-acrylic acid) hydrogels. *Reactive and Functional Polymers*, 70(5), 294-300.

Liu, L., Barber, A. H., Nuriel, S., and Wagner, H. D. (2005). Mechanical properties of functionalized single-walled carbon-nanotube/poly(vinyl alcohol) nanocomposites. *Advanced Functional Materials*, 15(6), 975-980.

Liu, Y., Song, S., Liu, S., Zhu, X., and Wang, P. (2022). Application of nanomaterial in hydrogels related to wound healing. *Journal of Nanomaterials*, 2022, 4656037.

Liu, Y., Zhan, H., Skelton, S., and Lee, B. P. (2013). Marine adhesive containing nanocomposite hydrogel with enhanced materials and bioadhesive properties. *MRS Online Proceedings Library*, 1569(1), 33-38.

Long, Y., Wang, Z., Ding, H., Geng, J., and Bai, B. (2019). Investigation and characterization of a robust nanocomposite preformed particle gel for enhanced oil recovery. *Energy & Fuels*, 33(6), 5055-5066.

Loste, J., Lopez-Cuesta, J. M., Billon, L., Garay, H., and Save, M. (2019). Transparent polymer nanocomposites: an overview on their synthesis and advanced properties. *Progress in Polymer Science*, 89, 133-158.

Lu, B., Li, T., Zhao, H., Li, X., Gao, C., Zhang, S., and Xie, E. (2012). Graphene-based composite materials beneficial to wound healing. *Nanoscale*, 4(9), 2978-2982.

Ma, P. C., Siddiqui, N. A., Marom, G., and Kim, J. K. (2010). Dispersion and functionalization of carbon nanotubes for polymer-based nanocomposites: a review. *Composites Part A: Applied Science and Manufacturing*, 41(10), 1345-1367.

Ma, Y., Ge, Y., and Li, L. (2017). Advancement of multifunctional hybrid nanogel systems: construction and application in drug co-delivery and imaging technique. *Materials Science and Engineering: C*, 71, 1281-1292.

Magnin, D., Lefebvre, J., Chornet, E., and Dumitriu, S. (2004). Physicochemical and structural characterization of a polyionic matrix of interest in biotechnology, in the pharmaceutical and biomedical fields. *Carbohydrate Polymers*, 55(4), 437-453.

Mai, B., Jia, M., Liu, S., Sheng, Z., Li, M., Gao, Y., Wang, X., Liu, Q., and Wang, P. (2020). Smart hydrogel-based DVDMS/bFGF nanohybrids for antibacterial phototherapy with multiple damaging sites and accelerated wound healing. *ACS Applied Materials & Interfaces*, 12(9), 10156-10169.

Makuuchi, K. (2010). Critical review of radiation processing of hydrogel and polysaccharide. *Radiation Physics and Chemistry*, 79(3), 267-271.

Mallam, S., Horkay, F., Hecht, A. M., and Geissler, E. (1989). Scattering and swelling properties of inhomogeneous polyacrylamide gels. *Macromolecules*, 22(8), 3356-3361.

McCracken, J. M., Rauzan, B. M., Kjellman, J. C. E., Kandel, M. E., Liu, Y. H., Badea, A., Miller, L. A., Rogers, S. A., Popescu, G., and Nuzzo, R. G. (2019). 3D-printed hydrogel composites for predictive temporal (4D) cellular organizations and patterned biogenic mineralization. *Advanced Healthcare Materials*, 8(1), 1800788.

Meddahi-Pellé, A., Legrand, A., Marcellan, A., Louedec, L., Letourneur, D., and Leibler, L. (2014). Organ repair, hemostasis, and *in vivo* bonding of medical devices by aqueous solutions of nanoparticles. *Angewandte Chemie International Edition*, 53(25), 6369-6373.

Meenach, S. A., Anderson, A. A., Suthar, M., Anderson, K. W., and Hilt, J. Z. (2009). Biocompatibility analysis of magnetic hydrogel nanocomposites based on poly(N-isopropylacrylamide) and iron oxide. *Journal of Biomedical Materials Research Part A*, 91A(3), 903-909.

Mehrali, M., Thakur, A., Pennisi, C. P., Talebian, S., Arpanaei, A., Nikkhah, M., and Dolatshahi-Pirouz, A. (2017). Nanoreinforced hydrogels for tissue engineering: biomaterials that are compatible with load-bearing and electroactive tissues. *Advanced Materials*, 29(8), 1603612.

Mendes, B. B., Gómez-Florit, M., Hamilton, A. G., Detamore, M. S., Domingues, R. M. A., Reis, R. L., and Gomes, M. E. (2019). Human platelet lysate-based nanocomposite bioink for bioprinting hierarchical fibrillar structures. *Biofabrication*, 12(1), 015012.

Meng, Y., Cao, J., Chen, Y., Yu, Y., and Ye, L. (2020). 3D printing of a poly(vinyl alcohol)-based nano-composite hydrogel as an artificial cartilage replacement and the improvement mechanism of printing accuracy. *Journal of Materials Chemistry B*, 8(4), 677-690.

Merino, S., Martín, C., Kostarelos, K., Prato, M., and Vázquez, E. (2015). Nanocomposite hydrogels: 3D polymer-nanoparticle synergies for on-demand drug delivery. *ACS Nano*, 9(5), 4686-4697.

Miyazaki, S., Karino, T., Endo, H., Haraguchi, K., and Shibayama, M. (2006). Clay concentration dependence of microstructure in deformed poly(N-isopropylacrylamide)-clay nanocomposite gels. *Macromolecules*, 39(23), 8112-8120.

Montarnal, D., Capelot, M., Tournilhac, F., and Leibler, L. (2011). Silica-like malleable materials from permanent organic networks. *Science*, 334(6058), 965-968.

Moon, Y. E., Jung, G., Yun, J., and Kim, H. I. (2013). Poly(vinyl alcohol)/poly(acrylic acid)/TiO2/graphene oxide nanocomposite hydrogels for pH-sensitive photocatalytic degradation of organic pollutants. *Materials Science and Engineering: B*, 178(17), 1097-1103.

Nadernezhad, A., Caliskan, O. S., Topuz, F., Afghah, F., Erman, B., and Koc, B. (2019). Nanocomposite bioinks based on agarose and 2D nanosilicates with tunable flow properties and bioactivity for 3D bioprinting. *ACS Applied Bio Materials*, 2(2), 796-806.

Nejadnik, M. R., Yang, X., Bongio, M., Alghamdi, H. S., van den Beucken, J. J. J. P., Huysmans, M. C., Jansen, J. A., Hilborn, J., Ossipov, D., and Leeuwenburgh, S. C. G. (2014). Self-healing hybrid nanocomposites consisting of bisphosphonated hyaluronan and calcium phosphate nanoparticles. *Biomaterials*, 35(25), 6918-6929.

Oishi, M., Hayashi, H., Iijima, M., and Nagasaki, Y. (2007). Endosomal release and intracellular delivery of anticancer drugs using pH-sensitive PEGylated nanogels. *Journal of Materials Chemistry*, 17(35), 3720-3725.

Okajima, T., Harada, I., Nishio, K., and Hirotsu, S. (2000). Discontinuous crossover between fast and slow kinetics at the volume phase transition in poly-N-isopropylacrylamide gels. *Japanese Journal of Applied Physics*, 39(2), L875-L877.

Okumura, Y., and Ito, K. (2001). The polyrotaxane gel: a topological gel by figure-of-eight cross-links. *Advanced Materials*, 13(7), 485-487.

Olad, A., and Rashidzadeh, A. (2008). Preparation and anticorrosive properties of PANI/Na-MMT and PANI/O-MMT nanocomposites. *Progress in Organic Coatings*, 62(3), 293-298.

Omidi, M., Yadegari, A., and Tayebi, L. (2017). Wound dressing application of pH-sensitive carbon dots/chitosan hydrogel. *RSC Advances*, 7(18), 10638-10649.

Omidian, H., Rocca, J. G., and Park, K. (2005). Advances in superporous hydrogels. *Journal of Controlled Release*, 102(1), 3-12.

Orthman, J., Zhu, H. Y., and Lu, G. Q. (2003). Use of anion clay hydrotalcite to remove coloured organics from aqueous solutions. *Separation and Purification Technology*, 31(1), 53-59.

Oun, A. A., and Rhim, J. W. (2017). Carrageenan-based hydrogels and films: effect of ZnO and CuO nanoparticles on the physical, mechanical, and antimicrobial properties. *Food Hydrocolloids*, 67, 45-53.

Palioura, D., Armes, S. P., Anastasiadis, S. H., and Vamvakaki, M. (2007). Metal nanocrystals incorporated within pH-responsive microgel particles. *Langmuir*, 23(10), 5761-5768.

Pardo-Yissar, V., Gabai, R., Shipway, A. N., Bourenko, T., and Willner, I. (2001). Gold nanoparticle/hydrogel composites with solvent-switchable electronic properties. *Advanced Materials*, 13(17), 1320-1323.

Parisi-Amon, A., Lo, D. D., Montoro, D. T., Dewi, R. E., Longaker, M. T., and Heilshorn, S. C. (2017). Protein-nanoparticle hydrogels that self-assemble in response to peptide-based molecular recognition. *ACS Biomaterials Science & Engineering*, 3(5), 750-756.

Park, J., Kim, I. Y., Patel, M., Moon, H. J., Hwang, S. J., and Jeong, B. (2015). 2D and 3D hybrid systems for enhancement of chondrogenic differentiation of tonsil-derived mesenchymal stem cells. *Advanced Functional Materials*, 25(17), 2573-2582.

Pasqui, D., Atrei, A., Giani, G., De Cagna, M., and Barbucci, R. (2011). Metal oxide nanoparticles as cross-linkers in polymeric hybrid hydrogels. *Materials Letters*, 65(2), 392-395.

Paul, A., Manoharan, V., Krafft, D., Assmann, A., Uquillas, J. A., Shin, S. R., Hasan, A., Hussain, M. A., Memic, A., Gaharwar, A. K., and Khademhosseini, A. (2016). Nanoengineered biomimetic hydrogels for guiding human stem cell osteogenesis in three dimensional microenvironments. *Journal of Materials Chemistry B*, 4(20), 3544-3554.

Peak, C. W., Wilker, J. J., and Schmidt, G. (2013). A review on tough and sticky hydrogels. *Colloid and Polymer Science*, 291(9), 2031-2047.

Peppas, N. A., Hilt, J. Z., Khademhosseini, A., and Langer, R. (2006). Hydrogels in biology and medicine: from molecular principles to bionanotechnology. *Advanced Materials*, 18(11), 1345-1360.

Peppas, N. A., and Sahlin, J. J. (1989). A simple equation for the description of solute release. III. Coupling of diffusion and relaxation. *International Journal of Pharmaceutics*, 57(2), 169-172.

Pongjanyakul, T., and Puttipipatkhachorn, S. (2007). Xanthan-alginate composite gel beads: molecular interaction and *in vitro* characterization. *International Journal of Pharmaceutics*, 331(1), 61-71.

Qiu, L., Bulut Coskun, M., Tang, Y., Liu, J. Z., Alan, T., Ding, J., Truong, V. T., and Li, D. (2016). Ultrafast dynamic piezoresistive response of graphene-based cellular elastomers. *Advanced Materials*, 28(1), 194-200.

Quesada-Pérez, M., Maroto-Centeno, J. A., Forcada, J., and Hidalgo-Alvarez, R. (2011). Gel swelling theories: the classical formalism and recent approaches. *Soft Matter*, 7(22), 10536-10547.

Rajabi, N., Kharaziha, M., Emadi, R., Zarrabi, A., Mokhtari, H., and Salehi, S. (2020). An adhesive and injectable nanocomposite hydrogel of thiolated gelatin/gelatin methacrylate/Laponite® as a potential surgical sealant. *Journal of Colloid and Interface Science*, 564, 155-169.

Rezazadeh, B., Sirousazar, M., Abbasi-Chianeh, V., and Kheiri, F. (2020). Polymer-clay nanocomposite hydrogels for molecular irrigation application. *Journal of Applied Polymer Science*, 137(18), 48631.

Romero, G., and Moya, S. E. (2012). Synthesis of organic nanoparticles. *Frontiers of Nanoscience*, 115-141.

Rose, S., Prevoteau, A., Elzière, P., Hourdet, D., Marcellan, A., and Leibler, L. (2014). Nanoparticle solutions as adhesives for gels and biological tissues. *Nature*, 505(7483), 382-385.

Ruan, C., Zeng, K., and Grimes, C. A. (2003). A mass-sensitive pH sensor based on a stimuli-responsive polymer. *Analytica Chimica Acta*, 497(1), 123-131.

Santiago, F., Mucientes, A. E., Osorio, M., and Rivera, C. (2007). Preparation of composites and nanocomposites based on bentonite and poly(sodium acrylate). Effect of amount of bentonite on the swelling behaviour. *European Polymer Journal*, 43(1), 1-9.

Sengupta, R., Bandyopadhyay, A., Sabharwal, S., Chaki, T. K., and Bhowmick, A. K. (2005). Polyamide-6,6/*in situ* silica hybrid nanocomposites by sol-gel technique: synthesis, characterization and properties. *Polymer*, 46(10), 3343-3354.

Servant, A., Leon, V., Jasim, D., Methven, L., Limousin, P., Fernandez-Pacheco, E. V., Prato, M., and Kostarelos, K. (2014). Graphene-based electroresponsive scaffolds as polymeric implants for on-demand drug delivery. *Advanced Healthcare Materials*, 3(8), 1334-1343.

Servant, A., Methven, L., Williams, R. P., and Kostarelos, K. (2013). Electroresponsive polymer-carbon nanotube hydrogel hybrids for pulsatile drug delivery *in vivo. Advanced Healthcare Materials*, 2(6), 806-811.

Shao, C., Chang, H., Wang, M., Xu, F., and Yang, J. (2017). High-strength, tough, and self-healing nanocomposite physical hydrogels based on the synergistic effects of dynamic hydrogen bond and dual coordination bonds. *ACS Applied Materials & Interfaces*, 9(34), 28305-28318.

Shao, C., Wang, M., Chang, H., Xu, F., and Yang, J. (2017). A Self-healing cellulose nanocrystal-poly(ethylene glycol) nanocomposite hydrogel via Diels-Alder click reaction. *ACS Sustainable Chemistry & Engineering*, 5(7), 6167-6174.

Sharma, G., Bhogal, S., Naushad, M., Inamuddin, Kumar, A., and Stadler, F. J. (2017). Microwave assisted fabrication of La/Cu/Zr/carbon dots trimetallic nanocomposites with their adsorptional vs photocatalytic efficiency for remediation of persistent organic pollutants. *Journal of Photochemistry and Photobiology A: Chemistry*, 347, 235-243.

Sharma, G., Thakur, B., Naushad, M., Kumar, A., Stadler, F. J., Alfadul, S. M., and Mola, G. T. (2018). Applications of nanocomposite hydrogels for biomedical engineering and environmental protection. *Environmental Chemistry Letters*, 16(1), 113-146.

Shibayama, M., Karino, T., Miyazaki, S., Okabe, S., Takehisa, T., and Haraguchi, K. (2005). Small-angle neutron scattering study on uniaxially stretched poly(N-isopropylacrylamide)-clay nanocomposite gels. *Macromolecules*, 38(26), 10772-10781.

Shin, S., Kwak, H., and Hyun, J. (2018). Melanin nanoparticle-incorporated silk fibroin hydrogels for the enhancement of printing resolution in 3D-projection stereolithography of poly(ethylene glycol)-tetraacrylate bio-ink. *ACS Applied Materials & Interfaces*, 10(28), 23573-23582.

Sirousazar, M., Forough, M., Farhadi, K., Shaabani, Y., and Molaei, R. (2014). Hydrogels: properties, preparation, characterization and biomedical applications in tissue engineering, drug delivery and wound care, *Advanced Healthcare Materials*, 295-357.

Sirousazar, M., and Ghasemi, M. (2018). Nanocomposite hydrogel stoppers for preventing the natural gas leakage during the welding. *Nashrieh Shimi va Mohandesi Shimi Iran*, 37(2), 67-77.

Sirousazar, M., and Kokabi, M. (2012). Intelligent nanocomposite hydrogels, *Intelligent Nanomaterials*, 487-531.

Sperinde, J. J., and Griffith, L. G. (2000). Control and prediction of gelation kinetics in enzymatically cross-linked poly(ethylene glycol) hydrogels. *Macromolecules*, 33(15), 5476-5480.

Stornes, M., Shrestha, B., and Dias, R. S. (2018). pH-dependent polyelectrolyte bridging of charged nanoparticles. *The Journal of Physical Chemistry B*, 122(44), 10237-10246.

Strong, L. E., and West, J. L. (2011). Thermally responsive polymer-nanoparticle composites for biomedical applications. *WIREs Nanomedicine and Nanobiotechnology*, 3(3), 307-317.

Stuart, M. A. C., Huck, W. T. S., Genzer, J., Müller, M., Ober, C., Stamm, M., Sukhorukov, G. B., Szleifer, I., Tsukruk, V. V., Urban, M., Winnik, F., Zauscher, S., Luzinov, I., and Minko, S. (2010). Emerging applications of stimuli-responsive polymer materials. *Nature Materials*, 9(2), 101-113.

Su, E., and Okay, O. (2018). Hybrid cross-linked poly(2-acrylamido-2-methyl-1-propanesulfonic acid) hydrogels with tunable viscoelastic, mechanical and self-healing properties. *Reactive and Functional Polymers*, 123, 70-79.

Su, M., Liu, Y., Li, S., Fang, Z., He, B., Zhang, Y., Li, Y., and He, P. (2019). A rubber-like, underwater superoleophobic hydrogel for efficient oil/water separation. *Chemical Engineering Journal*, 361, 364-372.

Sun, J., Schmidt, B. V. K. J., Wang, X., and Shalom, M. (2017). Self-standing carbon nitride-based hydrogels with high photocatalytic activity. *ACS Applied Materials & Interfaces*, 9(3), 2029-2034.

Tai, Z., Yang, J., Qi, Y., Yan, X., and Xue, Q. (2013). Synthesis of a graphene oxide-polyacrylic acid nanocomposite hydrogel and its swelling and electroresponsive properties. *RSC Advances*, 3(31), 12751-12757.

Tamesue, S., Ohtani, M., Yamada, K., Ishida, Y., Spruell, J. M., Lynd, N. A., Hawker, C. J., and Aida, T. (2013). Linear versus dendritic molecular binders for hydrogel network formation with clay nanosheets: studies with ABA triblock copolyethers carrying guanidinium ion pendants. *Journal of the American Chemical Society*, 135(41), 15650-15655.

Tan, C., Cao, X., Wu, X. J., He, Q., Yang, J., Zhang, X., Chen, J., Zhao, W., Han, S., Nam, G. H., Sindoro, M., and Zhang, H. (2017). Recent advances in ultrathin two-dimensional nanomaterials. *Chemical Reviews*, 117(9), 6225-6331.

Tan, Y., Wang, D., Xu, H., Yang, Y., An, W., Yu, L., Xiao, Z., and Xu, S. (2018). A fast, reversible, and robust gradient nanocomposite hydrogel actuator with water-promoted thermal response. *Macromolecular Rapid Communications*, 39(8), e1700863.

Teja, A. S., and Koh, P. Y. (2009). Synthesis, properties, and applications of magnetic iron oxide nanoparticles. *Progress in Crystal Growth and Characterization of Materials*, 55(1), 22-45.

Thakur, T., Xavier, J. R., Cross, L., Jaiswal, M. K., Mondragon, E., Kaunas, R., and Gaharwar, A. K. (2016). Photocrosslinkable and elastomeric hydrogels for bone regeneration. *Journal of Biomedical Materials Research Part A*, 104(4), 879-88.

Tongwa, P., and Bai, B. (2014). Degradable nanocomposite preformed particle gel for chemical enhanced oil recovery applications. *Journal of Petroleum Science and Engineering*, 124, 35-45.

Turcheniuk, K., and Mochalin, V. N. (2017). Biomedical applications of nanodiamond (Review). *Nanotechnology*, 28(25), 252001.

Ullah, F., Othman, M. B. H., Javed, F., Ahmad, Z., and Akil, H. M. (2015). Classification, processing and application of hydrogels: a review. *Materials Science and Engineering: C*, 57, 414-433.

Vallés-Lluch, A., Poveda-Reyes, S., Amorós, P., Beltrán, D., and Monleón Pradas, M. (2013). Hyaluronic acid-silica nanohybrid gels. *Biomacromolecules*, 14(12), 4217-4225.

Varaprasad, K., Raghavendra, G. M., Jayaramudu, T., Yallapu, M. M., and Sadiku, R. (2017). A mini review on hydrogels classification and recent developments in miscellaneous applications. *Materials Science and Engineering: C*, 79, 958-971.

Vermonden, T., Censi, R., and Hennink, W. E. (2012). Hydrogels for protein delivery. *Chemical Reviews*, 112(5), 2853-88.

Wahid, F., Zhong, C., Wang, H. S., Hu, X. H., and Chu, L. Q. (2017). Recent advances in antimicrobial hydrogels containing metal ions and metals/metal oxide nanoparticles. *Polymers*, 9(12), 636.

Wang, C., Flynn, N. T., and Langer, R. (2004). Controlled structure and properties of thermoresponsive nanoparticle-hydrogel composites. *Advanced Materials*, 16(13), 1074-1079.

Wang, D., Song, J., Wang, J., Zhang, Z., Shen, Q., Wang, J., Guo, H., Wang, R., Xie, C., Lu, W., and Liu, M. (2020). Efficient gene delivery based on guanidyl-nucleic acid molecular interactions. *Advanced Functional Materials*, 30(45), 2004783.

Wang, J., Hu, H., Yang, Z., Wei, J., and Li, J. (2016). IPN hydrogel nanocomposites based on agarose and ZnO with antifouling and bactericidal properties. *Materials Science and Engineering: C*, 61, 376-386.

Wang, L., Geng, J., and Bai, B. (2018). Highly deformable nano-cross-linker-bridged nanocomposite hydrogels for water management of oil recovery. *Energy & Fuels*, 32(3), 3068-3076.

Wang, L. S., Du, C., Chung, J. E., and Kurisawa, M. (2012). Enzymatically cross-linked gelatin-phenol hydrogels with a broader stiffness range for osteogenic differentiation of human mesenchymal stem cells. *Acta Biomaterialia*, 8(5), 1826-1837.

Wang, N., Ma, M., Luo, Y., Liu, T., Zhou, P., Qi, S., Xu, Y., and Chen, H. (2018). Mesoporous silica nanoparticles-reinforced hydrogel scaffold together with pinacidil loading to improve stem cell adhesion. *ChemNanoMat*, 4(7), 631-641.

Wang, Q., Gao, Z., and Yu, K. (2017). Interfacial self-healing of nanocomposite hydrogels: theory and experiment. *Journal of the Mechanics and Physics of Solids*, 109, 288-306.

Wang, Q., Mynar, J. L., Yoshida, M., Lee, E., Lee, M., Okuro, K., Kinbara, K., and Aida, T. (2010). High-water-content mouldable hydrogels by mixing clay and a dendritic molecular binder. *Nature*, 463(7279), 339-343.

Wang, T., Zheng, S., Sun, W., Liu, X., Fu, S., and Tong, Z. (2014). Notch insensitive and self-healing PNIPAm-PAM-clay nanocomposite hydrogels. *Soft Matter*, 10(19), 3506-3512.

Wang, Y., Dong, A., Yuan, Z., and Chen, D. (2012). Fabrication and characterization of temperature-, pH- and magnetic-field-sensitive organic/inorganic hybrid poly (ethylene glycol)-based hydrogels. *Colloids and Surfaces A: Physicochemical and Engineering Aspects*, 415, 68-76.

Wang, Z., Zhao, J., Tang, W., Hu, L., Chen, X., Su, Y., Zou, C., Wang, J., Lu, W. W., Zhen, W., Zhang, R., Yang, D., and Peng, S. (2019). Multifunctional nanoengineered hydrogels consisting of black phosphorus nanosheets upregulate bone formation. *Small*, 15(41), 1901560.

Wei, S., Liu, X., Zhou, J., Zhang, J., Dong, A., Huang, P., Wang, W., and Deng, L. (2020). Dual-crosslinked nanocomposite hydrogels based on quaternized chitosan and clindamycin-loaded hyperbranched nanoparticles for potential antibacterial applications. *International Journal of Biological Macromolecules*, 155, 153-162.

Wirthl, D., Pichler, R., Drack, M., Kettlguber, G., Moser, R., Gerstmayr, R., Hartmann, F., Bradt, E., Kaltseis, R., Siket, C. M., Schausberger, S. E., Hild, S., Bauer, S., and Kaltenbrunner, M. (2017). Instant tough bonding of hydrogels for soft machines and electronics. *Science Advances*, 3(6), e1700053.

Wiśniewski, R., Śmieszek, E., and Kamińska, E. (1995). Three-dimensional solubility parameters: simple and effective determination of compatibility regions. *Progress in Organic Coatings*, 26(2), 265-274.

Wu, C. J., Wilker, J. J., and Schmidt, G. (2013). Robust and adhesive hydrogels from cross-linked poly(ethylene glycol) and silicate for biomedical use. *Macromolecular Bioscience*, 13(1), 59-66.

Wu, D., Xu, J., Chen, Y., Yi, M., and Wang, Q. (2018). Gum arabic: a promising candidate for the construction of physical hydrogels exhibiting highly stretchable, self-healing and tensility reinforcing performances. *Carbohydrate Polymers*, 181, 167-174.

Wu, S., Li, H., and Chen, J. P. (2004). Modeling investigation of volume variation kinetics of fast response hydrogels. *Journal of Macromolecular Science, Part C*, 44(2), 113-130.

Xu, Y., Wu, Q., Sun, Y., Bai, H., and Shi, G. (2010). Three-dimensional self-assembly of graphene oxide and DNA into multifunctional hydrogels. *ACS Nano*, 4(12), 7358-7362.

Yang, C. M., Lee, J., Lee, H., and Park, W. H. (2022). ZnO nanoparticle-embedded modified silk fibroin-tannin multifunctional hydrogel. *International Journal of Biological Macromolecules*, 210, 1-10.

Yang, J., Zhu, L., Yan, X., Wei, D., Qin, G., Liu, B., Liu, S., and Chen, Q. (2016). Hybrid nanocomposite hydrogels with high strength and excellent self-recovery performance. *RSC Advances*, 6(64), 59131-59140.

Yetisen, A. K., Naydenova, I., da Cruz Vasconcellos, F., Blyth, J., and Lowe, C. R. (2014). Holographic sensors: three-dimensional analyte-sensitive nanostructures and their applications. *Chemical Reviews*, 114(20), 10654-10696.

Yu, Y. H., Jen, C. C., Huang, H. Y., Wu, P. C., Huang, C. C., and Yeh, J. M. (2004). Preparation and properties of heterocyclically conjugated poly(3-hexylthiophene)-clay nanocomposite materials. *Journal of Applied Polymer Science*, 91(6), 3438-3446.

Yuan, L., Pan, M., Lei, M., Zhou, X., Hu, D., Liu, Q., Chen, Y., Li, W., and Qian, Z. (2020). A novel composite of micelles and hydrogel for improving skin delivery of hydrocortisone and application in atopic dermatitis therapy. *Applied Materials Today*, 19, 100593.

Yue, Y., Kurokawa, T., Haque, M. A., Nakajima, T., Nonoyama, T., Li, X., Kajiwara, I., and Gong, J. P. (2014). Mechano-actuated ultrafast full-colour switching in layered photonic hydrogels. *Nature Communications*, 5, 4659.

Zareie, C., Bahramian, A. R., Sefti, M. V., and Salehi, M. B. (2019). Network-gel strength relationship and performance improvement of polyacrylamide hydrogel using nano-silica; with regards to application in oil wells conditions. *Journal of Molecular Liquids*, 278, 512-520.

Zhao, Q. S., Ji, Q. X., Xing, K., Li, X. Y., Liu, C. S., and Chen, X. G. (2009). Preparation and characteristics of novel porous hydrogel films based on chitosan and glycerophosphate. *Carbohydrate Polymers*, 76(3), 410-416.

Zhao, W., Odelius, K., Edlund, U., Zhao, C., and Albertsson, A. C. (2015). *In situ* synthesis of magnetic field-responsive hemicellulose hydrogels for drug delivery. *Biomacromolecules*, 16(8), 2522-2528.

Zhao, X., Guo, B., Wu, H., Liang, Y., and Ma, P. X. (2018). Injectable antibacterial conductive nanocomposite cryogels with rapid shape recovery for noncompressible hemorrhage and wound healing. *Nature Communications*, 9(1), 2784.

Zhao, X., Huebsch, N., Mooney, D. J., and Suo, Z. (2010). Stress-relaxation behavior in gels with ionic and covalent crosslinks. *Journal of Applied Physics*, 107(6), 063509.

Zhong, M., Liu, Y. T., and Xie, X. M. (2015). Self-healable, super tough graphene oxide-poly(acrylic acid) nanocomposite hydrogels facilitated by dual cross-linking effects through dynamic ionic interactions. *Journal of Materials Chemistry B*, 3(19), 4001-4008.

Zhou, L., He, B., and Zhang, F. (2012). Facile one-pot synthesis of iron oxide nanoparticles cross-linked magnetic poly(vinyl alcohol) gel beads for drug delivery. *ACS Applied Materials & Interfaces*, 4(1), 192-199.

Zhu, H. Y., Fu, Y. Q., Jiang, R., Yao, J., Xiao, L., and Zeng, G. M. (2012). Novel magnetic chitosan/poly(vinyl alcohol) hydrogel beads: preparation, characterization and application for adsorption of dye from aqueous solution. *Bioresource Technology*, 105, 24-30.

Zhu, W., Lu, J., and Dai, L. (2018). Multifunctional pH-responsive sprayable hydrogel based on chitosan and lignin-based nanoparticles. *Particle & Particle Systems Characterization*, 35(12), 1800145.

Zolfaghari, R., Katbab, A. A., Nabavizadeh, J., Tabasi, R. Y., and Nejad, M. H. (2006). Preparation and characterization of nanocomposite hydrogels based on polyacrylamide for enhanced oil recovery applications. *Journal of Applied Polymer Science*, 100(3), 2096-2103.

Chapter 2

Self-Healing Mechanisms in Materials and Potential Applications of Their Nanocomposite Hydrogels

Aminur Rahman
and Abu Bin Imran*
Department of Chemistry, Faculty of Engineering,
Bangladesh University of Engineering and Technology, Dhaka, Bangladesh

Abstract

Over the last several decades, many self-healing materials, including nanocomposites, polymers, elastomers, hydrogels, and so on, have been developed using different approaches for various applications. A self-healing mechanism enables the integration of sufficient functionalities into the lateral chain of these products. Nanofillers boost the mechanical strength of self-healing nanocomposites; however, the self-healing properties are typically governed by the appropriate functionalization of these nanofillers. The functionalization and dispersion issues with nanofillers are still prevalent in self-healing formulations. Self-healing nanocomposites are employed in engineering applications such as coatings, sensors, actuators, soft robotics, and biomedical applications such as drug delivery, adhesives, tissue engineering, and wound healing. This chapter provides a detailed approach and a brief overview of achieving various self-healing materials and potential applications of self-healing nanocomposite hydrogels have also been discussed.

Keywords: self-healing, nanofiller, elastomer, covalent interactions, non-covalent interactions

2.1. Introduction

Living organisms can heal their injury to continue their active and regenerative functions. The capacity to sustain these metabolic activities determines their lifespan. The main role of synthetic materials is functionality; however, they lack metabolic properties. To increase their lifespan, self-healing has become the prime aspect of scientific and technological research. Malinskii et al., (1970) first reported comprehensive research on the self-healing of cracks in polyvinyl alcohol (PVA) specimens paved the way for the concept of self-healing. After that,

* Corresponding Author's Email: abimran@chem.buet.ac.bd.

In: Nanocomposite Hydrogels and their Emerging Applications
Editor: Mohammad Sirousazar
ISBN: 979-8-88697-675-5

self-healing became an appealing subject, and their research in thermoplastic and crosslinked polymer systems began (Wool and O'Connor, 1981; Wool, 1980). In the 1990s, a matrix cracking repair and filling cement was reported (Dry, 1994). White et al., (2001) published their work on observing the self-healing behavior of polymer-based materials, opening up the era of smart healable materials. Since then, self-healing materials have become a major area of interest.

Over the last decade, several successful methods for repairing mechanically damaged materials have been researched and recent developments have been summarized in several review articles (Hager, Greil, Leyens, van der Zwaag, and Schubert, 2010; Khan, Ahmed, and Rabnawaz, 2020; Murphy and Wudl, 2010; Qamar, Sottos, and Trask, 2020; Wool, 2008; Wu, Meure, and Solomon, 2008). Various materials, including nanostructured materials, have been effectively used for the self-healing concept (Barbero, Ford, and Mayugo, 2009; Frei et al., 2013; Kötteritzsch, Schubert, and Hager, 2013; Trask, Williams, and Bond, 2007). The use of nanofillers is extremely favorable not only for enhancing their mechanical characteristics by providing greater stiffness reinforcement but also for enhancing their functional and biological properties (Annabi et al., 2014; Lau and Kiick, 2015; Tee, Wang, Allen, and Bao, 2012).

Typical nanofillers include carbon nanotubes (CNTs) (Qi, Liu, Gao, and Mäder, 2013; Sahoo, Rana, Cho, Li, and Chan, 2010), graphene (Huang, Qi, Boey, and Zhang, 2012; Potts, Dreyer, Bielawski, and Ruoff, 2011), ceramic oxides (Mera, Navrotsky, Sen, Kleebe, and Riedel, 2013), nano silica (Engel and Kickelbick, 2013), and nanocellulose (Galland et al., 2014). With the addition of cellulose nano-whiskers, the tensile strength of PVA has been boosted by up to 60 times (Ding et al., 2018). Additionally, nanofillers have been used to promote self-healing characteristics. For instance, conducting nanofillers with high thermal conductivity (graphene, CNT, and metals) are utilized as nanoscale heaters to enhance self-healing properties in the polymer matrix (Bai and Shi, 2017; Dai, Zhou, Wang, Ding, and Yuan, 2018). There are huge specific interfacial areas in nanocomposite materials, generally more than 100 m^2/g. Optimum nanofiller dispersion and process rheology can only be achieved by carefully designing and managing the interfacial bond dynamics between polymer and nanofiller (including through particle and/or polymer functionalization).

The self-healing properties of nanocomposites and their applications are addressed thoroughly in this article. After a brief discussion of self-healing principles, we proceed on to an overview of various materials that may be used for self-healing. A review of nanocomposite self-healing chemistry is presented thereafter, followed by a discussion of useful applications for self-healing nanocomposite hydrogels.

2.2. Different Mechanisms of Self-Healing

Based on the healing mechanism, three kinds of self-healing materials exist capsule-based, vascular-based, and intrinsic-based (Imran, 2021; Imran, Seki, and Takeoka, 2010; Kessler, 2007). The following section will briefly introduce the mechanism of these materials.

2.2.1. Capsule Mechanism

As a general rule, the capsule-based technique comprises the healing material inside the capsules. A rupture in the system leads the capsules to explode, releasing the healing agent.

When damage occurs, a healing agent is released from the capsules; this causes the healing agent to travel to the damage site, where it may be restored. Microcapsules also contain the healing agent during storage, and then the capsules are dispersed evenly throughout the matrix. The healing agent doesn't interfere while it's being stored. The capsule wall is broken by the injury, which causes the healing agent to leak out and seal the fracture. In Figure 2.1a, the entire process is depicted.

2.2.2. Vascular Mechanism

This procedure is similar to a biological system, how plants and animals use vascular networks to transport healing ingredients to damaged areas. It uses a controlled sequence of network channels to build the material. As seen in Figure 2.1b, hollow tubes or systems with fiber-like features carry the healing agent. This approach is superior to the capsule mechanism because it continuously supplies the healing chemical, even if there are several damages to a particular region. This method transported more healing agents to the fracture than the capsule-based mechanism. Unlike capsule-based techniques, which may be readily implanted into any device but whose functionality can only be used once, this vascular mechanism can be employed several times.

2.2.3. Intrinsic Mechanism

Polymers and polymer composites have fundamental characteristic based on intrinsic mechanisms that facilitates fracture healing under particular circumstances and certain stimuli (such as light, heat, and pH). Because there are no impregnated capsules or vascular systems (hollow fibers) in intrinsic self-healing polymers, this process is superior to the other mechanisms. Therefore, the mechanical characteristics of the specimen remain unaffected. Figure 2.1c represents a simple illustration of the intrinsic process. Most of the investigated self-healing hydrogel, elastomer/polymer, and nanocomposite systems function on the intrinsic mechanism. This chapter, therefore, focuses on materials that have been developed by intrinsic mechanisms.

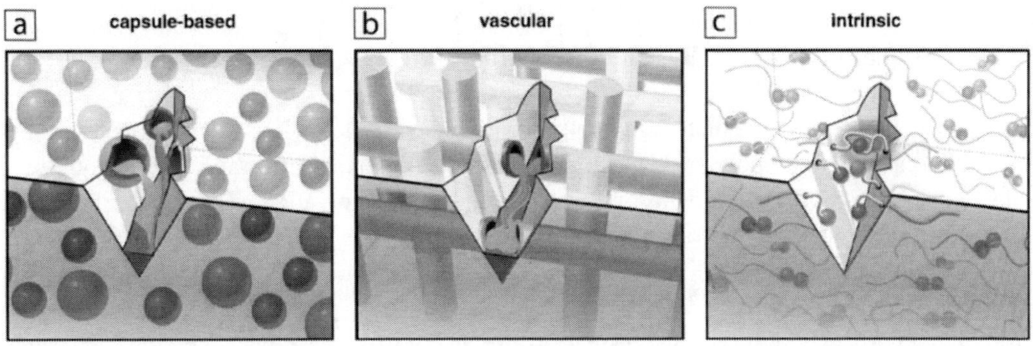

Figure 2.1. Different mechanisms for self-healing (a) capsule, (b) vascular, and (c) intrinsic (Isabel et al., 2020).

2.3. Synthesize and Self-Healing Mechanism of Intrinsic Self-Healing Materials

In intrinsic self-healing approaches, the polymer self-heals after damage by exploiting dynamic/reversible interactions under different physicochemical conditions. Physical interactions (reversible or non-covalent bonding) and chemical interactions (irreversible or covalent bonding) have been used to develop self-healing materials (Figure 2.2).

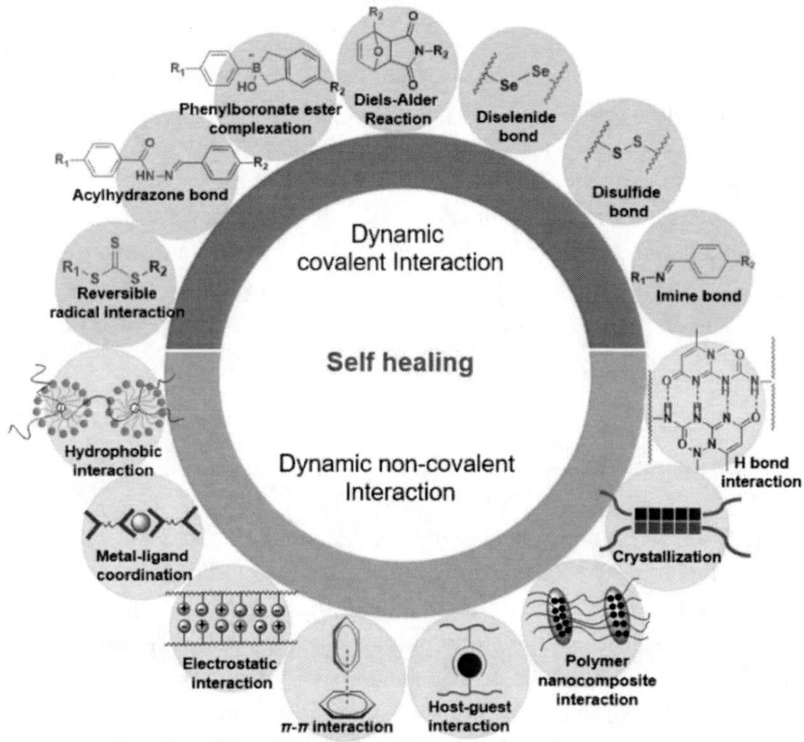

Figure 2.2. Typical examples of commonly used functionalities in covalent and non-covalent interactions (Khan et al., 2020).

2.3.1. Physical Interaction-Based Self-Healing Materials

Non-covalent interactions, such as hydrophobic interactions, host-guest interactions, hydrogen bonds, crystallization, polymer-nanocomposite interactions, electrostatic interaction, pi-pi interaction, and coordination bonds, are at the core of the preparation of physical self-healing gels, elastomers, and polymers.

2.3.1.1. Hydrogen Bonds

Although hydrogen bonds (H-bonds) are weaker than conventional covalent bonds, they can be used to create a wide variety of polymers with different mechanical characteristics because of their positional accuracy and affinity. Strength and network remodeling are influenced by concentration levels, H-bond strength, length, and stiffness of polymer chains. H-bonds have been used to create many successful self-repairing materials. These are multi-hydrogen directional bonds, such as thymine and 2,6-diaminotriaine (DAT) bonds, ureidopyrimidinone (Upy) quadruple H-bonds, and multi-urea are analogous to covalent bonds in terms of their strength and stability. Poly(methyl methacrylate) and two dynamic poly(acrylate amide) blocks have been used to create spherical core-shell nanoparticles that exhibit self-healing properties (Chen and Guan, 2014).

2.3.1.2. Metal-Ligand Coordination

Using N-heterocyclic carbenes (NHCs) and transition metals that can self-heal, complexes were designed to use this property and coordinate with transition metals (Williams, Boydston, and Bielawski, 2007). Heat will repair the damage due to the complex balance between metal and polymer, causing microcracks to flow and coordination bonds to rebuild. The high self-healing temperature results from the coordination bond's strength and lack of soft segments. Furthermore, when the block copolymer hydroxyethylenediamino triacetic acid sodium salt (HEEDTA) was added to Ru(II)-bis-terpyridine crosslinked block copolymers, the temperature variations freed Ru from the ligand, leading to bonding and debonding (Williams et al., 2007). On the other hand, Ru(II)-bis-terpyridine has a binding strength equivalent to a conventional covalent bond that necessitates large dissociation energy (Gohy, Lohmeijer, and Schubert, 2002).

There are many possible combinations of ligands and metal-ligand coordination that can be adjusted. Similar but organized networks of Zn(II)-bis-terpyridine and Fe(II)-bis-terpyridine will lead to self-heal, and heterogeneities will play a key role in network rearrangements. The optically healable supramolecular polymer with an amorphous poly(ethylene-co-butylene) core and 2,6-bis(1-methylbenzimidazolyl)pyridine (Mebip) ligands at the ends can be crosslinked with the help of Zn(II)-bis-terpyridine coordination (Burnworth et al., 2011). In phase-separated morphology, a hydrophobic section and a polar metal-ligand motif are present. Because of this, when ultraviolet (UV) light irradiates, metal-ligand motifs are electrically stimulated. This causes them to dissociate from one another and reduces their molecular mass and viscosity. However, this dissociation can be reversed by exposing them to ultraviolet light again. Macromonomers, which are formed when the hard phase is broken down into smaller fragments, are capable of fixing mechanical damage by returning to their initial morphology via coordination bonds between metals and ligands.

2.3.1.3. Pi-pi Stacking Interactions

Pi-pi-stacking interactions between end-capped pi-electron-deficient groups and pi-electron-rich aromatic backbone molecules have been used to make reversible self-healing supramolecular polymer networks that can be turned on and off by heat. End-capped polysiloxane (Burattini, Colquhoun, Greenland, and Hayes, 2009) or polyamide (Burattini et al., 2011) chains of chain-folding copolyimide (electron-poor) and pyrenyl (electron rich) were used to obtain this. By changing the spacer and changing the blend compositions, the glass transition temperature (Tg) of the network can be tuned to make it self-heal over a wide range of temperatures (about 50-100°C). Due to a flexible "soft" spacer, the stacking interactions would be broken when the material is heated, allowing the pyrenyl end-capped chains to separate from the copolyimide and flow. So, by rearranging the stacking, the damage can be fixed and mechanical strength will be restored.

Figure 2.3 shows how interactions between bis-pyrenyl end groups (compound 1) and naphthalene-diimide (compound 3) chains form the tweezer type. There are also non-tweezer form interactions that can be observed using mono-pyrenyl groups (compound 2). Compared to the [1 + 2] blend, the [1 + 3] blend exhibits higher tensile modulus and elongation, but it needs a higher healing temperature and more time.

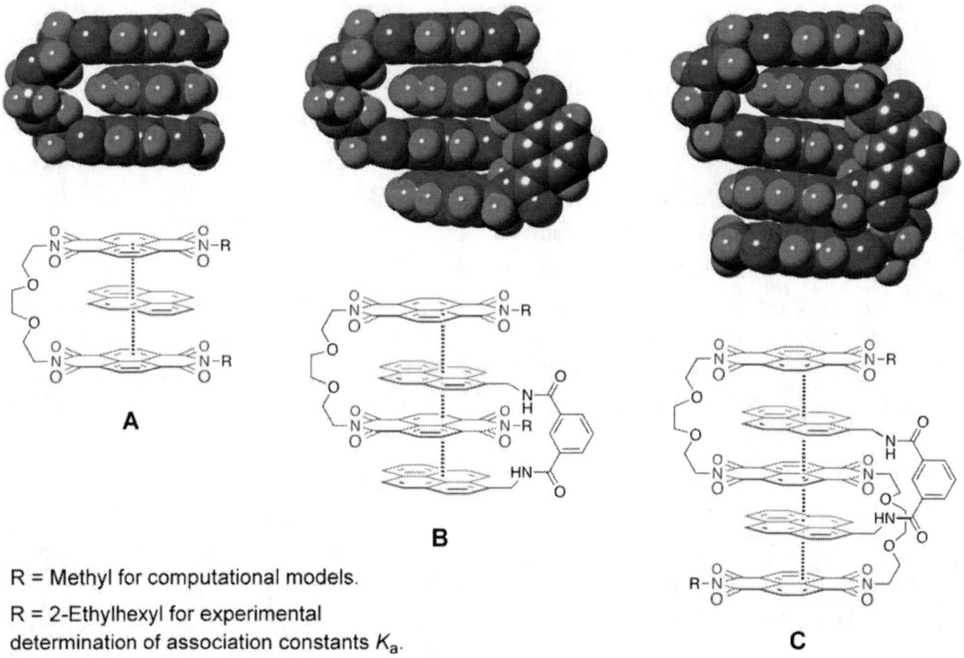

Figure 2.3. An energy-minimized model of the complexes produced between pyrene or tweezer-type derivative and naphthalene diimide oligomers (Burattini et al., 2010).

2.3.1.4. Ionic Interactions

Interactions between positive and negative charged polymer chain functional groups have produced self-healing properties in certain polymers (Eisenberg, Hird, and Moore, 1990). Polyelectrolytes derived from poly(acrylic acid) (PAA)/poly(allylamine hydrochloride) (PAH) pairs are typical examples (Reisch et al., 2014). Ultracentrifugation and extrusion in the presence of salt yielded compact polyelectrolyte complexes (CoPECs) that exhibit self-healing properties. For example, if the concentration of NaCl salt is increased, the self-healing capability increases, suggesting that salt disrupts interactions between polyelectrolyte pairs, resulting in higher chain mobility.

2.3.1.5. Host-Guest Interactions

The term "host-guest interactions" refers to the binding by weak interactions of two structurally diverse but dimensionally matched macromolecular units. Supramolecular hydrogels that are self-healable may be formed by synthetic host-guest pairs capable of reversible guest exchanges (Nakahata, Takashima, Yamaguchi, and Harada, 2011; Pluth and Raymond, 2007). Natural host-guest assemblies have tremendous biological significance. Cyclodextrin (CD), which has a cavity capable of trapping other molecules, may be utilized to make hydrogels by interacting with ferrocene-terminated polyacrylic acid (Fc) (Nakahata et al., 2011). Redox stimuli induced a sol-gel transition, and the dissociation and re-association of CD and ferrocene promoted self-healing properties (Figure 2.4). Supramolecular hydrogels produced from polyacrylamide modified with CD crosslinked by hydrophobic interaction with aliphatic guest groups also demonstrated self-healing properties (Kakuta et al., 2013).

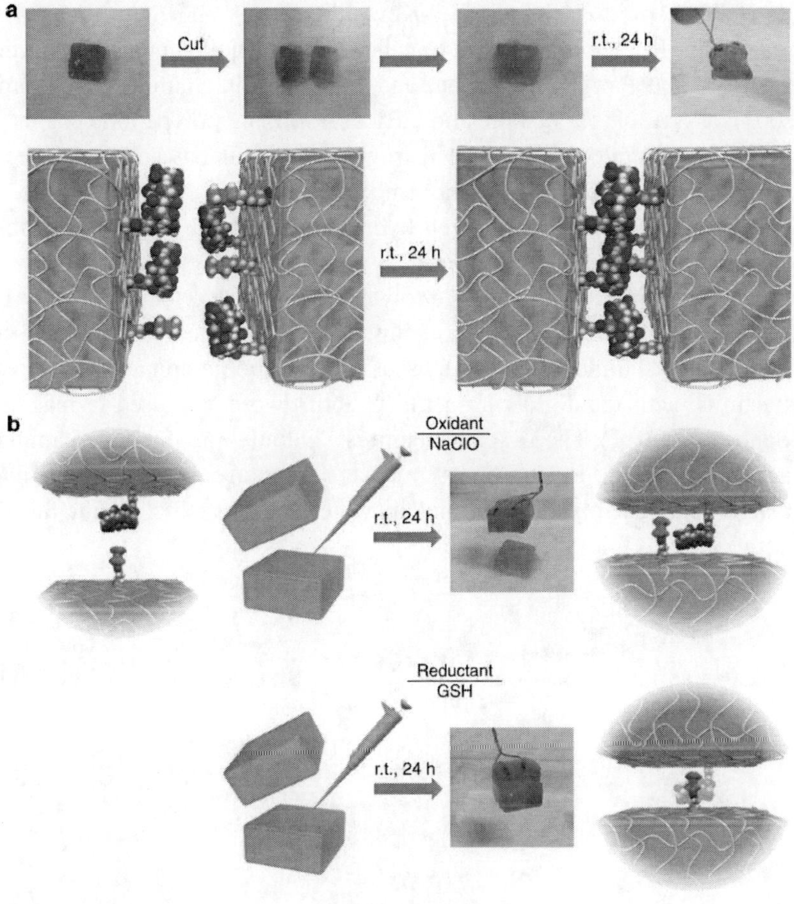

Figure 2.4. (a) Two parts of sliced PAA-6-β-CD/PAA-Fc hydrogel (1+1 wt.%) reunite and the crack adequately heals to create one gel and (b) healing experiment of redox-responsive PAA-6-β-CD / PAA-Fc hydrogel using oxidizing and reducing agents (Nakahata et al., 2011).

2.3.1.6. Hydrophobic Interactions

Hydrophobic interactions are an essential part of creating self-healing hydrogels. Hydrophobic monomer units in the polymer chains cause hydrophobic interactions between the chains, which leads to a temporary crosslink network. Micelle polymerization has recently been developed to make self-healing hydrogels that are held together by hydrophobic interactions between surfactant micelles (Tuncaboylu, Argun, Sahin, Sari, and Okay, 2012; Tuncaboylu, Sahin, Argun, Oppermann, and Okay, 2012; Tuncaboylu, Sari, Oppermann, and Okay, 2011). This system's hydrogels have self-healing properties because of the reversible dissociation and binding of hydrophobic crosslinks in networks. In the presence of NaCl, a micelle-based hydrogel may be formed by copolymerizing hydrophobic monomer stearyl methacrylate (C18) with hydrophilic acrylamide monomer (PAAm-co-C18) in a micellar solution of sodium dodecyl sulfate (SDS).

2.3.1.7. Crystallization

A non-covalent bond is formed between polymer chains with specific conformations, such as L-polylactide (L-PLA) acid (Hales, Barner-Kowollik, Davis, and Stenzel, 2004), α-helix or β-sheet peptide chain (Kotharangannagari, Sánchez-Ferrer, Ruokolainen, and Mezzenga, 2012; Silva et al., 2013), and PVA (Hyon, Cha, and Ikada, 1989; Stammen, Williams, Ku, and Guldberg, 2001), as a result of crystallization. By crystallizing polypeptide chains, the diblock copolypeptides may generate self-healing hydrogels in an aqueous solution. These hydrogels could be synthesized, for example, by self-assembly of diblock copolypeptide amphiphiles with two separate hydrophobicities, i.e., charged hydrophilic segment and hydrophobic segment, with regular architectures in dilute solutions (2-8 wt.%) (Nowak et al., 2002).

In this case, highly charged polyelectrolytes like poly(L-lysine) or poly(L-glutamate sodium salt) were used as hydrophilic domains. Poly(L-leucine) with rod-like a-helices or poly(L-valine) with crystalline sheets were used as hydrophobic domains. As repeating units, filamentous actin (F-actin) and globular actin (G-actin) were also used to make self-healing hydrogels (Sano et al., 2011). The G-actin subunit is a globular protein of 375 amino acid units. It can self-assemble into the F-actin because it has a structure of two helical filaments. The bismaleimide- polyethylene glycol (PEG) linkers were used as a crosslinker to fabricate F-actin hydrogel (Figure 2.5).

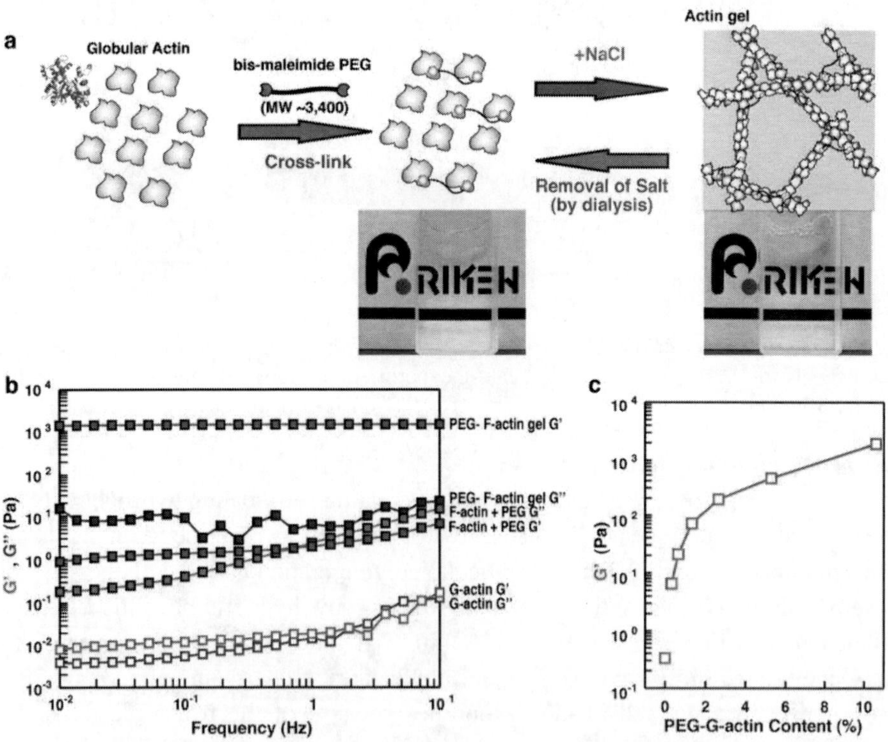

Figure 2.5. (a) A schematic of how to make a crosslinked F-actin hydrogel with bis-maleimide PEG (b) Storage (G′) and loss (G″) moduli (c) concentration dependence of the PEG-G-actin on the storage modulus (G′) of the PEG-F-actin gel (Sano et al., 2011).

By changing the electrolyte concentration or the rheological recovery test, the dynamic sol-gel transformation of F-actin hydrogels could be influenced. The hydrogel's G^0 instantly returns to its original value after removing the breaking strain (20-1000 percent). Extended recovery time is needed when a longer period of breaking strain is administered.

2.3.2. Chemical Interaction-Based Self-Healing Materials

A self-healing exchange reaction cannot be generated in ordinary chemical gels because the networks are permanently linked by stable covalent bonds. Dynamic covalent chemistry is a promising method for creating self-healing gels. Reversible dynamic covalent links in chemical self-healing gels may repair new covalent bonds surrounding the damaged area regardless of whether or not certain stimuli (heat, light, or pH) are used.

2.3.2.1. Phenylboronate Ester Complexations

A reversible boronate ester in water can form diol and boronic acid complexes, the stability of which depends on the pH value (Dowlut and Hall, 2006; Pal, 2021; Paugam, Valencia, Boggess, and Smith, 1994; Pizer and Babcock, 1977; Pizer and Ricatto, 1994). Roberts et al., (2007) came up with a self-healing hydrogel that is made of polymer-bound phenylboronic acid (PBA) and salicylhydroxamic acid (SHA) that binds with reversible boronate ester complexation.

2.3.2.2. Disulfide Bonds

A disulfide bond is a covalent bond that may be affected by changes in pH or redox potential as it is based on a dynamic exchange reaction between thiol and disulfide. Canadell et al., (2011) have invented a self-healing rubber from epoxy resins containing disulfide linkages, a dynamic disulfide polymer. However, only a few disulfide bond-dependent gels have been reported (Kemp, Go, and Jones, 2008; Lei, Xiang, Yuan, Rong, and Zhang, 2014; Messens and Collet, 2013; Nagy, 2013; Pleasants, Guo, and Rabenstein, 1989). By depositing a solution of a -SH functionalized star polymer (multi-arm star polymer chains with thiol (-SH) groups at the arm periphery) on silicon wafers with an oxidation catalyst (I_2 or $FeCl_3$), thin reversible organo-gel films crosslinked by S-S bonds have been synthesized.

2.3.2.3. Imine Bonds (Schiff Base)

Self-healing hydrogels could be made using imine bonds, also known as the Schiff base, as dynamic crosslinkers. Chitosan-PEG hydrogels may be self-healing because of aromatic Schiff base linkages (Yang et al., 2012; Zhang, Tao, Li, and Wei, 2011; Zhang et al., 2012). To design a hydrogel, the benzaldehyde-modified telechelic PEG was used to crosslink the amine groups of primary chitosan polymer chains.

2.3.2.4. Acylhydrazone Bonds

The acylhydrazone link is one of the imine bonds established by the aldehyde and hydrazine reaction utilized to make dynamic covalent polymers with multifaceted functions (Apostolides and Patrickios, 2018; Giuseppone, Fuks, and Lehn, 2006; Maeda, Otsuka, and Takahara, 2009; Ono, Fujii, Nobori, and Lehn, 2007a, 2007b; Skene and Lehn, 2004). There have lately been reports of self-healing gels made of acylhydrazone linkages. As an example, self-healing organogels crosslinked by acylhydrazone linkages via condensation of acylhydrazines have been synthesized at the two ends of polyethylene oxide (PEO) with aldehyde groups in tris[(4-formylphenoxy)methyl] ethane. With glacial acetic acid as a catalyst, it is possible to break and develop new acylhydrazone bonds in polymer networks. This is attributed to damaged gel plates sticking together after being held in contact for some time.

2.3.2.5. Reversible Radical Reactions

Gels based on reversible radical reactions show self-healing activity under light illumination. Nicolay and colleagues (2010) developed a self-healing device capable of repeating self-healing under UV irradiation by dynamic covalent reshuffling of trithiocarbonate (TTC) units. The self-healing organogel containing TTC units was created by employing n-butyl acrylate (BA) as the monomer and TTC as both a chain transfer agent and a crosslinker in anisole by reversible addition-fragmentation chain transfer polymerization (RAFT) (Figure 2.6). The self-healing properties of the organogel may be explained as follows. Bond homolysis in the nitrogen environment under UV irradiation formed carbon radicals by the TTC units at the reactive chain ends of the chain under reaction. The transport of carbon radicals across crosslinked networks allowed for chain rearrangement and the synthesis of new backbones based on reshuffling reactions.

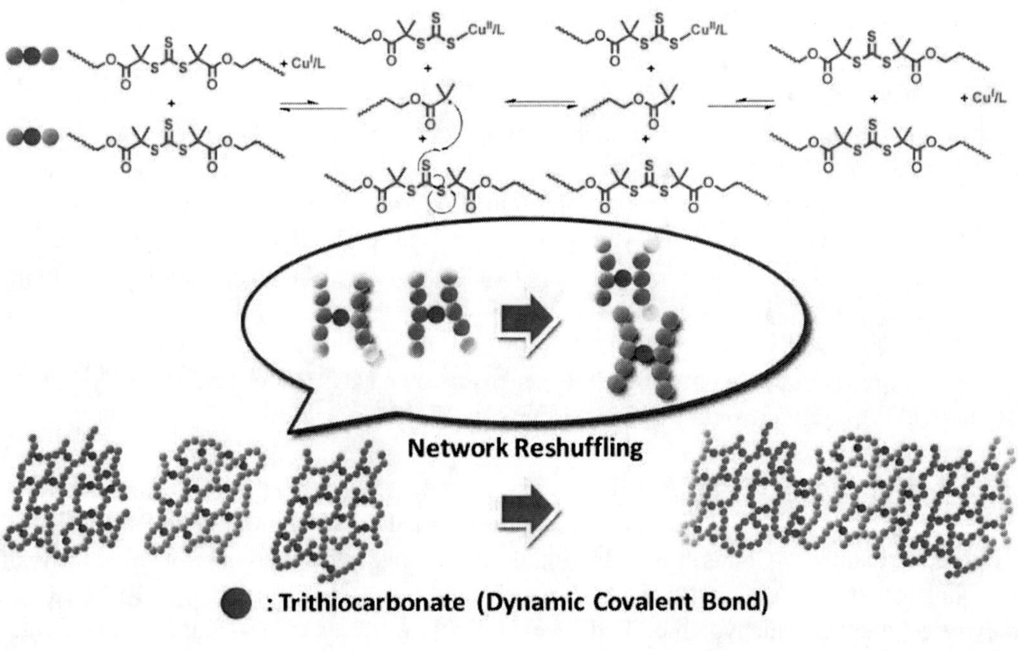

Figure 2.6. Self-healing mechanism of gels prepared with a trithiocarbonate crosslinker via reshuffling of reversible covalent bonds (Nicolay et al., 2010).

2.3.2.6. Diels-Alder Reactions

Diels-Alder (DA) reaction, one of the "click chemistry" reactions, has played a significant role in the design of diverse functional gels because it gives excellent selectivity and yields without side reactions and by-products (Bergman and Wudl, 2007; Gacal et al., 2006; Gandini, 2013; Gandini, Silvestre, and Coelho, 2013; Kavitha and Singha, 2009; Moses and Moorhouse, 2007; Ossipov and Hilborn, 2006). Additionally, the DA reaction is thermo-reversible, making it feasible to break the DA bonds produced by the diene and dienophile and re-establish equilibrium (Adzima, Kloxin, and Bowman, 2010; X. Chen et al., 2002; Luo et al., 2009; Peterson, Jensen, and Palmese, 2010; Schenzel, Moszner, and Barner-Kowollik, 2017). For the reasons stated above, the DA reaction has become one of the most well-known chemical reactions for producing self-healing gels. Kavitha and Singha (2009) synthesized a thermoreversible self-healing organogel film using poly(furfuryl methacrylate) (PFMA) as a diene and dienophile bismaleimide (BM) as a crosslinker.

2.4. Self-Healing Nanocomposites

Nanoparticles, nanotubes, and nanofibers can not contribute self-repairing properties to the self-healing properties of polymer matrix unless self-healing segments are intrinsically incorporated. However, if the interfacial areas between the polymer matrix and reinforcing rigid entities show the ability of bond breaking and reformation, the interfacial regions may fulfill that role. The surface may be altered in most nanoparticles, nanorods/nanowires, nanotubes, nanofibers, and graphene sheets to form reactive amine, carboxylic acid, and hydroxyl, UPy, thiol, furfuryl, cinnamoyl, anthracene, and pyrenyl groups (Imran and Shapter, 2015; Artel et al., 2013; Biyani, Foster, and Weder, 2013; Engel and Kickelbick, 2014; Froimowicz, Klinger, and Landfester, 2011; Vaiyapuri et al., 2012; Wang et al., 2013; Yadav and Cho, 2013; Yoonessi et al., 2012).

To generate reversible bonds with a polymer matrix, the polymer motif dictates the kind of modification to be used. Because of this, interfacial chemistry will be crucial for accomplishing reversible self-healing. The surface functionalization is most effective when the polymer matrix contains reactive groups that "match". In the reaction, interfacial bonding would boost both covalent and non-covalent interactions between the two components. Reforming bonding at the fiber-matrix interface can make repairs easier while the interface goes through a partial damage-repair period. However, a composite's overall mechanical integrity remains the same. When a load is removed, the reformed interface can return to a state of equilibrium. Polyhedral oligomeric silsesquioxane (POSS), with the empirical formula $R_n(SiO_{1.5})_n$, where n = 8, 10, or 12, is a good example of an organic-inorganic hybrid compound where the R groups on each cage silicone atom can be either reactive or non-reactive organic groups.

The cubic cage may be the lowest possible nanoscale silica particles (Cheng, Vautravers, Morris, and Cole-Hamilton, 2008; Liang, Li, Toghiani, Koo, and Pittman, 2006; Xue et al., 2012). Thermally healable POSS using DA reactions is shown in Figure 2.7 (Xu, Zhao, Wang, and Lin, 2013). Compared to nanoparticles functionalized with furfuryl groups, POSS composites have a modulus of 1.0 GPa and a yield strength of 118 MPa. Due to the molecular dispersion of the POSS nanocage system, self-healing of a rupture at 135-150 °C can be done effectively within 30 minutes. POSS may be functionalized with different stimulus-responsive reactive groups (Kango et al., 2013). Stimuli-responsive reactive moieties may be added to nanoparticle surfaces using "grafting to" or "grafting from" techniques. When alkane chains are added to nanoparticles, interfacial interactions may be enhanced, leading to better phase compatibility and mechanical characteristics in polymers.

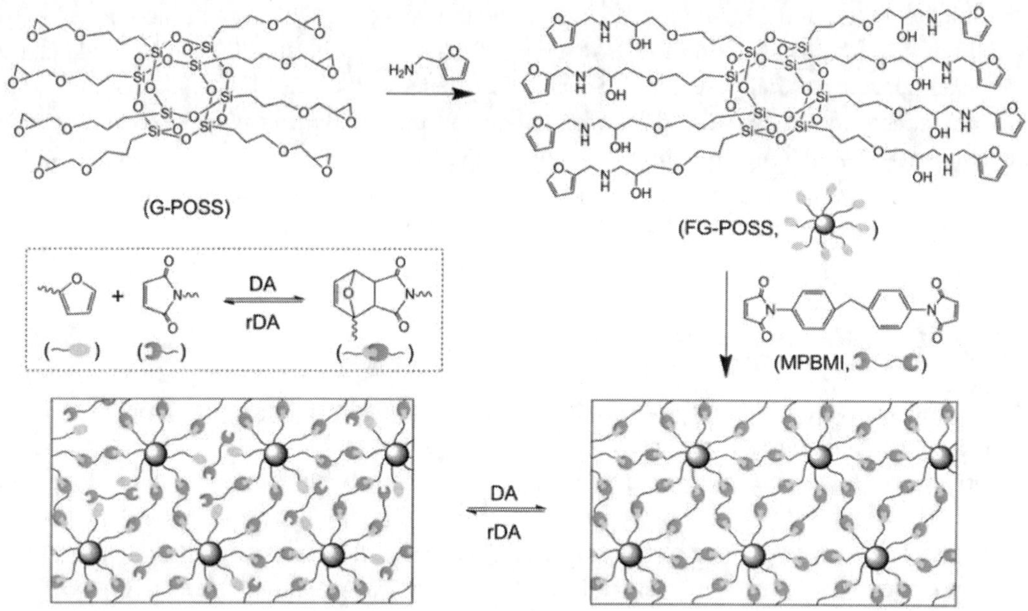

Figure 2.7. Schematic of a thermally healable POSS nanocomposite based on the DA reaction (Xu et al., 2013).

Still, because the interfacial interactions between nanoparticles and polymer matrices are usually made through physical interactions, there is a chance that will form aggregates when nanocomposites are subjected to changing temperatures or cyclic loads (Lee, Zhang, Emrick, and Crosby, 2006). Changing nanoparticles with sensitive functional groups like cinnamoyl, pyrenyl, and furfuryl and making reversible polymer matrix covalent linkage can lead to self-healing triggered by light or heat. Polydiimide interacts with functionalized gold nanoparticles (AuNPs) of pyrene and end-capped polyamide of pyrene (as electron donors) (Vaiyapuri, Greenland, Colquhoun, Elliott, and Hayes, 2013) (Figure 2.8).

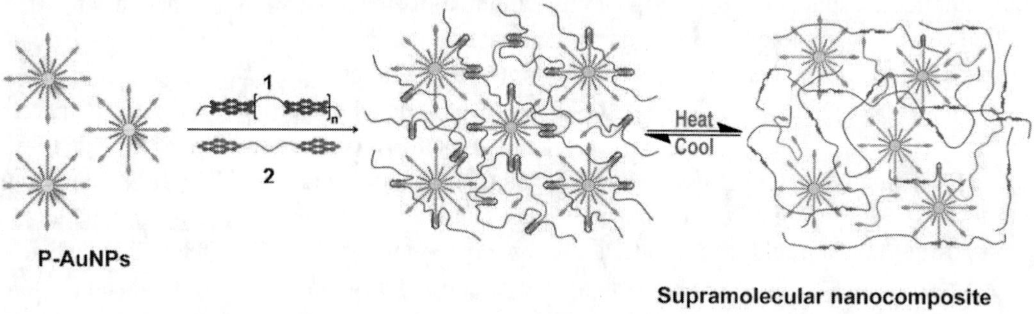

Figure 2.8. An illustration of the formation/dissociation of a supramolecular nanocomposite from chain-folding polymer 1, pyrene end-capped polymer 2, and the introduction of a pyrene functionalized AuNP (P-AuNP) filler (Vaiyapuri et al., 2013).

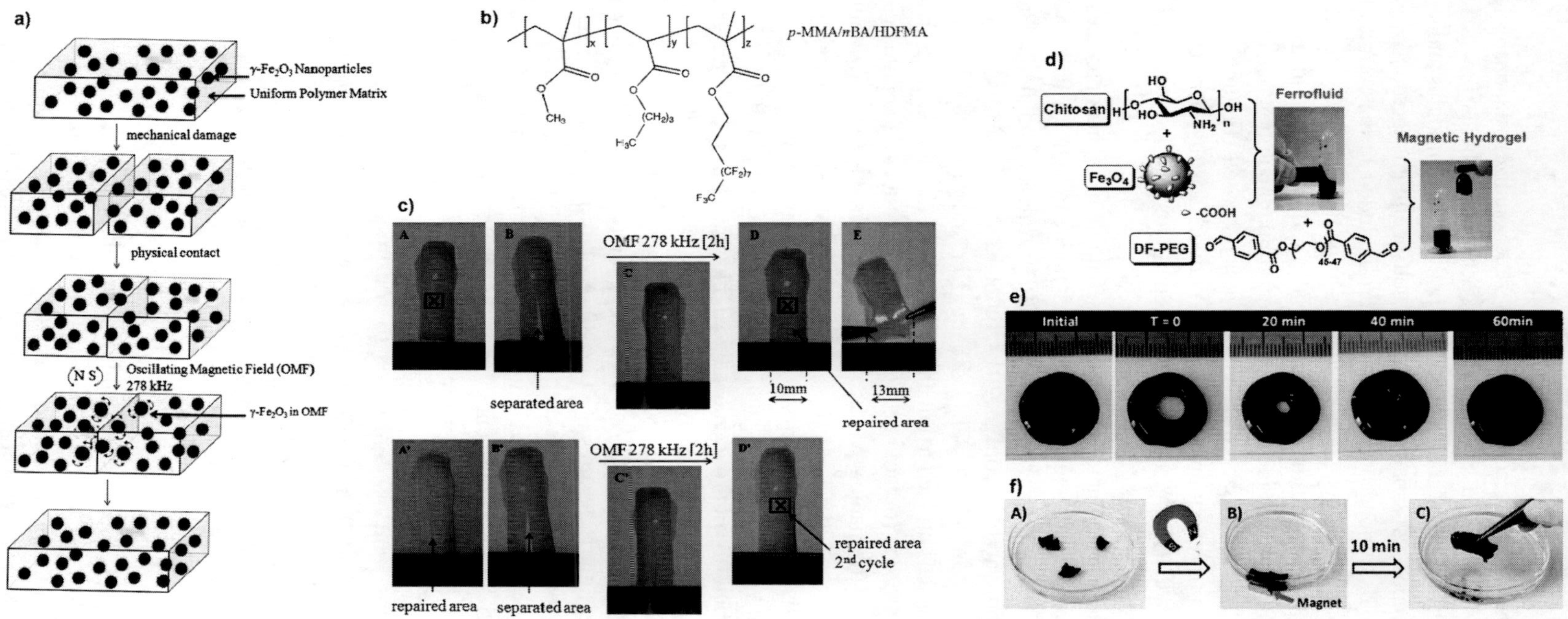

Figure 2.9. A thermoplastic polymer healing process for magnetic nanoparticles assisted by oscillatory nanoparticle movements in oscillating magnetic fields (Amaral et al., 2017).

Self-healing is formed by interactions between functionalized nanoparticles and the polymer matrix that can be reversed by changing temperatures. AuNPs functionalized with pyrene have uniform distribution and a linear increase in the tensile modulus and strength. The development of nanocomposites capable of remote self-healing is particularly intriguing. Electric and magnetic fields, among other variables, are included in the definition, as is the conversion of external stimuli to heat generated locally in the damaged area. As a result, polymer chains or segments move around more efficiently, and interfacial diffusion is sped up to help the polymer self-healing. The idea of using superparamagnetic Fe_2O_3 nanoparticles that move around in magnetic fields has been developed and proposed (Figure 2.9). The Fe_2O_3 nanoparticles are added to the thermoplastic polymer network for this purpose (Amaral and Pasparakis, 2017). The oscillating movements of nanoparticles translate magnetic frequency into heat and repeatedly replicate this procedure.

Similarly, surface plasmon resonance generation used noble metal nanoparticles like AuNP and AgNP to fix the damage remotely, which turns incident light (usually near-infrared spectrum) into localized heat in the area (Alessandri, 2010; Amendola et al., 2009; Amendola and Meneghetti, 2012; Cortie and McDonagh, 2011; Fiore, Rowan, and Weder, 2013; Giannini, Fernández-Domínguez, Heck, and Maier, 2011; Zhang, Xia, and Zhao, 2012). One of the most frequent types of nanomaterials utilized to enhance mechanical and self-healing capacities into composite materials is nanorods (organic or inorganic). Dispersions of cellulose nanorods in the polymer matrix improve mechanical power, whereas stacking interactions within the matrix governs self-healing (Fox et al., 2012). Phase separation occurred when nanorod concentration exceeded 10%, compromising mechanical strength and healing ability owing to inefficient interfacial interaction between nanorods and polymer matrix. Like POSS, hydroxyl-rich cellulose may be modified to improve mechanical characteristics and self-healing. The cellulose nanocrystals (CNC) with many hydroxyl groups react with UPy to produce strong UPy-UPy hydrogen-bonding interactions with polymer matrices containing UPy motifs. At 320-390 nm wavelength and 250-350 mW/cm^2, the UPy pattern absorbs UV light and converts it to heat, causing self-healing. Exposures of 20 to 80 seconds are required for repairs to take place. As a result, the idea of nanorods with optoelectronic capabilities (such as transition metal oxide and noble oxide nanorods) was brought up. Optoelectronic response of nanocomposites and remotely controlled healing efficiency may be developed using metal nanorods functionalized with functional photo-responsive groups (i.e., UPy, cinnamoyl, and anthracene) (Chen, Shao, Li, and Wang, 2013; Pearce, Melanko, and Salem, 2007).

Host-guest interactions were also employed to construct self-healing nanocomposites. Although the host-guest strategy may have certain drawbacks, one of its key advantages is its quick response. Three-component recognition cellulose nanorod-reinforced composite hydrogels have been designed using CB macrocyclic host molecules mixed with methyl viologen (MV)/naphthyl (Np) (Kim et al., 2000). A supramolecular composite hydrogel is made up of soft PVA-co-MV containing the first guest MV, hard CNC grafted with poly(dimethylaminoethyl methacrylate-co-naphthyl methacrylate) (CNC-g-PDMAEMA-co-NpMA) containing the second guest Np, and host CB (McKee et al., 2014). The MV and Np guest motifs bind to the host CB in aqueous conditions, resulting in a supramolecular network. Rheological studies demonstrated the most remarkable storage module of 14.3 kPa and the quick transitions from sol-to-gel as well as fast self-healing (approximately 6 s). These characteristic behaviors of nanocomposite hydrogels are particularly noteworthy in biomedical applications because of their ability to overcome the hydrogel's inherent fragility and quickly restore mechanical properties. It is well known that the distinctive physical features of CNTs and graphene nanosheets (GNS) include great mechanical strength, large surface area, and combination structures with exceptional electrical properties (Novoselov et al., 2004). When combined with other polymers, they acquire additional strength and integration (Stankovich et al., 2006). CNT and GNS were chosen because of their unique electromagnetic characteristics in developing nanocomposites for remotely triggered self-healing. Protonated Nafion SMP containing amide functionalized single-walled CNTs (SWNTs) evenly distributed by infrared (IR) radiation showed restoration characteristics (Kohlmeyer, Lor, and Chen, 2012).

The composite made of graphene-reinforced polyurethane (PUR) also demonstrated healing properties. Graphene's capacity to convert IR irradiation, electromagnetic wavelength, or electricity into heat, improved PUR chain mobility and re-entanglement at the damaged interfaces, resulting in multi-responsiveness in remote-triggered self-healing. Both CNT and GNS may be chemically modified to add carboxylic acid groups, accompanied by polymer matrix crosslinking or additional modifications with various functionalities to promote dispersion and self-healing responses (Huang et al., 2013).

The big limitation is the tendency to oxidize heavily, which reduces optical absorption and electrical conductivity. Following the modification of graphene oxide with phenyl isocyanate, phenylhydrazine reduces graphene oxide to restore near-infrared absorbance to preserve optoelectronic characteristics (Kim, Kim, Kim, Kwon, and Jeong, 2013). CNT and GNS can directly make potentially healing nanocomposites; however, unique manufacturing strategies may be helpful. When aligned-CNT sheets are prepared by dry drawing from a spinnable CNT array and wrapped around self-healing polymer fibers, an electrically conductive wire with a uniformly connected CNT sheet shell is obtained from the polymer fiber core (Sun et al., 2014). Increases in conductivity reduced electrical resistance and capacitance recovery were seen after self-healing the aligned CNT sheet.

In addition to nanoparticles, the most extensively utilized materials in producing fiber-reinforced healing composites include nanorods, CNTs, graphene nanosheets, carbon and glass fibers, and metal wires. When constructing fiber-reinforced self-healing composites, proper production procedures are essential for evenly distributing anisotropic components. For instance, when silver nanowire solution was dropped onto a substrate, it formed a silver nanowire-reinforced healable conductor, which was subsequently deposited with DA reactive groups on top of the silver nanowire layer and healed at 70 °C to yield the healable conductive composite (Gong et al., 2013). Several fiber/polymer interfacial healing methods have been reported to date. For example, maleimide groups were added to glass fiber so that it could bond with a furan-functionalized epoxy matrix after heat treatment (Peterson, Jensen, and Palmese, 2011). Microcapsules or nanocapsules filled with healing agents were placed on the surface of carbon or glass fibers along the same line and then combined into the polymer matrix. Once the interfacial damage has occurred, the capsule is broken and the healing agent is released. This re-bonds the fiber and polymer matrix in a repeatable manner (Jones, Blaiszik, White, and Sottos, 2013). Although microcapsule-based healing agents can only be used once, fiber/polymer matrix interfacial healing seems to have additional benefits.

2.5. Applications and Future of Self-Healing Nanocomposites

Self-healing nanocomposites are a big step forward in science and technology (Wu and Chen, 2016; Zhang, Liu, Pan, and Li, 2018). CNT-based self-healing nanocomposites are an excellent choice for making advanced coatings, sensors, electronics, solar cells, and biomedical devices. CNT-based nanocomposites have been used to make coatings that can fix themselves. Vijayan and coworkers came up with coatings made of epoxy and CNT that can self-heal. In these coatings, there were also self-healing nanocapsule healing agents. Their ability to self-heal is linked to the amount of nanofiller material and healing agent in the material (Ding et al., 2013; Wojtecki, Meador, and Rowan, 2011). Tatiya et al., (2013) also developed self-healing anti-corrosion coatings out of polyurethane and CNT. The coatings make the metal very resistant to corrosion. Microcapsules filled with diisocyanates were used in the coatings to make them self-healing (Wang, Xu, Li, Yang, and An, 2014).

Biomedical self-healing superhydrophobic coatings were invented by Li et al., (2010). CNT nanocomposites have been employed for piezoresistive pressure sensing applications. Wu et al., (2017) have designed autonomous self-healing nanocomposite pressure sensors based on CNTs. With 19.7 vol.% multi-walled CNT (MWCNT), the nanocomposite displays remarkable pressure-sensing efficiency. They have been tested for electrical conductivity utilizing the percolation and tunneling principle (Kim, Park, Okoli, and Zhang, 2009; Zhou, Zhu, and Yan, 2014). Polymer/CNT nanocomposites with self-healing properties have also been used in solar cells. Habisreutinger et al., (2014) employed a self-healing polymer/CNT as an organic hole transport medium in a solar cell. The amount of power solar cells could convert was 10.2%. But important improvements need to be made to self-healing polymer/CNT for a high-efficiency solar cell to be stable for a long time (Gupta et al., 2011; Yang and Urban, 2013).

A self-healing light-emitting device (LED) made of polymer and CNTs has been designed (Byrne and Gun'ko, 2010). Shi and coworkers made electroluminescent devices that can stretch and fix themselves by using polyurethane/CNT nanocomposite electrodes. An adequately oriented CNT was used to make the wrinkled structure, which makes it easier to stretch. The unit's brightness has remained stable throughout 10 cycles (Kwon et al., 2013). The LED industry has not extensively researched the self-healing characteristics of CNT polymers. Self-healing TPU and CNT nanocomposites made by melt blending have been examined by Ramoa and coworkers for their electromagnetic interference shielding effectiveness (EMI SE). With 10-15 wt.% CNT, the EMI SE nanocomposite was 20-22 dB. But there are not many reports on this subject (Menon, Madras, and Bose, 2019; Yang et al., 2018).

The biomedical field is important for self-healing polymer/CNT (Brochu, Craig, and Reichert, 2011; Shin et al., 2013). There are possible uses for improved mechanical, thermal, rheological, morphological, and biological properties in the biomedical field. Self-healing polymer/CNT nanocomposites have been used in drug delivery systems and tissue engineering scaffolds (Jing, Mi, Napiwocki, Peng, and Turng, 2017). Poly (3,4-ethylenedioxythiophene) and CNT were used to build a self-healable drug delivery system. Self-healable polymer/CNT has also been used as a drug carrier. Polycaprolactone/CNT nanocomposite scaffolds have been used to repair cardiac tissue. Researchers have examined the nanocomposites' thermal, crystalline, and biodegradation properties (Haraguchi, Uyama, and Tanimoto, 2011; Li, Yan, Yang, Chen, and Zeng, 2015). Still, more research needs to be done on self-healing polymer/CNT properties and effectiveness for drug delivery and tissue engineering. In short, research needs to be motivated to figure out how to deal with these materials' complicated molecular and structural designs.

All self-healing polymer/CNT nanocomposite hydrogels need to understand how chemical nature, morphology, nanostructuring, and network development affect the self-healing process. We designed a multifunctional double network nanocomposite self-healing hydrogel that can conduct electricity, have excellent mechanical properties, be bioactive, and heal wounds (Figure 2.10) (Rahman, Foyez, Susan, and Imran, 2020). Self-healing conducting polymer/CNT, on the other hand, needs to be looked into for energy storage devices like supercapacitors and actuators (Michel et al., 2012). E-textiles are also waiting for self-healing polymer/CNT fabrics to come out. Most importantly, these nanocomposites have not yet been used in aerospace or cars, but clean, self-repairing polymers have already been used in this field (Makhlouf, 2014; Périchaud et al., 2012). Lastly, future challenges of auto-repair materials need to be investigated for all of the above application areas.

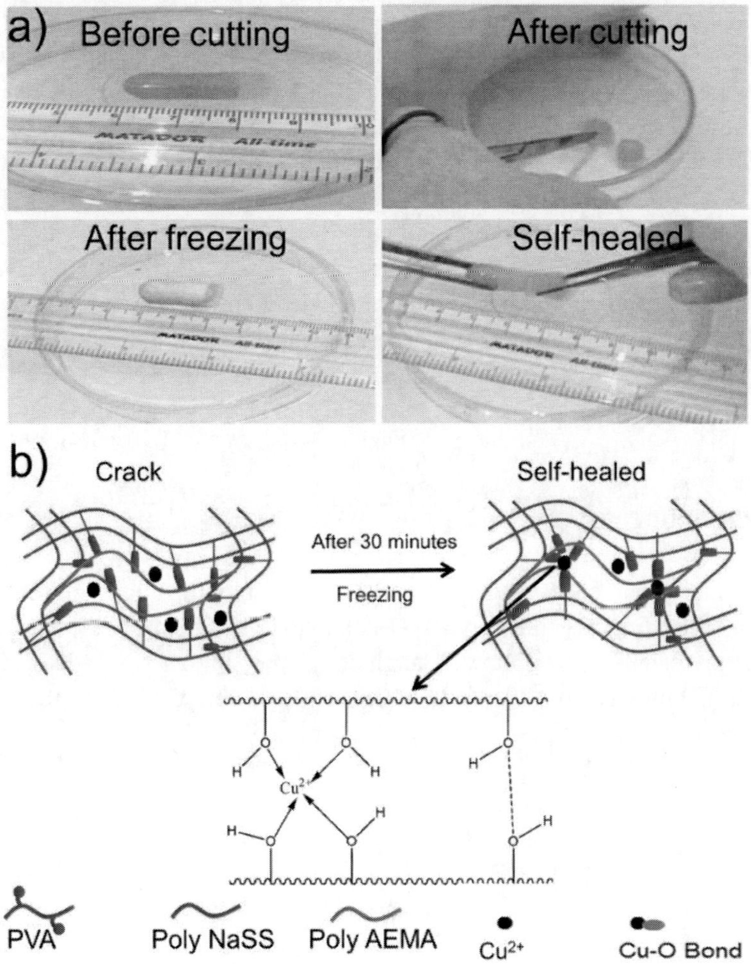

Figure 2.10. (a) The PVA-P(NaSS)-P(AEMA)@Cu-based self-healable and conductive double network nanocomposite hydrogel and (b) self-healing mechanism of the hydrogel (Rahman, Foyez, Susan, and Imran, 2020).

Conclusion

The potential of materials to mimic the self-healing processes of living organisms is critical to the sustainability of life and the efficient use of resources and energy. Technological requirements have investigated many intriguing alternatives in the recent decade. Synthetic polymers, including stimuli-responsive components, such as complicated covalent bonds and supramolecular chemistries, may be highly effective in 21st century material design. Heat, electromagnetic radiation, environmental changes, or the presence of atmospheric CO_2 and H_2O are all external stimuli that might help create favorable thermodynamic and kinetic conditions for self-healing. Even though most research on self-healing is focused on the chemical aspect, the physical aspect is equally important.

In this chapter, different examples are used to show how important heterogeneities and how they are distributed are to the self-heal of polymer materials. It will open the door for a new generation of materials that behave like living things by combining hard and soft segments into one polymer network or making them stronger with nanomaterials with attractive interfacial chemistries. A single reaction is often insufficient to promote network repairs, just as mechanical damage does not merely result in a single cleavage of chemical bonds. When physical and chemical processes are synergized, extraordinary self-healing characteristics are possible. It is tough to self-heal because of these inherent characteristics. Understanding mechanical and morphological features and molecular-level data are critical to understanding the chemical processes that occur during network healing. Chemical analysis and targeted mechanical study are required to improve our knowledge in this crucial field. Research in the future should apply the heterogeneity principle to produce fast stimulus responsiveness of rigid materials, as well as multi-level chemical interactions accomplished by coordinated chain movements and to repair several kinds of chemical entities in a single material.

Acknowledgments

A. B. Imran greatly appreciates the support of the Capacity Utilization Program under Special Allocation for Science and Technology (BS-182 and PHY'S-467) from the Ministry of Science and Technology, People's Republic of Bangladesh. The authors would like to express their gratitude to the Committee on Advanced Studies and Research (CASR) in BUET for providing financial support.

References

Adzima, B. J., Kloxin, C. J., and Bowman, C. N. (2010). Externally triggered healing of a thermoreversible covalent network via self-limited hysteresis heating. *Advanced Materials,* 22(25), 2784-2787.

Alessandri, I. (2010). Writing, self-healing, and self-erasing on conductive pressure-sensitive adhesives. *Small,* 6(15), 1679-1685.

Amaral, A. J. R., and Pasparakis, G. (2017). Stimuli responsive self-healing polymers: gels, elastomers and membranes. *Polymer Chemistry,* 8(42), 6464-6484.

Amendola, V., Dini, D., Polizzi, S., Shen, J., Kadish, K. M., Calvete, M. J. F., Hanack, M., and Meneghetti, M. (2009). Self-healing of gold nanoparticles in the presence of zinc phthalocyanines and their very efficient nonlinear absorption performances. *The Journal of Physical Chemistry C,* 113(20), 8688-8695.

Amendola, V., and Meneghetti, M. (2012). Advances in self-healing optical materials. *Journal of Materials Chemistry,* 22(47), 24501-24508.

Annabi, N., Tamayol, A., Uquillas, J. A., Akbari, M., Bertassoni, L. E., Cha, C., Camci-Unal, G., Dokmeci, M. R., Peppas N. A., and Khademhosseini, A. (2014). 25th anniversary article: rational design and applications of hydrogels in regenerative medicine. *Advanced Materials,* 26(1), 85-124.

Apostolides, D. E., and Patrickios, C. S. (2018). Dynamic covalent polymer hydrogels and organogels crosslinked through acylhydrazone bonds: synthesis, characterization and applications. *Polymer International,* 67(6), 627-649.

Artel, V., Cohen, R., Aped, I., Ronen, M., Gerber, D., and Sukenik, C. N. (2013). Controlled formation of thiol and disulfide interfaces. *Langmuir,* 29(1), 191-198.

Bai, J., and Shi, Z. (2017). Dynamically cross-linked elastomer hybrids with light-induced rapid and efficient self-healing ability and reprogrammable shape memory behavior. *ACS Applied Materials and Interfaces,* 9(32), 27213-27222.

Barbero, E. J., Ford, K. J., and Mayugo, J. A. (2009). Modeling self-healing of fiber-reinforced polymer-matrix composites with distributed damage. *Self-healing Materials*, doi: 10.1002/9783527625376.ch9.

Bergman, S. D., and Wudl, F. (2007). Re-mendable polymers. *Self Healing Materials*, 45-68.

Biyani, M. V., Foster, E. J., and Weder, C. (2013). Light-healable supramolecular nanocomposites based on modified cellulose nanocrystals. *ACS Macro Letters,* 2(3), 236-240.

Brochu, A. B. W., Craig, S. L., and Reichert, W. M. (2011). Self-healing biomaterials. *Journal of Biomedical Materials Research Part A,* 96(2), 492-506.

Burattini, S., Colquhoun, H. M., Greenland, B. W., and Hayes, W. (2009). A novel self-healing supramolecular polymer system. *Faraday Discussions,* 143, 251-264.

Burattini, S., Greenland, B. W., Hayes, W., Mackay, M. E., Rowan, S. J., and Colquhoun, H. M. (2011). A supramolecular polymer based on tweezer-type $\pi-\pi$ stacking interactions: molecular design for healability and enhanced toughness. *Chemistry of Materials,* 23(1), 6-8.

Burnworth, M., Tang, L., Kumpfer, J. R., Duncan, A. J., Beyer, F. L., Fiore, G. L., Rowan, S. J., and, Weder, C. (2011). Optically healable supramolecular polymers. *Nature,* 472(7343), 334-337.

Byrne, M. T., and Gun'ko, Y. K. (2010). Recent advances in research on carbon nanotube-polymer composites. *Advanced Materials,* 22(15), 1672-1688.

Canadell, J., Goossens, H., and Klumperman, B. (2011). Self-healing materials based on disulfide links. *Macromolecules,* 44(8), 2536-2541.

Chen, H., Shao, L., Li, Q., and Wang, J. (2013). Gold nanorods and their plasmonic properties. *Chemical Society Reviews,* 42(7), 2679-2724.

Chen, X., Dam, M. A., Ono, K., Mal, A., Shen, H., Nutt, S. R., Sheran, K., and Wudl, F. (2002). A thermally re-mendable cross-linked polymeric material. *Science,* 295(5560), 1698-1702.

Chen, Y., and Guan, Z. (2014). Multivalent hydrogen bonding block copolymers self-assemble into strong and tough self-healing materials. *Chemical Communications,* 50(74), 10868-10870.

Cheng, G., Vautravers, N. R., Morris, R. E., and Cole-Hamilton, D. J. (2008). Synthesis of functional cubes from octavinylsilsesquioxane (OVS). *Organic and Biomolecular Chemistry,* 6(24), 4662-4667.

Cortie, M. B., and McDonagh, A. M. (2011). Synthesis and optical properties of hybrid and alloy plasmonic nanoparticles. *Chemical Reviews,* 111(6), 3713-3735.

Dai, S., Zhou, X., Wang, S., Ding, J., and Yuan, N. (2018). A self-healing conductive and stretchable aligned carbon nanotube/hydrogel composite with a sandwich structure. *Nanoscale,* 10(41), 19360-19366.

Ding, F., Xu, W., Graff, G. L., Zhang, J., Sushko, M. L., Shao, Y., Engelhard, M. H., Nie, Z., Xiao, J., and Liu, X. (2013). Dendrite-free lithium deposition via self-healing electrostatic shield mechanism. *Journal of the American Chemical Society,* 135(11), 4450-4456.

Ding, Q., Xu, X., Yue, Y., Mei, C., Huang, C., Jiang, S., Wu, Q., and Han, J. (2018). Nanocellulose-mediated electroconductive self-healing hydrogels with high strength, plasticity, viscoelasticity, stretchability, and biocompatibility toward multifunctional applications. *ACS Applied Materials and Interfaces,* 10(33), 27987-28002.

Dowlut, M., and Hall, D. G. (2006). An improved class of sugar-binding boronic acids, soluble and capable of complexing glycosides in neutral water. *Journal of the American Chemical Society,* 128(13), 4226-4227.

Dry, C. (1994). Matrix cracking repair and filling using active and passive modes for smart timed release of chemicals from fibers into cement matrices. *Smart Materials and Structures,* 3(2), 118.

Eisenberg, A., Hird, B., and Moore, R. B. (1990). A new multiplet-cluster model for the morphology of random ionomers. *Macromolecules,* 23(18), 4098-4107.

Engel, T., and Kickelbick, G. (2013). Thermoreversible reactions on inorganic nanoparticle surfaces: Diels-Alder reactions on sterically crowded surfaces. *Chemistry of Materials,* 25(2), 149-157.

Engel, T., and Kickelbick, G. (2014). Self-healing nanocomposites from silica–polymer core-shell nanoparticles. *Polymer International,* 63(5), 915-923.

Fiore, G. L., Rowan, S. J., and Weder, C. (2013). Optically healable polymers. *Chemical Society Reviews,* 42(17), 7278-7288.

Fox, J., Wie, J. J., Greenland, B. W., Burattini, S., Hayes, W., Colquhoun, H. M., Mackay, M. E., and Rowan, S. J. (2012). High-strength, healable, supramolecular polymer nanocomposites. *Journal of the American Chemical Society,* 134(11), 5362-5368.

Frei, R., McWilliam, R., Derrick, B., Purvis, A., Tiwari, A., and Di Marzo Serugendo, G. (2013). Self-healing and self-repairing technologies. *The International Journal of Advanced Manufacturing Technology,* 69(5), 1033-1061.

Froimowicz, P., Klinger, D., and Landfester, K. (2011). Photoreactive nanoparticles as nanometric building blocks for the generation of self-healing hydrogel thin films. *Chemistry-A European Journal,* 17(44), 12465-12475.

Gacal, B., Durmaz, H., Tasdelen, M. A., Hizal, G., Tunca, U., Yagci, Y., and Demirel, A. L. (2006). Anthracene-maleimide-based Diels-Alder "click chemistry" as a novel route to graft copolymers. *Macromolecules,* 39(16), 5330-5336.

Galland, S., Leterrier, Y., Nardi, T., Plummer, C. J. G., Månson, J. A. E., and Berglund, L. A. (2014). UV-cured cellulose nanofiber composites with moisture durable oxygen barrier properties. *Journal of Applied Polymer Science,* 131(16), 40604.

Gandini, A. (2013). The furan/maleimide Diels-Alder reaction: a versatile click-unclick tool in macromolecular synthesis. *Progress in Polymer Science,* 38(1), 1-29.

Gandini, A., Silvestre, A., and Coelho, D. (2013). Reversible click chemistry at the service of macromolecular materials. Part 4: Diels-Alder non-linear polycondensations involving polyfunctional furan and maleimide monomers. *Polymer Chemistry,* 4(5), 1364-1371.

Giannini, V., Fernández-Domínguez, A. I., Heck, S. C., and Maier, S. A. (2011). Plasmonic nanoantennas: fundamentals and their use in controlling the radiative properties of nanoemitters. *Chemical Reviews,* 111(6), 3888-3912.

Giuseppone, N., Fuks, G., and Lehn, J. M. (2006). Tunable fluorene-based dynamers through constitutional dynamic chemistry. *Chemistry-A European Journal,* 12(6), 1723-1735.

Gohy, J. F., Lohmeijer, B. G. G., and Schubert, U. S. (2002). Reversible metallo-supramolecular block copolymer micelles containing a soft core. *Macromolecular Rapid Communications,* 23(9), 555-560.

Gong, C., Liang, J., Hu, W., Niu, X., Ma, S., Hahn, H. T., and Pei, Q. (2013). A healable, semitransparent silver nanowire-polymer composite conductor. *Advanced Materials,* 25(30), 4186-4191.

Gupta, V., Chaudhary, N., Srivastava, R., Sharma, G. D., Bhardwaj, R., and Chand, S. (2011). Luminscent graphene quantum dots for organic photovoltaic devices. *Journal of the American Chemical Society,* 133(26), 9960-9963.

Habisreutinger, S. N., Leijtens, T., Eperon, G. E., Stranks, S. D., Nicholas, R. J., and Snaith, H. J. (2014). Carbon nanotube/polymer composites as a highly stable hole collection layer in perovskite solar cells. *Nano Letters,* 14(10), 5561-5568.

Hager, M. D., Greil, P., Leyens, C., van der Zwaag, S., and Schubert, U. S. (2010). Self-healing materials. *Advanced Materials,* 22(47), 5424-5430.

Hales, M., Barner-Kowollik, C., Davis, T. P., and Stenzel, M. H. (2004). Shell-cross-linked vesicles synthesized from block copolymers of poly (D, L-lactide) and poly (N-isopropyl acrylamide) as thermoresponsive nanocontainers. *Langmuir,* 20(25), 10809-10817.

Haraguchi, K., Uyama, K., and Tanimoto, H. (2011). Self-healing in nanocomposite hydrogels. *Macromolecular Rapid Communications,* 32(16), 1253-1258.

Huang, L., Yi, N., Wu, Y., Zhang, Y., Zhang, Q., Huang, Y., Ma, Y., and Chen, Y. (2013). Multichannel and repeatable self-healing of mechanical enhanced graphene-thermoplastic polyurethane composites. *Advanced Materials,* 25(15), 2224-2228.

Huang, X., Qi, X., Boey, F., and Zhang, H. (2012). Graphene-based composites. *Chemical Society Reviews,* 41(2), 666-686.

Hyon, S. H., Cha, W. I., and Ikada, Y. (1989). Preparation of transparent poly (vinyl alcohol) hydrogel. *Polymer Bulletin,* 22(2), 119-122.

Imran, A. B. (2021). Design, development, characterization and application of smart polymeric hydrogel. *Manufacturing Systems: Recent Progress and Future Directions*, 187-218.

Imran, A. B., Seki, T., and Takeoka, Y. (2010). Recent advances in hydrogels in terms of fast stimuli responsiveness and superior mechanical performance. *Polymer Journal,* 42(11), 839-851.

Imran, A. B., and Shapter, J. (2015). *Innovations in Nanomaterials*, Nova Science Publishers, Inc.

Jing, X., Mi, H. Y., Napiwocki, B. N., Peng, X. F., and Turng, L. S. (2017). Mussel-inspired electroactive chitosan/graphene oxide composite hydrogel with rapid self-healing and recovery behavior for tissue engineering. *Carbon,* 125, 557-570.

Jones, A. R., Blaiszik, B. J., White, S. R., and Sottos, N. R. (2013). Full recovery of fiber/matrix interfacial bond strength using a microencapsulated solvent-based healing system. *Composites Science and Technology,* 79, 1-7.

Kakuta, T., Takashima, Y., Nakahata, M., Otsubo, M., Yamaguchi, H., and Harada, A. (2013). Preorganized hydrogel: self-healing properties of supramolecular hydrogels formed by polymerization of host-guest-monomers that contain cyclodextrins and hydrophobic guest groups. *Advanced Materials,* 25(20), 2849-2853.

Kango, S., Kalia, S., Celli, A., Njuguna, J., Habibi, Y., and Kumar, R. (2013). Surface modification of inorganic nanoparticles for development of organic-inorganic nanocomposites-A review. *Progress in Polymer Science,* 38(8), 1232-1261.

Kavitha, A. A., and Singha, N. K. (2009). "Click chemistry" in tailor-made polymethacrylates bearing reactive furfuryl functionality: a new class of self-healing polymeric material. *ACS Applied Materials and Interfaces,* 1(7), 1427-1436.

Kemp, M., Go, Y. M., and Jones, D. P. (2008). Nonequilibrium thermodynamics of thiol/disulfide redox systems: a perspective on redox systems biology. *Free Radical Biology and Medicine,* 44(6), 921-937.

Kessler, M. R. (2007). Self-healing: a new paradigm in materials design. *Proceedings of the Institution of Mechanical Engineers, Part G: Journal of Aerospace Engineering,* 221(4), 479-495.

Khan, A., Ahmed, N., and Rabnawaz, M. (2020). Covalent adaptable network and self-healing materials: current trends and future prospects in sustainability. *Polymers,* 12(9), 2027.

Kim, J., Jung, I. S., Kim, S. Y., Lee, E., Kang, J. K., Sakamoto, S., Yamaguchi, K., and Kim, K. (2000). New cucurbituril homologues: syntheses, isolation, characterization, and X-ray crystal structures of cucurbit [n] uril (n= 5, 7, and 8). *Journal of the American Chemical Society,* 122(3), 540-541.

Kim, J. T., Kim, B. K., Kim, E. Y., Kwon, S. H., and Jeong, H. M. (2013). Synthesis and properties of near IR induced self-healable polyurethane/graphene nanocomposites. *European Polymer Journal,* 49(12), 3889-3896.

Kim, M., Park, Y. B., Okoli, O. I., and Zhang, C. (2009). Processing, characterization, and modeling of carbon nanotube-reinforced multiscale composites. *Composites Science and Technology,* 69(3-4), 335-342.

Kohlmeyer, R. R., Lor, M., and Chen, J. (2012). Remote, local, and chemical programming of healable multishape memory polymer nanocomposites. *Nano Letters,* 12(6), 2757-2762.

Kotharangannagari, V. K., Sánchez-Ferrer, A., Ruokolainen, J., and Mezzenga, R. (2012). Thermoreversible gel-sol behavior of rod-coil-rod peptide-based triblock copolymers. *Macromolecules,* 45(4), 1982-1990.

Kötteritzsch, J., Schubert, U. S., and Hager, M. D. (2013). Triggered and self-healing systems using nanostructured materials. *Nanotechnology Reviews,* 2(6), 699-723.

Kwon, W., Do, S., Lee, J., Hwang, S., Kim, J. K., and Rhee, S. W. (2013). Freestanding luminescent films of nitrogen-rich carbon nanodots toward large-scale phosphor-based white-light-emitting devices. *Chemistry of Materials,* 25(9), 1893-1899.

Lau, H. K., and Kiick, K. L. (2015). Opportunities for multicomponent hybrid hydrogels in biomedical applications. *Biomacromolecules,* 16(1), 28-42.

Lee, J. Y., Zhang, Q., Emrick, T., and Crosby, A. J. (2006). Nanoparticle alignment and repulsion during failure of glassy polymer nanocomposites. *Macromolecules,* 39(21), 7392-7396.

Lei, Z. Q., Xiang, H. P., Yuan, Y. J., Rong, M. Z., and Zhang, M. Q. (2014). Room-temperature self-healable and remoldable cross-linked polymer based on the dynamic exchange of disulfide bonds. *Chemistry of Materials,* 26(6), 2038-2046.

Li, L., Yan, B., Yang, J., Chen, L., and Zeng, H. (2015). Novel mussel-inspired injectable self-healing hydrogel with anti-biofouling property. *Advanced Materials,* 27(7), 1294-1299.

Li, Y., Li, L., and Sun, J. (2010). Bioinspired self-healing superhydrophobic coatings. *Angewandte Chemie,* 122(35), 6265-6269.

Liang, K., Li, G., Toghiani, H., Koo, J. H., and Pittman, C. U. (2006). Cyanate ester/polyhedral oligomeric silsesquioxane (POSS) nanocomposites: synthesis and characterization. *Chemistry of Materials,* 18(2), 301-312.

Luo, X., Ou, R., Eberly, D. E., Singhal, A., Viratyaporn, W., and Mather, P. T. (2009). A thermoplastic/thermoset blend exhibiting thermal mending and reversible adhesion. *ACS Applied Materials and Interfaces,* 1(3), 612-620.

Maeda, T., Otsuka, H., and Takahara, A. (2009). Dynamic covalent polymers: reorganizable polymers with dynamic covalent bonds. *Progress in Polymer Science,* 34(7), 581-604.

Makhlouf, A. S. H. (2014). Protective coatings for automotive, aerospace and military applications: current prospects and future trends. *Handbook of Smart Coatings for Materials Protection*, 121-131.

Malinskii, Y. M., Prokopenko, V. V., Ivanova, N. A., and Kargin, V. A. (1970). Investigation of self-healing of cracks in polymers. *Polymer Mechanics,* 6(2), 240-244.

McKee, J. R., Appel, E. A., Seitsonen, J., Kontturi, E., Scherman, O. A., and Ikkala, O. (2014). Healable, stable and stiff hydrogels: combining conflicting properties using dynamic and selective three-component recognition with reinforcing cellulose nanorods. *Advanced Functional Materials,* 24(18), 2706-2713.

Menon, A. V., Madras, G., and Bose, S. (2019). Light weight, ultrathin, and "thermally-clickable" self-healing MWNT patch as electromagnetic interference suppressor. *Chemical Engineering Journal,* 366, 72-82.

Mera, G., Navrotsky, A., Sen, S., Kleebe, H. J., and Riedel, R. (2013). Polymer-derived SiCN and SiOC ceramics-structure and energetics at the nanoscale. *Journal of Materials Chemistry A,* 1(12), 3826-3836.

Messens, J., and Collet, J. F. (2013). Thiol-disulfide exchange in signaling: disulfide bonds as a switch. *Antioxidants and redox signaling*, 18, 1594-1596.

Michel, S., Chu, B. T. T., Grimm, S., Nüesch, F. A., Borgschulte, A., and Opris, D. M. (2012). Self-healing electrodes for dielectric elastomer actuators. *Journal of Materials Chemistry,* 22(38), 20736-20741.

Moses, J. E., and Moorhouse, A. D. (2007). The growing applications of click chemistry. *Chemical Society Reviews,* 36(8), 1249-1262.

Murphy, E. B., and Wudl, F. (2010). The world of smart healable materials. *Progress in Polymer Science,* 35(1-2), 223-251.

Nagy, P. (2013). Kinetics and mechanisms of thiol-disulfide exchange covering direct substitution and thiol oxidation-mediated pathways. *Antioxidants and Redox Signaling,* 18(13), 1623-1641.

Nakahata, M., Takashima, Y., Yamaguchi, H., and Harada, A. (2011). Redox-responsive self-healing materials formed from host-guest polymers. *Nature Communications,* 2(1), 1-6.

Nicolay, R., Kamada, J., Van Wassen, A., and Matyjaszewski, K. (2010). Responsive gels based on a dynamic covalent trithiocarbonate cross-linker. *Macromolecules,* 43(9), 4355-4361.

Novoselov, K. S., Geim, A. K., Morozov, S. V., Jiang, D. E., Zhang, Y., Dubonos, S. V., Grigorieva, I. V., and Firsov, A. A. (2004). Electric field effect in atomically thin carbon films. *Science,* 306(5696), 666-669.

Nowak, A. P., Breedveld, V., Pakstis, L., Ozbas, B., Pine, D. J., Pochan, D., and Deming, T. J. (2002). Rapidly recovering hydrogel scaffolds from self-assembling diblock copolypeptide amphiphiles. *Nature,* 417(6887), 424-428.

Ono, T., Fujii, S., Nobori, T., and Lehn, J. M. (2007a). Optodynamers: expression of color and fluorescence at the interface between two films of different dynamic polymers. *Chemical Communications,* 42, 4360-4362.

Ono, T., Fujii, S., Nobori, T., and Lehn, J. M. (2007b). Soft-to-hard transformation of the mechanical properties of dynamic covalent polymers through component incorporation. *Chemical Communications,* 1, 46-48.

Ossipov, D. A., and Hilborn, J. (2006). Poly (vinyl alcohol)-based hydrogels formed by "click chemistry." *Macromolecules,* 39(5), 1709-1718.

Pal, K. (2021). *Sensors for Stretchable Electronics in Nanotechnology*, CRC Press.

Paugam, M. F., Valencia, L. S., Boggess, B., and Smith, B. D. (1994). Selective dopamine transport using a crown boronic acid. *Journal of the American Chemical Society,* 116(24), 11203-11204.

Pearce, M. E., Melanko, J. B., and Salem, A. K. (2007). Multifunctional nanorods for biomedical applications. *Pharmaceutical Research,* 24(12), 2335-2352.

Périchaud, A. A., Iskakov, R. M., Kurbatov, A., Akhmetov, T. Z., Prokohdko, O. Y., Razumovskaya, I. V., Bazhenov, S. L., Apel, P. Y., Voytekunas, V. Y., and Abadie, M. J. M. (2012). Auto-reparation of polyimide film coatings for aerospace applications challenges and perspectives. *High Performance Polymers-Polyimides Based-From Chemistry to Applications,* 215-244.

Peterson, A. M., Jensen, R. E., and Palmese, G. R. (2010). Room-temperature healing of a thermosetting polymer using the Diels-Alder reaction. *ACS Applied Materials and Interfaces,* 2(4), 1141-1149.

Peterson, A. M., Jensen, R. E., and Palmese, G. R. (2011). Thermoreversible and remendable glass-polymer interface for fiber-reinforced composites. *Composites Science and Technology,* 71(5), 586-592.

Pizer, R., and Babcock, L. (1977). Mechanism of the complexation of boron acids with catechol and substituted catechols. *Inorganic Chemistry,* 16(7), 1677-1681.

Pizer, R., and Ricatto, P. J. (1994). Thermodynamics of several 1:1 and 1:2 complexation reactions of the borate ion with bidentate ligands. 11B NMR spectroscopic studies1. *Inorganic Chemistry,* 33(11), 2402-2406.

Pleasants, J. C., Guo, W., and Rabenstein, D. L. (1989). A comparative study of the kinetics of selenol/diselenide and thiol/disulfide exchange reactions. *Journal of the American Chemical Society,* 111(17), 6553-6558.

Pluth, M. D., and Raymond, K. N. (2007). Reversible guest exchange mechanisms in supramolecular host-guest assemblies. *Chemical Society Reviews,* 36(2), 161-171.

Potts, J. R., Dreyer, D. R., Bielawski, C. W., and Ruoff, R. S. (2011). Graphene-based polymer nanocomposites. *Polymer,* 52(1), 5-25.

Qamar, I. P. S., Sottos, N. R., and Trask, R. S. (2020). Grand challenges in the design and manufacture of vascular self-healing. *Multifunctional Materials,* 3(1), 013001.

Qi, H., Liu, J., Gao, S., and Mäder, E. (2013). Multifunctional films composed of carbon nanotubes and cellulose regenerated from alkaline-urea solution. *Journal of Materials Chemistry A,* 1(6), 2161-2168.

Rahman, A., Foyez, T., Susan, M. A. B. H., and Imran, A. B. (2020). Self-healable and conductive double-network hydrogels with bioactive properties. *Macromolecular Chemistry and Physics,* 221(17), 2000207.

Reisch, A., Roger, E., Phoeung, T., Antheaume, C., Orthlieb, C., Boulmedais, F., Lavalle, P., Schlenoff, J. B., Frisch, B., and Schaaf, P. (2014). On the benefits of rubbing salt in the cut: self healing of saloplastic PAA/PAH compact polyelectrolyte complexes. *Advanced Materials,* 26(16), 2547-2551.

Roberts, M. C., Hanson, M. C., Massey, A. P., Karren, E. A., and Kiser, P. F. (2007). Dynamically restructuring hydrogel networks formed with reversible covalent crosslinks. *Advanced Materials,* 19(18), 2503-2507.

Sahoo, N. G., Rana, S., Cho, J. W., Li, L., and Chan, S. H. (2010). Polymer nanocomposites based on functionalized carbon nanotubes. *Progress in Polymer Science,* 35(7), 837-867.

Sano, K. I., Kawamura, R., Tominaga, T., Oda, N., Ijiro, K., and Osada, Y. (2011). Self-repairing filamentous actin hydrogel with hierarchical structure. *Biomacromolecules,* 12(12), 4173-4177.

Schenzel, A. M., Moszner, N., and Barner-Kowollik, C. (2017). Self-reporting dynamic covalent polycarbonate networks. *Polymer Chemistry,* 8(2), 414-420.

Shin, S. R., Jung, S. M., Zalabany, M., Kim, K., Zorlutuna, P., Kim, S. B., Nikkhah, M., Khabiry, M., Azize, M., Kong, J., and Wan, K. T. (2013). Carbon-nanotube-embedded hydrogel sheets for engineering cardiac constructs and bioactuators. *ACS Nano,* 7(3), 2369-2380.

Silva, D., Natalello, A., Sanii, B., Vasita, R., Saracino, G., Zuckermann, R. N., Doglia, S. M., and Gelain, F. (2013). Synthesis and characterization of designed BMHP1-derived self-assembling peptides for tissue engineering applications. *Nanoscale,* 5(2), 704-718.

Skene, W. G., and Lehn, J. M. P. (2004). Dynamers: polyacylhydrazone reversible covalent polymers, component exchange, and constitutional diversity. *Proceedings of the National Academy of Sciences,* 101(22), 8270-8275.

Stammen, J. A., Williams, S., Ku, D. N., and Guldberg, R. E. (2001). Mechanical properties of a novel PVA hydrogel in shear and unconfined compression. *Biomaterials,* 22(8), 799-806.

Stankovich, S., Dikin, D. A., Dommett, G. H. B., Kohlhaas, K. M., Zimney, E. J., Stach, E. A., Piner, R. D., Nguyen, S. T., and Ruoff, R. S. (2006). Graphene-based composite materials. *Nature,* 442(7100), 282-286.

Sun, H., You, X., Jiang, Y., Guan, G., Fang, X., Deng, J., Chen, P., Luo, Y., and Peng, H. (2014). Self-healable electrically conducting wires for wearable microelectronics. *Angewandte Chemie,* 126(36), 9680-9685.

Tatiya, P. D., Hedaoo, R. K., Mahulikar, P. P., and Gite, V. V. (2013). Novel polyurea microcapsules using dendritic functional monomer: synthesis, characterization, and its use in self-healing and anticorrosive polyurethane coatings. *Industrial and Engineering Chemistry Research,* 52(4), 1562-1570.

Tee, B. C. K., Wang, C., Allen, R., and Bao, Z. (2012). An electrically and mechanically self-healing composite with pressure-and flexion-sensitive properties for electronic skin applications. *Nature Nanotechnology,* 7(12), 825-832.

Trask, R. S., Williams, H. R., and Bond, I. P. (2007). Self-healing polymer composites: mimicking nature to enhance performance. *Bioinspiration and Biomimetics,* 2(1), P1.

Tuncaboylu, D. C., Argun, A., Sahin, M., Sari, M., and Okay, O. (2012). Structure optimization of self-healing hydrogels formed via hydrophobic interactions. *Polymer,* 53(24), 5513-5522.

Tuncaboylu, D. C., Sahin, M., Argun, A., Oppermann, W., and Okay, O. (2012). Dynamics and large strain behavior of self-healing hydrogels with and without surfactants. *Macromolecules,* 45(4), 1991-2000.

Tuncaboylu, D. C., Sari, M., Oppermann, W., and Okay, O. (2011). Tough and self-healing hydrogels formed via hydrophobic interactions. *Macromolecules,* 44(12), 4997-5005.

Vaiyapuri, R., Greenland, B. W., Colquhoun, H. M., Elliott, J. M., and Hayes, W. (2013). Molecular recognition between functionalized gold nanoparticles and healable, supramolecular polymer blends-a route to property enhancement. *Polymer Chemistry,* 4(18), 4902-4909.

Vaiyapuri, R., Greenland, B. W., Rowan, S. J., Colquhoun, H. M., Elliott, J. M., and Hayes, W. (2012). Thermoresponsive supramolecular polymer network comprising pyrene-functionalized gold nanoparticles and a chain-folding polydiimide. *Macromolecules,* 45(13), 5567-5574.

Wang, C., Liu, N., Allen, R., Tok, J. B. H., Wu, Y., Zhang, F., Chen, Y., and Bao, Z. (2013). A rapid and efficient self-healing thermo-reversible elastomer crosslinked with graphene oxide. *Advanced Materials,* 25(40), 5785-5790.

Wang, W., Xu, L., Li, X., Yang, Y., and An, E. (2014). Self-healing properties of protective coatings containing isophorone diisocyanate microcapsules on carbon steel surfaces. *Corrosion Science,* 80, 528-535.

White, S. R., Sottos, N. R., Geubelle, P. H., Moore, J. S., Kessler, M. R., Sriram, S. R., Brown, E. N., and Viswanathan, S. (2001). Autonomic healing of polymer composites. *Nature,* 409(6822), 794-797.

Williams, K. A., Boydston, A. J., and Bielawski, C. W. (2007). Towards electrically conductive, self-healing materials. *Journal of the Royal Society Interface,* 4(13), 359-362.

Wojtecki, R. J., Meador, M. A., and Rowan, S. J. (2011). Using the dynamic bond to access macroscopically responsive structurally dynamic polymers. *Nature Materials,* 10(1), 14-27.

Wool, R., and O'Connor, K. M. (1981). A theory crack healing in polymers. *Journal of Applied Physics,* 52(10), 5953-5963.

Wool, R. P. (1980). Crack healing in semicrystalline polymers, block copolymers and filled elastomers, *Adhesion and Adsorption of Polymers*, 341-362.

Wool, R. P. (2008). Self-healing materials: a review. *Soft Matter,* 4(3), 400-418.

Wu, D. Y., Meure, S., and Solomon, D. (2008). Self-healing polymeric materials: a review of recent developments. *Progress in Polymer Science,* 33(5), 479-522.

Wu, T., and Chen, B. (2016). Synthesis of multiwalled carbon nanotube-reinforced polyborosiloxane nanocomposites with mechanically adaptive and self-healing capabilities for flexible conductors. *ACS Applied Materials and Interfaces,* 8(36), 24071-24078.

Wu, T., and Chen, B. (2017). Autonomous self-healing multiwalled carbon nanotube nanocomposites with piezoresistive effect. *RSC Advances,* 7(33), 20422-20429.

Xu, Z., Zhao, Y., Wang, X., and Lin, T. (2013). A thermally healable polyhedral oligomeric silsesquioxane (POSS) nanocomposite based on Diels-Alder chemistry. *Chemical Communications,* 49(60), 6755-6757.

Xue, Y., Liu, Y., Lu, F., Qu, J., Chen, H., and Dai, L. (2012). Functionalization of graphene oxide with polyhedral oligomeric silsesquioxane (POSS) for multifunctional applications. *The Journal of Physical Chemistry Letters,* 3(12), 1607-1612.

Yadav, S. K., and Cho, J. W. (2013). Functionalized graphene nanoplatelets for enhanced mechanical and thermal properties of polyurethane nanocomposites. *Applied Surface Science,* 266, 360-367.

Yang, B., Zhang, Y., Zhang, X., Tao, L., Li, S., and Wei, Y. (2012). Facilely prepared inexpensive and biocompatible self-healing hydrogel: a new injectable cell therapy carrier. *Polymer Chemistry,* 3(12), 3235-3238.

Yang, W., Shao, B., Liu, T., Zhang, Y., Huang, R., Chen, F., and Fu, Q. (2018). Robust and mechanically and electrically self-healing hydrogel for efficient electromagnetic interference shielding. *ACS Applied Materials and Interfaces,* 10(9), 8245-8257.

Yang, Y., and Urban, M. W. (2013). Self-healing polymeric materials. *Chemical Society Reviews,* 42(17), 7446-7467.

Yoonessi, M., Shi, Y., Scheiman, D. A., Lebron-Colon, M., Tigelaar, D. M., Weiss, R. A., and Meador, M. A. (2012). Graphene polyimide nanocomposites; thermal, mechanical, and high-temperature shape memory effects. *ACS Nano,* 6(9), 7644-7655.

Zhang, H., Xia, H., and Zhao, Y. (2012). Optically triggered and spatially controllable shape-memory polymer-gold nanoparticle composite materials. *Journal of Materials Chemistry,* 22(3), 845-849.

Zhang, Q., Liu, L., Pan, C., and Li, D. (2018). Review of recent achievements in self-healing conductive materials and their applications. *Journal of Materials Science,* 53(1), 27-46.

Zhang, Y., Tao, L., Li, S., and Wei, Y. (2011). Synthesis of multiresponsive and dynamic chitosan-based hydrogels for controlled release of bioactive molecules. *Biomacromolecules,* 12(8), 2894-2901.

Zhang, Y., Yang, B., Zhang, X., Xu, L., Tao, L., Li, S., and Wei, Y. (2012). A magnetic self-healing hydrogel. *Chemical Communications,* 48(74), 9305-9307.

Zhou, S., Zhu, H., and Yan, Z. (2014). Materials, theories and experiments of microcapsule self-healing method - a review. *Tunneling and Underground Construction*, 195-204.

Chapter 3

Nanocomposite Hydrogels for Tissue Regeneration Applications

Jingan Li[1,*] and Kun Zhang[2]

[1] School of Materials Science and Engineering,
Zhengzhou University, Zhengzhou, China
[2] School of Life Science,
Zhengzhou University, Zhengzhou, China

Abstract

Hydrogels are widely used in many fields due to their unique properties such as flexibility, conductivity, biocompatibility, environmental responsiveness, and injectable properties. In the field of biomedicine, hydrogels are usually applied in tissue repair and regeneration, surface modification, drug controlled release, and many other directions. Wherein, nanocomposite hydrogels, with their unique size, surface, and quantum tunneling effects, have great influences on drug delivery, targeted therapy, and accurate diagnosis. The shape, composition, size, dispersion coefficient, and charge of hydrogel affect its interaction with cells and tissues. These are very interesting experimental phenomena. In this chapter, the recent study about nanocomposite hydrogels and other related hydrogels for biomedical applications from our research group in recent years will be introduced.

Keywords: hydrogels, biomedical application, tissue engineering, biomaterials, drug controlled release

3.1. Introduction

Since the development of human civilization, a large number of catastrophic events, such as war, disease, and accidents, have caused serious damage to human tissues. Human's pursuit of tissue repair and regeneration has never stopped since ancient times, such as drug treatment, surgical repair, organ transplantation, prosthesis replacement, and so on. However, limited by scientific cognition and technical means, many tissue wounds cannot be properly treated. Some

* Corresponding Author's Email: lijingan@zzu.edu.cn.

In: Nanocomposite Hydrogels and their Emerging Applications
Editor: Mohammad Sirousazar
ISBN: 979-8-88697-675-5

chronic and malignant diseases, such as diabetes foot, myocardial infarction, and cancer, cause cell disorder, tissue damage, and organ dysfunction, which need to be properly treated and improved. The treatment of these diseases and injuries with traditional drugs and biomedical materials is not enough to meet the needs of patients. Therefore, it is urgent to develop more simple and effective materials with higher biosafety levels. Compared with the traditional treatment methods such as oral administration, injection of drugs or surgical implantation, *in situ* drug delivery, minimally invasive implantation, drug release, targeted drug delivery, and other treatment methods occupy more and more people's vision with their excellent functions.

Accordingly, these treatment methods also put forward higher and higher requirements for materials, such as porous structure suitable for cell migration and growth, drug loading, elastic modulus matching the lesion, moisture content and biodegradation rate matching the tissue, and biological responsiveness (e.g., pH, temperature, time, and enzyme responsiveness). Hydrogels are undoubtedly the best of these new materials, which are widely used in tissue engineering, drug controlled release, medical device processing, and other fields. Among them, nano hydrogels have been widely studied in specific fields because of their unique advantages, such as size, charge, and surface chemical effects. Over the past decade, our team has done a series of research in the above fields, found some basic scientific phenomena, and explored some scientific laws. Some of the research results have been tentatively applied in clinical practice in the form of non-profit cooperation, benefiting some patients. In this chapter, we will systematically introduce these studies.

3.2. Hydrogels and Nanocomposite Hydrogels for Cardiovascular Application

Cardiovascular disease (CVD) is the leading cause of death worldwide all the time (Guan, Li, Zhang, and Li, 2021). Stent intervention is generally applied as an effective method for treating CVD (Li, 2021). However, the current stents are limited by several complications of inflammation, hyperplasia, delayed endothelialization, and late thrombosis (Li, 2020). The coating strategy of the stent can significantly improve the multiple biological functions of the surface to prevent or alleviate the above-mentioned concurrent symptoms (Li, De Rosa, Wang, and Zhang, 2020). Hydrogels and nanocomposite hydrogels can be used as coating materials for cardiovascular stents for CVD treatment due to their excellent biocompatibility and drug-loading performance (Tong, Sheng, Hou, Iqbal, Chen, and Li, 2022).

3.2.1. Coatings of Cardiovascular Stents

Hyaluronic acid is (HA) a natural hydrogel with wide applications in tissue engineering due to its unique features (Figure 3.1) (Yasin, Ren, Li, Sheng, Cao, and Zhang, 2022). As early as 2013, our team began to conduct systematic research on the behavior and function of HA micropattern regulating vascular endothelial cells (EC) and smooth muscle cells (SMC) on the surface of cardiovascular materials (Li, Zhang, Wu, Zhang, Yang, Tu, and Huang, 2015; Li, Zhang, Yang, Wu, Chen, Zhao, Li, and Huang, 2013). Our study found that HA micro-tripes can regulate the phenotypic contraction of SMC, and this function will be enhanced when fluid shear stress is applied along the direction of micro-stripes. However, some peer-reviewed experts suggest that SMC in the coronary artery does not extend and grow along the blood flow, but grows perpendicular to the blood flow direction in the circular vessel wall.

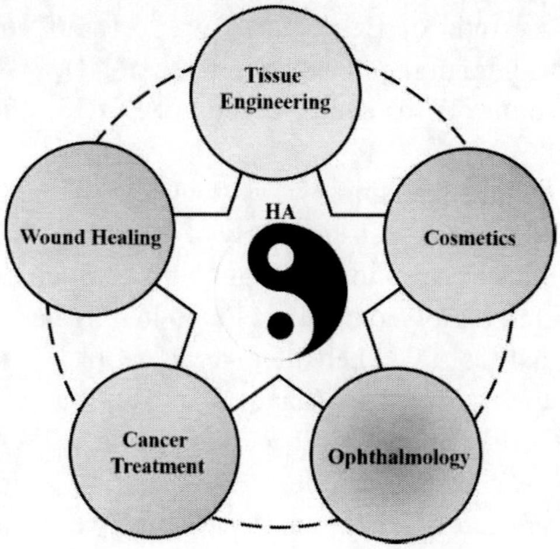

Figure 3.1. The applications of HA natural hydrogels as popular biomaterials (Yasin, Ren, Li, Sheng, Cao, and Zhang, 2022).

When supplementing the relevant experiments, we were surprised to find that when the fluid shear stress perpendicular to the HA micro-stripe direction was applied, although the SMC grew slowly, its phenotypic shrinkage function was much higher than that when the fluid shear stress parallel to the HA micro-stripe direction. The regulation of SMC contractile phenotype requires a wide range of HA micro-stripe scales, ranging from a few microns to hundreds of microns. However, EC has strict restrictions on the size of the HA micro-stripe. In our early work, 40, 25, and 10 μm-wide HA micro-stripes were used to simulate human blood flow shear stress on the regulation of morphologies and functional factors release of EC. The results showed that the HA micro-stripe with 25 μm width was close to the effect of human blood flow shear stress 15 dyn/cm^2 on EC morphological index and the regulation of functional factors release, such as nitric oxide (NO) and fibronectin (Fn).

SMC and EC are the main constituent cells of vascular mesomembrane and intima respectively (Li, Zhang, Yang, Wu, Chen, Zhao, Li, and Huang, 2013). Therefore, the co-culture of SMC and EC is an important cellular basis for vascular tissue repair and reconstruction. For a long time, SMC and EC have been unable to imitate the co-culture model under physiological conditions for many reasons, such as loss of cell phenotype and function, and EC/SMC cannot be contacted directly. To solve these problems, we controlled the SMC to contractile phenotype using the HA micro-pattern with 25 μm width and building the EC/SMC co-culture model by adding hyaluronidase to hydrolyze high molecular weight HA (HMW-HA) into low molecular weight HA (LMW-HA) to induce EC adhesion and growth (Li, Li, Zhang, Liao, Yang, Maitz, and Huang, 2013). In this co-culture model, it was verified that the contractile SMC could promote NO and prostacyclin (PGI_2) release from EC.

Yet, another EC/SMC co-culture model was built by coating type IV collagen (ColIV) onto the HA patterned SMC and inducing EC adhesion and growth (Li, Zhang, Xu, Chen, Yang, Zhao, Zhao, and Huang, 2014). The EC in this new co-culture model was proved to release more NO compared with the EC in the hyaluronidase model. Nevertheless, EC in both co-culture models could not completely cover the SMC as EC in the vessel wall. By dissecting the femoral artery of dogs, we found that the number of EC and SMC in the vessel was not equal, because the size of SMC was generally larger than EC. Further studies found that in the ColIV co-culture model, when the ratio of EC to SMC was 4:1, EC could completely cover the contractile SMC, and SMC benign regulated the release of NO, PGI_2, thrombomodulin (TM), tissue pathway factor inhibitor (TPFI) and E-selectin of EC (Li, Zhang, Wu, Liao, Yang, and Huang, 2015).

Based on the above understanding, we speculated that a stent coating with good biocompatibility may be prepared by self-assembly of HA and Col IV mixture. In the first step of the experiment, we were surprised to find that Col IV, which originally took 24 hours to dissolve in acetic acid buffer, dissolved in HA solution in only half an hour, suggesting a very good integration of HA and Col IV. Thereafter, the HA/Col IV coating presented excellent functions of pro-endothelialization (Li, Zhang, Wu, He, Yang, and Huang, 2015), anti-hyperplasia (Li, Zhang, Chen, Liu, Yang, Zhao, and Huang, 2014), anti-coagulant, and anti-inflammation (Li, Zhang, Ma, Wu, Yang, He, and Huang, 2016). In spite of that, the specific parameters of HA as a stent coating are not clear. Thus, we conjugated HA onto the stent materials via depositing poly-dopamine (PDA) as the bonding layer which may combine the HA by electrostatic interaction (Wu, Li, Zhang, He, Yang, Zou, and Huang, 2016).

Our data suggested that when the concentration of the HA solution was 2 mg/ml, the PDA/HA coating showed stronger blood compatibility, cell compatibility, and histocompatibility. In the follow-up study, we covalently grafted HA with different MW (4×10^3, 1×10^5, and 5×10^5 Dalton) onto the surface of stainless steel stent to study their multi-functions (Li, Wu, Zhang, He, Zou, Luo, Fan, Yang, Zhao, and Huang, 2017). The systematic biological evaluation confirmed that when the MW of HA in the coating was 1×10^5 Dalton, the stent possessed stronger biocompatibility. In order to explore whether this rule is also applicable to biodegradable magnesium (Mg) alloy vascular stents, we conjugated HA with different MW (4×10^3, 1×10^5, 5×10^5, and 1×10^6 Dalton) onto the surface of NaOH-treated Mg-Zn-Y-Nd alloy surface via PDA adhesion to investigate their corrosion resistance and biocompatibility (Li, Chen, Zhang, and Guan, 2020).

Our data showed that the coatings composed of HA with 1×10^5 and 1×10^6 Dalton could endow the Mg alloy with better corrosion resistance. Wherein, the HA with 1×10^5 Dalton coated Mg-Zn-Y-Nd alloy displayed better biocompatibility. Obviously, the HA coatings on the Mg alloy showed consistent trends on the stainless steel. In addition, the HA coatings may control the degradation products release of the Mg alloy, and it has been proven that the size and concentration of the degradation products release of the Mg alloy may influence the macrophages phenotypes, further affect EC growth and function (Figure 3.2) (Hou, Witte, Li, and Guan, 2022). This research also found that the Mg ion could enter into EC to regulate its function.

The theory of "Spatial and temporal order of functions" proposed in our previous publications (Li, Zhang, and Hou, 2020; Li, and Hou, 2022; Li, Hou, and Zhang, 2021), revealed the truth that macrophage phenotype may influence SMC contractile status and further affect the endothelial monolayer formation (Figures 3.3 and 3.4).

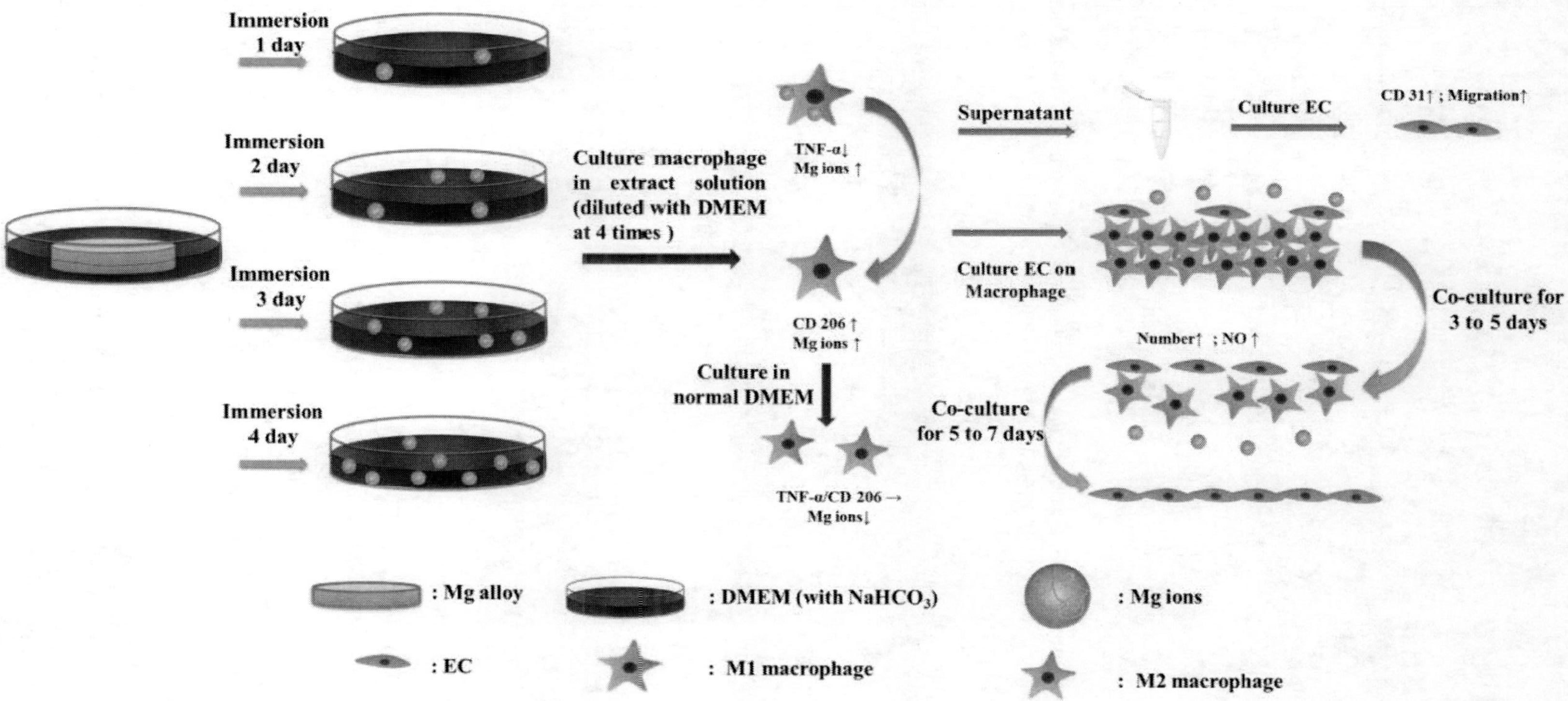

Figure 3.2. Mg alloy degradation products regulating macrophages and EC behaviors (Hou, Witte, Li, and Guan, 2022).

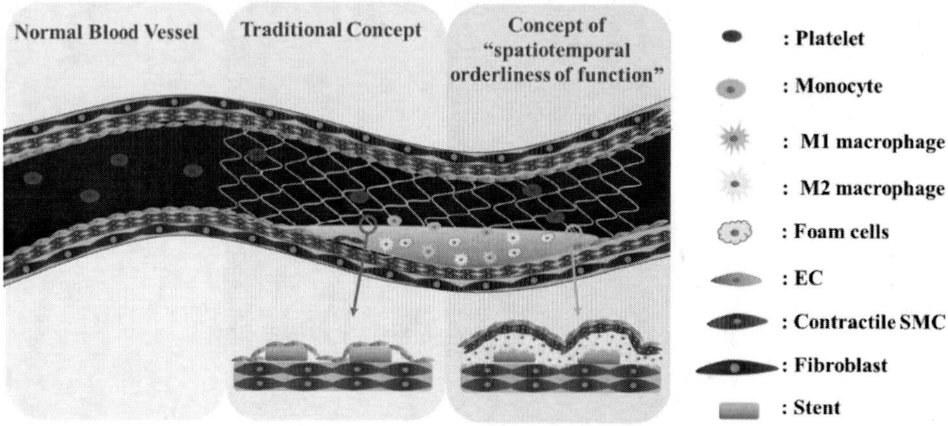

Figure 3.3. The diagram of the traditional concept and the concept of "spatiotemporal orderliness of function" for the endothelialization on the cardiovascular stents in comparison (Li, Hou, and Zhang, 2021).

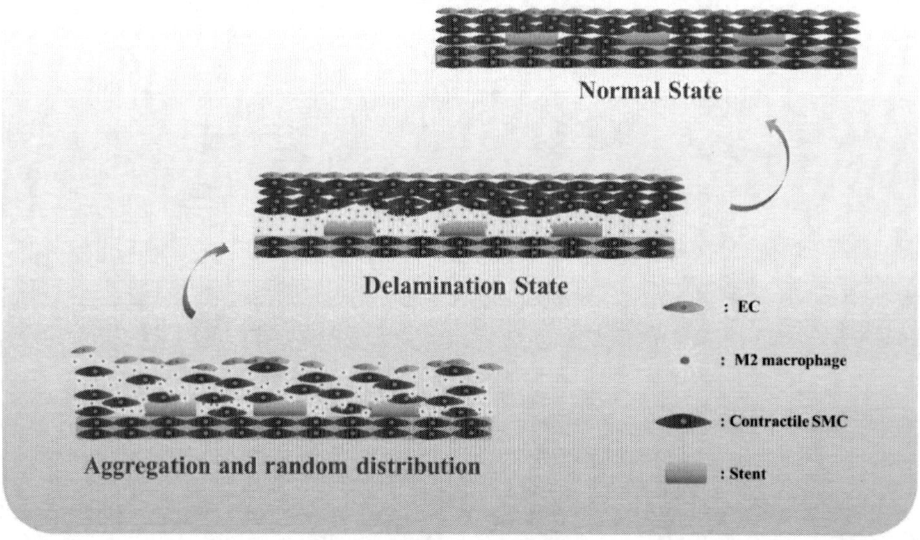

Figure 3.4. The evolution of EC, contractile SMC, and M2 macrophages distributed in the regenerated tissues according to the concept of "spatiotemporal orderliness of function" for the endothelialization of cardiovascular stents (Li, Hou, and Zhang, 2021).

In the additional study, we prepared HA nanoparticles with different sizes, all of which could combine with the Mg ion. The small particle could carry the Mg ion into EC to enhance their NO release and CD31 expression, suggesting enhanced functionality. The big particle prevents the carried Mg ion into the EC, further reducing their NO release and CD31 expression compared to the mall particle group, indicating weakened functions (Wang, Zhu, Wang, Chang, Wang, Li, and Guan, 2020). Based on the knowledge gained in this study, we prepared these small HA particles onto the Mg-Zn-Y-Nd alloy surface and further strengthened the biocompatibility (Li, Wang, Sheng, Liu, Xue, Tong, and Guan, 2021).

Although a large number of studies have shown that the use of extracellular matrix (ECM) components to modify the surface of vascular scaffolds can significantly improve their anticoagulant, anti-proliferative, and pro-endothelial functions, it may be that the full-component ECM coating has a better effect. Tu and coworkers prepared a full component ECM coating on the metals by the method of culturing EC to secrete ECM and then removing cells, and the EC-ECM coating inhibited platelet adhesion and promoted EC adhesion. They further fabricated the SMC-ECM coating in the same way, and the differences between the two ECM coatings were one of them improved EC growth and the other one promoted functional factor release from EC (Li, Zhang, and Huang, 2017). However, the EC which secreted ECM on the surface was not in a physiological state, so many functions were missing, including the secretion function of ECM (Li, Zhang, and Huang, 2017). Thus, we controlled the EC to their physiological state by HA micro-stripe and obtained the ECM/HAP coatings by the method reported by Tu and coworkers (Li, Zhang, Wu, Zhang, Yang, Tu, and Huang, 2015; Zou, Luo, Han, Li, Yang, Li, and Huang, 2019). The ECM/HAP coatings showed stronger blood compatibility, anti-inflammation, anti-hyperplasia, pro-endothelialization functions, and histocompatibility than the previous one.

In the previous work, it has been reported that the HA micro-stripes also possess the ability of regulating SMC to contractile phenotype. Therefore, the SMC-ECM/EC-ECM double layer was prepared by controlling SMC and EC to their physiological state by HA micro-stripe, and removing the cells in sequence (Han, Luo, Zou, Li, Zhang, Yang, and Huang, 2019). This structure is also very similar to the vascular base membrane, which is crucial for maintaining the functions of the blood vessel wall. The SMC-ECM/EC-ECM double layer showed more complete biological functions compared to the single SMC-ECM and EC-ECM. However, this double-layer ECM is not easy to be prepared onto biodegradable materials or materials with irregular shapes. Then, we distributed this nature-inspired ECM into solution by ultrasonic oscillation method, and re-conjugated the ECM onto biodegradable Mg alloy, endowing better corrosion resistance and biological functions (Liu, Chen, Zhang, Li, and Guan, 2022).

Although HA has powerful biological functions as a coating for cardiovascular stents, its hyaluronidase stability in extraction limits its application, that is, HA coating is easily decomposed by hyaluronidase *in vivo*. To solve this problem, we modified the HA molecule by sulfonation and found that the stability of HA against hyaluronidase increased with the increase of sulfonation degree (Yu, Zhu, Hou, Li, and Guan, 2020). In addition, with the increase of sulfonation degree, the function of sulfonated hyaluronic acid (S-HA) in inhibiting macrophage adhesion and activation, inhibiting SMC proliferation, and promoting EC growth and migration was gradually enhanced. Thus, the S-HA was conjugated onto the ZE21B alloy which was pre-treated with hydrofluoric acid and deposited by PDA to create a multi-functional coating (Yu, Zhu, Dong, Zhang, Li, and Guan, 2022). This coating endowed the ZE21B alloy with a better ability of anticoagulation, promoting EC growth, controlling SMC to the contractile phenotype, and regulating macrophages to the M2 phenotype. Additionally, it also promised the ZE21B alloy stronger corrosion resistance.

Another molecule with the sulfonic acid group, chondroitin sulfate (CS), came into our view in 2017 as a good coating material for surface modification of cardiovascular stents (Li, Zou, Zhang, Luo, Yang, Jing, Zhang, Cui, and Huang, 2017). There are two subtypes of natural CS, CSA, and CSC, which usually mix together in nature to play a role. In 2017, we found the PDA/HD-CS coating prepared by the CS mixture significantly improved EC adhesion/ proliferation and inhibited EC apoptosis simultaneously on the stainless steel surface. In the beagle dog femoral artery implantation model, the PDA/HD-CS group also showed a strong function to regulate the conversion of macrophages to M2 phenotype and the contraction of SMC in the new tissues, finally forming the endothelial monolayer. However, which subtypes, CSA or CSC played a more crucial role in surface functionalization was not clear at that time.

Hence, CSA and CSC were conjugated onto the PDA/HD coated stents separately, and the results showed that both CSA and CSC contributed to anticoagulation, capturing endothelial progenitor cells (EPC), improving EC growth and NO release, but CSC showed stronger functions on these aspects (Zou, Li, Kou, Luo, and Yang, 2021). These two subtypes showed no significant difference in suppressing SMC proliferation, but both of them owned better ability compared to stainless steel and PDA/HD in this aspect. In a white rabbit model of iliac artery implantation, CSA did not form an endothelial monolayer in the new tissue on the stent surface for one month, while CSC played a more significant role in this regard. In addition, we also compared another rare CS subtype, CSB with CSA as a cardiovascular stent coating (Li, Li, Zou, Kou, Hou, Yasin, and Zhang, 2022). The CSB coating showed no significant difference in inhibiting platelet adhesion compared to CSA but presented the advantages of anti-inflammation, anti-hyperplasia, pro-endothelialization, and dynamic blood compatibility.

3.2.2. Thermosensitive Hydrogel for Myocardial Tissue Repair

Myocardial tissue repair is a major challenge in heart disease treatment, and a large number of studies have proved that angiogenesis plays an important role in it (Chen, Ye, Wang, Li, and Chen, 2022; Lee, Parthiban, Jin, Knowles and Kim, 2021). At present, injectable hydrogels, as a cutting-edge and effective scaffold for tissue engineering, have attracted extensive attention. Based on previous studies, we developed a temperature-sensitive hydrogel containing double nanoparticles (He, Zhao, Su, Zhang, Wang, Luo, Gao, Li, and Yang, 2019). The hydrogel was liquid at room temperature and could form a gel when the temperature reached 37°C. Its elastic modulus meets the requirements of myocardial tissue. Its internal porous wetting environment is conducive to the growth of EC, mesenchymal stem cells (MSC), and other cells related to vascular regeneration and myocardial repair. In order to induce MSC to migrate to the hydrogel, we loaded the nanoparticles of MSC chemokine SDF-1 with the same charge in the hydrogel to facilitate the release of chemokine.

In order to accelerate the growth and rapid vascularization of the homing MSC and EC, we loaded the hydrogel with nanoparticles of endothelial growth factor VEGF, which is opposite to its charge. The experiment in the chorioallantoic membrane model showed that the temperature-sensitive hydrogel containing double nanoparticles could accelerate vascular regeneration. This study became the most-read paper after publication in 2019. Then, we introduced RGD into the hydrogel which may increase the MSC differentiation, and the results showed that the new hydrogels regulated the SMC, EC, and MSC growth in orders (Ding, Zhao, Liu, Wang, Gao, Li, and Yang, 2020). The rat myocardial infarction model results showed that the ultrasonic electrocardiogram, ejection fraction calculation, and shortening fraction of the SD rats' myocardial infarction model after injecting the new hydrogel presented no difference with the healthy rats, suggesting a strong curative effect. This study was honored with Athanasiou ABME Student Award in 2021.

3.3. Hydrogels and Nanocomposite Hydrogels for Wound Healing

3.3.1. Injectable Hydrogels for Treating Traumatic Brain Injury

Traumatic brain injury (TBI) affects people worldwide and is characterized by high mortality and disability (Prabhakar, Khan, Grewal, and Singh, 2022). Several previous research demonstrated that stem cell replacement therapy may have potential applications in the treatment of nervous system damage. However, after stem cell transplantation, there are some defects that seriously limit their therapeutic effect, including inflammatory reaction, mechanical damage, and the rapid flow of cerebrospinal fluid at the lesion, all of which will cause the loss or very low vital ratio of transplanted stem cells.

To solve these problems, we designed an injectable hydrogel with sodium alginate and HA to carry the stem cells as a tissue engineering scaffold (Zhang, Shi, Zhou, Xing, Ma, Li, Zhang, Yao, Wang, Li, Li, and Guan, 2018). This SA/HA hydrogel could be gel within 5-8 minutes after preparation, which not only left enough injection time but also ensured rapid gelation without being washed away by cerebrospinal fluid. Its water content was about 90%, which was slightly higher than that of brain tissue, while subsequent animal experiments showed that it did not cause a serious inflammatory reaction. Its elastic modulus was equivalent to that of brain tissue to ensure that the implanted hydrogel will not cause stress damage to brain tissue. Its 7-day degradation rate was about 50%, and then it degraded slowly, the degradation rate was less than 60% in 28 days. The gelled SA/HA hydrogel was in the shape of transparent jelly, and the microstructure is a porous structure with a pore diameter of 50-100 μm. Compared with traditional SA hydrogels, SA/HA hydrogels could load more stem cells, and the survival rate of loaded stem cells was higher. Due to a more uniform pore structure, stem cells were more evenly distributed in SA/HA hydrogels. The injection experiment of the SD rat brain injury model showed that the SA/HA group SD rats got a lower mNSS score after injection for 35 days, suggesting a better health status.

A series of further behavioral experiments also confirmed that SD rats in the SA/HA group had a better memory. Immunohistochemical and immunofluorescence tests after tissue sections showed that there were more stem cells in SA/HA group at the implantation site, and more stem cells differentiated into nerve cells. Based on the above work, we prepared another hydrogel with SA and type I collagen (ColI), which had very similar functions to the SA/HA hydrogel (Zhou, Zhang, Ma, Liu, Yao, Li, Wang, and Guan, 2019).

3.3.2. Hydrogels for Repairing Skin Injury

As the body's first line of defense against external stimuli, the skin not only protects the body from foreign germs but also senses stimuli and maintains homeostasis. At the same time, skin is vulnerable to acute trauma, high temperature, chemicals, and other factors, which can cause serious damage. Once severe skin damage occurs, homeostasis of the body would be affected, and the open wound significantly increases the risk of infection (Cui, Li, Guan, Zhang, Zhang, and Li, 2022; Su, Li, Liang, Zhang, and Li, 2021). Skin lesions can be divided into acute and chronic; wherein acute wounds are always neutrally alkaline (pH = 6.5-8.5), while chronic wounds are alkaline (pH = 7.2-8.9) (Liang, Cui, Li, Guan, Zhang, and Li, 2021). While acute wounds (such as burns, penetrating wounds, and blunt wounds) tend to heal quickly, chronic wounds (such as venous leg ulcers, pressure sores, and diabetic foot ulcers) are difficult to heal and are susceptible to inflammation and infection (Liang, Cui, Li, Guan, Zhang, and Li, 2021). The key to treating skin damage is to provide the outermost barrier, the epidermis, to prevent infection, stop bleeding, moisturize, and reduce pain (Liang, Cui, Li, Guan, Zhang, and Li, 2021). However, epidermal coverage alone does not solve the problems of graft fragility, wound contraction, and scarring. Thus, providing strength and elasticity for epidermal transplantation is necessary to promote skin regeneration and wound healing (Liang, Cui, Li, Guan, Zhang, and Li, 2021).

In 2020, we cooperated with Shanghai Junshi Biosciences Co., Ltd. to load benlysta with injectable hydrogel prepared by SA (Wang, Guan, Zhang, and Li, 2020). The benlysta was then a new drug that had just been approved by the US Food and Drug Administration (FDA) for treating systemic lupus erythematosus (SLE) in children aged 5 years and above. Our experiments results indicated that the benlysta-loaded SA hydrogel could significantly inhibit the inflammatory response and promote the growth of fibroblasts and skin cells, and later cells are the main constituent cells of skin tissue, and their rapid growth implies the rapid regeneration of skin tissue. In addition, the benlysta-loaded SA hydrogel also accelerated the EC growth, which is the main cellular component of capillaries, and its rapid growth is conducive to the regeneration of capillaries, thus accelerating the transportation and metabolism of nutrients in the process of skin repair.

In 2021, for the timely treatment of acute skin injury, we prepared a new injectable *in situ* gel-forming hydrogel (Guan, Zhang, Cui, Liang, Li, and Guan, 2022). The hydrogel was stabilized by a dynamic Schiff base and consisted of gelatin, oxidized dextran, and apolipoprotein. The gel-forming time of the hydrogel was about 5-12 minutes, the moisture content was about 95%, and the elastic modulus was equivalent to that of normal human skin. It could be completely degraded in a simulated *in vivo* microenvironment for about 4-7 days. After gelatinization, the surface and interior of the hydrogel showed different porous structures, with a surface pore size of about 40 μm and an internal pore size of about 160 μm. *In vitro* biological tests showed that the hydrogel had a low hemolysis rate and good compatibility with fibroblasts and skin cells. In addition, it also had effective hemostatic properties. *In vivo* studies indicated that the hydrogel could reduce wound inflammation, promote angiogenesis, and then promote skin tissue regeneration.

In the same year, another multi-functional hydrogel was prepared with carboxymethyl chitosan, HA, and dopamine via crosslinking by horseradish peroxidase and hydrogen peroxide (Cui, Li, Guan, Zhang, Zhang, and Li, 2022). This kind of hydrogel had the characteristics of high adhesion, oxidation resistance, and hemostasis. It is an ideal material to promote the repair of skin damage. The results of the full-thickness skin injury model in mice showed that the hydrogel could significantly improve angiogenesis and cell proliferation at the injured location, and inhibit inflammation. At the same time, through literature search and review, we found that nano hydrogels can promote vascular regeneration and have potential application value in skin tissue repair and regeneration (Cui, Liang, Liu, Zhang, and Li, 2020). However, there are few reports on direct research in this field.

3.3.3. Hydrogels for Preventing Puncture Site Bleeding

In 2021, our collaborator, Professor Hualong Bai from the First Affiliated Hospital of Zhengzhou University, hopes to improve the biological performance of the puncture needle to inhibit bleeding during the puncture process. His initial idea was to prepare CD34 antibody coating on the puncture needle to accelerate the growth of endothelial cells and endothelial progenitor cells in the lesion after puncture, to achieve the purpose of tissue repair and inhibit bleeding. But the key problem is how to make the CD34 antibody relatively stable and temporarily fixed on the puncture needle. So, we tried to load the CD34 antibody into the HA solution with a molecular weight of 1×10^5 Dalton, and use the double-sided adhesive function of PDA to graft the HA loaded with CD34 antibody onto the needle by the chemical combination (Wei, Xu, Wang, Li, Sun, Xie, Xing, Bai, Kan, Li, and Dardik, 2021). In this way, the drug release function of HA can be used to accurately deliver the CD34 antibody to the lesion during puncture. The results of animal experiments showed that the HA coating was used as a carrier, which not only significantly inhibited bleeding during puncture but also obtained additional anti-inflammatory, anti-hyperplasia, rapid endothelialization, and accelerated angiogenesis. After that, this technique was used to stop bleeding at different puncture sites and achieved good results (Xu, Wang, Wei, Sun, Bai, and Li, 2021).

3.4. Hydrogels and Nanocomposite Hydrogels for Vascular Patch

Professor Bai is a vascular surgeon. He has made a long-term academic visit to Yale University, which is famous in this field, and tried to promote these more advanced clinical technologies and concepts in China. The author of this chapter, Professor Jingan Li, first met him through the introduction of Professor Shaokang Guan, vice president of Zhengzhou University, to exchange the preparation process of magnesium alloy vascular stent surface coating. Professor Bai originally thought that Professor Li was just a pure material scientist who did not understand biomedicine, let alone vascular structure. However, through in-depth communication, Professor Bai found that Professor Li was not only familiar with the general knowledge of the biomedical field, but also had his own unique views on vascular tissue regeneration, even cell differentiation, dedifferentiation, and other very professional subfields. In Professor Li's own words, he was a so-called pure biomaterial researcher.

At that time, Professor Bai's research group was trying to use the acellular-treated autologous or allogeneic vascular tissue as a vascular patch for peripheral vascular repair and had achieved good results. After acellular treatment, the vascular tissue reduced the immune rejection to a certain extent. However, due to a large number of proteins, the immune rejection could not be completely eliminated. The thickening of the vascular wall caused by the hyperplasia after implantation was also an important factor limiting the technology. Their initial strategy was to reduce the neointimal thickness as the goal, and heparin was selected as the coating material of the acellular vascular tissue patch due to the proliferation inhibition function. However, heparin as a coating material has precise requirements for the dosage and grafting method. Excessive use or release will cause serious bleeding. After careful consideration, we decided to load heparin into HA to prepare a drug sustained-release coating, which can not only ensure sufficient cumulative drug dosage but also ensure that the heparin release per unit time is not excessive (Bai, Wang, Li, Liu, Wang, Sun, Wei, Wang, Li, and Dardik, 2020). Compared with the wrinkle morphology of the acellular vascular tissue, the HA-heparin composite coating had a smoother surface and a uniform porous structure, which is a typical feature of the drug release coating. After implantation, the patch not only reduced the intimal thickness but also expressed more vascular intimal specific factor vWF, which proved that a good endothelial function was obtained. In addition, the introduction of HA coating further reduced the immune response, which was due to the fact that HA can protect other proteins from immune attack.

Another problem raised by Professor Bai was that the patch stimulates tumor growth after repairing blood vessels. Therefore, an anti-tumor drug needed to be loaded on the patch. Thus, we chose a powerful tumor suppressant, programmed death-1 (PD-1), and loaded it into the HA coating (Bai, Wang, Li, Sun, Wei, Wang, Wang, Xing, Li, and Dardik, 2021). The results of animal experiments showed that the patch with drug-loaded coating significantly inhibited the tumor and further reduced the thickness of neointima. Then, this technique was extended to many different samples and achieved good efficacy consistently (Sun, Wang, Liu, Li, Wei, Xu, Qiao, Wang, Fu, Bai, and Li, 2021). In addition, we also loaded the TGF β 1 into the HA hydrogel to create a novel drug delivery system that deceased aneurysm and pseudoaneurysm formation and progression by securing the drug in place between the medium and adventitia in the aneurysm and pseudoaneurysm to make it release slowly (Bai, Sun, Wei, Xie, Li, Xu, Wang, Liu, Zhang, Wu, Wang, Xing, Wang, and Li, 2021).

Inspired by the superelasticity of the swim bladder, Professor Bai and our research group performed acellular treatment on it and prepared HA coating loaded with rapamycin on the surface (Bai, Sun, Wu, Wei, Xie, Wang, Hou, Li, Dardik, and Li, 2021). The experimental results showed that the treated swim bladder suture did not rupture on the blood vessel and showed excellent biocompatibility. This experiment fully proves that the swim bladder can be used as a potential vascular patch material after fine treatment. Using the same method, we treated the leaves with acellular treatment, obtained a transparent tough membrane, and prepared rapamycin-loaded HA coating on its surface (Bai, Xie, Wang, Li, Sun, Wei, Wang, Wu, Bai, and Li, 2021). Different from the study of the swim bladder, rapamycin on the leaf surface was first loaded into PLGA nanoparticles before loading into HA, which was not only conducive to the further release of rapamycin but also conducive to the rapid entry of the drug into the cell, thus enhancing the drug efficacy (Figure 3.5). The scanning electron microscope (SEM) images depicted that the microscopic morphology of the leaf surface was very smooth, but after the acellular treatment, it presented a uniform grid structure, which was quite different from that before the acellular treatment. After the drug-loaded HA coating treatment, the surface morphology was again becoming smooth, and the previous morphology was restored. Therefore, this process can also be regarded as a bionic technology that endows leaves with multiple functions.

Animal experiments showed that the implanted leaf patch not only promoted tissue regeneration and reduced the thickness of neointima, but also showed significant CD31 expression on one side of the intima, which proved that there was a trend of endothelial monolayer formation. This study fully proved that the leaves can also be used as a sufficient spare material for vascular patches after fine treatment such as acellular and coating design. In the follow-up study, we conducted acellular and coating treatment on onion skin, sawdust, and other materials. The drugs loaded included various anti-inflammatory and antioxidant drugs, tannic acid, and even Professor Li's collection of red wine from Europe and China also contributed one bottle each (in the field of food, red wine, and raw onion have a variety of health preservation and chronic disease treatment effects). It has been proved that these treated and optimized materials can be used as potential substitutes for high-quality vascular patches.

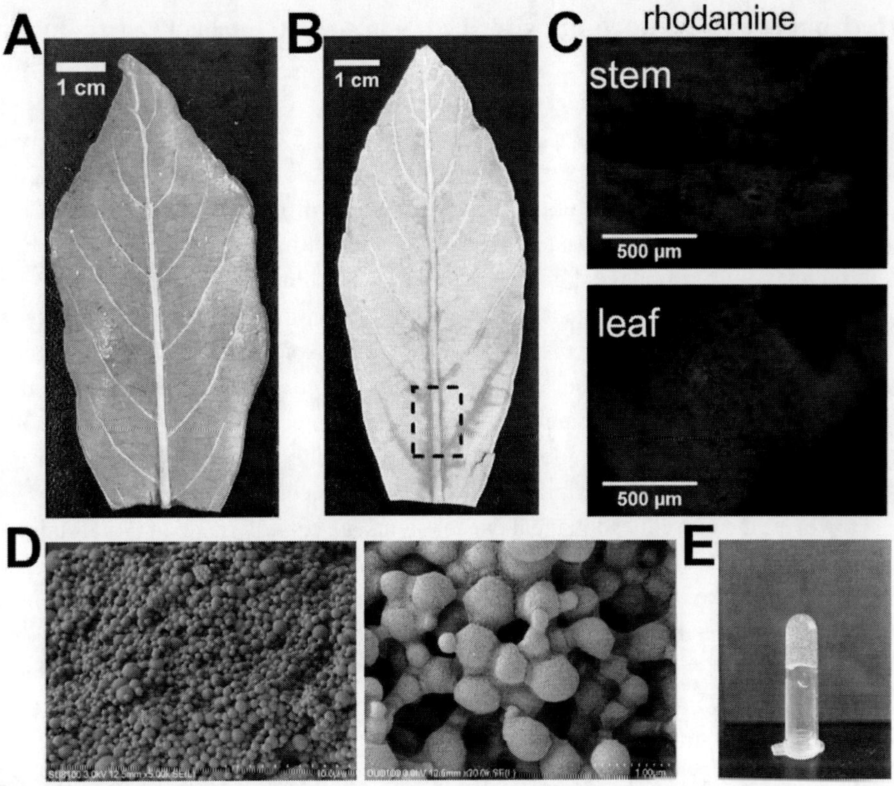

Figure 3.5. Photographs showing the plant leaf before and after rhodamine water perfusion: (A) photograph showing the plant leaf before rhodamine water perfusion, (B) photograph showing the decellularized leaf perfused with rhodamine water (ruler marks: 1 mm), (C) immunofluorescence photographs showing the leaf perfused with rhodamine water (scale bar: 500 µm; n = 3), (D) SEM image showing the PLGA-based nanoparticles, and (E) photograph showing the liquid hydrogel that solidified in 10 min (Bai, Xie, Wang, Li, Sun, Wei, Wang, Wu, Bai, and Li, 2021).

In order to develop a functional sequential biological patch, Professor Bai loaded three groups of different drugs into the SA/HA hydrogel we developed before and sewed three hydrogels loaded with different drugs together (Wei, Li, He, Shu, Dardik, and Bai, 2022). The outermost hydrogel was loaded with Necrostatin-1, an inhibitor of cell necrosis. The middle layer hydrogel was loaded with rapamycin and SB431542 to inhibit proliferation and inflammation. The inner hydrogel was loaded with heparin and CD34 antibody for anticoagulation and endothelialization. All drugs were pre-loaded onto PLGA nanoparticles and then dispersed into hydrogels. Compared with the acellular thoracic artery patch and bovine pericardial patch, the three-layer hydrogel suture bio-patch formed thinner neointima 14 days after implantation. Compared with the hydrogel patch without the drug, the drug-loaded three-layer hydrogel showed a better function of inhibiting hyperplasia and inflammation. Although the preparation technology of this drug-loaded three-layer hydrogel patch is still rough and inconvenient for industrialized mass production, the direction of its sequential function is correct. This concept can provide direction for the design and clinical transformation of hydrogels.

3.5. Hydrogels and Nanocomposite Hydrogels for Cancer Treatment

Cancer is a kind of disease with very high morbidity and mortality rate. Talking about cancer and panic really happens around us. Esophageal cancer ranks among the top five deadly malignant tumors, and the 5-year survival rate of patients is not more than 20%. Esophageal cancer has different incidence factors in different regions of the world, but they are all related to eating habits. In the west, it is related to drinking and getting used to cold food. The patients are often patients with esophageal adenocarcinoma. In China, especially in Henan Province, patients often suffer from esophageal squamous cell carcinoma because people have the habit of eating while it is hot, especially eating stewed noodles, spicy soup, and other hot delicacies. Once esophageal cancer enters the advanced stage, the cure rate is very low, and surgery, chemotherapy, and radiotherapy cannot guarantee that cancer will not metastasize. Palliative therapy with stent intervention can not only prolong patients' life and improve their quality of life but also greatly reduce the possibility of esophageal cancer metastasis to the lungs and other organs. However, the current esophageal cancer stents are mainly made of 317L stainless steel.

Although they have good biocompatibility and mechanical properties to ensure sufficient support and low cytotoxicity after implantation, esophageal cancer cells will also climb to the surface of the stents, resulting in restenosis in the stents. Therefore, it is particularly important to prepare a coating on the surface of esophageal cancer stents that can inhibit the adhesion and growth of cancer cells. In 2017, we covalently grafted polyethyleneimine (PEI) onto the surface of 317L stainless steel through the double-sided adhesive function of PDA (Zhang, Bai, Wang, Li, Guan, and Li, 2017). The results of apoptosis experiments showed that the PDA/PEI coatings not only inhibited the adhesion and growth of esophageal squamous cell carcinoma cells but also significantly killed pathological fibroblasts, epithelial cells, and other cells involved in stent restenosis. This work was also authorized by the Chinese invention patent. Subsequent studies found that the treatment effect of PDA/PEI coating loaded with pentafluorouracil was better. It could inhibit both squamous cell carcinoma and adenocarcinoma cells, and not only inhibit the pathological cells on the surface of the scaffold but also kill the pathological cells in the lumen. The killing mechanism of drug-loaded coating on different pathological cells was finally revealed.

Another problem with the 317L stainless steel stent is that although it has good mechanical properties and strong support, its shortcomings are also caused by excessive support and hardness, which will cause damage to the normal tissue at the lesion and unbearable pain. In addition, if stent restenosis occurs, the stent needs to be removed for secondary implantation. The additional damage caused by the adhesion between the stent and the tissue is often intolerable or even fatal. Therefore, it is very important to develop a biodegradable esophageal cancer stent to avoid secondary surgery. Magnesium alloy has come into our view as a common material for biodegradable vascular stents. Magnesium alloy has good mechanical properties and can be used for stent processing. Its excellent plasticity determines that the supporting force of magnesium alloy stent can be adjusted by the material composition and crystal structure. In addition, the degraded alkaline products can neutralize the acidic environment of the cancer lesion, and the degraded hydrogen can also reduce the inflammatory environment. Therefore, we chose the Mg-Zn-Y-Nd alloy developed by the Key Laboratory of Advanced Magnesium Alloys in Henan Province of China to compare with 317L stainless steel (Wang, Zhang, Li, Liu, and Guan, 2020). The experimental results showed that Mg-Zn-Y-Nd alloy not only had appropriate mechanical properties and excellent degradability, but also significantly inhibited the adhesion and growth of esophageal cancer cells, pathological fibroblasts, esophageal epithelial cells, and macrophages, suggesting a potential biodegradable stent material for esophageal cancer.

In 2017, a study reported that doctors prescribed disulfiram for drinking cancer patients. After taking it for a period of time, some patients' cancer could be improved. Further study found that the combined use of disulfiram and copper ion could significantly improve the anti-cancer effect. Our own experience also tells a story: Mr. Shichang Liu, a Professor at Qiqihar University, suffered from advanced pancreatic cancer, and his doctor diagnosed that he could not live for half a year. He contacted us to ask if we knew the team about malaria treatment for advanced cancer and whether we could introduce him as a patient volunteer because he couldn't stand the pain. Unfortunately, we don't, thus we suggested that he try to use disulfiram and copper gluconate under the guidance of his doctor, which made him live for another year with high quality, and finally, he died of a cerebral hemorrhage after an accidental fall in a walk. This case made us continue our efforts in the research and development of the low-cost anti-cancer drug disulfiram. In the following study, we loaded the disulfiram and copper ions into HA nanoparticles to enhance the targeted drug delivery function (Xu, Zhang, Liang, Gao, Li, and Guan, 2021). The use of HA nanoparticles further optimizes blood compatibility in the drug supply process. Therefore, this technology is expected to solve the moderate complications of stroke caused by disulfiram in anti-cancer applications.

Colorectal cancer is a kind of killer threatening human health and life safety. Unlike other cancers that prefer weak acidic environments, colorectal cancer is often in a more alkaline environment. However, most oral drugs used to treat colorectal cancer have to undergo reciprocating changes in different acid-base environments. There are also great differences in apoptosis and inflammatory protein expression between colorectal cancer and normal mucosa adjacent to cancer (Sun, Xue, Yasin, He, Chai, Li, and Zhang, 2022). Therefore, it is an arduous task to select an acid and alkali resistant hydrogel that can target colorectal cancer cells as a drug delivery material.

It has been reported that colorectal cancer cells mainly rely on the enriched Cu ion environment for proliferation, diffusion, and metastasis, so removing the Cu ion from the lesion and reducing its active concentration can inhibit the further development of malignant tumors. Therefore, we prepared a new material with SA hydrogel microspheres as the core and PEI and SA as the outer surface respectively by means of core gelation and layer-by-layer wrapping (Xu, Su, Cui, Zhang, and Li, 2019). After testing, the SA/PEI/SA hydrogel microspheres had extremely high acid and alkali resistance and good corrosion resistance in the simulated pH environment of the esophagus, stomach, and colorectal. At the same time, the dense surface of the microspheres was full of nanopores, which could ensure the combination with Cu ions without being involved in colorectal cancer cells. Finally, the captured Cu ions can be discharged from the body through the patient's excretion. Cell experiments *in vitro* showed that SA/PEI/SA hydrogel microspheres had a high inhibitory effect on colorectal cancer cells. This inhibitory effect came from the deprivation of Cu ions and the carboxyl group of the hydrogel itself.

Conclusion

Due to the urgent need for biomedicine and tissue regeneration, hydrogel and nanocomposite hydrogel technology have developed rapidly in the past decades. Various new hydrogels have been deeply studied and widely used in cardiovascular stent coatings, injectable tissue engineering scaffolds, wound repair dressings, vascular surgical patches, targeted cancer therapy, etc. In particular, nanocomposite hydrogels have attracted much attention and favor because of their unique size, charge, and targeting effects. However, there are still many difficulties in their clinical transformation. For example, the effects of nanocomposite hydrogels in different animals are different. However, more difficulties and challenges come from non-scientific fields such as market share, medical promotion, the number of demand groups, and conversion costs. In combination with the existing research reviewed in this chapter, with chemical synthesis and modification as the means and *in situ* sequential release of the integration of diagnosis and treatment functions as the goal, the smaller-scale nanocomposite hydrogels (less than 10 nm in diameter) will be more systematically studied in the field of tissue regeneration in the future and then obtain broader applications.

Funding

In terms of financial assistance, this research was funded by the National Natural Science Foundation of China, Grant Number U2004164; the Key Scientific and Technological Research Projects in Henan Province, Grant Number 222102310234.

References

Bai, H., Sun, P., Wu, H., Wei, S., Xie, B., Wang, W., Hou, Y., Li, J., Dardik, A., and Li, Z. (2021). The application of tissue-engineered fish swim bladder vascular graft. *Communications Biology*, 4, 1153.

Bai, H., Sun, P., Wei, S., Xie, B., Li, M., Xu, Y., Wang, W., Liu, Y., Zhang, L., Wu, H., Wang, Z., Xing, Y., Wang, Z., and Li, J. (2021). A novel intramural TGF β 1 hydrogel delivery method to decrease murine abdominal aortic aneurysm and rat aortic pseudoaneurysm formation and progression. *Biomedicine & Pharmacotherapy*, 137, 111296.

Bai, H., Wang, Z., Li, M., Liu, Y., Wang, W., Sun, P., Wei, S., Wang, Z., Li, J., and Dardik, A. (2020). Hyaluronic acid-heparin conjugated decellularized human great saphenous vein patch decrease neointimal thickness. *Journal of Biomedical Materials Research: Part B - Applied Biomaterials*, 108(6), 2417-2425.

Bai, H., Wang, Z., Li, M., Sun, P., Wei, S., Wang, W., Wang, Z., Xing, Y., Li, J., and Dardik, A. (2021). Inhibition of programmed death-1 decreases neointimal hyperplasia after patch angioplasty. *Journal of Biomedical Materials Research: Part B - Applied Biomaterials*, 109(2), 269-278.

Bai, H., Xie, B., Wang, Z., Li, M., Sun, P., Wei, S., Wang, W., Wang, W., Bai, L., and Li, J. (2021). Application of the tissue-engineered plant scaffold as a vascular patch. *ACS Omega*, 6(17), 11595-11601.

Chen, Y., Ye, M., Wang, X., Li, W., and Chen, W. (2022). Functionalized gelatin/strontium hydrogel bearing endothelial progenitor cells for accelerating angiogenesis in wound tissue regeneration. *Biomaterials Advances*, 136, 212803.

Cui, L., Li, J. K., Guan, S., Zhang, K. X., Zhang, K., and Li, J. (2022). Injectable multifunctional CMC/HA-DA hydrogel for repairing skin injury. *Materials Today Bio*, 14, 100257.

Cui, L., Liang, J., Liu, H., Zhang, K., and Li, J. (2020). Nanomaterials for angiogenesis in skin tissue engineering. *Tissue Engineering Part B: Reviews*, 26(3), 203-216.

Ding, Y., Zhao, A., Liu, T., Wang, Y., Gao, Y., Li, J., and Yang, P. (2020). An injectable nanocomposite hydrogel for potential application of vascularization and tissue repair. *Annals of Biomedical Engineering*, 48(5), 1511-1523.

Guan, S., Li, J. K., Zhang, K., and Li, J. (2021). Environmentally responsive hydrogels for repair of cardiovascular tissue. *Heart Failure Reviews*, 26(5), 1273-1285.

Guan, S., Zhang, K., Cui, L., Liang, J., Li, J., and Guan, F. (2022). Injectable gelatin/oxidized dextran hydrogel loaded with apocynin for skin tissue regeneration. *Biomaterials Advances*, 133, 112604.

Han, C., Luo, X., Zou, D., Li, J., Zhang, K., Yang, P., and Huang, N. (2019). Nature-inspired extracellular matrix coating produced by micro-patterned smooth muscle and endothelial cells endows cardiovascular materials better biocompatibility. *Biomaterials Science*, 2019(7), 2686-2701.

He, D., Zhao, A., Su, H., Zhang, Y., Wang, Y., Luo, D., Gao, Y., Li, J., and Yang, P. (2019) An injectable scaffold based on temperature responsive hydrogel and factors loaded nano-particles for potential application of vascularization in tissue engineering. *Journal of Biomedical Materials Research: Part A*, 107(10), 2123-2134.

Hou, Y., Witte, F., Li, J., and Guan, S. (2022). The increased ratio of Mg^{2+}/Ca^{2+} from degrading magnesium alloys directs macrophage fate for functionalized growth of endothelial cells. *Smart Materials in Medicine*, 3, 188-198.

Lee, J. H., Parthiban, P., Jin, G. Z., Knowles, J. C., and Kim, H. W. (2021). Materials roles for promoting angiogenesis in tissue regeneration. *Progress in Materials Science*, 117, 100732.

Li, J. (2021). Advanced Biomaterials for Drug Delivery and Tissue Regeneration. *Current Drug Delivery*, 18(7), 834-835.

Li, J. (2020). Advanced Healthcare Biomaterials. *Advanced Materials Letters*, 11(10), 20101560.

Li, J., Chen, L., Zhang, X., and Guan, S. (2020). Enhancing biocompatibility and corrosion resistance of biodegradable Mg-Zn-Y-Nd alloy by preparing PDA/HA coating for potential application of cardiovascular biomaterials. *Materials Science & Engineering C*, 109, 110607.

Li, J., De Rosa, S., Wang, J., and Zhang, K. (2020). Biomaterials development, modification, and potential application for interventional cardiology. *BioMed Research International*, 2020, 4890483.

Li, J., and Hou, Y. (2022). Surface modification of metallic biomaterials for cardiovascular cells regulation and biocompatibility improvement. *Nanoscale Engineering of Biomaterial: Properties and Application*, 207-225.

Li, J., Hou, Y., and Zhang, K. (2021). Inspiration from new discoveries in vascular endothelial cell growth: a bold conjecture about mature cell growth. *Current Approaches in Science and Technology* Research, 4, 36-44.

Li, J., Li, G., Zhang, K., Liao, Y., Yang, P., Maitz, M. F., and Huang, N. (2013). Co-culture of vascular endothelial cells and smooth muscle cells by hyaluronic acid micro-pattern on titanium surface. *Applied Surface Science*, 273, 24-31.

Li, J., Li, W., Zou, D., Kou, F., Hou, Y., Yasin, A., and Zhang, K. (2022). Comparison of conjugating chondroitin sulfate A and B on amine-rich surface: for deeper understanding on directing cardiovascular cells fate. *Composites Part B: Engineering*, 228, 109430.

Li, J., Wang, S., Sheng, Y., Liu, C., Xue, Z., Tong, P., and Guan, S. (2021). Designing HA/PEI nanoparticle composite coating on biodegradable Mg-Zn-Y-Nd alloy to direct cardiovascular cells fate. *Smart Materials in Medicine*, 2, 124-133.

Li, J., Wu, F., Zhang, K., He, Z., Zou, D., Luo, X., Fan, Y., Yang, P., Zhao, A., and Huang, N. (2017). Controlling molecular weight of hyaluronic acid conjugated on amine-rich surface: towards better multifunctional biomaterials for cardiovascular implants. *ACS Applied Materials & Interfaces*, 9(36), 30343-30358.

Li, J., Zhang, K., Chen, H., Liu, T., Yang, P., Zhao, Y., and Huang, N. (2014). A novel coating of type IV collagen and hyaluronic acid on stent material-titanium for promoting smooth muscle cells contractile phenotype. *Materials Science and Engineering C: Materials for Biological Applications*, 38, 235-243.

Li, J., Zhang, K., and Hou, Y. (2020). From selective cardiovascular cells adhesion to regulating "spatiotemporal orderliness of function": understanding based on biomaterials surface modification with functional molecules. *Current Topics in Medicine and Medical Research*, 5, 155-161.

Li, J., Zhang, K., and Huang, N. (2017). Engineering cardiovascular implant surfaces to create a vascular endothelial growth microenvironment. *Biotechnology Journal*, 12(12), 1600401

Li, J., Zhang, K., Ma, W., Wu, F., Yang, P., He, Z., and Huang, N. (2016). Investigation of enhanced hemocompatibility and tissue compatibility associated with multi-functional coating based on hyaluronic acid and type IV collagen. *Regenerative Biomaterials*, 3(3), 149-157.

Li, J., Zhang, K., Wu, F., He, Z., Yang, P., and Huang, N. (2015). Constructing bio-functional layers of hyaluronan and type IV collagen on titanium surface for improving endothelialization. *Journal of Materials Science*, 50, 3226-3236.

Li, J., Zhang, K., Wu, J., Liao, Y., Yang, P., and Huang, N. (2015). Co-culture of endothelial cells and patterned smooth muscle cells on titanium: Construction with high density of endothelial cells and low density of smooth muscle cells. *Biochemical and Biophysical Research Communications*, 456(2), 555-561.

Li, J., Zhang, K., Wu, J., Zhang, L., Yang, P., Tu, Q., and Huang, N. (2015). Tailoring of the titanium surface by preparing cardiovascular endothelial extracellular matrix layer on the hyaluronic acid micro-pattern for improving biocompatibility. *Colloids and Surfaces B: Biointerfaces*, 128, 201-210.

Li, J., Zhang, K., Xu, Y., Chen, J., Yang, P., Zhao, Y., Zhao, A., and Huang, N. (2014). A novel co-culture models of human vascular endothelial cells and smooth muscle cells by hyaluronic acid micro-pattern on titanium surface. *Journal of Biomedical Materials Research: Part A*, 102(6), 1950-1960.

Li, J., Zhang, K., Yang, P., Wu, L., Chen, J., Zhao, A., Li, G., and Huang, N. (2013). Research of smooth muscle cells response to fluid flow shear stress by hyaluronic acid micro-pattern on a titanium surface. *Experimental Cell Research*, 319(17), 2663-2672.

Li, J., Zou, D., Zhang, K., Luo, X., Yang, P., Jing, Y., Zhang, Y., Cui, G., and Huang, N. (2017). Strong multi-functions based on conjugating chondroitin sulfate on amine-rich surface direct vascular cells fate for cardiovascular implanted devices. *Journal of Materials Chemistry B*, 2017(5), 8299-8313.

Liang, J., Cui, L., Li, J. K., Guan, S., Zhang, K., and Li, J. (2021). Aloe vera: a medicinal plant used in skin wound healing. *Tissue Engineering Parts B: Reviews*, 27(5), 455-474.

Liu, C., Chen, L., Zhang, K., Li, J., and Guan, S. (2022). Tailoring ZE21B alloy with nature-inspired extracellular matrix secreted by micro-patterned smooth muscle cells and endothelial cells to promote surface biocompatibility. *International Journal of Molecular Sciences*, 23(6), 3180.

Prabhakar, N. K., Khan, H., Grewal, A. K., and Singh, T. G. (2022). Intervention of neuroinflammation in the traumatic brain injury trajectory: *in vivo* and clinical approaches. *International Immunopharmacology*, 108, 108902.

Su, J., Li, J. K., Liang, J., Zhang, K., and Li, J. (2021). Hydrogel preparation methods and biomaterials for wound dressing. *Life*, 11(10), 1016.

Sun, P., Wang, Z., Liu, W., Li, M., Wei, S., Xu, Y., Qiao, Z., Wang, W., Fu, Y., Bai, H., and Li, J. (2021). Programmed death-1 mediates venous neointimal hyperplasia in humans and rats. *Aging-US*, 13(12), 16656-16666.

Sun, X., Xue, Z., Yasin, A., He, Y., Chai, Y., Li, J., and Zhang, K. (2022). Colorectal cancer and adjacent normal mucosa differ in apoptotic and inflammatory protein expression. *Engineered Regeneration*, 2, 279-287.

Tong, P., Sheng, Y., Hou, R., Iqbal, M., Chen, L., and Li, J. (2022). Recent progress on coatings of biomedical magnesium alloy. *Smart Materials in Medicine*, 3, 104-116.

Wang, S., Zhang, X., Li, J., Liu, C., and Guan, S. (2020). Investigation of Mg-Zn-Y-Nd alloy for potential application of biodegradable esophageal stent material. *Bioactive Materials*, 5(1), 1-8.

Wang, X., Guan, S., Zhang, K., and Li, J. (2020). Benlysta loaded sodium alginate hydrogel and its selective functions on promoting skin cell growth and inhibiting inflammation. *ACS Omega*, 5(18), 10395-10400.

Wang, Z., Zhu, S., Wang, L., Chang, L., Wang, J., Li, J., and Guan, S. (2020). Preparing a novel magnesium-doped hyaluronan/ polyethyleneimine nanoparticle to improve endothelial functionalization. *IET Nanobiotechnology*, 14(2), 142-147.

Wei, S., Li, J., He, H., Shu, C., Dardik, A., and Bai, H. (2022). A three-layered hydrogel patch with hierarchy releasing of PLGA nanoparticle drugs decrease neointimal hyperplasia. *Smart Materials in Medicine*, 3, 139-147.

Wei, S., Xu, Y., Wang, Z., Li, M., Sun, P., Xie, B., Xing, Y., Bai, H., Kan, Q., Li, J., and Dardik, A. (2021). Hydrogel-coated needles prevent puncture site bleeding. *Acta Biomaterialia*, 128, 305-313.

Wu, F., Li, J., Zhang, K., He, Z., Yang, P., Zou, D., and Huang, N. (2016). Multi-functional coating based on hyaluronic acid and dopamine conjugate for potential application on surface modification of cardiovascular implanted devices. *ACS Applied Materials & Interfaces*, 8(1), 109-121.

Xu, R. Su, C., Cui, L., Zhang, K., and Li, J. (2019). Preparing sodium alginate/polyethyleneimine spheres for potential application of killing tumor cells by reducing the concentration of copper ions in the lesions of colon cancer. *Materials*, 12(9), 1570.

Xu, R., Zhang, K., Liang, J., Gao, F., Li, J., and Guan, F. (2021). Hyaluronic acid/polyethyleneimine nanoparticles loaded with copper ion and disulfiram for esophageal cancer. *Carbohydrate Polymers*, 261, 117846.

Xu, Y., Wang, Z., Wei, S., Sun, P., Bai, H., and Li, J. (2021). Hydrogel-coated needles prevent puncture site bleeding in arteriovenous fistula and arteriovenous grafts in rats. *Biomedicine & Pharmacotherapy*, 143, 112113.

Yasin, A., Ren, Y., Li, J., Sheng, Y., Cao, C., and Zhang, K. (2022). Advances in hyaluronic acid for biomedical application. *Frontiers in Bioengineering and Biotechnology*, 10, 910290.

Yu, Y., Zhu, S., Dong, H., Zhang, X., Li, J., and Guan, S. (2021). A novel MgF2/PDA/S-HA coating on the biodegradable ZE21B alloy for better multi-functions on cardiovascular application. *Journal of Magnesium and Alloys*, In press, doi: 10.1016/j.jma.2021.06.015.

Yu, Y., Zhu, S., Hou, Y., Li, J., and Guan, S. (2020). Sulfur contents in sulfonated hyaluronic acid direct the cardiovascular cells fate. *ACS Applied Materials & Interfaces*, 12(41), 46827-46836.

Zhang, K., Bai, Y., Wang, X., Li, Q., Guan, F., and Li, J. (2017). Surface modification of esophageal stent materials by a polyethylenimine layer aiming at anti-cancer function. *Journal of Materials Science: Materials in Medicine*, 28(8), 125.

Zhang, K., Shi, Z., Zhou, J., Xing, Q., Ma, S., Li, Q., Zhang, Y., Yao, M., Wang, X., Li, Q., Li, J., and Guan, F. (2018). Potential application of an injectable hydrogel scaffold loaded with mesenchymal stem cells for treating traumatic brain injury. *Journal of Materials Chemistry B*, 2018(6), 2982-2992.

Zhou, J., Zhang, K., Ma, S., Liu, T., Yao, M., Li, J., Wang, X., and Guan, F. (2019). Preparing an injectable hydrogel with sodium alginate and Type I collagen to create better MSCs growth microenvironment. *e-Polymers*, 19(1), 87-91.

Zou, D., Li, J., Kou, F., Luo, X., and Yang, P. (2021). Reveal crucial subtype of natural chondroitin sulfate on the functionalized coatings for cardiovascular implants. *Journal of Materials Science & Technology*, 91, 67-77.

Zou, D., Luo, X., Han, C., Li, J., Yang, P., Li, Q., and Huang, N. (2019). Preparation of a biomimetic ECM surface on cardiovascular biomaterials via a novel layer-by-layer decellularization for better biocompatibility. *Materials Science and Engineering C: Materials for Biological Applications*, 96, 509-521.

Chapter 4

Nanocomposite Hydrogels: Their Types and Applications in Tissue Engineering

Hamna Yasin
Zubaida Yousaf*
Arusa Aftab
Nadia Riaz
and Afifa Younas
Department of Botany,
Lahore College for Women University Lahore, Lahore, Pakistan

Abstract

Tissue engineering is a unique division of study that enables the creation of physiologically functional constructions for drug development, disease modeling, and the therapy of diseased or damaged organs. Hydrogel is a three-dimensional crosslinked polymeric network with distinct structure and behavior. Nanocomposite hydrogels have evolved as a new kind of composite material in recent decades, attracting significant interest, investment in research, and development around the world. Because of their unique attributes such as biocompatibility, low toxicity, high water content, and acceptable mechanical strength, these hydrogels are frequently employed in the field of medicine for drug delivery, wound healing, tissue engineering, and cancer therapy. Diverse nanoparticles have been incorporated into the hydrogel network to generate nanocomposite hydrogels. Natural and synthetic polymers were used to create nanocomposite hydrogels with improved mechanical and reactive qualities, resulting in significant potential and advancements for a variety of biological applications. This chapter will begin with an introduction, synthesis of nanocomposite hydrogels from plants, types of nanocomposite hydrogels, and recent developments of nanocomposite hydrogels with an emphasis on tissue engineering.

Keywords: nanocomposite hydrogels, drug delivery, tissue engineering, polymers, biomaterials

---

* Corresponding Author's Email: zubaida.yousaf@lcwu.edu.pk.

In: Nanocomposite Hydrogels and their Emerging Applications
Editor: Mohammad Sirousazar
ISBN: 979-8-88697-675-5

4.1. Introduction

Hydrogels have been described as two or multi-component systems made of a 3-dimensional network of polymeric chains and water that fills the spaces between macromolecules. Such structures contain diverse amounts of water in an equilibrium state reliant on the features of the polymer along with the variety and density of the network joints; typically in the swollen state, the mass fraction of water in a hydrogel is much higher than the mass section of polymer. As in real life, it is usual to use synthetic polymers which are soluble in water in their uncrosslinked state to obtain substantial levels of swelling. Several "classical" chemical processes can be used to create hydrogels. It includes single-step processes, like polymerization and simultaneous crosslinking of multi-dimensional monomers, as well as multi-step processes such as the combination of reactive group-containing polymer molecules and their subsequent crosslinking; possibly also by reacting polymers with appropriate crosslinking agents. Polymer engineers may develop and construct polymer networks with molecular-scale control over the structure, such as crosslinking density, and with specialized properties, such as biodegradation, mechanical strength, and chemical and biological reactivity to stimuli (Burkert et al., 2007).

Organic/inorganic nanocomposites are useful constituents made of immiscible organic and inorganic parts that can be used to construct intricate nanometer-scale structures. For several applications, polymer/clay nanocomposites have been successfully created and have undergone significant study (Okada and Usuki, 2006). By applying the idea of nanocomposite to the realm of soft hydrogel materials, Haraguchi and Takehisa (2002), for the first time, announced the development of "nanocomposite hydrogels," which had distinctive organic-inorganic network structures. The exceptional mechanical, optical, and swelling/deswelling capabilities of the nanocomposite hydrogels allowed them to outperform the restrictions of traditional chemically crosslinked hydrogels. The creation of nanocomposite hydrogels was made possible by permitting clay platelets to function as multifaceted crosslinkers in the production of polymer networks, rather than merely incorporating clay nanoparticles into a chemically crosslinked network. Nanocomposite hydrogels have garnered a lot of attention due to their outstanding qualities and are thought to be a new kind of hydrogel (Okada and Usuki, 2006).

4.2. Synthesis of Nanocomposite Hydrogels

Due to the simplicity and adaptability of the synthetic process, nanocomposite hydrogels can be created using a diversity of molds in an extensive range of shapes, including large blocks, pieces, tiny films, rods, hollow tubes, spheres, bellows, and irregular sheets (Haraguchi, 2007). This is caused by the reaction solution's simple injection and the ensuing gel's extremely high toughness. Additionally, porous nanocomposite hydrogels with a low density, and microgels with diameters of several hundred nm could be created (Zhang et al., 2007). Furthermore, direct replica moldings have been used to prepare micrometer-scale, surface-patterned nanocomposite hydrogel films. Here in the following sections, the synthesis methods for two typical nanocomposite hydrogels prepared based on natural products, i.e., *Linum usitatissimum* and guar gum are described.

4.2.1. *Linum Usitatissimum*-Loaded Nanocomposite Hydrogels

Linum usitatissimum, commonly called Flax, is a dicotyledonous plant that is commonly grown for its fiber, oil, and therapeutic properties in temperate climatic zones (Shim et al., 2014). Figure 4.1a shows a typical photograph of the *Linum usitatissimum* plant. An oilseed known as flaxseed is utilized extensively in both commercial and herbal products. It has a variety of fascinating biological properties. It is a vital source of alkaloids, cyanogenic glycosides, polysaccharides, cyclic peptides, and linolenic acid. Flaxseed extracts have a variety of biological functions that are valuable for tissue engineering applications, such as antioxidant and anti-inflammatory actions (Slavova et al., 2016).

Figure 4.1. Typical photographs of the (a) *Linum usitatissimum* and (b) *Cyamopsis tetragonoloba* plants.

Table 4.1. Taxonomy of *Linum usitatissimum* plant

Kingdom:	Plantae
Division:	Magnoliophyta
Class:	Magnoliopsida
Order:	Malpighiales
Family:	Linaceae
Genus:	*Linum*
Species:	*Linum usitatissimum*
Botanical name:	*Linum usitatissimum*

The aerial part (flower, leave, stem, petals, shoots, etc.) of the plant *Linum usitatissimum* has been used to create, describe, and use a nanocomposite hydrogel based on alginate (Alg) and nano-hydroxyapatite (nHA) as a bone tissue engineering framework (Mohammadpour et al., 2022). nHA was created via a wet chemical method/precipitation process and added to an alginate hydrogel as filler by physically crosslinking it. A blast crusher measuring 1mm was used to smash the plant material. To reduce the unit size and increase extraction yields, the fine powder was once more processed in an electric mill just before extraction. The powder created in this way was extracted using an electric stirrer for one hour at room temperature. Plant powder measuring 500 g was dissolved in 10 L of a 50/50 (v/v) H_2O/methanol solution. To get rid of the methanol, the resulting solution was evaporated at 40 °C under a vacuum, after being filtered through a Büchner funnel. The XAD 16-N resin was treated with crude extract (4 L). After removing non-recoverable compounds with 2 L of H_2O, the resin was washed with 2 L of an 80/20% ethanol and water solution to recover the desired compounds. After vacuum-assisted evaporation of this later fraction, which contained polyphenolic compounds, the desiccated matter of the purified extract from the aerial section of *Linum usitatissimum* was obtained. Scanning electron microscopy, dynamic light scattering, and zeta potential characterizations showed that the synthesized material has a nanometric-scale plate-like shape. The created nanocomposite features an interconnected porous architecture. The porosity ranged from 80-90%, and the typical hole size was in the 100-200 nm range. According to the *Linum usitatissimum* release test, approximately 90% of the loaded medication was released within 12 hours, monitored by a steady release over the course of 48 hours. The nanocomposite has considerable antioxidant activity that supports bone repair, according to *in vitro* tests (Mohammadpour et al., 2022).

4.2.2. Guar Gum-Based Nanocomposite Hydrogels

Cluster beans, also known as guar beans or *Cyamopsis tetragonoloba*, which is an annual legume crop grown in tropical regions that is the source of abundant solvable dietary fiber known as guar gum. The appearance of the *Cyamopsis tetragonoloba* plant is shown in Figure 4.1b. Guar is an annual, deep-rooted legume that can withstand droughts and is planted throughout the summer. The Sanskrit terms gau and ahaar are thought to have given rise to the name guar. The storage polysaccharide known as guar gum assists to prevent the desiccation of the seed in the endosperm of cluster beans (Overbeeke et al., 1990). Guar gum has a variety of uses in the food business because of its hydrophilic gel-forming abilities. The United States, Australia, India, Pakistan, and Africa are the main countries where the cluster bean is grown.

Table 4.2. Taxonomy of *Cyamopsis tetragonoloba* plant

Kingdom:	Plantae
Clade:	Tracheophytes
Clade:	Angiosperms
Clade:	Eudicots
Clade:	Rosids
Order:	Fabales
Family:	Fabaceae
Subfamily:	Faboideae
Genus:	Cyamopsis
Species:	C. tetragonoloba
Binomial name:	Cyamopsis tetragonoloba (L.) Taub

From guar gum-g-polyacrylamidoglycolic acid, a dual perspective nanocomposite hydrogel network has been created using sodium borohydride, which quickly and easily causes crosslinking and the formation of silver nuclei by a reduction reaction (Palem et al., 2020). Utilizing graft polymer as a template, the composite hydrogels were created using the previously reported methodology (Prasad et al., 2012). The creation of the composite hydrogels took place in just one step. The graft polymer was first dissolved in 100 mL of double distilled water using a 100 rpm magnetic stirrer until it was homogeneous, yielding a 2% graft polymer solution. The next step was to add 30 mL of graft polymer solution (2%) to a 100 mL beaker that already contained 10 mL of newly made 0.001 M silver nitrate ($AgNO_3$) solution. After 30 minutes of stirring, 3 mL of 0.01 M sodium borohydride ($NaBH_4$) was added to the reaction mixture, which led to the production of the crosslinked graft polymer (GGAA) and its silver nanocomposite (GGAA@SNC) hydrogels. The nanocomposite hydrogels were then cleaned by submerging them in double distilled water for 24 hours. The sustained release of the anticancer medicine 5-fluorouracil (5-FU) was explored using the crosslinked graft polymer GGAA and GGAA@SNC hydrogels, and their antibacterial activities were assessed. The crosslinked graft polymer hydrogels showed tremendous antibacterial activity towards *Bacillus subtilis* and *Salmonella ebony* using the zone of inhibition test (Palem et al., 2020).

4.3. Properties of Nanocomposite Hydrogels

Nanocomposite hydrogels are soft and elastic materials with surprising mechanical properties. The nanocomposite hydrogels can endure significant deformation, including stretching, bending, knotting, and compression (Haraguchi, 2007). The extensibility of a nanocomposite hydrogel is dependent on the type of nanomaterials and polymer employed and can generally be extended to more than 1000% of its initial length. In the case of clay-based nanocomposite hydrogels, the elongation at break is nearly consistent irrespective of clay and polymer contents, thus, by merely switching the crosslinker from a chemical one to clay, fragile hydrogels are radically transformed into flexible nanocomposite hydrogels. When the network is not properly built, elongation at break increases significantly (by more than 2000%). The ability of nanocomposite hydrogels to recover from substantial strains is also very time-dependent and varies depending on the clay/polymer system and composition (Haraguchi et al., 2003; Haraguchi and Li, 2006). Contrary to conventional hydrogels, which easily fracture under slight pressures, nanocomposite hydrogels often endure up to 90% compression. Strength and compression moduli rise almost linearly with clay content. This is a result of clay platelets forming a micro-complex structure with higher concentrations of chemical crosslinks (Haraguchi, 2007).

Swelling is one of the most important properties of nanocomposite hydrogels which occurs by the spontaneous absorption of water from aqueous solutions due to the hydrophilic and crosslinked nature of nanocomposite hydrogels. Some nanocomposite hydrogels can change their swelling ability in response to environmental conditions (temperature, pH, light, pressure, ionic strength, etc.), which are known as smart or intelligent nanocomposite hydrogels. The type of nanoparticles and polymer and nanoparticles/polymer ratio affect the swelling ability of nanocomposite hydrogels. In some cases, compared to conventional hydrogels, nanocomposite hydrogels show a higher level of equilibrium swelling in water (Xu et al., 2007). Unusual behaviors have been seen in the swelling behavior of nanocomposite hydrogels, including the emergence of a highest in the swelling curve and a striking increase in the degree of equilibrium swelling as a result of post-treatment in particular nanocomposite hydrogels with the conservation of strong motorized characteristics. When nanoparticles (e.g., clay particles) and entangled polymer chains are rearranged during swelling or post-treatment, both occurrences are explained (Zhang et al., 2006b). Among the smart nanocomposite hydrogels, temperature-sensitive nanocomposite hydrogels have received a lot of attention (Kopecek and Yang, 2007). Poly (N-isopropyl acrylamide) is a well-known temperature-sensitive hydrogel that has a number of significant drawbacks, including limited change in volume during swelling, a sluggish deswelling rate, and poor mechanical characteristics. These constraints have been shown to be simultaneously overcome in poly (N-isopropyl acrylamide) nanocomposite hydrogels with a large change in volume, high sluggish deswelling rate, and outstanding mechanical characteristics. It has been shown that by changing the composition of the poly (N-isopropyl acrylamide) nanocomposite hydrogel, there is a possibility to control the thermo-sensitivity of the network in a wide range (Miyazaki et al., 2006).

4.4. Types of Nanocomposite Hydrogels

There are different ways to classify nanocomposite hydrogels. One of the popular methods is the classification of nanocomposite hydrogels based on the type of constituent materials. Therefore, nanocomposite hydrogels can be classified as carbon-based, polymer-based, inorganic-based, and metallic-based nanocomposite hydrogels, as briefly introduced in Table 4.3.

Table 4.3. Types of nanocomposite hydrogels

Nanocomposite Hydrogels

Carbon-based nanocomposite hydrogels	Polymer-based nanocomposite hydrogels	Inorganic-based nanocomposite hydrogels	Metallic-based nanocomposite hydrogels
• Carbon nanotubes (CNTs) • Graphene/Graphene oxide (GO)	• Dendrimers/Hyperbranched polymers • Liposomes • Nanogels (Nanosized hydrogels)	• Hydroxyapatite • Nanoclays • Glass-ceramics • Calcium phosphate • Wollastonite	• Magnetic nanoparticles • Anti-microbial effect of metallic nanoparticles

4.4.1. Carbon-Based Nanocomposite Hydrogels

There are possible uses for carbon-based nanomaterials in biomedicine like as carbon nanotubes (CNTs), graphene, fullerene (C60), and nanodiamonds. Particularly because of their strong electrical conductivity, mechanical toughness, and optical characteristics, CNTs and graphene are frequently used. Carbon-based nanomaterials are SP2 hybridized, hollow, cylindrical tubes made of carbon (graphite) with a high aspect ratio (1000). A very small fraction, or yet the lack of chemically active hanging bonds, is just where perfect (defect-free)

CNTs vary from graphite. The variability of particular distances caused by the van der Waals forces is a characteristic of carbon-based nanomaterials.

Carbon-based nanomaterials with these qualities make excellent candidates for creating electrically conductive tissues like nerve, muscle, and heart tissues (Gaharwar et al., 2014). The strength of the C bond is what causes CNTs' stiffness. Additionally, CNTs can serve as a nanoscale heat sink that keeps the composite material from melting from deterioration and overheating. The attributes of optical CNTs are significantly impacted by symmetry, the number, and whether the caps (CNT ends) are present or not of its component carbon atoms (Kumar et al., 2017). Carbon-based materials' limited ability to interact with hydrophilic polymers and hydrogels is a result of their hydrophobic character. To improve their dispersion, CNT surfaces are often changed with a range of polar groups like NH_2, OH, and COOH or grafted with various polymeric chains. Within the nanocomposite hydrogel network, CNTs can act as reinforcing elements and, in several situations, make the hydrogel reactive to outside electrical/thermal stimuli (Gaharwar et al., 2014). Figure 4.2 shows how graphene could roll up to form CNT.

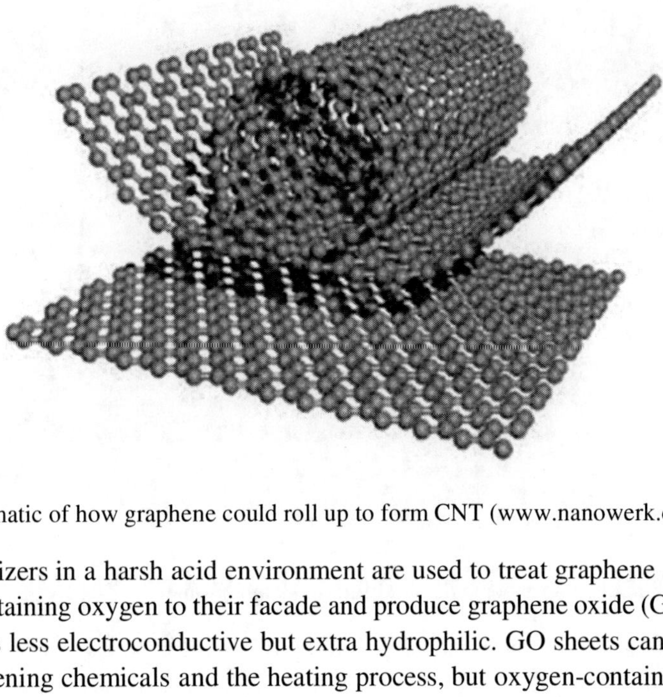

Figure 4.2. Schematic of how graphene could roll up to form CNT (www.nanowerk.com).

Strong oxidizers in a harsh acid environment are used to treat graphene sheets in order to bind groups containing oxygen to their facade and produce graphene oxide (GO). Compared to graphene, GO is less electroconductive but extra hydrophilic. GO sheets can be reduced once more using lessening chemicals and the heating process, but oxygen-containing residuals still are present. Paul et al., (2014) delivered pro-angiogenic human vascular endothelial growth factor plasmid DNA (pDNAVEGF) to injured cardiac tissues using gelatin methacrylate (GelMA) impregnated with functionalized GO nanosheets.

4.4.2. Polymer-Based Nanocomposite Hydrogels

Examples of nanoparticles with the capacity to entrap hydrophobic or hydrophilic medicines, proteins, genes, and other bioactive substances include different types of polymers and polymeric particles. Due to their capacity for regulated drug release, these hybrid hydrogels have attracted interest in the pharmaceutical and biological fields (Joshi and Grinstaff, 2008). This skill is a result of these polymeric nanoparticles' ability to respond to stimuli. When something changes suddenly in response to a slight alteration in an environmental factor like an increase or decrease in temperature, it is said to be stimuli-responsive. For pH-responsive systems, ionizable polymers with Pka between 3 to 10 are suitable. The ionization state of weak acids/bases, changes in response to pH changes. Whilst this ionizable group is connected to the polymer structure, the swelling behavior of nanocomposite hydrogels undergoes a conformational change (Schmaljohann, 2006). Polymer-based nanocomposite hydrogels can be classified as dendrimers, liposomes, and nanogels.

Due to the many functional groups on their perimeter and their greatly branched porous structure, dendrimers have a high degree of reactivity and loading efficiency, as it is shown in Figure 4.3. The dendrimer concentration has an impact on the hydrogels' stiffness, degradation characteristics, and hydration kinetics. The resulting dendrimer-containing nanocomposite hydrogels exhibit significant stress absorption that is essential for tissue engineering applications (Solouk et al., 2011).

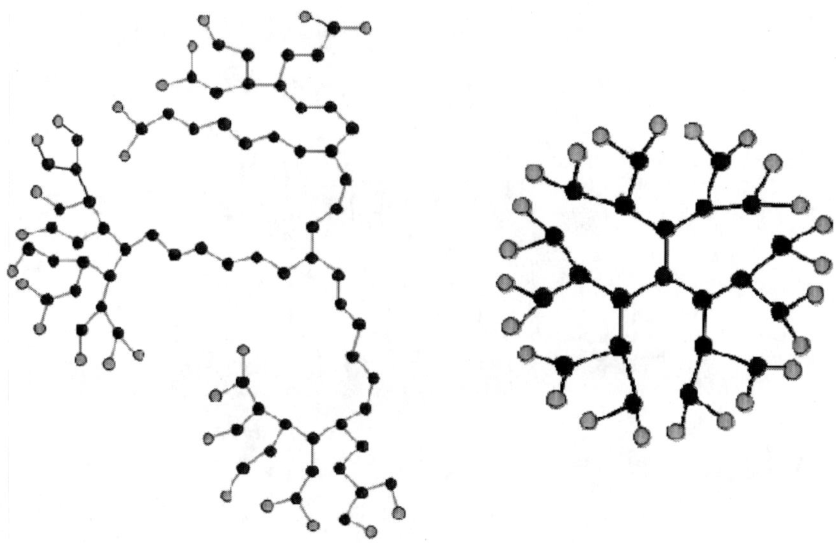

Figure 4.3. Hyperbranched polymers and dendrimers (Dalakoglou et al., 2012).

A tiny bubble known as a liposome is formed of the same substance as a cell membrane made of phospholipids. When the membrane is disturbed, it can self-assemble, either as bilayers or as monolayers, to form tiny spheres that are smaller than a typical cell (Deb et al., 2019). Liposomes are the bilayer cellular structures. The core and surface of liposomes are hydrophilic, but the outer layer is hydrophobic. As a result, they can include both hydrophobic/hydrophilic molecules, including medicines, bioactive substances, DNA, RNA, and genes. To carry the molecules to the position of action, drug-loaded liposomes could be combined with the substances such as degradable polymers functioning as a carrier in the system. These mixtures can then be crosslinked to create nanocomposite hydrogels (Gupta and Sharma, 2011).

Hydrogels with a nanoscale core, or nanogels, have the ability to encapsulate bioactive substances. Due to their nano-dimensions, they react to microenvironmental factors like temperature or pH. Depending on how they are crosslinked, nanogels are prepared using a variety of techniques. They are capable of being chemically crosslinked through different types of bonds and hydrophobic interactions (physically). For instance, mono-dispersive polymer micelles can self-assemble in H_2O using amphiphilic block copolymers. By covalently connecting the different polymer chains in these polymer micelles either at the core or shell, nanogels are created (Sasaki and Akiyoshi, 2010).

4.4.3. Inorganic-Based Nanocomposite Hydrogels

Most inorganic nanoparticles, include silica, calcium phosphate, bioactive glasses, silica, nHA, and synthetic silicate nanoparticles. The necessary minerals are wollastonite, glass, and ceramic, for the body's typical tissue processes. These are various implantable devices that frequently use various types of ceramics materials, especially for uses involving bones. A range of types of chemical bonds can bond ceramics together such as the direct covalent-ionic bond, electrostatic chemical bond, hydrogen bonding, van der Waals bonding of organic constituents, and complex combination of different types of bonding (Hench et al., 1971). Inorganic-based nanocomposite hydrogels can be prepared using some inorganic nanomaterials including hydroxyapatite, nanoclays, glass-ceramics, calcium phosphate, and wollastonite.

Mineral hydroxyapatite is present in bones and teeth. It is mostly employed as an implant covering to encourage bone ingrowth or as filler to replace damaged bone. They do, however, have several disadvantages, including a low *in vivo* bioresorption rate, a weak stimulating effect on the formation of new tissues, low resistance to crack, and low fatigue stability in the physiological environment. Adding silicon or uorine to the hydroxyapatite lattice is one way to get around these problems. Uorine can be used to enhance the biodegradation characteristics, protein adsorption, and adherence of the coating to the metal substrate (Zhang et al., 2006a). Tissue-adhesive nanocomposite hydrogels with improved properties are produced by combining linear and branched polymers with nanoclays. High surface interactions are caused by the anisotropic, plate-like, and elevated aspect-ratio morphology of nanoclay. Owing to the non-covalent connections between the polymer chains and the nanoparticles, the polymer chains can absorb and desorbs on the silicate surfaces (Schexnailder and Schmidt, 2009).

Glass is an inert, unstructured substance having an atomic structure that is randomly ordered. The random structure of glass can be rearranged into a prearranged structure of ceramic that is more even than other materials. It is referred to as a glass-ceramic (Deubener et al., 2018). Calcium phosphate comes in a variety of forms that can perform various functions. In the skeleton and teeth, hydroxyapatite, a form of calcium phosphate, can be found. In diet goods, calcium phosphate is utilized in other forms. However, it is not recommended to take hydroxyapatite as a food supplement. Therefore, calcium phosphate nanoparticles encapsulated in hydrogels may be an excellent option for promoting bone development (Straub, 2007). Various composite materials have been filled with wollastonite ($CaSiO_3$), a calcium silicate, to enhance their mechanical qualities. Given its high bioactivity and biocompatibility, it shows potential as a material for dental roots and artificial bones. These inorganic bioactive nanoparticles can be combined with organic or synthetic polymeric hydrogels to create reinforced nanocomposite hydrogels (Aza et al., 1998).

4.4.4. Metallic-Based Nanocomposite Hydrogels

For biomedical purposes, nanocomposite hydrogels can be made using a variety of metallic nanoparticles, such as Au, Ag, and metal oxide, like iron oxide nanoparticles (Fe_3O_4 and Fe_2O_3), titania (TiO_2), alumina, and zirconia. These nanoparticles exhibit appropriate substantial characteristics, like magnetic properties (iron oxides), electrical and thermal conductivity (Au nanorods), and antibacterial characteristics (Ag nanoparticles). Therefore, nanocomposite hydrogels containing metal or metal oxide are used as imaging agents or in drug delivery systems, switchable conductive scaffolds sensors, actuators, and electronics (Schexnailder and Schmidt, 2009).

Due to their toxicity toward bacteria and physiological stability, metals can be employed as antimicrobial agents. There are various ways to create polymer/metal nanocomposite hydrogels, including direct metal nanofiller addition to a hydrogel matrix or *in situ* creation of the nanoparticle within a hydrogel. These nanocomposites' antibacterial action is based on the stages describing how bacteria vanish in a biological environment. Antibacterial silver nanoparticles are harmless and have a low surface binding affinity. These nanoparticles can form and develop in the open gaps between the crosslinked hydrogel networks when they are inflated. The antibacterial effectiveness of the silver-loaded nanocomposite hydrogels can be influenced by the size of the silver nanoparticles. Because they could easily break free of the hydrogel network and interrelate with bacteria, it has been shown that the smaller nanoparticles improve the antibacterial activity of nanocomposite hydrogels. Even though silver nanoparticles can break free from the hydrogel network, they do not exhibit strong antibacterial properties, especially if they are huge in size (Murali et al., 2007).

When exposed to exterior magnetic fields, magnetic nanoparticles contained within a nanocomposite hydrogel network produce high temperatures. As a result, the temperature of the adjacent matrix rises over the polymer's inferior critical solution temperature, which causes the polymer chains to change from coil to globule. As a result, the active agent or drug compounds embedded within the nanocomposite hydrogel network could be released. Due to their ability to remotely interact with external magnetic fields, magnetic-based nanocomposite hydrogels can be employed for different applications (Gaharwar et al., 2009).

4.5. Applications of Nanocomposite Hydrogels in Tissue Engineering

In tissue engineering, scaffolds aid in cell metabolite and nutrition transfer along with cell adhesion, proliferation, and differentiation. Although hydrogels can serve as scaffolds for tissue regeneration, they are often unusable due to their weak mechanical properties. Clay and CNTs are two examples of different nanoparticles that can be added to increase the mechanical and functional properties of hydrogels to meet tissue engineering requirements. For instance, a nanocomposite hydrogel made of sodium alginate and CNTs demonstrated high biocompatibility, good inflammatory response, and non-cytotoxic. Because of its improved qualities, nanocomposite hydrogel makes for good scaffolding materials (Kawaguchi et al., 2006). In tissue engineering techniques, scaffolds play essential roles in promoting cell adhesion, migration, development, and even differentiation. For successful bone tissue engineering, resemblance to the natural structure of bone is a requirement. Collagen and hydroxyapatite form the biocomposite of bone, with the former providing the 3D structure, flexibility, and restorability. In light of this, 3D nanocomposites containing hydroxyapatite have attracted considerable interest for tissue engineering of the bone (Mondal et al., 2018).

4.5.1. Bone Tissue Engineering

A type of rigid connective tissue made up of cells, fibers, and the matrix is bone tissue. Numerous biomaterial scaffolds for bone regeneration have been developed recently as a result of the serious clinical issues associated with localized bone defects caused by trauma, infection, or inherited disorders. The elastic modulus of cancellous bone is between 0.1 and 0.5 GPa, and the compressive strength is between 4.0 and 12.0 MPa. The elastic modulus of compact bone is between 12.0 and 18.0 GPa, and the compressive potency is between 130 and 180 MPa. Therefore, in order to support new tissues, scaffolds must have mechanical properties that are similar to those of actual bone (Zhang et al., 2007). The following requirements should be met in addition to an acceptable scaffold for bone tissue engineering including biocompatibility, suitable pore size, osteoconductivity, and biodegradability. The holding capabilities of bone tissue are significantly influenced by hydroxyapatite nanoparticles, mineral-based nanomaterials that serve as nucleation sites for the deposition of the hard phase of bone. Additionally, they regulate the growth and regeneration of new tissue by triggering extracellular matrix proteins like fibronectin, osteocalcin, etc. (Wilson et al., 2005). Despite hydroxyapatite's widespread use in bone tissue engineering, its non-degradable nature has always been a challenge. Nanocomposite hydrogels for bone tissue engineering can also be created by tricalcium phosphate, a bioactive mineral with easily degradable characteristics.

4.5.2. Cartilage Tissue Engineering

Chondrocytes and extracellular matrix make up the majority of the avascular and lymphatic tissue known as articular cartilage. It offers a lubricating surface that can withstand heavy loading for an extended period of time. This makes it possible to apply the least amount of pressure to the subchondral bones beneath, absorb vibrations, reduce friction articulation, and encourage joint movement throughout muscular motion (Zuo et al., 2015). These days, cartilage abnormalities and cartilage degeneration are major issues in orthopedics clinics due to the rise in sports injuries and the population's hastened aging. However, because of its low-down chondrocyte activity and deficiency of vascularity, the self-repairing capacity of cartilage tissue is constrained. Additionally, the cartilage regeneration process creates fibrocartilage tissue, which differs from native tissue in terms of mechanical and biological qualities (Farokhi et al., 2020).

Hybrid hydrogels' mechanical properties can be significantly enhanced by combining stiff nanoparticles and supple polymer strands. Consequently, substantial research has been done on nanocomposite hydrogels for cartilage tissue engineering. In the realm of cartilage restoration, natural polymer scaffolds including alginic acid, hyaluronic acid, collagen, and gelatin are being quickly developed as tissue substitutes, however, they are constrained by weak mechanical qualities. Nanocomposite hydrogels proved to have promising biomaterial for cartilage tissue engineering. Piluso and colleagues (2019) created 3D nanocomposite hydrogels with starch nanocrystals inserted in the gelatin matrix to boost the involuntary properties of these scaffolds.

4.5.3. Skeletal Tissue Engineering

Muscle tissue, which includes skeletal muscles, is one of the body's most prevalent and intricately arranged tissues. In order to prevent removing a muscle flap at the donor site, skeletal muscle tissue engineering aims to replicate the natural structure and function of muscle *in vitro* and transplant this tissue in the injured location. Given that scaffold flexibility and strength would control myogenic cell activity in skeletal muscle tissue engineering, such scaffolding systems would need to be mechanically robust. It has been shown that conductive nanocomposite hydrogels make excellent 3D biomaterials for skeletal muscle tissue engineering. Carbon nanoparticles are one of the conductive sources for biomaterials to achieve conductivity in the domain of skeletal muscle regeneration. Two of the most popular carbon nanomaterials for skeletal muscle regeneration are graphene and CNTs (Dong et al., 2020). Patel et al., (2018) created graphene-polysaccharide nanocomposite hydrogel scaffolds by combining graphene with chitosan and gellan gum. Scaffolds' solubility, electrical conductivity, tensile strength, and toughness were all improved by adding graphene to them.

4.5.4. Myocardial Therapy

Due to advancements in technology, there is a magnificent increase in the survival rate of the patient after a heart attack. Recent research suggests that heart function may be improved by stem cell transplantation through a variety of processes, including cardiac differentiation, neoangiogenesis, and paracrine effects (Chong et al., 2014). Despite encouraging outcomes, the use of stem cell therapy is still restricted by logistical, financial, and scheduling constraints, especially when separating cells from older patients in clinical settings. Heart on a chip, also known as *in vitro* miniaturized strong and sick cardiac tissue model, has established substantial attention in recent years. In order to accomplish this, tissue engineering is incoming a new phase in which it combines the knowledge of micro-scale technology with the biology of stem cells, especially induced pluripotent stem cells. These bioengineered cardiac structures are also useful in functional tissue replacement therapy as cardiac patches (Bhatia and Ingber, 2014).

A traditional method involves combining cardiac cells with hydrogels and growing them in perfusable bioreactors that stimulate mechanically and electrically. However, the prevalent biomaterials' weak conductivity causes cardiac constructions to perform less than optimally. This issue might be resolved by including electroconductive nanomaterials. It has been demonstrated that CNT-reinforced hybrid GelMA hydrogel has improved mechanical and electrical properties (Shin et al., 2013). The resultant nanocomposite hydrogel dramatically enhanced the electrophysiology, maturation, and alignment of cardiomyocytes. It is possible to construct a correct electroconductive network and rapid beating behavior within the cardiac cells by creating such biomimetic nanocomposite hydrogels. Dvir and coworkers showed incorporation of gold nanowires and nanoparticles into hydrogels derived from biomaterials can greatly improve the electrical message between adjacent heart cells in a 3D microenvironment (Shevach et al., 2014). As a result, the nearby cardiac cells responded to electrical pacing synchronously and produced appropriate excitation coupled to contraction with each beat. Thus, it can be said that engineering nanocomposite hydrogels, which accurately mimic the heart cell niche and bio-inspired microarchitecture, are necessary for the fabrication of functional cardiac tissues.

4.5.5. Nerve Tissue Engineering

The nervous system played a significant role in controlling various functions of the body. It is divided into two parts i.e., central nervous system abbreviated as CNS and the peripheral nervous system abbreviated as PNS. Nerve tissue engineering aims to create nerve graft substitutes for the long-term treatment of nerve injuries and functional improvement. These nerve grafts should, in theory, offer stable mechanical and electrical surroundings that can act as a link for regenerating axons (Schmidt and Leach, 2003). Recent research has demonstrated that carbon-based nanomaterials may help with the creation of nanocomposite hydrogels for the engineering of nerve tissue. Hybrid hydrogels based on graphene might be a good choice for promoting nerve regeneration (Huang et al., 2017). Nanocomposite hydrogels made of graphene and polyurethane were created by Huang et al., (2017). These polyurethane hydrogels using graphene as a substrate might promote neural stem cell neuronal development and proliferation. Studies specifically demonstrated that polypyrrole (PPy) and CNT may increase the electrical conductivity of hydrogels without having a negative impact on cell viability. By electrically stimulating them, these polyethylene glycol nanocomposite hydrogels could improve mean neuritis length and total neuritis outgrowth, whereas PPy-hyaluronic acid nanocomposite hydrogels may greatly aid in the neuronal development of human embryonic neural stem cells. In addition to carbon-based nanomaterials, hydrogels based on regular and artificial polymers are intriguing for the manufacturing of neural tissue (Imaninezhad et al., 2018). Some important research works performed in past years regarding the application of nanocomposite hydrogels in biomedicine and tissue engineering are summarized in Table 4.4.

Table 4.4. Some important works regarding the application of nanocomposite hydrogels in biomedicine and tissue engineering

Type of Nanoparticles	Main Features	Application
Inorganic Nanoparticles	• Emphasized the potential of – Laponite – High aspect ratio – Interacts with chemical entities – Simply functionalized – Readily degraded in the body – Chitosan/montmorillonite-based films – Improved tensile strength – Amended Young's modulus – Enhanced thermal stability • Less rate of moisture vapor transmission • Physical crosslinkers • Enhanced cell compatibility • Developed fragility Greater volume and rapid thermo-responsibility	• Drug delivery • Bio-imaging • Wound dressing • Hard tissue engineering • Hemostatic agents for emergency conditions • Bone tissue engineering

Table 4.4. (Continued)

Type of Nanoparticles	Main Features	Application
Polymeric Nanoparticles	• Responsive to pH • Nanogels that can bear pollutants are incorporated • Liposome is crosslinked • Maleimide-functionalized crosslinkers • Novel chemical degradation • Collagen-grafted hydrogel based on polyhedral oligomeric silsesquioxane (POSS) nanoparticles • Better cell adhesion • Improved surface properties	• Drug delivery
Carbonic Nanoparticles	– Based on Alginate – Functionalized with GO – Appropriate proliferation of intermolecular hydrogen bonds – Emphasized the mechanisms of antibacterial activity in graphene-based materials • Direct interaction with bacterial membranes – Covering – Photothermal ablation – Based on PGS – CNTS as both physical and covalent Crosslinker • Crosslinking in a non-reversible manner • Greater electrical properties • Advanced maturation of cultured muscle cells	• Articular tissues • Antibacterial systems • Packaging • Wound dressing • Musculoskeletal muscles production • Myocardial gene therapy
Metallic Nanoparticles	• Instinctively resilient – Elastomeric • Chitosan-g-poly(acrylamide)/Zn – Ofloxacin made of this nanoparticle – Showed pH-dependent drug release – Showed good antibacterial activity – Based on Alginate material – Gold nanowires connect the pore walls – Enhanced electrical connection between cardiac cells • Denser and better-aligned tissues • Based on Hemicellulose • Responsive to Magnetic field • They combine the renewability of hemicellulose and the magnetism of Fe_3O_4	• Drug delivery • Antibacterial drug carriers • Cardiac patches

Conclusion

Nanocomposite hydrogels open up a new era of advanced biomaterials for their use in different applications, e.g., tissue engineering. Nanocomposite hydrogels synthesized from plants have been employed for the treatment of various therapies. While there are still many obstacles in the development of biomimetic nanoscaffold materials. Future research will focus heavily on biologically or chemically modified composite nanomaterials and injectable or biomimetic materials in the process of developing biomaterials. To learn more about the clinical efficacy of scaffold-based tissue-engineered cartilage healing, randomized clinical trials and extended follow-up periods are required.

Hydrogel and nanocomposite hydrogel scaffolds have been used frequently in animal experiments, but not in human clinical studies. With the advancement of nanotechnology and thorough study of the microenvironment of cartilage, nanocomposite hydrogels have found extensive use in tissue engineering due to their biocompatibility, ease of preparation, high drug loading efficiency, multiple stimulus responses, controllable drug release, and improved mechanical properties. As a result, nanoparticles and hydrogels are extremely versatile and will be studied more in the future. The performance of nanocomposite hydrogels can be greatly enhanced by constructing ultrastructures or dispersing nanoparticles in hydrogels to create composite materials using template molding, self-assembly, microfluidics, and 3D printing technologies.

References

Aza, P. N., Guitian, F., Aza, D. S., and Valle, F. J. (1998). Analytical control of wollastonite for biomedical applications by use of atomic absorption spectrometry and inductively coupled plasma atomic emission spectrometry. *Analyst*, 123, 681-685.

Bhatia, S. N., and Ingber, D. E. (2014). Microfluidic organs-on-chips. *Nature Biotechnology,* 32(8), 760-772.

Burkert, S., Schmidt, T., Gohs, U., Dorschner, H., and Arndt, K. F. (2007). Cross-linking of poly(n-vinyl pyrrolidone) films by electron beam irradiation. *Radiation Physics and Chemistry*, 76(8-9), 1324-1328.

Chong, J. J., Yang, X., Don, C. W., Minami, E., Liu, Y. W., Weyers, J. J., Mahoney, W. M., Biber, B. V., Cook, S. M., Palpant, N. J., Gantz, J. A., Fugate, J. A., Muskheli, V., Gough, G. M., Vogel, K. W., Astley, C. A., Hotchkiss, C. E., Baldessari, A., Pabon, L., Reinecke, H., Gill, E. A., Nelson, V., Kiem, H. P., Laflamme, M. A., and Murry, C. E. (2014). Human embryonic-stem-cell-derived cardiomyocytes regenerate non-human primate hearts. *Nature*, 510(7504), 273-277.

Dalakoglou, G., Karatasos, K., Lyulin, S., Larin, S., Darinskii, A., and Lyulin, A. (2012). Conformational effects in non-stoichiometric complexes of two hyperbranched molecules with a linear polyelectrolyte. *Polymers*, 4(1), 240-255.

Deb, P. K., Al-Attraqchi, O., Chandrasekaran, B., Paradkar, A., and Tekade, R. K. (2019). Protein/peptide drug delivery systems: practical considerations in pharmaceutical product development. *Basic Fundamentals of Drug Delivery*, 651-684.

Deubener, J., Allix, M., Davis, M. J., Duran, A., Höche, T., Honma, T., Komatsu, T., Krüger, S., Mitra, I., Müller, R., Nakane, S., Pascual, M. J., Schmelzer, J. W. P., Zanotto, E. D., and Zhou, S. (2018). Updated definition of glass-ceramics, *Journal of Non-Crystalline Solids,* 501, 3-10.

Dong, R., Ma, P. X., Guo, B. (2020). Conductive biomaterials for muscle tissue engineering. *Biomaterials*, 229, 119584.

Dvir, T., Timko, B. P., Brigham, M. D., Naik, S. R., Karajanagi, S. S., Levy, O., Jin, H., Parker, K. K., Langer, R., and Kohane, D. S. (2011). Nanowired three-dimensional cardiac patches. *Nature Nanotechnology*, 6(11), 720-725.

Farokhi, M., Jonidi Shariatzadeh, F., Solouk, A., and Mirzadeh, H. (2020). Alginate based scaffolds for cartilage tissue engineering: a review. *International Journal of Polymeric Materials and Polymeric Biomaterials*, 69(4), 230-247.

Gaharwar, A. K., Avery, R. K., Assmann, A., Paul, A., McKinley, G. H., Khademhosseini, A., and Olsen, B. D. (2014). Shear-thinning nanocomposite hydrogels for the treatment of hemorrhage. *ACS Nano*, 8(10), 9833-9842.

Gaharwar, A. K., Wong, J. E., Müller-Schulte, D., Bahadur, D., and Richtering, W. (2009). Magnetic nanoparticles encapsulated within a thermoresponsive polymer. *Journal of Nanoscience and Nanotechnology*, 9(9), 5355-5361

Gupta, M., and Sharma, V. (2011). Targeted drug delivery system: a Review. *Research Journal of Chemical Sciences*, 1(2), 135-138.

Haraguchi, K. (2007). Nanocomposite hydrogels. *Current Opinion in Solid State and Materials Science*, 11(3-4), 47-54.

Haraguchi, K., and Li, H. J. (2006). Mechanical properties and structure of polymer-clay nanocomposite gels with high clay content. *Macromolecules,* 39:1898-1905.

Haraguchi, K., and Takehisa, T. (2002). Nanocomposite hydrogels: a unique organic-inorganic network structure with extraordinary mechanical, optical, and swelling/de-swelling properties. *Advanced Materials,* 14, 1120-1124.

Haraguchi, K., Farnworth, R., Ohbayashi, A., and Takehisa, T. (2003). Compositional effects on mechanical properties of nanocomposite hydrogels composed of poly(N,N-dimethylacrylamide) and clay. *Macromolecules*, 36, 5732-5741.

Hench, L. L., Splinter, R. J., Allen, W. C., and Greenlee, T. K. (1971). Bonding mechanisms at the interface of ceramic prosthetic materials. *Journal of Biomedical Materials Research*, 5(6), 117-141.

Huang, C. T., Shrestha, L. K., Ariga, K., and Hsu, S. H. (2017). A graphene-polyurethane composite hydrogel as a potential bioink for 3D bioprinting and differentiation of neural stem cells. *Journal of Materials Chemistry B*, 5(44), 8854-8864.

Imaninezhad, K., Pemberton, F. L., Xu, K., Kalinowski, R., Bera., and Zustiak, S. P. (2018). Directed and enhanced neurite outgrowth following exogenous electrical stimulation on carbon nanotube-hydrogel composites. *Journal of Neural Engineering,* 15(5), 056034.

Joshi, N., and Grinstaff, M. (2008). Applications of dendrimers in tissue engineering. *Current topics in Medicinal Chemistry*, 8(14), 1225-1236.

Kawaguchi, M., Fukushima, T., Hayakawa, T., Nakashima, N., Inoue, Y., Takeda, S., Okamura, K., and Taniguchi, K. (2006). Preparation of carbon nanotube-alginate nanocomposite gel for tissue engineering. *Dental Materials Journal,* 25, 719-725.

Kopecek, J., and Yang, J. (2007). Hydrogels as smart biomaterials. *Polymer International,* 56, 1078-1098.

Kumar, S., Rani, R., Dilbaghi, N., Tankeshwar, K., and Kim, K. H. (2017). Carbon nanotubes: a novel material for multifaceted applications in human healthcare. *Chemical Society Reviews*, 46(1), 158-196.

Miyazaki, S., Karino, T., Endo, H., Haraguchi, K., and Shibayama, M. (2006). Clay concentration dependence of microstructure in deformed poly(N-isopropylacrylamide)-clay nanocomposite gels. *Macromolecules*, 39, 8112-8120.

Mohammadpour, M., Samadian, H., Moradi, N., Izadi, Z., Eftekhari, M., Hamidi, M., Shavandi, A., Quéro, A., Petit, E., Delattre, C., and Elboutachfaiti, R. (2022). Fabrication and characterization of nanocomposite hydrogel based on alginate/nano-hydroxyapatite loaded with *Linum usitatissimum* extract as a bone tissue engineering scaffold. *Marine Drugs*, 20(1), 20.

Mohan, Y. M., Lee, K., Premkumar, T., and Geckeler, K. E. (2007). Hydrogel networks as nanoreactors: a novel approach to silver nanoparticles for antibacterial applications. *Polymer*, 48(1), 158-164.

Mondal, S., Dorozhkin, S. V., and Pal, U. (2018). Recent progress on fabrication and drug delivery applications of nanostructured hydroxyapatite. *Wiley Interdisciplinary Reviews: Nanomedicines and Nanobiotechnology,* 10, e1504.

Okada, K., and Usuki, A. (2006). Twenty years of polymer-clay nanocomposites. *Macromolecular Materials and Engineering*, 291, 1449-1476.

Overbeeke, N., Termorshuizen, G., Giuseppin, M., Underwood, D, and Verrips C. (1990). Secretion of the alpha galactosidase from *Cyamopsis tetragonoloba* (guar) by Bacillus subtilis. *Applied and Environmental Microbiology*, 56(5), 1429-1434.

Palem, R., Shimoga, G., Rao, K., Lee, S., and Kang, T. (2020). Guar gum graft polymer-based silver nanocomposite hydrogels: synthesis, characterization and its biomedical applications. *Journal of Polymer Research,* 27, 68.

Patel, A., Xue, Y. E., Hartley, R., Sant, V., Eles, J. R., Cui, X. T., Stolz, D. B., and Sant, S. (2018). Hierarchically aligned fibrous hydrogel films through microfluidic self-assembly of graphene and polysaccharides. *Biotechnology and Bioengineering,* 115, 2654-2667.

Paul, A., Hasan, A., Kindi, H. A., Gaharwar, A. K., Rao, V. T., Nikkhah, M., and Khademhosseini, A. (2014). Injectable graphene oxide/hydrogel-based angiogenic gene delivery system for vasculogenesis and cardiac repair. *ACS Nano*, *8*(8), 8050-8062.

Prasad, S. S., Rao, K. M., Reddy, P. R. S., Reddy, N. S., Rao, K. S. V. K., and Subha, M. C. S. (2012) Synthesis and characterisation of guar gum-g-poly(acrylamidoglycolic acid) by redox initiator. *Indian Journal of Advances in Chemical Sciences,* 1, 28-32.

Sasaki, Y., and Akiyoshi, K. (2010). Nanogel engineering for new nanobiomaterials: from chaperoning engineering to biomedical applications. *The Chemical Record*, 10(6), 366-376.

Schexnailder, P., and Schmidt, G. (2009). Nanocomposite polymer hydrogels. *Colloid and Polymer Science*, 287(1), 1-11.

Schmaljohann, D. (2006). Thermo-and pH-responsive polymers in drug delivery. *Advanced Drug Delivery Reviews*, 58(15), 1655-1670.

Schmidt. C. E., and Leach, J. B. (2003). Neural tissue engineering: strategies for repair and regeneration. *Annual Review of Biomedical Engineering*, 5, 293-347.

Shevach, M., Fleischer, S., Shapira, A., and Dvir, T. (2014). Gold nanoparticle-decellularized matrix hybrids for cardiac tissue engineering. *Nano Letters,* 14(10), 5792-5796.

Shim, Y. Y., Gui, B., Wang, Y., and Reaney, M. J. T. (2014). Flaxseed (*Linum usitatissimum* L.) bioactive compounds and peptide nomenclature: a review. *Trends in Food Science and Technol*ogy, 38(1), 5-20.

Shin, S. R., Jung, S. M., Zalabany, M., Kim, K., Zorlutuna, P., Kim, S. B., Nikkhah, M., Khabiry, M., Azize, M., Kong, J., and Wan, K. T. (2013). Carbon-nanotube-embedded hydrogel sheets for engineering cardiac constructs and bioactuators. *ACS Nano* 7(3), 2369-2380.

Slavova, K., Karama, A., Kancheva, V., and Amarowicz, R. (2016). Antioxidant activity of flaxseed extracts in lipid systems. *Molecules*, 21(1), 17.

Solouk, A., Mirzadeh, H., Shokrgozar, M. A., Solati-Hashjin, M., Najarian, S., and Seifalian, A. M. (2011). The study of collagen immobilization on a novel nanocomposite to enhance cell adhesion and growth. *Iranian Biomedical Journal*, 15, 6-14.

Song, L., Zhu, M., Chen, Y., and Haraguchi, K. (2008). Surface-patterning of nanocomposite hydrogel film by direct replica molding and subsequent change in pattern size. *Polymer Journal*, 40(9), 800-805.

Straub, D. A. (2007). Calcium supplementation in clinical practice: a review of forms, doses, and indications. *Nutrition in Clinical Practice*, 22(3), 286-296.

Wilson, C. J., Clegg, R. E., Leavesley, D. I., and Pearcy, M. J. (2005). Mediation of biomaterial-cell interactions by adsorbed proteins: a review. *Tissue Engineering,* 11(1-2), 1-18.

Xu, K., Wang, J., Xiang, S., Chen, Q., Yue, Y., Su, X., Song, C., and Wang, P. (2007). Polyampholytes superabsorbent nanocomposites with excellent gel strength. *Composites Science and Technology*, 67, 3480-3486.

Zhang, O., Zha, L., Ma, J., and Liang, B. (2007). A novel route to the preparation of poly (N-isopropylacrylamide) microgels by using inorganic clay as a cross-linker. *Macromolecular Rapid Communication*, 28, 116-120.

Zhang, Q., Xue, R., Li, X., Wang, Z., Zhou, Z., and Chen, L. (2014). Silk sericin/poly (NIPAM/LMSH) nanocomposite hydrogels: rapid thermo-responsibility and good carrier for cell proliferation. *Pure and Applied Chemistry*, 86(5), 721-731.

Zhang, S., Xianting, Z., Yongsheng, W., Kui, C., and Wenjian, W. (2006a). Adhesion strength of sol-gel derived fluoridated hydroxyapatite coatings. *Surface and Coatings Technology*, 200(22-23), 6350-6354.

Zhang, W., Liu, Y., Zhu, M., Zhang, Y., Liu, X., Yu, H., Jiang, Y., Chen, Y., Kuckling, D. and Adler, H. J. P. (2006b). Surprising conversion of nanocomposite hydrogels with high mechanical strength by posttreatment: from a low swelling ratio to an ultrahigh swelling ratio. *Journal of Polymer Sciences Part A: Polymer Chemistry*, 44, 6640-6645.

Zuo, Y., Liu, X., Wei, D., Sun, J., Xiao, W., Zhao, H., Guo, L., Wei, Q., Fan, H., and Zhang, X. (2015). Photo-cross-linkable methacrylated gelatin and hydroxyapatite hybrid hydrogel for modularly engineering biomimetic osteon. *ACS Applied Materials & Interfaces*, 7(19), 10386-10394.

Chapter 5

Fundamentals of Nanocomposite Hydrogels and Their Applications in Cancer Therapy

Renad AlAnsari
Bushra Hasan
and G. Roshan Deen*
Materials for Medicine Research Group, School of Medicine,
The Royal College of Surgeons in Ireland (RCSI),
Medical University of Bahrain, Kingdom of Bahrain

Abstract

Nanocomposite hydrogels are multicomponent systems with excellent properties in terms of mechanical strength and biocompatibility. The presence of metallic nanoparticles, inorganic materials, polymer micelles, or liposomes in the matrix of hydrogel allows the manipulation of the properties of these materials. Nanocomposite hydrogels have a variety of biomedical applications such as in drug delivery, tissue engineering, cell therapy, cancer vaccines, and antimicrobial coatings. Drug delivery systems based on nanocomposite hydrogels (including stimuli-responsive systems) have fewer side effects than traditional chemotherapy and offer direct delivery of therapeutics to tumors and programmed biodegradation. The fundamental concepts of these materials and their applications in cancer therapy are summarized in this chapter.

Keywords: nanocomposites, hydrogels, nanoparticles, cancer therapy

5.1. Introduction

Advances in material synthesis and nanotechnology have opened the door for valuable and diverse range of applications in all branches of medicine and engineering. Among the vast materials developed, hydrogels and nanocomposite hydrogels have become the spotlight for biomedical applications impacting day-to-day life. The importance of these materials is clear

* Corresponding Author's Email: rdeen@rcsi.com.

In: Nanocomposite Hydrogels and their Emerging Applications
Editor: Mohammad Sirousazar
ISBN: 979-8-88697-675-5

from the exponential growth in research publications over the last 60 years, with the total number of publications exceeding 100,000 by 2020 (Correa et al., 2021). Hydrogels are three-dimensional (3D) crosslinked polymer networks that can retain a significant amount of water or physiological fluid without being dissolved in the medium. The structural integrity of the swollen macromolecular network is maintained by either physical or chemical crosslinking. Chemically crosslinked hydrogels have permanent crosslink junctions or nodes through the formation of covalent bonds, while physically crosslinked hydrogels have transient or non-permanent junctions. These non-permanent junctions are due to the entanglement of polymer chains, ionic interactions, hydrophobic interactions, or hydrogen bonding (Deen and Mah, 2016; Peppas and Bures, 2000). The physical interactions are non-permanent due to reversible interactions between the polymeric materials. The structures of chemically crosslinked and physically crosslinked hydrogels are illustrated in Figure 5.1.

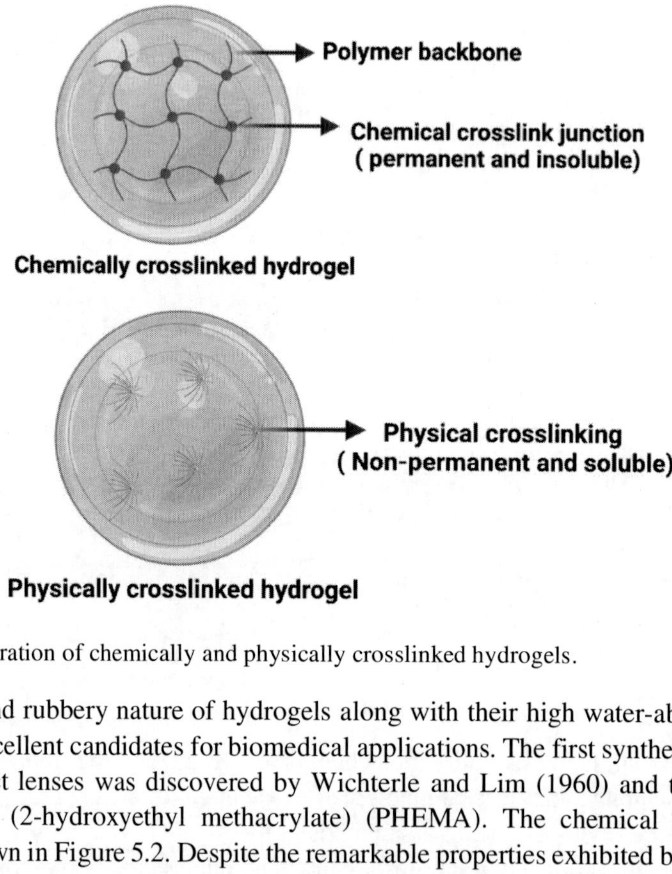

Figure 5.1. Illustration of chemically and physically crosslinked hydrogels.

The soft and rubbery nature of hydrogels along with their high water-absorbing capacity makes them excellent candidates for biomedical applications. The first synthetic hydrogel used to make contact lenses was discovered by Wichterle and Lim (1960) and this material was based on poly (2-hydroxyethyl methacrylate) (PHEMA). The chemical structure of this polymer is shown in Figure 5.2. Despite the remarkable properties exhibited by hydrogels, they suffer from low mechanical strength, hence preventing their applications in regenerative medicine (Haraguchi, 2007).

In recent years, nanocomposite or multicomponent hybrid hydrogels have received much research attention for targeted and versatile applications in medical and non-medical fields. The approach of introducing a second component (at a nanoscale) in the hydrogel (multicomponent approach) increases the mechanical and stimuli-responsive properties of the material. The concept of the multicomponent approach in the synthesis of nanocomposite hydrogels is shown in Figure 5.3. The improved properties allow the material to be used in applications that demand high mechanical strength, good compression tolerance, and elasticity, such as in tissue and cartilage regeneration using polymer-based scaffolds.

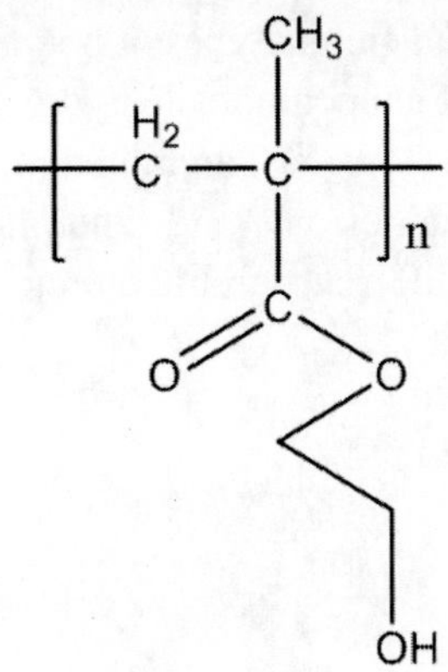

Figure 5.2. Chemical structure of poly (2-hydroxyethyl methacrylate) (PHEMA).

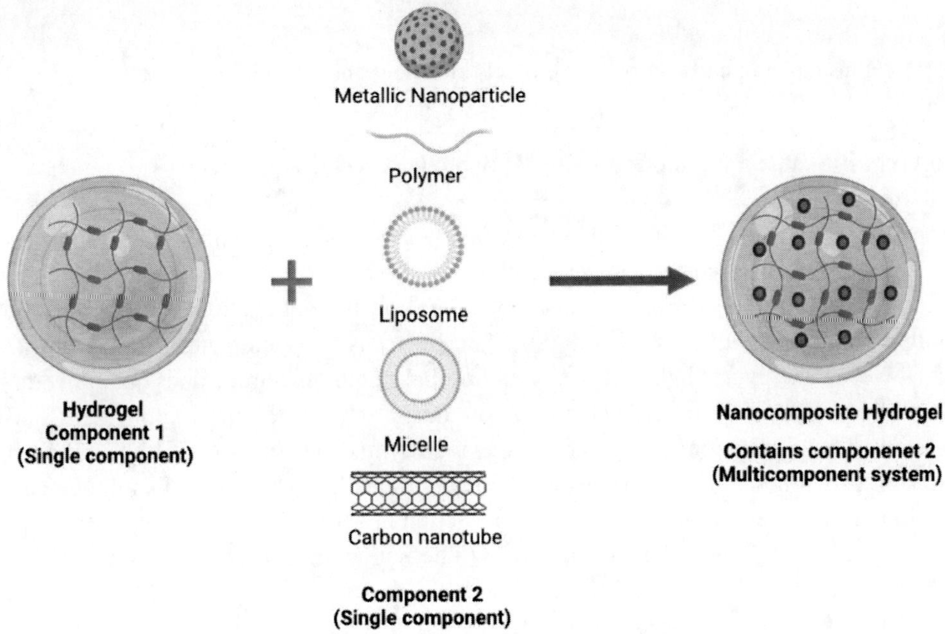

Figure 5.3. Illustration of multicomponent approach in the synthesis of nanocomposite hydrogels.

For a wider range of biomedical applications of nanocomposite hydrogels, important properties of the hydrogel and the incorporated materials need to be fully understood in terms of material properties, self-assembly and interactions, chemical functionality, diffusion rates, rheological properties, etc. In this chapter, the fundamentals of nanocomposites, and their applications in cancer diagnosis and therapy are described. An overview of other medical applications and future prospects of these materials are also presented.

5.2. Types of Nanocomposite Hydrogels

There are two main types of nanocomposite hydrogels, organic-based (including carbon-based) and organic-inorganic-based. Many different types of polymers and metallic nanoparticles have been widely used in the synthesis of nanocomposite hydrogels for a wide range of applications. The incorporation of polymers (in various architectures) and nanoparticles enhances the mechanical, stimuli-responsive, adhesion, and morphological properties of the final nanocomposite hydrogels. The different architectures of nanocomposite hydrogels are illustrated in Figure 5.4.

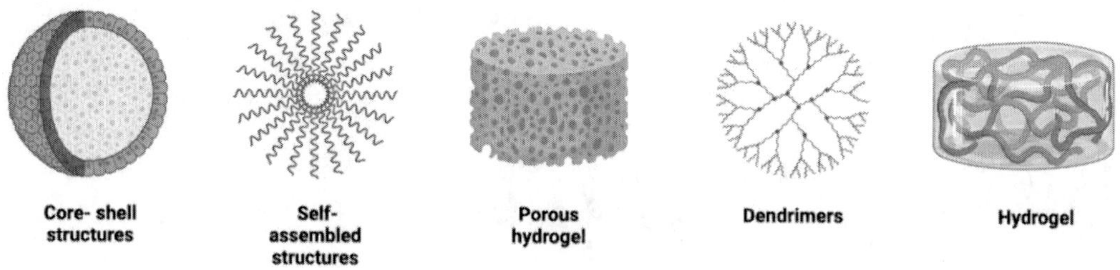

Figure 5.4. Illustration of nanocomposite hydrogels in various architectures.

5.2.1. Organic-Based Nanocomposite Hydrogels (Carbon-Based)

Carbon nanotubes, Buckminster fullerenes (buckyballs), nanodiamonds, and graphenes are carbon-based nanomaterials that have been widely used in the synthesis and fabrication of nanocomposite hydrogels for biomedical applications (Kumar et al., 2017). Carbon nanotubes are hollow structures with a high aspect ratio that can act as nano templates for ions of different sizes. However, the hydrophobic nature of carbon-based nanomaterials does not allow them to interact with hydrophilic functional groups of the polymer, thereby limiting their uniform dispersion. This limitation is overcome by modifying the surface of the carbon-based nanomaterials with polar functional groups (Gaharwar et al., 2015). Depending on the polymer matrix of the hydrogel, the addition of a small fraction of carbon nanotubes provides additional mechanical, electrical, and thermal properties. The nanocomposite hydrogels based on poly (glycerol sebacate) and carbon nanotubes show a significant improvement in compression and tensile strength, with no changes to the elastic modulus of the polymer matrix (Gaharwar et al., 2015).

Graphene oxide is another carbon-based material used in the synthesis and fabrication of nanocomposite hydrogels for targeted drug delivery of therapeutics, genes, and growth factors. These materials when incorporated with cationic polymers such as PEI, greatly improve the effective delivery of genes. Nanocomposite hydrogels based on gold nanoparticles, graphene oxide, and polymers are an interesting class of materials that have applications in bioimaging, biosensing, and photo-thermal applications (Turcheniuk et al., 2015). Despite the improved properties, the long-term cytotoxicity issues of these materials require further studies.

5.2.2. Organic-Based Nanocomposite Hydrogels (Polymer-Based)

Polymer-based nanocomposite hydrogels contain hyperbranched polymers (dendrimers), micelles, nanogels, core-shell gels, or liposomes dispersed with a polymer matrix either through direct dispersion or self-assembly processes. These hybrid materials are sometimes called "plum-pudding gels". These nanocomposite materials have gained widespread pharmaceutical and biomedical applications. The incorporation of stimuli-responsive polymers as the second component enhances the stimuli-responsive property of the nanocomposite hydrogel. External stimuli include pH, temperature, light intensity, biomolecules, enzymes, pressure, sound, and magnetic field (Oral and Peppas, 2004).

Due to the presence of a large number of functional groups, dendrimer-based nanocomposite hydrogels show a high loading capacity for therapeutic drugs and good mechanical properties. The enhanced stress-holding capacity of these hydrogels makes them suitable candidates for applications in regenerative medicine (e,g. cartilage tissue engineering). Nanocomposite hydrogel films based on dendrimers exhibit improved properties in terms of wettability and morphology. These properties are vital for cell adhesion and proliferation in regenerative medicine. The nanostructured internal morphology (nanopores or nanochannels) of the nanocomposite hydrogels improves the mechanical strength and adhesion of fibroblasts (Zhang et al., 2013).

Micro and nanogels are crosslinked polymeric materials in the dimensions range of 10-1000 nm (Deen et al., 2014). Due to the large surface area to volume ratio, therapeutic drug-loaded nanogels can be injected intravenously thereby targeting the site of delivery. By this approach, the limitations of macroscopic hydrogels can be overcome. In comparison to other drug delivery systems viz. liposomes and polymersomes, nanogel-based nanocomposites are easily taken up by cells thereby promoting the effective delivery of therapeutic agents by intracellular means. Hybrid nanogels containing ionizable functional groups such as tertiary amines and carboxylic acids have been used for stimuli-responsive drug delivery in response to changes in external pH conditions.

One major concern with these systems in biomedical applications is hemocompatibility. This includes coagulation of blood, agglomeration of erythrocytes, and activation of platelets (Huang et al., 2014). Nanogel-based nanocomposites with cationic polymers, often exhibit higher toxic properties than systems containing anionic or neutral polymers. The toxicity and biocompatibility are overcome by adding a protective third component around the cationic polymers.

Liposome-based nanocomposite hydrogels have been developed as biodegradable site-specific drug delivery systems. Liposomes are spherical vesicles that are composed of phospholipid bilayers. The core of the layer is hydrophilic, and the shell is hydrophobic in nature. This property allows the liposomes to encapsulate hydrophilic drugs in the core and hydrophobic drugs in the shell. Micellar aggregate-based nanocomposite hydrogels have also been developed. Encapsulation of the drug dexamethasone acetate in micelles of poly (ethylene glycol) and poly (ε-caprolactone) and dispersing in a hydrogel construct has been developed for cancer treatment and to improve single-dose therapies.

5.2.3. Organic-Inorganic-Based Nanocomposite Hydrogels (Metallic Nanoparticles-Polymer)

Metallic nanoparticle-incorporated nanocomposite hydrogels are an interesting class of nanocomposites that has found a wide range of biomedical applications including the diagnosis and treatment of cancer. Nanoparticles of gold (Au), silver (Ag), platinum (Pt), palladium (Pd), copper (Cu), nickel (Ni), selenium (Se), and oxides of metals such as zinc oxide (ZnO), iron oxide (Fe_2O_3 and Fe_3O_4), titanium dioxide (TiO_2), etc., have been used in the synthesis and fabrication of metallic nanoparticle-based nanocomposite hydrogels (Rafieian et al., 2019). The metallic nanoparticles have unique electrical, thermal, magnetic, and anti-microbial properties, and when incorporated in hydrogels they provide the nanocomposite with superior features for applications in suitable and wearable electronics, and conductive scaffolds (Gaharwar et al., 2014). The surface of nanoparticles can be surface functionalized to enhance their interaction with polymer chains of the hydrogel matrix, for sustained release of therapeutic agents from the nanocomposite hydrogel. The incorporation of silver, copper, and selenium nanoparticles in hydrogels renders the nanocomposite with antimicrobial properties (antibacterial, antiviral, and antifungal). The size and shape of the nanoparticles play a vital role in enhancing these properties of the nanocomposite. The smaller the size of the spherical nanoparticles, the better the antimicrobial properties. The small size of the nanoparticles allows easier diffusion from the hydrogel matrix and better interaction with the microbe (Mohan et al., 2007).

Nanocomposite hydrogels containing magnetic nanoparticles have important biomedical applications such as imaging, diagnosis, bioactivation, and delivery of cancer therapeutics. When subjected to an external magnetic field, the magnetic nanoparticles dispersed in the hydrogel matrix generate local heating. If the hydrogel matrix is composed of a thermal-responsive polymer (e.g., poly (*N*-isopropyl acrylamide) (PNIPAM)), the local heating of the magnetic nanoparticles causes reversible shrinkage of the nanocomposite. By modulating the heating, the release of therapeutic agents at the site of action can be controlled. This type of magnetic-responsive nanocomposite hydrogels can be synthesized through the proper choice of polymeric materials that have physiologically relevant lower critical solution temperatures (LCST). The LCST is defined as the critical solution temperature at which the polymer undergoes phase separation through the coil-to-globule transition. Nanocomposite hydrogel scaffolds based on alginate and gold nanowires have been developed for the treatment of damaged cardiac tissues (Oviar et al., 2011). The incorporation of gold nanowires in the alginate nanocomposite improves the electrical conductivity for bridging and synchronizing the cardiac cells. The nanowires act as reinforcing agents and improve the mechanical strength of the nanocomposite.

5.2.4. Organic-Inorganic-Based Nanocomposite Hydrogels (Ceramic-Polymer)

Inorganic nanomaterials such as nano-hydroxyapatite, silicates (nanoclays), calcium phosphate, and wollastonite incorporated nanocomposite hydrogels are widely used in bone-related applications, mainly in implantable devices. Hydroxyapatite is a mineral widely found in bones. Protein-based nanocomposite hydrogels containing nano-hydroxyapatite are used in soft tissue engineering and regenerative medicine. The presence of this inorganic material in the hydrogel matrix stimulates cell growth thereby making the material bioactive.

Nanocomposite hydrogels with tissue-adhesive properties are synthesized using silicate nanoparticles (nanoclays). The morphology of the silicate particles is planar and anisotropic with a high aspect ratio. These features allow better surface interactions with the hydrogel matrix thereby improving the mechanical strength of the material. Nanocomposite hydrogels based on chitosan sulfate containing exfoliated montmorillonite nanolayers have been developed for wound dressing applications. These materials showed better cell attachment for fibroblast cells with no cytotoxicity.

Bio-modified clay (chitosan-based containing montmorillonite), and gelatin-based nanocomposites containing nanoclays have been developed for the treatment of hemorrhages. Without the use of any exogenous growth factors, nanoclays can induce osteogenic differentiation of human mesenchymal cells. Poly (glycerol sebacate)-based nanocomposite hydrogels containing 2D nanoclays show good mechanical stiffness and elasticity with cell growth-enhancing features. These properties arise due to stronger interactions between the nanoclays and the polymer chain. The long-term cytotoxicity of silicate nanoparticles needs to be investigated in detail. Wollastonite is a calcium silicate that is a filler material. It is used as a filler to improve the mechanical properties of materials. Its good biocompatibility and bioactivity make it a good candidate for materials used to make artificial bones and dental roots. A few important types of nanocomposite hydrogels and their biomedical applications are summarized in Table 5.1 (Rafieian et al., 2019).

Table 5.1. Nanocomposite hydrogels and their applications

Type of Nanoparticle	Hydrogel Matrix	Properties	Applications
Polymer	Pullulan and cholesterol	Stimuli-responsive	Drug delivery
Polymer	Crosslinked maleimide	Chemical degradation	Drug delivery
Polymer	Dendrimers	High drug loading	Drug delivery
Metallic	Thermoplastic	High mechanical strength	Anti-bacterial drug carriers
Metallic	Hemicellulose	Magnetic-responsive and degradation	Drug delivery
Metallic	Chitosan and polyacrylamide	pH-responsive and antibacterial	Drug delivery
Inorganic	Chitosan and montmorillonite	High mechanical strength and thermal stability	Wound dressing and tissue engineering
Inorganic	Gelatin	Mineralization and improved stiffness	Tissue engineering
Inorganic	Silk sericin	Enhanced cell differentiation	Bone tissue engineering

5.2.5. Injectable Nanocomposite Hydrogels and Considerations

Injectable nanocomposite hydrogels are formed through *in situ* gelation of the components through physical interactions (secondary bonding). The polymer solutions are directly injected into tissues where they solidify and form a gel through hydrogen bonding, hydrophobic interactions, and dispersion interactions. Injectable hydrogels are widely preferred in medical applications as they do not require invasive surgical procedures. However, the kinetics of gelation need to be carefully considered for medical applications. The gelation can be very fast with the nanocomposite forming a solid gel with the syringe, or gel very slowly leading to poor mixing of the components leading to the loss of therapeutic agents.

Characterizing the rheological or flow properties of injectable nanocomposite hydrogels is a prerequisite for any biomedical applications. Injectability, viscosity, and applied pressure are all interrelated, and these factors need to be considered for clinical applications, as the materials are normally injected using a needle or a catheter. The relationship between the material viscosity and the pressure is modeled using steady-state flow models. For polymer solutions that can undergo physical crosslinking, the viscosity (η) is related by a power law :

$$\eta = K\gamma^{n-1} \quad (5\text{-}1)$$

where K is the consistency index, and n is the shear-thinning parameter. For a Newtonian fluid with constant viscosity $n = 1$, and for a shear-thinning fluid (decrease in viscosity with increase in shear rate) $n < 1$. Based on the power law, the injection pressure (P), for nanocomposite hydrogels (physically crosslinked) is given by the following equation (Paxton et al., 2017; Hernandez et al., 2021):

$$P = \left(\frac{3n+1}{n}\right)^n K \left(\frac{Q}{\pi}\right)^n \frac{2l}{R^{3n+1}} \quad (5\text{-}2)$$

where Q is the flow rate, R is the radius of the needle or catheter, and l is the length of the needle or catheter. Injectable nanocomposite hydrogels based on synthetic polymers and hydroxyapatite nanocrystals show excellent properties for the mineralization of bones, growth of osteoblasts, and vascular infiltration. Similar composite materials using triblock copolymers, collagen, and hydroxyapatite have shown bio-mimetic properties in the regeneration of osteoblast cells.

5.3. Synthesis Strategies of Nanocomposite Hydrogels

Many methods are currently available in the synthesis and fabrication of nanocomposite hydrogels. The strategies successfully incorporate both organic and inorganic nanomaterials in hydrogels as fillers to improve the mechanical strength of the hydrogel. The components can also be crosslinked to manipulate the network structure of the final material. Natural and synthetic polymers have been used in the development of a variety of nanocomposites such as chitosan, collagen, gelatin, carrageenan, carboxy methyl cellulose, polyvinyl alcohol, PHEMA, poly (lactic acid glycolic acid) (PLGA), PNIPAM, poly (ethylene imine) (PEI), peptides, etc. These polymers provide the microenvironment and an array of molecular structures for the incorporation of nanomaterials. More durable and complex nanocomposite hydrogels are prepared through self-assembly processes involving cooperative interactions (physical interactions). The self-assembled structures act as nano templates for physical entrapment leading to homogenous distribution of nanomaterials in the hydrogel matrix.

In the regeneration of bone tissue, nanocomposite hydrogels containing titanium, silver, and gold nanoparticles dispersed in natural polymers or neutral synthetic polymers are widely used. The properties of the nanocomposite hydrogels are dependent on the type of polymers, size, and shape of nanoparticles (polydispersity), concentration, degree of crosslinking, and degree of dispersion. A few important approaches in the synthesis of nanocomposite hydrogels are presented in this section.

5.3.1. Blending Method

This is the simplest method involved in the synthesis of nanocomposite hydrogels. In this method, the polymers, crosslinking agents, polymerization initiators, and nanomaterials are mixed, and the polymerization reaction is initiated either using light or heat to form the hydrogel. The uniform distribution of the nanomaterial in the hydrogel can be tuned by varying the speed of stirring, the concentration of monomers, the types of monomers, and the time of reaction. The main disadvantage of this method is the inhomogeneous distribution of nanomaterials in the polymer matrix, and this leads to nanocomposite hydrogels with weak mechanical strength (Miwa et al., 2014).

5.3.2. Grafting Method

In this method, the surface of the nanomaterial is functionalized to achieve covalent interactions with the polymer chains of the hydrogel. By this method the problems associated with the blending method can be avoided, leading to products with good mechanical strength and homogenous distribution of nanomaterials. Nanocomposite hydrogels based on surface-modified iron oxide nanoparticles and polydimethylsiloxane (PDMS) prepared by the grafting method exhibit high mechanical strength and stability (Hu et al., 2019).

5.3.3. Freeze-Thaw Method

Some physically crosslinked nanocomposite hydrogels are prepared using the freeze-thaw method. In this method, the nanocomposite mixture is frozen and thawed several times to allow physical crosslinking between the components leading to gelation. Physical crosslinking occurs through the formation of secondary forces such as hydrogen bonds, hydrophobic interactions, and dispersion interactions. The main disadvantages of this method are poor mechanical strength and loss of polymer network structure, in some cases.

5.3.4. *In Situ* Method

In this method, the monomers, nanoparticles, and crosslinking agents are mixed, and the polymerization reaction is initiated using either light, heat, or redox reactions. As hydrogel formation occurs, the nanostructures get trapped between the polymer networks. The polymer network acts as a stabilizing agent preventing the aggregation and inhomogeneous distribution of nanostructures in the final product. By this method, uniform distribution of nanomaterials can be achieved leading to the formation of nanocomposites with good mechanical properties. Many nanocomposite hydrogels containing magnetic nanoparticles are synthesized by the *in situ* method (Wang et al., 2018).

5.3.5. Powder Method

This method was developed by Deen and Hara (2005) to prepare ionomer-based nanocomposites containing lead sulfide (PbS) nanoparticles. Ionomers are co-polymers containing a small fraction of monomers that contain ionizable chemical functional groups. An example of an ionomer is the copolymer of methyl methacrylate containing various percentages of methacrylic acid (PMMA-co-PMA). In this method, the ionomer is mixed with metal salt instead of the nanoparticle. The solution is freeze-dried to obtain a powder which is then reduced using suitable reducing methods. The powder containing the nanoparticle is fabricated using compression molding. Due to the presence of a small fraction of ionizable units, the distribution of nanomaterial in the nanocomposite is homogeneous. By this method, nanocomposite materials with excellent tensile strength and stability can be synthesized and fabricated.

5.4. Fundamental Requirements of Nanocomposite Hydrogels for Medical Applications

Hydrogels have made significant impacts and contributions to biomedical applications. The translation of these materials to clinical applications is not a simple task, and various fundamental factors need to be considered mainly for the sustained delivery of therapeutics in cancer and diabetes therapies (Nourouzi et al., 2016).

5.4.1. Basis of Drug Delivery Using Nanocomposite Hydrogels

Drug delivery applications using hydrogels and nanocomposite hydrogels are fundamentally based on the mass transport of the cargo (drugs, genes, nucleic acids, proteins, etc.). The mass transport is dependent on the internal morphology and mesh size (ξ) of the nanocomposite hydrogel and the hydrodynamic diameter (D_h) of the material to be delivered (Figure 5.5). The mesh size or correlation length is the linear distance between two adjacent crosslink points in a hydrogel matrix. These characteristic features have important implications in the delivery of the cargo either through the process of diffusion or erosion (extended-release), or both. The mesh size of the nanocomposite hydrogel can be controlled by the degree of crosslinking and through the incorporation of nanostructures within the matrix.

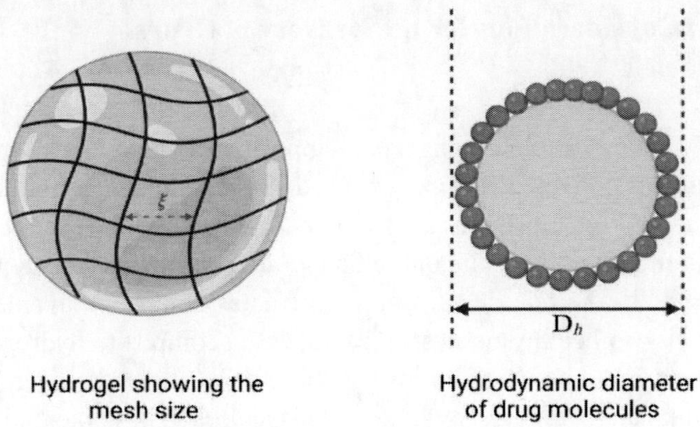

Figure 5.5. Illustration of network mesh size of hydrogels and hydrodynamic diameter of drug molecules.

If the size of the drug is much smaller than the mesh size ($D_h < \xi$), the drug diffuses freely out of the nanocomposite leading to the characteristic "burst" release, just after administration. In this case, diffusion is the dominant mechanism, and the drug is released completely from the nanocomposite in a very short time frame. It must be pointed out that the nature of the hydrogel matrix and chemical functionality of the drug, and the diffusion can be affected leading to hindered diffusion (Calgue and Phillips, 1996). The important polymer architectural features and drug molecules are illustrated in Figure 5.6. If the size of the drug is comparable to the mesh size ($D_h \sim \xi$), the drug diffuses significantly slower. If the size of the drug is larger than the mesh size ($D_h > \xi$), the drug is completely trapped in the matrix of the nanocomposite and cannot diffuse out. In this case, the release of the drug is dependent on many factors such as erosion of the gel, swelling, and mechanical disturbances. Here, the dominant mechanism of drug release is erosion-controlled (Li and Mooney, 2016). Nanocomposite hydrogels for biomedical applications are designed to degrade or surface erode through various mechanisms including enzymatic hydrolysis and stimuli-responsive degradation. Thus, understanding the mechanism of diffusion and erosion of nanocomposite hydrogels are important prerequisites in the design, development, and fabrication of such systems for site-specific drug delivery applications.

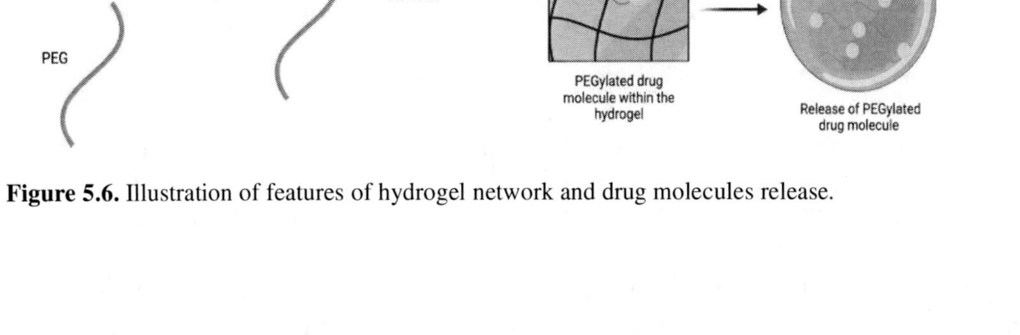

Figure 5.6. Illustration of features of hydrogel network and drug molecules release.

5.4.2. Important Considerations for the Delivery of Drugs, Proteins, and Nucleic Acids

In many drug delivery applications using nanocomposite hydrogels, small molecules and drugs smaller than 15 nm are delivered (Burczak et al., 1994). Controlling the release of such small molecules from the hydrogel matrix is a challenge for sustained release. The main aim of any drug delivery system is to deliver the entire active drug to the target site without any loss in concentration. As such, the release of drugs at off-sites should be avoided to prevent drug overdose and toxicity to healthy tissues. Injectable nanocomposite hydrogels can be directly administered at the target site using an invasive approach, therefore overcoming the toxicity associated with bulk nanocomposite hydrogels. This approach has potential in the treatment of injured tissues.

The mesh size of the hydrogel matrix can be controlled by varying the degree of crosslinking thereby regulating the drug release kinetics. However, highly crosslinked hydrogels are brittle and show poor mechanical strength, which is an undesirable material property for drug delivery applications. This can be overcome through the interaction of the drug with the matrix of the hydrogel (drug-polymer interactions). The interactions are physical (secondary bonding) such as hydrophobic, hydrogen bonding, electrostatic, and dispersion (van der Waals). Polyelectrolyte-incorporated nanocomposite hydrogels can control the release of small molecules that are oppositely charged, thereby reducing the undesirable burst release (Sivakumaran et al., 2011). Nanocomposite hydrogels based on cyclodextrin have been shown to prevent the burst release of small molecules through host-guest interactions. The use of pH- and temperature-responsive nanocomposite hydrogels in the sustained release of doxorubicin (hydrophilic drug) and paclitaxel (hydrophobic drug) is promising in reducing the toxicity associated with common chemotherapy. The solubility of the drugs plays an important role in influencing the mechanical properties and release kinetics (Zhao et al., 2011).

Not all small molecules can have significant interactions with the polymer, and template methods are used to enhance the polymer-drug interactions. In this approach, the nanocomposite hydrogels are synthesized using a molecular template which is then removed (Sellergren, 1997). The structure (cavity) left behind is used to bind the small molecules on the polymer network. An alternative method to enhance the polymer-drug interaction is through the conjugation of the drug with another polymer like polyethylene glycol (PEG). The process of conjugation is termed PEGylation. This method has been used in conjugating the chemotherapy drug camptothecin with PEG to increase the overall size of the cargo (Yu et al., 2008). By this method, the circulatory half-life and biodistribution can be extended and altered. The method of PEGylation of camptothecin with a triblock copolymer of poly (lactic acid-co-glycolic acid) and PEG had been developed as an injectable nanocomposite hydrogel for cancer therapy. In the long run, similar hydrogel systems have the potential to replace the long infusions of chemotherapy and mitigate toxic side effects.

Peptides and therapeutic proteins are valuable in the treatment of various conditions such as cancer, diabetes, arthritis, and infectious diseases. Protein-based drugs are often unstable with a short shelf-life. Further, these drugs require repeated administration for maximum benefit. The stabilization of protein-based drugs can be improved through encapsulation using hydrogels. Proteins with molar mass above 100 kDa have better sustained release profiles. A high-dose delivery of a clinically approved drug, bevacizumab (Avastin), in cancer therapy has been achieved using this approach. For the delivery of smaller proteins like growth factors, hormones, and cytokines, the method of PEGylation and nanoencapsulation are preferred for sustained release and to prevent burst release.

Nanocomposite hydrogels used for the delivery of DNA and RNA, are based on electrostatic interactions with supramolecular building blocks. These types of nanocomposites poses cell penetrating abilities and can protect the molecules from degradation. Thermosensitive nanocomposite hydrogel based on PLGA-PEG-PLGA has been used to deliver shRNA for the tumor oncogene PLK1 by the method of complexation. Nucleic acid particles that are assembled by electrostatic interactions can alter gene expression. This concept was used to deliver DNA using hydrogels based on methacrylate to encode pro-angiogenic growth factors (Paul et al., 2014). By chemically modifying the nucleic acids, they can be loaded as particles in hydrogels for effective delivery. Through chemical modification and polymer conjugation gene expressions can be altered in tissues. This feature has been demonstrated using supramolecular nanocomposite hydrogels based on cholesterol and mRNA in altering the gene expression in cardiac tissues (Jung et al., 2010). Overall, many properties of hydrogel materials (degradation, biocompatibility, mechanical strength, diffusion, etc.) and nanoparticles (morphology, size, self-assembly, surface functionality, etc.) need to be carefully evaluated for the successful and targeted delivery of molecules at the site of action. To attain nanocomposites with tailored functionality, the interface between hydrogel and nanomaterial can be manipulated with existing polymerization techniques and the choice of nanomaterials.

5.5. Applications of Nanocomposite Hydrogels in Cancer Therapy

Nanocomposite hydrogels have found immense medical applications due to their excellent properties that resemble natural tissues. They are used in drug and gene delivery, regeneration of tissues, cancer diagnosis, cancer immunotherapy, coatings for medical devices, etc. In this section, the application of nanocomposite hydrogels mainly in cancer therapy is presented.

5.5.1. Background on Cancer and Cancer Therapy

In recent years, cancer is the most deleterious disease and is one of the leading causes of death with an estimated more than 10 million mortality per year at a global scale (Zaimy et al., 2017). Cancer is a disease caused by uncontrolled cell division and proliferation through infiltration and metastasis leading to the disruption of normal cell functions. The five most prevalent types of cancer are lung, breast, colorectal, prostate, and liver cancer. The mortality rates of these types of cancer vary from country to country. The highest mortality rates due to lung, colorectal, breast, and prostate cancer are observed in high-income countries, while low-income countries have mortality rates for liver, stomach, esophageal and cervical cancer (Torr et al., 2014).

Lung cancer is characterized by the uncontrolled division and proliferation of lung cells and is one of the leading causes of death. There are two types of lung cancer viz. non-small cell lung cancer (NSCLC) and small cell lung cancer (SCLC), and their prevalence is about 85% and 15%, respectively (Amreddy et al., 2017). On a global scale, breast cancer is considered to have the highest mortality rate in women, with about 685,000 deaths reported in 2021 (Nieto et al., 2022). This type of cancer is a challenge to the global health system due to the residual and recurrent nature of the tumor cells. Some of the most effective treatments for breast cancer are radiotherapy and chemotherapy after tumor resection. However, these treatment methods are often associated with severe systemic side effects owing to the nature and dosage of the drugs used.

Colon cancer is the third most diagnosed cancer in the world and is characterized by the formation and proliferation of benign polyps (clumps of cells) in the colon. The first line of treatment for this type of cancer is a systemic therapy which includes chemotherapy and immunotherapy. Prostate cancer found in males is characterized by uncontrolled cell division in the prostate gland, and current treatment methods include radiation therapy and radical prostatectomy (Dvoracek, 1998).

Liver cancer is the leading cause of death globally and is the fifth in the US. The two forms of liver cancer, hepatocellular carcinoma (HCC) and cholangiocarcinoma (CCA) account for 80-90% and 10-15%, respectively in adults. It is reported that liver cancer is more prevalent among adults in developing countries as the population is associated with at least one of the following risk factors: fatty liver disease, alcohol-related cirrhosis, obesity, smoking, unhealthy lifestyles, and hepatitis B and C viruses. The current treatment for liver cancer is transarterial chemoembolization (TACE) and kinase inhibitor treatment using sorafenib. Sorafenib is an inhibitor that is also used as a chemotherapeutic drug in the treatment of late-stage liver cancer (Ma et al., 2022). Given the adverse toxic effects associated with current chemotherapy, radiation therapy, immune therapy, and surgery for cancer treatment, alternative methods with fewer side effects are always being sought.

5.5.2. Nanocomposite Hydrogels in Cancer Therapy and Their Advantages

Nanocomposite hydrogels are promising materials for cancer diagnosis and therapy as they can address the shortcomings of current immunotherapy. These materials are well-suited for the effective and targeted delivery of cancer therapeutics to the affected cells. The normal barrier associated with dose-limiting toxicity in immunotherapy can be overcome using these materials. In addition, hydrogels limit the stimulation of the immune system in the microenvironment of tumor cells thereby making the therapies more tolerable. Further, the internal structure of nanocomposite hydrogel can be easily tuned to function as scaffolds and drug depots which can enhance the growth of endogenous immune cells. In cancer therapies, this material property is very promising as it serves as a template for creating an immunosuppressive environment within the microenvironment of the tumor (Binniewies et al., 2018).

Delivery of tumor-targeting antibodies using nanocomposite hydrogels as depot materials has been utilized. After surgical removal of the tumor, the cavity left behind can be directly delivered with antibodies and drugs. Injectable nanocomposite hydrogels have been used to deliver monoclonal antibodies (Herceptin) directly to the surgical site using anti-monoclonal HER2 antibodies-breast cancer models (Chen et al., 2019). This direct delivery of Herceptin at high concentrations using hydrogels as delivery materials prevents the recurrence of the tumor. It must be pointed out that HER2 monoclonal antibodies express well on healthy cells leading to very toxic side effects on cardiac cells. The use of hydrogels as drug delivery devices prevents the diffusion of antibodies into the systemic circulation and prevents cardiotoxicity. This is seen as a major advantage of using hydrogel-based materials for cancer therapy. A wide range of nanocomposite hydrogels has been developed for the delivery of cancer therapeutics such as doxorubicin, docetaxel, carboplatin, mitoxantrone, and leucovorin. The hydrogel or macromolecular carrier is based on crosslinked polymers, block copolymers (di- and tri-block), nanogels, and liposomes. Some examples of the nanocomposite systems are, lecithin with cholesterol-modified PEG, PEG modified with starch and dithiodipropionic acid, PEG-PLGA-PEG, chitosan with hydroxyapatite, hyaluronic acid with calcium and chitosan, PEGylated liposomes, and polyvinyl alcohol with montmorillonite.

5.5.3. Stimuli-Responsive Nanocomposite Hydrogels and On-Demand Therapeutic Effects in Cancer Therapy

Stimuli-responsive nanocomposite hydrogels offer better features in terms of on-demand delivery of therapeutics in response to changes in external stimuli. Magnetic field-responsive nanocomposite hydrogels containing magnetic nanoparticles are of particular interest in cancer diagnosis and cancer therapies. Iron oxide nanoparticles with paramagnetic properties are commonly used in magnetic-responsive nanocomposite hydrogels. Upon exposure to the magnetic field and magnetic dipole is created causing the iron oxide nanoparticles to vibrate. This vibration generates heat (localized heating) by the mechanism of thermal ablation. The heat generated is termed hyperthermia or overheating in general and is utilized in promoting the therapeutic efficacy of cancer drugs. Based on the targeted region of the tumor, hyperthermia is categorized into three categories, (i) diathermia, (ii) moderate hyperthermia, and (iii) thermal ablation. Diathermia refers to the low-temperature range with a maximum of 41°C, moderate hyperthermia falls in the mid-temperature range between 41-46 °C, and thermal ablation corresponds to the high-temperature range between 46-51 °C. A variety of biomedical applications use hyperthermia effects. For example, diathermia is usually used in physiotherapy and the treatment of rheumatic diseases. Thermal ablation is used to cause necrosis and carbonization of cancer cells. The types of hyperthermia and their biomedical applications are illustrated in Figure 5.7.

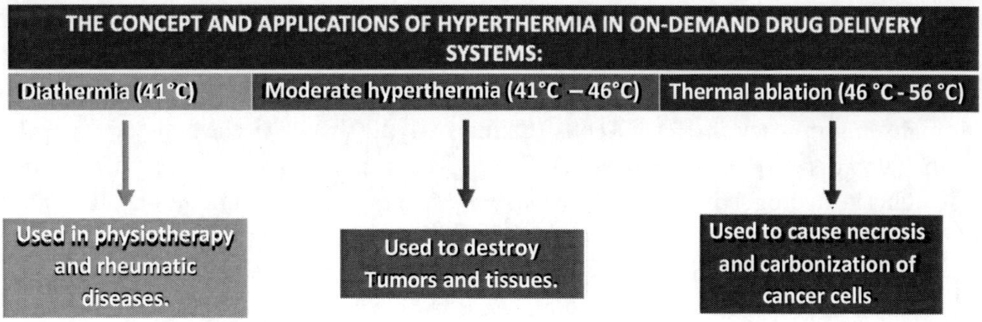

Figure 5.7. Types of hyperthermia, the temperature range, and biomedical applications.

Hyperthermia cancer therapy is used alongside with chemotherapy and radiation therapy to target and destroy tumors. This therapy relies on the strength of the external magnetic field, length of exposure, and targeted region of cancer cells. Since its introduction in 1957, this type of therapy has been effective in treating pancreatic cancer and glioblastoma. Further, hyperthermia therapy is a non-invasive treatment method, and the oscillating external magnetic field and the radiation emitted is felt only by the nanoparticles and not the body of the patient.

Given the extremely small size of the nanoparticles, they can pass through cell membranes and the blood-brain barrier and are useful in the treatment of glioblastoma. However, such diffusions in relation to the size of nanoparticles are still controversial. Magnetic-responsive nanocomposite hydrogels greatly control tumor growth due to therapeutic efficacy, vascularization, and oxygen supply, which are considered advantageous in cancer treatment (Ganguly and Margel, 2021). Magnetic-responsive nanocomposite hydrogels associated with thermal-responsive hydrogel matrix offer better control in the release of cancer therapeutics. The polymer matrix responds to hyperthermia caused by the nanoparticles and undergoes a phase transition from a swollen state to a shrunk state controlling the drug release. By proper choice of polymers, the drug release at the site of action can be modulated, promoting on-demand drug release. This method of drug release based on hyperthermia and phase transitions promotes synergistic cytotoxicity through chemo and thermotherapeutics (Kasinski et al., 2020). Hyperthermia stimulated drug delivery system is schematically illustrated in Figure 5.8.

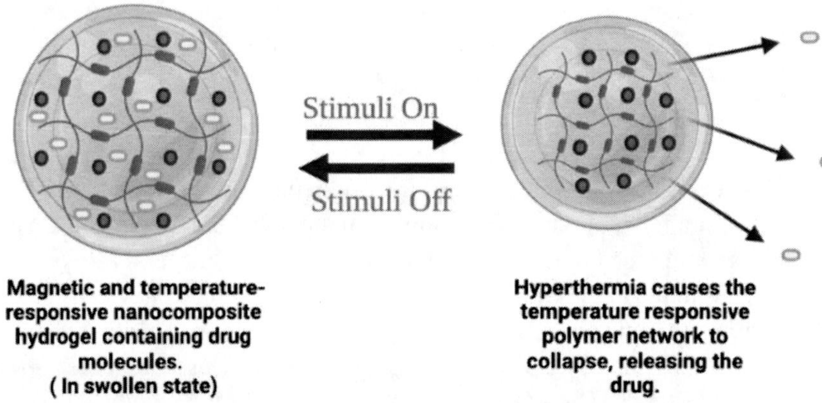

Figure 5.8. Illustration of magnetic and temperature-responsive nanocomposite hydrogels in drug delivery.

Cisplatin-loaded magnetic and thermal responsive nanocomposite hydrogels based on PNIPAM, dimethylamide, and Fe_3O_4 demonstrated the super magnetic features of the material with high drug efficacy in HepG2 cells (Salimi et al., 2018). The chemotherapy agent and immune-suppressant, methotrexate (MTX) loaded in magnetic and pH-responsive nanocomposite hydrogel (in the form of beads) in response to changes in external pH conditions was found to suppress the proliferation of HepG2 cells. Magnetic-responsive and pH-responsive nanocomposite hydrogels containing graphene oxide nanosheets were found to possess superior properties such as high magnetic hypothermia, biocompatibility, and thermal stability. This material is used in trans arterial embolization therapy and reduces postoperative recurrences of tumors in the liver. Injectable magnetic-responsive nanocomposite hydrogels based on silk fibroin containing iron oxide nanotubes have demonstrated the ability to be injected directly into liver tumors by ultrasound-guided interventional treatment (Qian et al., 2020).

Ultrasound-responsive nanocomposite hydrogels are used for the release of drugs in tumor microenvironments (Chen and Du, 2013). Ultrasound has various effects on the mechanism of drug targeting and delivery. High-intensity focused ultrasound produces localized heating through acoustic vibrations, and the heat generated is used in cancer therapy to destruct the tumor cells. Ultrasound induces cavitation which improves the permeability of cell membranes which facilitates the direct delivery of cancer therapeutics to the tumor cells. Nanocomposite hydrogels based on supramolecular structures containing mesoporous silica nanoparticles have been used as theranostic agents for the efficient release of cancer drugs and in imaging-guided therapy.

Active targeting using various ligands as targeting chemical moieties is a method to deliver cancer therapeutics with high efficacy. The large surface area of nanoparticles present in the nanocomposite allows interaction between ligands which in turn interacts with over-expressed receptors or antigens of tumor cells (Rosenblum et al., 2018). Such interactions mediate endocytosis and specific delivery of nanoparticles (or drug-functionalized nanoparticles) at the target site. Active targeting is widely used in nanomedicine for the successful delivery of drugs for liver cancer therapy. Short peptides, galactose, glycyrrhetinic acid, folate, transferring, antibodies and aptamers are currently used as ligands in liver cancer therapy.

Combination therapy which combines more than one therapy is a valuable platform in the treatment of cancer. This method is used to deliver anticancer drugs with high efficacy and to prevent the recurrence of tumors. Nanocomposite hydrogels based on PEGylated gold nanorods and polymeric micelles, loaded with paclitaxel have been developed for direct injection into the tumor, in combination therapy. The gold nanorods are activated using laser irradiation and are restricted to the tumor cells with the healthy cells remaining unaffected. The loaded drug is released (sustained release) at the same time to destruct any residual tumor cells (Zhang et al., 2016). Polypeptide-based injectable nanocomposites containing hollow gold nanorods were used in chemo-photothermal therapy to deliver doxorubicin (DOX) to liver tumor cells. Thermo-sensitive nanocomposite hydrogels containing a therapeutic radionuclide (colloid of Tin and Rhenium-188) have been developed for *in situ* gelation at body temperature and showed successful delivery of DOX to liver tumor cells (Peng et al., 2013). The residence time of the hydrogel containing the radionuclide colloid was significantly longer in the tumor preventing tumor growth. In an effort to study the effect of drug combinations in chemotherapy the drugs, DOX, cisplatin, and cyclophosphamide were combined with inflammatory GMCSF cytokines in injectable hydrogels based on the natural polymer chitosan. The best synergy was obtained for cyclophosphamide with the cytokines and showed good anticancer effects on cervical cell models.

Reactive oxygen species (ROS) sensitive hydrogels have been developed to treat cold or poor immunogenic tumors. These types of tumors are less likely to provoke the immune system of the body. This is because myeloid-derived suppressor cells (MDSCs) which are pro-tumor myeloid cells and tumor-associated macrophages (TAMs) that are present in the tumor aid a tolerable microenvironment within the tumor. The production of ROS is significantly high in tumor cells and this feature is used in the development of ROS-sensitive nanocomposite hydrogels. When placed in this environment, the hydrogel responds to changes in ROS and beings to degrade releasing the loaded therapeutics. Repolarization of the tumor microenvironment occurs as the ROS reacts with the chemical components of the hydrogel (Li et al., 2016). The delivery of a chemotherapy drug, gemcitabine, using ROS-sensitive nanocomposite hydrogels has also been studied. The drug was found to greatly suppress the myeloid-derived suppressor cells and tumor-associated macrophages in the tumor. Moreover, it also increased the intratumoral levels of $CD8^+$, $CD4^+$, and T-cells, indicating the benefits of

ROS-sensitive hydrogels in the treatment of cold tumors. However, an increase in the expression of PD-L1 and PD-1 was also observed which are inhibitors to immune responses (Wang et al., 2018). By introducing the antibody aPD-1 as an encapsulation within the hydrogel, the overexpression was reduced. The overexpression of indoleamine-pyrrole 2,3-dioxygenase (IDO) in tumors suppresses the immune response of T-cells. ROS-sensitive hydrogels containing the drug D1MT were used as delivery systems to inhibit IDO overexpression in tumor cells (Yu. et al., 2018). Further, higher levels of tumor-infiltrating lymphocytes and lower levels of ROS were observed. The use of ROS-sensitive nanocomposite hydrogel drug delivery systems greatly improves the efficacy of the drug.

The treatment of cancers that have low levels of neoantigens is a medical challenge. Neoantigens are specific antigens that allow the immune system to detect cancer cells. When their level is low, the immune system of the body struggles to provoke an immune response against cancer. ROS-sensitive hydrogels have been used to overcome this defect by promoting the production of higher levels of neoantigens. The hypomethylating agent Zeb-HMA was delivered using ROS-sensitive hydrogels to cancer cells with low levels of neoantigens (Ruan et al., 2019). The hypomethylating agent altered some genetic factors leading to the expression of silenced genes that are responsible for producing the neoantigens.

Stimuli-responsive nanocomposite hydrogels are also used to improve tumor imaging. The metallic nanoparticles present in the hydrogels enhance contrast factors leading to a better imaging of tumor cells. Many hydrogel systems are already in the market as radiographic imaging agents. Iodinated TracelT™ hydrogel systems are used in clinics as radiographic markers of soft tissues. This hydrogel has a longer residence time and it presents in the target tissues up to three months. This characteristic provides a high-contrast image in image-guided radiotherapy. Another important application of stimuli-responsive nanocomposite hydrogels in cancer radiotherapy is shielding. These materials are used to shield healthy and sensitive tissues during radiotherapy, making radiotherapy more effective and safer. During radiotherapy, the healthy tissue that surrounds the targeted tumor is at risk of exposure leading to radiation damage (Correa et al., 2021). The SpaceOAR hydrogel system has been developed to prevent radiation damage to healthy cells during radiotherapy of prostate cancer. This hydrogel is radiopaque and acts as a physical barrier between prostate cancer cells and healthy cells. The hydrogel also tends to stay in the location of administration thereby protecting the healthy cells. The features make these hydrogels useful in the treatment of other types of cancers such as pancreatic and rectal cancer. The formulation of hydrogels can be manipulated for safer and more effective cancer radiotherapy.

Combination therapy is further beneficial in suppressing the recurrence of tumors, and these benefits cannot be obtained using a single therapy alone, for example, photothermal therapy. Through proper design of stimuli-responsive and biodegradable nanocomposite hydrogels, more precise on-demand drug delivery can be attained in radiotherapy and chemotherapy for cancer treatment. The development of the next generation of such nanocomposite hydrogels is in progress to release different drugs at a specific time frame. This is an active area of research and the preliminary data obtained is promising (Wang et al., 2018; Li et al., 2016). The injectable hydrogel composites that are in clinical trials since early 2020 for the treatment of rectal, pancreatic, prostate, colorectal, bladder, and head and neck cancers are based on PEG and mucoadhesive polymeric materials. A few important nanocomposite hydrogels used in combination therapy in the treatment of liver cancer are summarized in Table 5.2.

Table 5.2. Some utilized nanocomposite hydrogels in liver cancer therapy

Nanocomposite Hydrogel	Drug	Cancer Cell Line	Refs.
PCL-PEG-PCL with rhenium and tin colloid	DOX	BNL-Luc	(Peng et al., 2013)
Polypeptide-based with gold nanorods	DOX	HepG2	(Jin et al., 2019)
PCL-PEG-PCL with multi-walled carbon nanotubes	Rhodamine B and DOX	BEL-7402	(Wei et al., 2018)
Chitosan-based with PEGylated gold nanorods	Paclitaxel	HEPG2	(Zhang et al., 2016)
Lactobionic acid and carboxymethylated curdlan	Retinoic acid	HEPG2	(Na et al., 2000)

5.5.4. Limitations of Nanocomposite Hydrogels in Cancer Therapy

Although nanocomposite hydrogels offer better properties than conventional hydrogels, their direct applications in cancer therapy are still a challenge. Materials for drug delivery applications need to be robust with excellent mechanical properties such as high tensile strength, elastic modulus, compression strength, and fracture strength. If these parameters are not considered, premature dissolution or fracture of the material will occur at the site of application. The internal morphology of the nanocomposite hydrogel plays an important part in the drug loading and drug release capability of the drug delivery system. Large internal pore size limits drug loading capacity and promotes the burst release of the drug (Tsirigotis-Maniecka et al., 2021). Treatments using chemically crosslinked nanocomposite hydrogels are invasive and the material containing the drug must be implanted through surgical procedures.

All these factors limit the clinical applications of these materials. However, new nanocomposite hydrogels with improved mechanical strength, stimuli-responsive properties, biocompatibility, and biodegradability have been developed for cancer therapy and few are already in clinical trials as mentioned earlier. When developing a treatment plan for cancer therapy using hydrogels, the complex anatomical structures around the tumor such as arteries, veins, nerves, and the lymphatic system needs careful consideration to avoid toxic effects. Apart from this, the immunological response, biodistribution, and biodegradability must also be considered for the efficient and safe treatment of cancer using all types of hydrogels. Advances in materials synthesis strategies have allowed the development of new hydrogels and hybrids with tailored properties for targeted and safer applications, in cancer therapy. As new and sophisticated materials continue to emerge, they may unlock even better, targeted, and safe treatments for all types of cancer in the future.

Conclusion

The excellent features of nanocomposites such as high mechanical strength, porosity, biocompatibility, biodegradation, and thermal stability make them appealing in diverse medical applications. Stimuli-responsive nanocomposite hydrogels that respond to changes in external stimuli are important materials in cancer diagnosis and therapy. These materials when used in combination therapy offer higher efficacy of the therapeutics and high levels of safety. The main advantage of nanocomposite hydrogels in drug delivery strategies is the localized treatment to the vicinity of administration. However, for certain types of drugs used in chemotherapy, the tolerability of the drug needs to be evaluated in greater detail for toxicity. The emergence of new nanocomposite hydrogels, in particular, stimuli-responsive systems, opens the door for target-specific and safe applications in cancer combination therapy.

References

Amreddy, N., Babu, A., Muralidharan, R., Munshi, A., and Ramesh, R. (2017). Polymeric nanoparticle-mediated gene delivery for lung cancer treatment. *Topics in Current Chemistry*, 375, 1-23.

Binniewies, M., Roberts, E. W., Kersten, K., Chan, V. Fearon, D. F., Merad, M., Coussens, L. M., Gabrilovich, D. I., Ostrand-Rosenberg, S., Hedrick, C. C., Vonderheide, R. H., Pittet, M. J., Jain, R. K., Zou, W., Howcroft, T. K., Woodhouse, E. C., Weinberg, R. A., and Krummel, M. F. (2018). Understanding the tumour immune microenvironment (TIME) for effective therapy. *Nature Medicine*, 24, 541-550.

Burczak, K., Fujisato, T., Hatad, M., and Ikada, Y. (1994). Protein permeation through poly (vinyl alcohol) hydrogel membranes. *Biomaterials*, 15, 231-238.

Calgue, D. S., and Phillips, R. J. (1996). Hindered diffusion of spherical macromolecules through dilute fibrous media. *Physics of Fluids*, 8, 1720-1729.

Chen, W., and Du, J. (2013). Ultrasound and pH dually responsive polymer vesicles for anticancer drug delivery. *Scientific Reports*, 3, 2161-2169.

Chen, X., Wang, M., Yang, X., Wang, Y., Yu, L., Sun, J., and Ding, J. (2019). Injectable hydrogels for the sustained delivery of a HER2-targeted antibody for preventing local relapse of HER2+ breast cancer after breast conserving surgery. *Theranostics*, 9, 6080-6098.

Correa, S., Grosskopf, A., Hernandez, H., Chan, D., Yu, A., Stapleton, L. M., and Appel, E. A. (2021). Translational applications of hydrogels. *Chemical Reviews*, 121, 11385-11457.

Deen, G. R., Alsted T., Richtering, W., and Pedersen, J. S. (2011). Synthesis and characterization of nanogels of poly (N-isopropylacrylamide) by a combination of light and small angle X-ray scattering. *Physical Chemistry Chemical Physics*, 13, 3108-3114.

Deen, G. R., and Hara, M. (2005). Preparation and characterization of PbS nanoclusters made by using a powder method on ionomers. *Polymer*, 46, 10883-10889.

Deen, G. R., and Mah, C. H. (2016). Influence of external stimuli on the network properties of cationic poly (N-acryloyl-N'-propyl piperazine) hydrogels. *Polymer*, 89, 55-68.

Dvoracek, J. (1998). Adenocarcinoma of the prostate. *Casopis lekaru ceskych*, 137, 515-521.

Gaharwar, A. K., Patel, A., Dolatshahi-Pirous, A., Zhang, H., Rangarajan, K., Iviglia, G., Shin, S. R., Hussain, M. A., and Khademhosseini, A. (2015). Elastomeric nanocomposite hydrogels made from Poly (glycerol sebacate) chemically crosslinked with carbon nanotubes. *Biomaterials Science*, 111, 46-58.

Gaharwar, A. K., Peppas, N. A., and Khademhossini, A. (2014). Nanocomposite hydrogels for biomedical applications. *Biotechnology and Bioengineering*, 111, 441-453.

Ganguly, S., and Margel, S. (2021). Design of magnetic hydrogels for hyperthermia and drug delivery. *Polymers*. 13, 4259.

Haraguchi, K. (2007). Nanocomposite hydrogels. *Current Opinion in Solid State and Materials Science*, 11, 47-54.

Hernandez, L. H., Souza, J., and Appel, E. A. (2021). A quantitative description for designing the extrudability of shear-thinning physical hydrogels. *Macromolecular Biosciences*, 21, 2000295-2000300.

Hu, X., Nian. G., Liang, X., Wu, L., Yin, T., Lu, H., Qu, S., and Yang, W. (2019). Adhesive tough magnetic hydrogels with high Fe_3O_4 content. *ACS Applied Materials and Interfaces*. 11, 10292-10300.

Jin, R. M., Yang, J., Zhao, D. H., Hou, X., Li, C. Q., Chen, W., Zhao, Y. D., Yin, Z. Y., and Liu, B. (2019). Hollow gold nanoshells-incorporated injectable genetically engineered hydrogel for sustained chemo-photothermal therapy of tumor. *Journal of Nanobiotechnology*, 17, 1-16.

Jung, S., Lee, S. H., Mok, H., Chung, H. J., and Park, T. G. (2010). Gene silencing efficiency of siRNA-PEGconjugates: effect of PEGylation site and PEG molecular weight. *Journal of Controlled Release*, 144, 306-313.

Kasinski, A., Zielinska-Pisklak, M., Oledzka, E., and Sobczak, M. (2020). Smart hydrogels. Synthetic stimuli-responsive antitumor drug release systems. *International Journal of Nanomedicine*, 15, 4541-4572.

Kumar, S., Rani, R., Dilbarghi, N., Tankeshwar, K., and Kim, K. H. (2017). Carbon nanotubes: a novel material for multifaceted applications in human healthcare. *Chemical Society Reviews*, 46, 158-196.

Li, J., and Mooney, D. J. (2016). Designing hydrogels for controlled drug delivery. *Nature Review Materials*, 1, 16071-16079.

Li, Y., Fang, M., Zhang, J., Wang, J., Song, Y., Shi, J., Li, W., Wu, G., Ren, J., Wang, Z., Zou, W., and Wang, L. (2016). Hydrogel duel delivered celecoxib and anti-PD-1 synergistically improve antitumor immunity. *Oncolimmunology*, 5, e1074374.

Ma, J., Wang, B., Shao, H., Zhang, S., Chen, X., Li, F., and Liang, W. (2022). Hydrogels for localized chemotherapy of liver cancer: a possible strategy for improved and safe liver cancer treatment. *Drug Delivery*, 29, 1457-1476.

Miwa, Y., Zinchenko, A., Lopatina, L. I., Sergeyev, V. G., and Murata, S. (2014). Size control of gold nanoparticles synthesized in a DNA hydrogel. *Polymer International*, 63, 1566-1571.

Mohan, M. Y., Lee, K., Premkumar T., and Geckeler, K. E. (2007). Hydrogels as nanoreactors - a novel approach to silver nanoparticles for antibacterial applications. *Polymer*, 48, 158-164.

Na, K., Park. K. H., Kim. S. W., and Bae, Y. H. (2000). Self-assembled hydrogel nanoparticles from curdlan derivatives: characterization, anti-cancer drug release and interaction with a hepatoma cell line (HepG2). *Journal of Controlled Release*, 69, 225-236.

Nieto, C., Vega, M., Rodríguez, V., Pérez-Esteban, P., and del Valle, E. (2022). Biodegradable gellan gum hydrogels loaded with paclitaxel for HER2+ breast cancer local therapy. *Carbohydrate Polymers*, 294, 119732-119747.

Norouzi, M., Nazzari, B., and Miller, D. W. (2016). Injectable hydrogel-based drug delivery systems for local cancer therapy. *Drug Discovery Today*, 21, 1835-1849.

Oral, E., and Peppas, N. A. (2004). Responsive and recognitive hydrogels using star polymers. *Journal of Biomedical Materials Research Part A*, 68A, 439-447.

Ovir, T., Timko, B. P., Brigham, M. D., Naik, S. R., Karajanagi, S. S., Levy, O., Jin, H., Parker, K. K., Langer, R., and Kahone, D. S. (2011). Nanowired three-dimensional cardiac patches. *Nature Nanotechnology*, 6, 720-725.

Paul, A., Hasan, A., Kindi, H. A., Gaharawar, A. K., Rao, V. T., Nikkhah, M., Shin, S. R., Krafft, D., Dokemesi, M. R., Shum-Tim, D., and Khademhosseini, A. (2014). Injectable graphene oxide /hydrogel-based angiogenic gene delivery systems for vasculogenic and cardiac repair. *ACS Nano*, 8, 8050-8062.

Paxton, N., Smolan, W., Bock, T., Melchels, F., Groll. J., and Jungst, T. (2017). Proposal to assess printability of bioinks for extrusion-based bioprinting and evaluation of rheological properties governing bio printability. *Biofabrication*, 9, 044107.

Peng, C. L., Shih, Y. H., Linag, K. S., Chiang, P. F., Yeh, C. H., Tang, T. C., Yao, C. J., Lee, S. Y., Luo, T. Y., and Shieh, M. J. (2013). Development of *in situ* forming thermosensitive hydrogel for radiotherapy combined with chemotherapy in a mouse model of hepatocellular carcinoma. *Molecular Pharmaceutics*, 10, 1854-1864.

Peppas, N. A., Bures, P., Leobandung, W., and Ichikawa, S. (2000). Hydrogels in pharmaceutical formulations. *European Journal of Pharmacy and Biopharmacy*, 50, 27-46.

Qian, K. Y., Song, Y., Yan, X., Dong, L., Xue, J., Xu, Y., Wang, B., Cao, B., Hou, Q., Peng, W., Hu, J., Jiang, K., Chen, S., Wang, H., and Lu, Y. (2020). Injectable ferrimagnetic silk fibroin hydrogel for magnetic hyperthermia ablation of deep tumor. *Biomaterials*, 259, 234-247.

Rafieian, S., Mirzadeh, H., Mahdavi, H., and Masoumi, M. E. (2019). A review on nanocomposite hydrogels and their biomedical applications. *Science and Engineering Composite Materials*, 26, 154-174.

Rosenblum, D., Joshi, N., Tao, W., Karp, J. M., and Peer, D. (2018). Progress and challenges towards targeted delivery of cancer therapeutics. *Nature Communications*, 9, 1-12.

Ruan, H., Hu, Q., Wen, D., Chen, Q., Chen, G., Lu, Y., Wang, J., Cheng, H., Lu, W., and Gu, Z. (2019). A dual-bioresponsive drug-delivery depot for combination of epigenetic modulation and immune checkpoint blockade. *Advanced Materials,* 31, 1806957.

Salimi, F., Dilmaghani, K. A., Alzadeh, E., Akbarzadeh, A., and Davaran, S. (2018). Enhancing cisplatin delivery to hepatocellular carcinoma HepG2 cells using dual sensitive smart nanocomposite. *Artificial Cells, Nanomedicine, and Biotechnology*, 46, 949-958.

Sellergren, B. (1997). Noncovalent molecular imprinting antibody-like molecular recognition in polymeric network materials. *Trends in Analytical Chemistry*, 16, 310-320.

Sivakumaran, D., Maitland, D., and Hoare, T. (2011). Injectable microgel-hydrogel composites for prolonged small-molecule drug delivery. *Biomacromolecules*, 12, 4112-4120.

Torre, L., Siegel, R., Ward, E., and Jemal, A. (2016). Global cancer incidence and mortality rates and trends - an update. *Cancer Epidemiology, Biomarkers & Prevention*, 25, 16 - 27.

Tsirigotis- Maniecka, M., Szyk-Warazyńska, L., Maniecki, Ł., Szczesna, W., Warzynski, P., and Wilk, K. A. (2021). Tailoring the composition of hydrogel particles for the controlled delivery of phytopharmaceuticals. *European Polymer Journal,* 151, 110429-110438.

Turchenuik, K., Boukherroub, R., and Szuneritis, S. (2015). Gold-graphene nanocomposites for sensing and biomedical applications. *Journal of Materials Chemistry B*, 3, 4301-4324.

Wang, C., Wang, J., Zhang, X., Yu, S., Wen, D., Hu, Q., Ye, Y., Bomba, H., Hu, X., Liu, Z., Dotti, G., and Gu, Z. (2018). Situ formed reactive oxygen species-responsive scaffold with gemcitabine and checkpoint inhibitor for combination therapy. *Science Translation Med*icine, 429, 3682-3690.

Wang, Y., Li, B., Xu, F., Han, Z., Wei, D., Jia, D., and Zhou, Y. (2018). Tough magnetic chitosan hydrogel nanocomposites for remotely stimulated drug release. *Biomacromolecules,* 19, 3351-3360.

Wei, C., Dong, X., and Zhang, Y. (2018). Simultaneous fluorescence imaging monitoring of the programmed release of dual drugs from a hydrogel-carbon nanotube delivery system. *Sensors and Actuators B,* 3, 264-275.

Wichterle, O., and Lim, D. (1960). Hydrophilic gels for biological use. *Nature*, 185, 117-118.

Yu, S. Wang, C., Yu, J., Wang, J., Lu, Y., Zhang, Y., Zhang, X., Hu, Q., Sun, W., He, C., Chen, X., and Gu, Z. (2018). Injectable bioresponsive gel, depot for enhanced immune checkpoint blockade. *Advanced Materials*, 28, 1801527-1801538.

Zaimy, M. A., Saffarazadeh, N., Mohammadi, A., Pourghadamyari, H., Izadi, P., Sarli, A., Paschepani, S. R., Azizi, H., Torkamandi, S., and Bazzaz, J. T. (2017). New methods in the diagnosis of cancer and gene therapy of cancer based on nanoparticles. *Cancer Gene Therapy*, 24, 233-243.

Zhang, N., Xu, X., Zhang, X., Ding, Q., Xue, L., Mo, R., and Zhang, C. (2016). Nanocomposite hydrogel incorporating gold nanorods and paclitaxel-loaded chitosan micelles for combination photothermal-chemotherapy. *International Journal of Pharmacy*, 497, 210-221.

Zhang, H., Patel, A., Gaharwar, A. K., Mihaila, S. M., Iviglia, G., Mukundan, S., Bae, H., Yang, H., and Khademhosseini, A. (2013). Hyperbranched polyester hydrogels with controlled drug release and cell adhesion properties. *Biomacromolecules*, 14, 1299-1310.

Zhao, L., Zhu, L., Liu, F., Liu, C., Shan, D., Wang, Q., Zhang, C., Li, J., Liu, J., Qu, X., and Yang, Z. (2011). pH triggered injectable amphiphilic hydrogel containing doxorubicin and paclitaxel. *International Journal of Pharmacy*, 410, 83-91.

Chapter 6

Injectable Nanocomposite Hydrogels as an Emerging Platform for Cancer Therapy

Swayamprakash Patel*
Department of Pharmaceutics & Pharmaceutical Technology,
Ramanbhai Patel College of Pharmacy,
Charotar University of Science and Technology (CHARUSAT), Anand, India

Abstract

Nanotechnology has successfully lefts its footprints in the diagnosis, treatment, and mitigation of the oncological conditions of patients. Myriads of research work have been published that suggest nanotechnology-based formulations which may rein the situation in cancer patients. The present chapter has focused on the hydrogel-based injectable nanocomposite as one of the emerging approaches to cancer therapy. Patents are accentuated more to identify innovative formulations of injectable nanocomposite hydrogels. A comprehensive review of existing research on hydrogel-based injectable nanocomposite including a complete patent landscaping is incorporated in the current chapter.

Keywords: injectable nanocomposite hydrogel, synergistic therapy, tumor

6.1. Introduction

6.1.1. Injectable Nanocomposite Hydrogels

Nanocomposite hydrogel, abbreviatedly known as NC hydrogel or NCH, is basically a composite of nanomaterial and hydrogel. The term composite is generally utilized for the material made up of two or more individual components without intimately dissolving each individual of them. The resulting composite material may not have the physical or chemical properties of each individual component, but they must possess either totally different or elevated physicochemical properties as a composite material itself. Any type of nanomaterial

* Corresponding Author's Email: swayamprakashpatel.ph@charusat.ac.in.

In: Nanocomposite Hydrogels and their Emerging Applications
Editor: Mohammad Sirousazar
ISBN: 979-8-88697-675-5

can be embedded within the matrix of the hydrogel. However, the most studied nanomaterials for NHC are carbon-based nanomaterials (e.g., carbon nanotube, carbon nanosheets, and graphene oxide (GO)), metallic nanomaterials (e.g., ferric oxide magnetic nanoparticles and gold nanoparticles), polymeric nanoparticles, radioactive nanoparticles, etc. A gamut of natural and synthetic polymers has been studied for the development of injectable NHC for cancer therapy. The most important parameter for the selection of the hydrogel matrix is its biocompatibility and ability to reach and hold the payload at the tumor site for a longer duration. A few exemplary studies on injectable NCHs with their application in cancer therapy are listed in Table 6.1.

Table 6.1. Injectable NCHs with application in cancer treatment: Hydrogels and respective payloads

Hydrogel	Source	Payload	Refs.
Chitosan	Natural	Hydroxy Appetite	(Chen et al., 2016; Huang et al., 2011; Mishra et al., 2011)
		Gold Nanosheet	(Qin et al., 2018)
		Carbon Nanotube	(Gholizadeh et al., 2017)
Collagen or Gelatine	Natural	Hydroxy Appetite	(Van et al., 2016; Ma et al., 2016; Pelin et al., 2009)
		Carbon Nanotube	(Sun et al., 2017)
		Magnetic Nanoparticle	(Antman-Passig and Shefi, 2016)
Hyaluronic Acid (HA)	Natural	Gold Nanosheet	(Song et al., 2015)
Cyclodextrin	Natural	Gold Nanorods	(Xu et al., 2017)
Poly(N-isopropyl acrylamide) (PNIPAAm)	Synthetic	Polydopamine (PDA) Nanoparticles	(Xu et al., 2017; Wang, Zhao, and Yuan, 2020)
(Methoxy poly (ethylene glycol)-b-poly (ε-caprolactone-co-1,4, 8-trioxa [4.6] spiro-9-un-decanone) (mPECT)	Synthetic	Paclitaxel and Gold Nanorods	(Yin et al., 2017)
Poly(2-(diethylamino) ethyl methacrylate-co-2-hydroxyethyl methacrylate) (DMAEMA-co-HEMA)	Synthetic	PDA Nanoparticles	(Wang, Zhao, and Yuan, 2020)
Poly(NIPAAm-co-sulfobetaine methacrylate (PNS)	Synthetic	PDA Nanoparticle	(Zheng et al., 2020)
Modified Alginate	Synthetic	Molybdenum Disulfide (MoS2)-Bovine Serum Albumin (BSA) Nanoflakes	(Qi et al., 2022)
Poly(ethylene glycol-maleate-citrate) (PEGMC)	Synthetic	5-fluorouracil (5-FU), Doxorubicin (DOX)	(Gonsalves et al., 2021)

6.1.2. Challenges of Injectable NCHs

6.1.2.1. Sterilization

Being an injectable product, injectable NCH must fulfill certain requirements of parenteral preparations. Undoubtedly, the first and foremost requirement is sterility. Nevertheless, hydrogel possesses remarkable applications, sterilization of hydrogel is always impeding steps toward the development of injectable hydrogels. Sterilization of hydrogel with conventional practices may induce crosslinking, degradation, leaking of payload, or change in its three-dimensional structural properties. In the comprehensive review, Galante et al., (2018), have discussed the consequences of conventional sterilization practices on various hydrogel systems. In the case of injectable NCHs, dry or wet heat sterilization may induce irreversible gel-sol or sol-gel transformation, which renders the product useless. Study shows that the Ag nanoparticles (NPs)-loaded NCH can be sterilized with an autoclave. However, dry heat sterilization makes the product vulnerable to heat (Rafael et al., 2019).

Irradiation may have a similar impact on sterilization through heating (Escudero-Castellanos et al., 2016). Clarification or filtration process can easily remove or reduce the concentration of the payload embedded in the form of nanomaterial from the NCH. Sterilization through the chemical agent can disturb the delicate balance of composition and may induce or deduce the crosslinking of the NCH. It may also alter the delicate composition of the payload. In one study, compared to gamma irradiation and ethylene oxide treatment, ethylene oxide gives satisfactory results for polyvinyl alcohol (PVA) hydrogel, in terms of immune response (Pohan et al., 2020). Removal of chemical agents after sterilization also seems impractical. However, ethanol treatment is yet considered the best choice for the sterilization of hydrogel-based materials (Huebsch, Gilbert, and Healy, 2005; Escudero-Castellanos et al., 2016). Conclusively, each hydrogel system requires a separate optimized sterilization technique.

6.1.2.2. Pyrogenicity and Biocompatibility

Being injectable, NCHs must have biocompatibility and biodegradability and must not show any pyrogenic activity. Few hydrogel polymers like PVA, collagen, and HA do not show any pyrogenic responses (Peng, Peng, and Shen, 2011; Kim, Choi, and Lee, 2010). Most of the time, for the development of injectable NCH, biocompatible and biodegradable components are selected with minimum possibilities of biological shock. However, a critical review on the SpaceOAR Hydrogel® system revealed anaphylactic responses in few cases (Aminsharifi et al., 2019).

6.1.3. Aspect of Cancer for NHCs

Localized tumors are the best choice of condition to possibly treat with injectable NCHs. The localized tumor can be best treated with chemotherapy and surgical removal of the tumor. However, anatomical and physiological complications, sometimes, do not allow surgical removal of the tumor. Recurrence of the tumor after surgical removal is significant due to residual tumor cells. Antineoplastic agents are the only hope for the treatment of unremovable and recurrent tumors. Radiation and heating can bolster the cytotoxic impact of chemotherapeutic agents. Nano-pharmaceuticals have well-proven applications in the treatment of cancer. Myriads of nanomaterials have been studied to the date for effective treatment of cancer. The major advantage of nanomaterials is their capability for drug targeting and, thus, the low magnitude of adverse effects. Consequently, less amount of dose as well as the rare requirement of adverse effect management lowers the cost of treatment.

A tremendous amount of research work has been carried out to date for the effective targeting of nanomaterials to the tumor site. Many of these nanomaterials achieve drug targeting to a remarkable extent. Satisfactory targeting, nevertheless, is yet not possible. These nanomaterials are mostly introduced through intravenous routes for targeting the tumor. This approach leads to many complications.

Peripheral circulation of nanomaterials leads, before reaching the target site, allows the reticuloendothelial system to engulf the nanomaterials considering them as foreign particles. Unintended distribution and entrapment of nanomaterials reduce the concentration of chemotherapeutic agents at the target site.

Hydrogels have tremendous applications in the field of drug delivery systems. The combined effects of hydrogel and nano-pharmaceuticals as NCH can revolutionize the treatment strategies for cancer. Hydrogel allows releasing nano-pharmaceuticals in a controlled manner.

Stimuli-responsive gel-to-sol and sol-to-gel transformations allow to development of NCH-based systems for localized delivery around the tumor. *In situ* gelation technology allows injection of the NCH within or around the tumor as localized drug delivery. Extremely poor delivery of nano-pharmaceuticals in the form of infusion/injection at the tumor sites due to reticuloendothelial system (RES) clearance, inability to inject residual tumor sites after surgery, the limited blood supply to the tumor, deterred by other tissues, etc. can be addressed substantially by injectable NCHs (Xu et al., 2016).

As shown in Figure 6.1, localized injection of nanomaterial-embedded hydrogel in its sol state concentrates the payload in the extreme vicinity of the tumor. Under the influence of various types of stimuli, NCH transforms immediately from sol to gel state and confirms its presence around the tumor for a longer duration. The physicochemical property of the payload, interaction type of nanomaterial with the NCH matrix, and conformation properties of the NCH determine the residence time of NCH around the tumor. Controlled release of antineoplastic agents from NCH and nanomaterial matrix creates a continuous accumulation of drug inside the tumor cells and eventually lead to cell death. Multi-platform therapy can also be possible with injectable NCHs. Hydrogel embedded with radioisotopes, anticancer agents, and heat-boosting particles can strike tumor with three different approaches. Radiation and cytotoxic effects of radioisotopes and anticancer agents, respectively result in cell death or prevent its proliferation.

On another hand, near-infrared (NIR) stimulation for hydrogel heats up the metallic nanoparticles within the hydrogel. This localized heat acts against the tumor and bolsters the action of tumor destruction.

6.1.4. Stimuli-Responsive Injectable NCHs

Localized delivery of nanomaterial (payload) embedded NCH around the tumor is only possible due to the ability of the NCH to model its 3D conformation against various stimuli. Without stimuli-responsive hydrogel, it is nearly impossible to hold NCH around the tumor.

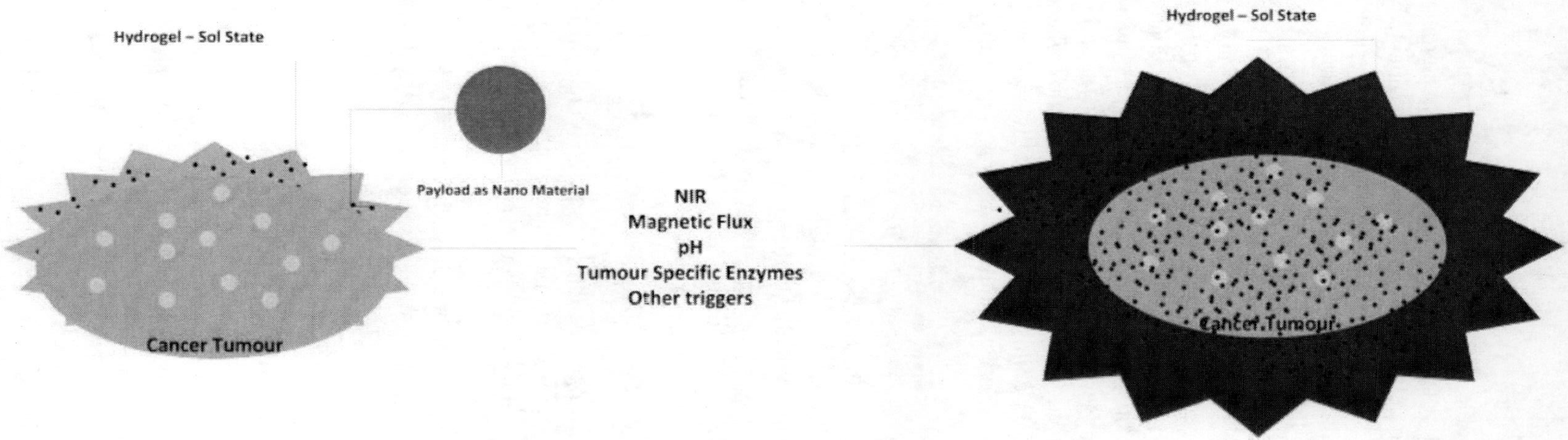

Figure 6.1. Working principle of injectable NCH against cancer.

6.1.4.1. Light as Stimulus

Light is perhaps the most acceptable stimulus because of its properties of non-invasive, high spatial/temporal resolution, and low contamination. Light-sensitive groups must be covalently or non-covalently intervened into gelators to produce light-controllable NCHs. These groups are azobenzenes, diarylethenes, and spiropyrans, photo-reversible dimerization groups like coumarins, anthracenes, pyrimidine derivatives, and photolabile groups (Li et al., 2022).

6.1.4.2. Thermal Stimulus

Thermo-sensitive NCH changes its structural configuration by varying temperature. This response occurs at lower and upper critical solution temperatures of the hydrogel, LCST and UCST, respectively. In the LCST hydrogel system, the hydrogel responds as phase separation at and above LCST, and vice versa. While in the case of UCST, the hydrogel system forms a sing phase at and above UCST, and vice versa. The most common thermo-responsive hydrogels studied for their biomedical application are PNIPAAm, butyl methacrylate, propyl acrylic acid, Pluronic F127 (PF127), glycol chitosan, HA, polyethylene oxide, methylcellulose, carboxymethyl cellulose (CMC), gelatine, chitosan, dextran, poly (ε-lysine), α-cyclodextrin, and xyloglucan.

6.1.4.3. Magnetic or Electrical Field Stimuli

Iron oxide nanoparticles with paramagnetic characteristics are typically integrated into NCHs to create magnetic-sensitive hydrogels. When exposed to a magnetic field, they vibrate and can significantly raise local temperatures, enhancing therapeutic efficacy through thermal-ablation mechanisms. Furthermore, these systems are frequently connected to thermo-sensitive hydrogels, which release drugs when the temperature rises, resulting in a synergistic effect between thermal and chemotherapeutic cytotoxicity.

6.1.4.4. pH Stimulus

Another common stimulus for hydrogel regulation is pH. To create pH-responsive hydrogel networks, a variety of crosslinks that create or disintegrate at particular pH values can be exploited, such as Schiff base crosslinks, acylhydrazone crosslinks, boronate-catechol ligands, and ionizable groups.

Most studied pH-responsive hydrogels are polyacrylic acid (PAA), poly(l-glutamic acid), albumin, methacrylate, Poly(2-(dimethylamino)ethyl methacrylate) (PDMAEMA), chitosan, and polyethylene oxide.

6.2. Current Research on Injectable NCHs

The present section embraces various research and patent literature to understand the contribution of NCHs in cancer therapy. The topic is more accentuated on injectable NCHs in the treatment of cancers. A preliminary search for the patent and non-patent literature revealed a huge number of publications on injectable NCHs for the specific treatment of cancers. Non-patent literatures were more than 700 cases on PubMed® (as of August 2022), and more than 50 patents were found in PatSeer Pro®, a commercial patent search and analysis tool. To this substantial number of literature and to precisely reduce the task of reviewing, the "text-mining" technique was adopted (Patel et al., 2021, 2022a, 2022b). Data from PubMed® and PatSeer Pro® were collected for the non-patent and patent literature which were subjected to the text-mining for further study and analysis of individual technical clusters. Search terminology for PubMed® was "nanocomposite hydrogel injectable". On another hand, the search string for the PatSeer Pro® was ((nanocomposite) and (hydrogel) and (inject*)) for the title, abstract, and claim fields only. The present section covers a detailed review of more than 700 research/review articles along with more than 50 patents. Text-mining and analysis was performed using Orange Canvas program version 3.32.0. The process of text-mining is briefly depicted in Figure 6.2.

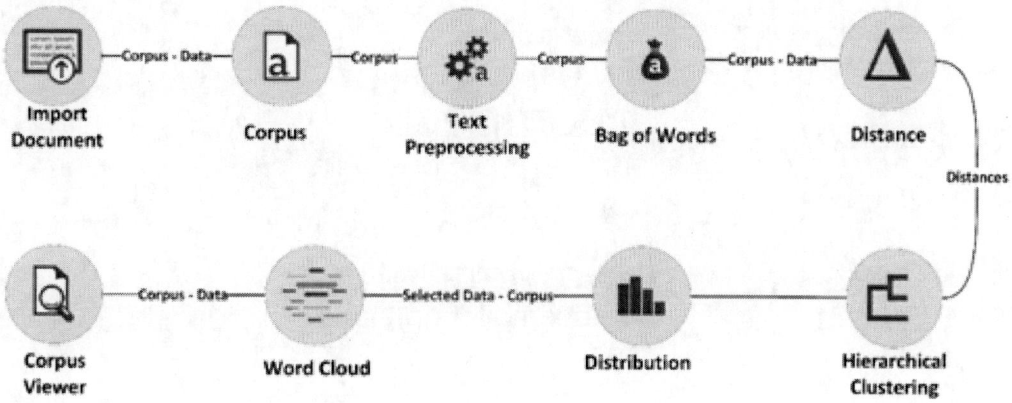

Figure 6.2. Workflow for text-mining of scientific literature[1].

6.3. Injectable NCHs in Non-Patent Literature

The PubMed® retrieval tool of the Orange Canvas program retrieved 742 literatures (Section, Title, and Abstract) with the search term "nanocomposite hydrogel injectable". This includes the data from the year 2006 to 2022. Literatures focused on "cancer" were filtered out using the filter option for further cluster formation and analysis. Text-mining has revealed plenty of different technological clusters. Each of these clusters represents the domain of injectable NCH in the treatment of cancer. Selected research work from each cluster is discussed in an elaborative manner.

6.3.1. mPECT-Based Injectable NCHs

As shown in Figure 6.3, the word cloud for the first cluster shows the highest occurrence of the word "mPECT" and embraces few research works. mPECT is an abbreviated form for a novel controlled release polymer, i.e., (methoxy poly (ethylene glycol)-b-poly(ε-caprolactone-co-

[1] Corpus - Collection of text data (Title, Abstract, and Claim); Pre-process text - Remove nontechnical and grammatical words, punctuations, assents, etc.; Bag of word - breaks sentences in words; Distance – Calculate occurrence of common word in document and accordingly calculate distance/relevance among document; Hierarchical cClustering – Clusters relevant document; Word cloud – Generate list of common words in cluster.

1,4,8-trioxa[4.6]spiro-9-un-decanone). mPECT is an outstanding amphiphilic block polymer that has the capability to self-assemble into the nanoparticle and entrap hydrophobic drug molecules. mPECT nanoparticles can be loaded in the saline at a higher concentration of up to 40 wt.% (Yin et al., 2017).

Figure 6.3. Word cloud generated for cluster 1.

In one of the outstanding works by Don et al., the paclitaxel (PTX)-mPECT NPs and the gold (Au) nanorod (NR)-mPECT loaded NCH have been presented. The NCH containing PTX-mPECT NPs and AU-NR-mPECT was prepared using cyclodextrin hydrogel (Liu et al., 2019), (Figure 6.4).

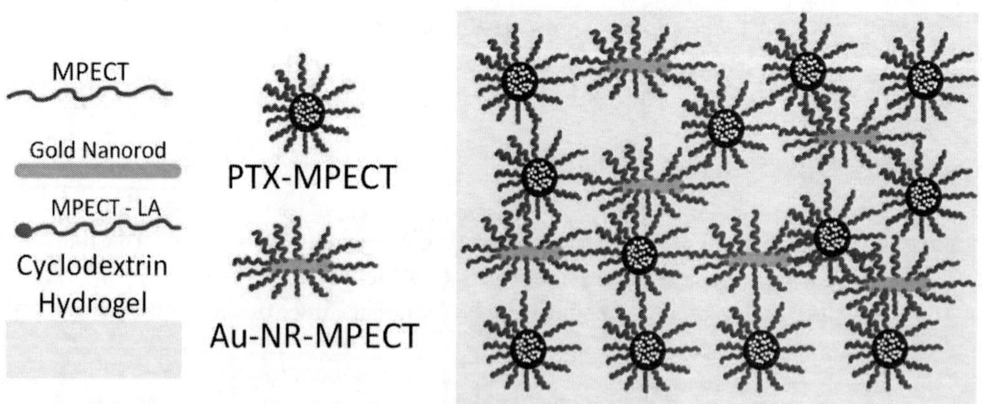

Figure 6.4. NCH prepared from PTX-mPECT NPs and Au-NR-mPECT.

The hydrogel composite was prepared through host-guest inclusion. α-cyclodextrin is proposed to be attached to the PEG (PEG) chains on the surface of PTX-mPECT and Au-NR-mPECT. The overall concept of this formulation is localized injection of PTX-mPECT and Au-NR-mPECT in the tumor which eventually gets gelatinized under influence of body temperature. The controlled release of the paclitaxel and infrared radiated thermal impact of gold nanoparticles suppresses the tumor progression. *In vivo* study for the treatment of induced breast cancer in rats was carried out to showcase the effectiveness of the concept and formulation. A similar concept was also demonstrated in another work, wherein the DOX and siBcl-2 nanoparticles were clubbed with the α-cyclodextrin as an NCH for controlled delivery of the DOX and siRNA for curbing the progression of the tumor. mPECT and cyclodextrin have collectively opened a newer dimension for the successful delivery of nanocomposites through the hydrogel inclusion at the site of the tumor by injection.

6.3.2. PDA-Based Injectable NCHs

One of the biggest advantages of NCH is its capability for *in situ* transformation from sol state to gel state. Most of the injectable hydrogel is in the sol state and can be converted into the gel state after injection at the site of application. This sol-to-gel transformation is best achievable due to thermosensitive polymers. However, this stimulus-response can be modulated in various ways by the incorporation of other materials or nanoparticles. PDA (polydopamine) nanoparticles while incorporated into a hydrogel, like PNIPAAm, change the behavior of hydrogel under exerting the stimulus. In presence of PDA, PNIPAAm gets influenced by the NIR radiation. So, PDA/PNIPAAm hydrogel exerts the phase transitions and volume changes under influence of NIR radiation. Specifically for cancer, PDA nanoparticles also aid photothermal response under NIR radiation.

In one of the works by Wang et al., (2020), NCH based on modified DMAEMA-co-HEMA containing PDA NPs was developed for the controlled delivery of the DOX. Under NIR irradiation, induced localized hyperthermia due to PDA NPs, the hydrophilicity-hydrophobicity transition of hydrogel occurs. This transition leads to the squeezing of the DOX. This work presented the best example for controlling the localized release of anticancer drug nanoparticles in the vicinity of the tumor under the NIR stimulation. In a similar interesting work, Zheng et al., (2020) developed an injectable composite hydrogel of poly (NIPAAm-co-sulfobetaine methacrylate) (PNS) containing anticancer drug and PDA NPs. The combined action of chemotherapy and the photothermal effect of PDA NPs have a synergistic effect on tumor regression. However, interestingly, the zwitterionic PNS hydrogel system avoids the formation of biofouling (adsorption of proteins and other components) around the surface of the hydrogel. So, PNS hydrogel containing PDA NPs and chemotherapeutic drugs seems even more effective than non-zwitterionic hydrogel.

6.3.3. Synergy of Injectable NCH and Photothermal Effect

Many research articles state the synergistic impact of injectable NCH containing drug-nano-pharmaceuticals and nanoparticles with a photothermal effect. This is only possible due to the injectability of hydrogel-containing nanomaterials. Localized release of drug as well as localized thermal effect synergistically reduces the tumor progression. The self-healing ability and sheer thinning properties of hydrogel play an important role in the successful development of injectable NCH (Liu and Hsu, 2018; Zheng et al., 2022; Chen et al., 2019). Qi et al., (2021) published their results with successful tumor regression due to the synergistic effect of the GO nanoparticles and needle-like hydroxyapatite (HAP). Aldehyde-modified PEG and CMC-based hydrogel with self-healing capability can carry a high concentration of GO NP and HAP. This PEG-CMC/GO/HAP had a synergistic impact on the treatment of breast cancer in mice. In another similar work, Qi et al., (2022) employed molybdenum disulfide (MoS_2)-bovine serum albumin (BSA) nanoflakes to exhibit synergistic anticancer impact under NIR radiation of 808 nm. Schiff base bond formation between oxidized sodium alginate (OSA) and hydroxypropyl chitosan (HPC) resulted in OSA-HPC injectable hydrogel.

Triple-action synergistic anticancer activity has been also projected through dual drug incorporation in the injectable hydrogel. Xiao et al., (2022) incorporated glucose oxidase (GOx)-conjugated polyacrylic PAA-stabilized iron oxide nanoparticles (GPI NPs) and Toll-like receptor 7/8 agonist resiquimod (R848) in sodium alginate-based hydrogel system. In starvation therapy, localized accumulation of GOx catalyzes the conversion of glucose into gluconic acid and hydrogen peroxide. This paucity of glucose starves the tumor cell to death. The chemo-dynamic treatment disrupts the tumor by using Fenton/Fenton-like reactions. The acidic condition of lysosomes promotes the disruption of hydrogen peroxide in cytotoxic reactive species. This process is further boosted by several folds in the presence of Fe^{+2} and Fe^{+3}. The resiquimod (R848), a toll-like receptor 7/8 agonist, may change the tumor microenvironment's (TME) functional orientation by repolarizing the M2-like phenotype toward an M1-like phenotype. By boosting the release of M1-related proinflammatory cytokines, such as tumor necrosis factor (TNF), interleukin-6 (IL-6), and IL-12, this transition may effectively lower the proportions of tumor-associated macrophages (TAMs), relieve the tumor-supportive microenvironment, and further impede the advancement of the tumor. The photothermal effect is combined with the ^{121}I-induced brachytherapy in one of the research works (Wu et al., 2022). PEG-coated gold nanorods induce photothermal impact under NIR radiation. While RGDY peptide labeled with ^{121}I exsert its cytotoxic effect through radioactivity.

Several research works with different photothermal aiding nanoparticle embedded hydrogels showed the synergistic impact on tumor cells. This photothermal aiding nanoparticle includes carbon-based NPs, black phosphorus nanosheets, semiconductor nanostructure, and organic NPs (Wang et al., 2016a, 2016b; Zhou et al., 2015; Zhang et al., 2015; Yang et al., 2018; Xing et al., 2018; Shao et al., 2018; Wu et al., 2018a). Innovatively, the dendrimer-encapsulated platinum nanoparticles are immobilized through the alginate-calcium hydrogel system.

Localized injection of this dendrimer encapsulated platinum with alginate hydrogel serves its purpose for efficiently concentrating metal nanoparticles around the tumor. On-demand injection of chelating agents around the hydrogel will initiate the gel-to-sol transformation of immobilized alginate.

The chelating agent forms chelates with Ca^{+2} ions thus reducing the viscosity of the alginate hydrogel. This process allows dendrimer-encapsulated platinum oxides to enter tumor cells. The photothermal effect under a high-intensity NIR beam causes tumor cells to die under influence of heat.

6.3.4. pH-Responsive Injectable NCHs

Like photothermal stimulus, pH-responsive attributes of the hydrogel and polymer can be used to develop injectable NCH with nanoparticles. A pH-responsive hydrogel composed of nanoparticles of the poly (D,L-lactic acid-co-glycolic acid) (PLGA) as the core containing the DOX and 5-fluorouracil (5-FU) with the carboxy methyl chitosan as the shell has been prepared. Nanoparticles were embedded in a hydrogel system composed of poly (ethylene glycol-maleate-citrate) (PEGMC) polymer. PEGMC consists of PEG, maleic acid, and citric acid which are most compatible with the skin environment. The environment within and in the vicinity of the cancer cells offers low pH that triggers the release of 5-FU and DOX from PLGA-CMC NPs in a controlled manner. This formulation was claimed to be most effective against the reoccurrence of skin cancer after surgery (Gonsalves et al., 2021).

An ultrasensitive pH-responsive injectable NCH was prepared to deliver self-healing hydrogel in the area with minor pH changes, like tumor and wound. (Wu et al., 2020). The sol-to-gel transformation occurs only at a pH change of 0.2. This supersensitive pH-responsive hydrogel is based on the Schiff base reaction between aldehyde-containing copolymers and amine-modified silica nanoparticles. Using an aromatic aldehyde-functionalized monomer, 4-formylbenzoate ethyl methacrylate (FBEMA), and 2-methacryloyloxyethyl phosphorylcholine (MPC), the copolymer was created. 3-Aminopropyl triethoxysilane (APTES)-modified silica nanoparticles (ASNP) system was employed as a drug delivery platform to modify the mechanical properties of the hydrogel.

An injectable NCH with multiple functionalities of pH responsiveness, tumor targeting, fluorescent probe, and enzyme-triggered drug release was developed and presented by Chen and Liu (2016). The DOX-loaded HA and a fluorescein isothiocyanate (FITC)-conjugated mesoporous silica nanocomposite (MSN) hydrogel was proposed for multiple actions of tumor targeting and controlled drug release under influence of pH change. The paclitaxel-loaded dual responsive nanogel was presented in one of the intriguing works (Chen et al., 2014). The methoxy poly(ethylene glycol) 2000 (mPEG2000)-isopropylideneglycerol polymer carries both the pH-responsive and hydrophilic moieties. Spontaneous formation of nano-micelles entraps the paclitaxel and can form a gel at body temperature. Under the influence of a mild acidic environment, pH-responsive gel deteriorates and releases the paclitaxel in a controlled manner.

6.3.5. Magnetic Field-Responsive Injectable NCHs

Hydrogel made up of α-cyclodextrin combined with PEGylated iron oxide can transform their gel-sol-gel transformation under an alternating current magnetic field (ACMF). The influence of ACMF gel-to-sol transformation allows the hydrogel to cover irregular contours of the postoperative wound in the breast cancer. The heat generated under ACMF causes thermal damage to the cancer cell. The DOX or paclitaxel incorporated in the hydrogel releases in a controlled manner in absence of ACMF (Wu et al., 2018b). This kind of arrangement allows complete coverage of the irregular surface of the tissue, thermal treatment of cancer, and controlled release of chemotherapeutics. In another similar work, Bacillus Calmette-Guerin (BCG) was targeted to the bladder for the treatment of cancer using magnetic nanoparticles. As per the study, the hydrogel system can lock BCG within the bladder for longer action (Zhang et al., 2013).

6.3.6. Prodrug- and Vaccine-Loaded Injectable NCH

An interesting strategy for the delivery of anticancer drugs to the target cells in the form of prodrugs was developed and comprehensively studied by Ono et al., (2021). An enzyme-prodrug therapy system was formulated using NCH comprised of PLGA-PEG-PLGA copolymer, laponite, and β-galactosidase (β-gal). 5-FU-β-gal prodrug remains well preserved inside the hydrogel matrix for longer periods and gets activated to 5-FU as soon as enters the tumor cell. In a synergistic approach of prodrug and photothermal effect, an NCH based on DOX and gold nanorods has been produced. Gold nanorods and adamantane-modified DOX (AD-DOX) prodrug were embedded in the polymer matrix of poly (NIPAm-co-MPCD). In presence of an acidic environment around the tumor cell, the acid labile hydrazone bond between adamantane and DOX undergoes cleavage and releases DOX. NIR influences the gold nanorods and creates a localized thermal impact on tumor cells.

Ultrasound-mediated gel-to-sol transformation of hydrogels can be employed for the on-demand delivery of payload. Tumor vaccine can be delivered on-demand when embedded inside the hydrogel. Hydrogel-based vaccine delivery and experiments have been set up with mice (Meng et al., 2021). Ultrasound-mediated gel-to-sol transformation results in burst release of vaccine followed by sol-to-gel transformation. Hydrogel acts as a vaccine reservoir and can be utilized as an on-demand switch on/off system for booster vaccine delivery.

6.4. Injectable NCHs in Patent Literature

Patent literatures on injectable NCH for the treatment of tumor were identified using the search string ((nanocomposite) and (hydrogel) and (inject*)) in the PatSeer Pro® patent search and analysis tool. More than fifty patents on NCH were found, however, out of these very few patents have focused on the treatment of tumors. In this section, these innovative and intriguing approaches are discussed in detail.

IN201931004799A is an Indian patent application submitted by the Department of Biosciences and Bioengineering, Indian Institute of Technology (IIT), Guwahati. It discloses the silk hydrogel-based NCH for targeted and controlled release of chemotherapeutics. The silk fibroin hydrogel was prepared from the blend of cocoon SF protein of mulberry *Bombyx mori* (BM) and gland SF protein of non-mulberry *Antheraea assama* (AA) without the aid of any crosslinkers. The patent has claimed stimuli-responsive and non-stimuli-responsive nanoparticles which are embedded in the SF hydrogel. These nanoparticles are selected from the group consisting of carbon nanotube, gold nanoparticles, GO NPs, polypyrrole NPs, (as stimuli-responsive NPs), mulberry silk fibroin NPs, and non-mulberry silk fibroin NPs (as non-stimuli responsive NPs). Nanoparticles were decorated with functionalizing agents like folic acid, RGD peptides, and aptamers for targeted delivery of chemotherapeutics. The patent has more specifically disclosed the single-walled carbon nanotubes functionalized with folic acid (FA) and loaded with DOX. The inventor has also disclosed Bombyx mori nanoparticles (nonfunctionalized) which are loaded with cisplatin. University of Sungkyunkwan, Korea, has a patent (KR20090118629A) in their portfolio that discloses the NCH for effective delivery of the tumor necrosis factor-related apoptosis-inducing ligand (TRAIL).

TRAIL is known to induce the self-killing of various types of cancer cells and over-activated immune cells and not affect normal cells. The PEG-TRAIL/HA NCH was claimed for targeted delivery of TRAIL for the treatment of cancer and rheumatoid arthritis. The HA has not only affinity towards the cancer site, but the accumulation of hyaluronidase at the cancerous site deteriorates the HA and releases TRAIL in a controlled manner. The binding of PEG with the N-terminus of TRAIL avoids hepatic clearance. The patent discloses that the HA is conjugated with PEG/TRAIL by electrostatic attraction of HA toward the negatively charged PEG/TRAIL. In a Chinese patent application (CN114425087A), radio isotope-based NCH was disclosed for targeted delivery. The broad scope of the claims has disclosed many components in the patent application (Table 6.2).

Table 6.2. Therapeutic function groups lined with the nanocarrier and broadly claimed in the CN114425087A patent

Photothermal	Magnetocaloric	Sonodynamic	Chemotherapy	Radio Isotopes
Polydopamine, noble metal nanomaterial, metal chalcogenide nanoparticles, azobenzene, o-nitrobenzyl derivatives, quantum dots, carbon derivatives, upconversion nanoparticles, scintillation nanoparticles, afterglow luminescent nanoparticles, photosensitizer, organic dye	Magnetic core-shell nanoparticles, magnetic liposomes, iron-containing porous metal nanocapsules	Acridine compounds, dyes, and perfluorocarbon nano-emulsions	Alkylating drugs, antimetabolites, antibiotic drugs, plant antineoplastic drugs, hormone drugs, and miscellaneous drugs	^{211}At ^{225}Ac ^{213}Bi ^{188}Re ^{90}Y ^{123}I ^{125}I ^{131}I ^{64}Cu ^{67}At

The concept of injectable NCH for the delivery of various therapeutic agents is adopted in one of the Korean patent applications (KR20100093048A) for the successful delivery of flavonoids. The catechin flavonoids have been loaded within nano-micelles for embedding in hydrogel and delivering the same to the cancer cells. In another invention (CN108144074B), radioiodine-labeled melanin nanoparticles were complexed with the four-armed PEG thiol group of the formula: $C(CH)_2O(CH_2CH_2O)\ nCH_2CH_2SH)_4$ (n=5 to 100). One of the Indian patents from Amity University (IN201911011659A) has disclosed their invention for graphene quantum dots-based hydrogel composite for controlled delivery of drugs at the targeted site. Hydrogel matrix was developed by single-step free radical aqueous co-polymerization of

Acrylic acid (AAc) and 2-(dimethylamino)ethylmethacrylate (DMAEMA) with filler concentration ranging from 0.01-0.3% weight/volume without the use of any chemical crosslinkers. Due to their capacity for pH-sensitive drug delivery of a biologically active agent for site-specific and sustained drug delivery, the NCH was provided in this invention as a suitable drug carrier.

One intriguing invention in an Indian patent application (IN201911039612A) describes the method for the development of CMC-cl-poly (lactic acid- co-itaconic acid) /ZnO-Ag nanocomposite for localized delivery of the anticancer agent. Both ZnO-Ag nanoparticle and CMC-cl-poly (lactic acid-co-itaconic acid) hydrogel matrix were developed by the coprecipitation method, individually, followed by composite preparation. Patent application EP3749288A1 has disclosed the method for delivering miRNA to targeted cancerous cells through NCH. The matrix of hydrogel was prepared from two amphiphilic cationic b-hairpin peptides of amino acid sequence VKVKVKVKV D PPTKVKVKVKV-NH 2 and VLTKVKTKV D PPTKVEVKVLV-NH 2. The nucleic acid molecule is conjugated with the first peptide which is unfolded in the hydrogel matrix. miRNA-peptide nanoparticles and the second peptide (for hydrogel) are self-assembled in presence of salt at pH=7.4 to develop NCH. In the line of thermosensitive hydrogel matrix, a US patent application (US9017726B2) has disclosed a newer hydrogel matrix of poly(organophosphazene) containing ferric oxide as nanomaterial.

This NCH can be utilized as a magnetic resonance imaging (MRI) contrast agent and can also trigger the gel-sol transformation of the hydrogel matrix under NIR radiation. This novel poly (organophosphazene) hydrogel matrix is revealed in the patent application through the formula shown in Figure 6.5.

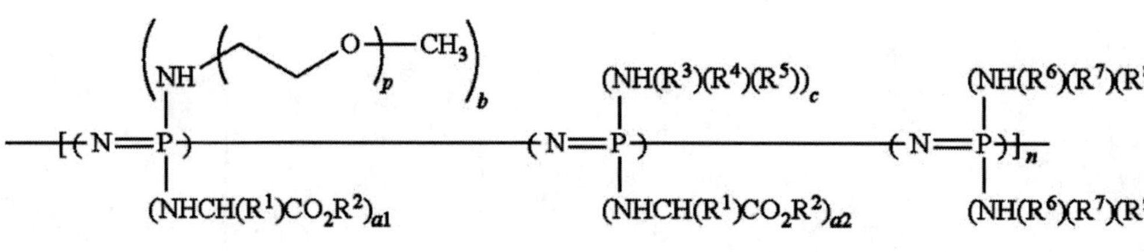

Figure 6.5. The formula of poly (organophosphazene) (US Patent US9017726B2).

Gambogic acid micelle hydrogel for the treatment of oral, lung, liver, colon, breast, stomach, pancreatic, and prostate cancers, and brain tumor, osteosarcoma, hematologic tumor, and melanoma, has been claimed in the Chinese patent application CN113952290A. Copolymerized acrylamide and acryloyl glycinamide in a water phase has been employed as a hydrogel matrix for the incorporation of dopamine-coated gold nanoparticles in patent application CN109503862B. The invention also discloses the process of preparing hydrogel matrix and nanoparticles which can go gel-to-sol transformation under NIR radiation. On the same line, the US2016184429A1 patent has disclosed an innovative approach to reduce the endotoxin level of chitosan for its utilization as a thermosensitive hydrogel matrix for the interleukin-12 in the treatment of bladder cancer.

In an interesting patent application from the Massachusetts Institute of Technology (WO2016134232A1), a theranostic nanoprobe has been disclosed to hamper multi-drug resistance during the chemotherapy of cancer. Many times, multi-drug resistance in cancer was developed due to the expression of energy-dependent transporters in cancerous cells. These transporters elevate the efflux of chemotherapeutics from the cell. This lead to a decrease in the concentration of drug inside the cancerous cell. The phosphoglycoprotein multidrug resistance protein 1 (MRPl-ABCCl) is often associated with resistance to a broad spectrum of anticancer drugs. The present invention has disclosed an NCH system containing gold nanoparticles as a probe. This prob detects the over-expression of MDR proteins and blocks the efflux of chemotherapeutic agents. This theranostic nanoprobe consists of a gold nanoparticle functionalized with at least one DNA-hairpin labeled with a fluorophore, and at least one anchor labeled with a quencher. The chemotherapeutic agent is included in another DNA-hairpin. WO2012070029A1 patent has developed an innovative process for the development of nano-micelles loaded with antineoplastic agents and antibodies like anti-MUC16, anti-MUC1, or anti-MUC4. These nano-micelles are further embedded with pH- and thermo-responsive hydrogel. The mucin antigens on cancer cells are the target of the nanomicelles. The drug-loaded nanomicelles can enter the tumor cell and deliver the chemotherapeutic medications because the anti-mucin antibodies can be absorbed by the tumor cells.

The hydrogel matrix was developed using the NIPAAm (a temperature-sensitive monomer), 2-(dimethylamino)ethyl methacrylate (DMAEMA), and poly (methacrylic acid) (PMAA), as pH-sensitive systems. In one of the Japanese patent applications (JP6387009B2), the injectable NCH for capturing cancer cells in the metastasis stage was developed and disclosed in detail.

The formulation contains nanoparticles with chemotherapeutic agents and a scaffold structure to entrap metastasis cancerous cells embedded in the hydrogel.

6.5. FDA-Approved Injectable Hydrogels and NCHs

Few of the injectable hydrogel and NCH for the treatment of cancer have been approved by the US Food and Drug Administration (FDA), which some of them are discussed herewith. Eligard® is a product developed and patented by Tolmar Pharmaceutical Inc. (US8226598B2). It is a dual injection delivery system of polymeric leuprolide acetate powder and a controlled release Atrigel® hydrogel delivery system. Eligard® is a subcutaneous injection of leuprolide acetate for the treatment of prostate cancer. Eligard® is available in four different strengths for one-month, three-months, four-months, and six-months controlled release. Under the aseptic conditions, both prefilled injections are mixed within injection only (patented technology). This mixture of hydrogel (in sol state) and polymeric particles of leuprolide acetate must be injected subcutaneously within thirty minutes of mixing (Dunn and Yarborough, 2012).

Lupron Depot® is another FDA-approved injectable hydrogel carrying an antineoplastic payload. However, it delivers microparticles of leuprolide acetate rather than nanoparticles. This product is developed by AbbVie Inc. and patented its single injection dual chamber technology for delivering the hydrogel. Sequential pushing of the plunger within a single barrel allows the mixing of the microsphere of leuprolide acetate with the hydrogel system.

Somatulin Depot® is a hexagonally arranged nanotube reservoir system for lanreotide for the treatment of gastro-entero-pancreatic neuroendocrine tumors. It is an NCH system for subcutaneous injection.

AGuIX® theranostic and treatment strategy is based on injectable gadolinium to detect tumors through MRI. High-intensity NIR beam can induce localized heat inside the tumor as the treatment strategy. NanoBiotix® technology with hafnium oxide nanoparticles works on the principle of localized radiation. These products are yet under clinical trial and are soon to be expected in the market.

Conclusion

FDA-approved hydrogels and increasing numbers of hydrogel-based marketed products shall open doors to injectable NCH systems. Clinically, injectable NCH has proven its significant impact compared to micro or macro payload hydrogel system. Though real injectable NCH has to pass many regulatory challenges to get successful marketing approval. Analytical method development, stability, delivery system, and sterilization issues are a few of the challenging components that may still hinder successful product development. In essence, injectable NCH possessed the highest curative possibilities for tumors. More research works must be focused on the analytical and scale-up challenges.

References

Aminsharifi A, Kotamarti S, Silver D and Schulman A. (2019). Major complications and adverse events related to the injection of the SpaceOAR hydrogel system before radiotherapy for prostate cancer: review of the manufacturer and user facility device experience database. *Journal of Endourology*, 33(10), 868-871.

Antman-Passig M and Shefi O. (2016). Remote magnetic orientation of 3D collagen hydrogels for directed neuronal regeneration. *Nano Letters*, 16(4), 2567-2573.

Chen D, Yu H, Sun K, Liu W and Wang H. (2014). Dual thermoresponsive and pH-responsive self-assembled micellar nanogel for anticancer drug delivery. *Drug Delivery*, 21(4), 258-64.

Chen H, Cheng R, Zhao X, Zhang Y, Tam A, Yan Y, Shen H, Zhang YS, Qi J, Feng Y, Liu L, Pan G, Cui W and Deng L. (2019). An injectable self-healing coordinative hydrogel with antibacterial and angiogenic properties for diabetic skin wound repair. *NPG Asia Materials*, 11, 3.

Chen X and Liu Z. (2016). A pH-responsive hydrogel based on a tumor-targeting mesoporous silica nanocomposite for sustained cancer labeling and therapy. *Macromolecular Rapid Communications*, 37(18), 1533-9.

Chen Y, Zhang F, Fu Q, Liu Y, Wang Z and Qi N. (2016). *In vitro* proliferation and osteogenic differentiation of human dental pulp stem cells in injectable thermo-sensitive chitosan/β-glycerophosphate/ hydroxyapatite hydrogel. *Journal of Biomaterials Applications*, 31(3), 317-327.

Dunn RL and Yarborough CL. (2012). Coupling syringe system and methods for obtaining a mixed composition. *US Patent*, US8226598B2.

Escudero-Castellanos A, Ocampo-García BE, Domínguez-García MV, Flores-Estrada J and Flores-Merino MV. (2016). Hydrogels based on poly(ethylene glycol) as scaffolds for tissue engineering application: biocompatibility assessment and effect of the sterilization process. *Journal of Materials Science: Materials in Medicine*, 27(12), 176.

Galante R, Pinto TJ, Colaco R and Serro AP. (2018). Sterilization of hydrogels for biomedical applications: A review. *Journal of Biomedical Materials Research Part B: Applied Biomaterials*, 106(6), 2472-2492.

Gholizadeh S, Moztarzadeh F, Haghighipour N, Ghazizadeh L, Baghbani F, Shokrgozar MA and Allahyari Z. (2017). Preparation and characterization of novel functionalized multiwalled carbon nanotubes/ chitosan/β-Glycerophosphate scaffolds for bone tissue engineering. *International Journal of Biological Macromolecules*, 97, 365-372.

Gonsalves A, Tambe P, Le D, Thakore D, Wadajkar AS, Yang J, Nguyen KT and Menon JU. (2021). Synthesis and characterization of a novel pH-responsive drug-releasing nanocomposite hydrogel for skin cancer therapy and wound healing. *Journal of Materials Chemistry B*, 9(46), 9533-9546.

Huang Z, Yu B, Feng Q, Li S, Chen Y and Luo L. (2011). *In situ*-forming chitosan/nano-hydroxyapatite/collagen gel for the delivery of bone marrow mesenchymal stem cells. *Carbohydrate Polymers*, 85(1), 261-267.

Huebsch N, Gilbert M and Healy KE. (2005). Analysis of sterilization protocols for peptide-modified hydrogels. *Journal of Biomedical Materials Research Part B: Applied Biomaterials*, 74B(1), 440-447.

Kim J, Choi J and Lee DY. (2010). Pyrogenicity of hyaluronic acid hydrogel cross-linked by divinyl sulfone for soft tissue augmentation. *Natural Science*, 2(7), 764-768.

Li Z, Zhou Y, Li T, Zhang J and Tian H. (2022). Stimuli-responsive hydrogels: fabrication and biomedical applications. *View*, 3(2), 20200112.

Liu M, Huang P, Wang W, Feng Z, Zhang J, Deng L and Dong A. (2019). An injectable nanocomposite hydrogel co-constructed with gold nanorods and paclitaxel-loaded nanoparticles for local chemo-photothermal synergetic cancer therapy. *Journal of Materials Chemistry B*, 7(16), 2667-2677.

Liu Y and Hsu SH. (2018). Synthesis and biomedical applications of self-healing hydrogels. *Frontiers in Chemistry*, 6, 449.

Ma X, He Z, Han F, Zhong Z, Chen L and Li B. (2016). Preparation of collagen/hydroxyapatite/alendronate hybrid hydrogels as potential scaffolds for bone regeneration. *Colloids and Surfaces B: Biointerfaces*, 143, 81-87.

Meng Z, Zhang Y, She J, Zhou X, Xu J, Han X, Wang C, Zhu M and Liu Z. (2021). Ultrasound-mediated remotely controlled nanovaccine delivery for tumor vaccination and individualized cancer immunotherapy. *Nano Letters*, 21(3), 1228-1237.

Mishra D, Bhunia B, Banerjee I, Datta P, Dhara S and Maiti TK. (2011). Enzymatically crosslinked carboxymethyl–chitosan/gelatin/nano-hydroxyapatite injectable gels for *in situ* bone tissue engineering application. *Materials Science and Engineering: C*, 31(7), 1295-1304.

Ono K, Hashimoto H, Katayama T, Ueda N and Nagahama K. (2021). Injectable biocatalytic nanocomposite hydrogel factories for focal enzyme-prodrug cancer therapy. *Biomacromolecules*, 22(10), 4217-4227.

Patel S, Patel A, Patel M, Shah U, Patel M, Solanki N and Patel S. (2021). Review and analysis of massively registered clinical trials of COVID-19 using the text mining approach. *Reviews on Recent Clinical Trials*, 16(3), 242-257.

Patel S, Patel A, Shah U, Patel M, Solanki N, Patel M and Patel S. (2022a). Data analysis and mapping of research interest in clinical trials of tuberculosis by text mining platform of artificial intelligence using open-source tool orange canvas. *The Chinese Journal of Artificial Intelligence*, 1(1), 3-17.

Patel S, Patel MS, Patel AD, Shah UH, Patel MM, Solanki ND and Patel MJ. (2022b). Probiotic formulations: A patent landscaping using the text mining approach. *Current Microbiology*, 79(5), 152.

Pelin IM, Maier SS, Chitanu GC and Bulacovschi V. (2009). Preparation and characterization of a hydroxyapatite–collagen composite as component for injectable bone substitute. *Materials Science and Engineering: C*, 29(7), 2188-2194.

Peng Z, Peng Z and Shen Y. (2011). Study on biological safety of polyvinyl alcohol/collagen hydrogel as a tissue substitute (II). *Journal of Macromolecular Science, Part A*, 48(8), 632-636.

Pohan G, Mattiassi S, Yao Y, Zaw AM, Anderson DE, Cutiongco MF, Hinds MT and Yim EK. (2020). Effect of ethylene oxide sterilization on polyvinyl alcohol hydrogel compared with gamma radiation. *Tissue Engineering Part A*, 26(19-20), 1077-1090.

Qi Y, Yuan Y, Qian Z, Ma X, Yuan W and Song Y. (2022). Injectable and self-healing polysaccharide hydrogel loading molybdenum disulfide nanoflakes for synergistic photothermal-photodynamic therapy of breast cancer. *Macromolecular Bioscience*, 22(9), e2200161.

Qi Y, Qian Z, Yuan W and Li Z. (2021). Injectable and self-healing nanocomposite hydrogel loading needle-like nano-hydroxyapatite and graphene oxide for synergistic tumour proliferation inhibition and photothermal therapy. *Journal of Materials Chemistry B*, 9(47), 9734-9743.

Qin H, Wang J, Wang T, Gao X, Wan Q and Pei X. (2018). Preparation and characterization of chitosan/β-glycerophosphate thermal-sensitive hydrogel reinforced by graphene oxide. *Frontiers in Chemistry*, 6, 565.

Rafael D, Andrade F, Martinez-Trucharte F, Basas J, Seras-Franzoso J, Palau M, Gomis X, Pérez-Burgos M, Blanco A, López-Fernández A, Vélez R, Abasolo I, Aguirre M, Gavaldà J and Schwartz S. (2019).

Sterilization procedure for temperature-sensitive hydrogels loaded with silver nanoparticles for clinical applications. *Nanomaterials*, 9(3), 380.

Shao J, Ruan C, Xie H, Li Z, Wang H, Chu PK and Yu XF. (2018). Black-phosphorus-incorporated hydrogel as a sprayable and biodegradable photothermal platform for postsurgical treatment of cancer. *Advanced Science*, 5(5), 1700848.

Song F, Hu W, Xiao L, Cao Z, Li X, Zhang C, Liao L and Liu L. (2015). Enzymatically cross-linked hyaluronic acid/graphene oxide nanocomposite hydrogel with pH-responsive release. *Journal of Biomaterials Science, Polymer Edition*, 26(6), 339-352.

Sun H, Zhou J, Huang Z, Qu L, Lin N, Liang C, Dai R, Tang L and Tian F. (2017). Carbon nanotube-incorporated collagen hydrogels improve cell alignment and the performance of cardiac constructs. *International Journal of Nanomedicine*, 12, 3109.

Van TD, Tran NQ, Nguyen DH, Nguyen CK, Tran DL and Nguyen PT. (2016). Injectable hydrogel composite based gelatin-PEG and biphasic calcium phosphate nanoparticles for bone regeneration. *Journal of Electronic Materials*, 45(5), 2415-2422.

Wang C, Wang X, Dong K, Luo J, Zhang Q and Cheng Y. (2016a). Injectable and responsively degradable hydrogel for personalized photothermal therapy. *Biomaterials*, 104, 129-37.

Wang C, Zhao N and Yuan W. (2020). NIR/thermoresponsive injectable self-healing hydrogels containing polydopamine nanoparticles for efficient synergistic cancer thermochemotherapy. *ACS Applied Materials & Interfaces*, 12(8), 9118-9131.

Wang X, Wang C, Zhang Q and Cheng Y. (2016b). Near infrared light-responsive and injectable supramolecular hydrogels for on-demand drug delivery. *Chemical Communications*, 52(5), 978-981.

Wu C, Zhao J, Hu F, Zheng Y, Yang H, Pan S, Shi S, Chen X and Wang S. (2018a). Design of injectable agar-based composite hydrogel for multi-mode tumor therapy. *Carbohydrate Polymers*, 180, 112-121.

Wu H, Song L, Chen L, Zhang W, Chen Y, Zang F, Chen H, Ma M, Gu N and Zhang Y. (2018b). Injectable magnetic supramolecular hydrogel with magnetocaloric liquid-conformal property prevents post-operative recurrence in a breast cancer model. *Acta Biomaterialia*, 74, 302-311.

Wu M, Chen J, Huang W, Yan B, Peng Q, Liu J, Chen L and Zeng H. (2020). Injectable and self-healing nanocomposite hydrogels with ultrasensitive pH-responsiveness and tunable mechanical properties: implications for controlled drug delivery. *Biomacromolecules*, 21(6), 2409-2420.

Wu Y, Yao Y, Zhang J, Gui H, Liu J and Liu J. (2022). Tumor-targeted injectable double-network hydrogel for prevention of breast cancer recurrence and wound infection via synergistic photothermal and brachytherapy. *Advanced Science*, 24(9), 2200681.

Xiao T, Zhu J, Jia L, Che H, Liu J, Deckers J, van Hest JCM and Shi X. (2022). Injectable alginate hydrogels for synergistic tumor combination therapy through repolarization of tumor-associated macrophages. *Journal of Controlled Release*, 348, 239-249.

Xing C, Chen S, Qiu M, Liang X, Liu Q, Zou Q, Li Z, Xie Z, Wang D, Dong B, Liu L, Fan D and Zhang H. (2018). Conceptually novel black phosphorus/cellulose hydrogels as promising photothermal agents for effective cancer therapy. *Advanced Healthcare Materials*, 7(7), e1701510.

Xu S, Yin L, Xiang Y, Deng H, Deng L, Fan H, Tang H, Zhang J and Dong A. (2016). Supramolecular hydrogel from nanoparticles and cyclodextrins for local and sustained nanoparticle delivery. *Macromolecular Bioscience*, 16(8), 1188-99.

Xu X, Huang Z, Huang Z, Zhang X, He S, Sun X, Shen Y, Yan M and Zhao C. (2017). Injectable, NIR/pH-responsive nanocomposite hydrogel as long-acting implant for chemophotothermal synergistic cancer therapy. *ACS Applied Materials & Interfaces*, 9(24), 20361-20375.

Yang G, Wan X, Gu Z, Zeng X and Tang J. (2018). Near infrared photothermal-responsive poly(vinyl alcohol)/black phosphorus composite hydrogels with excellent on-demand drug release capacity. *Journal of Materials Chemistry B*, 6(11), 1622-1632.

Yin L, Xu S, Feng Z, Deng H, Zhang J, Gao H, Deng L, Tang H and Dong A. (2017). Supramolecular hydrogel based on high-solid-content mPECT nanoparticles and cyclodextrins for local and sustained drug delivery. *Biomaterials Science*, 5(4), 698-706.

Zhang D, Sun P, Li P, Xue A, Zhang X, Zhang H and Jin X. (2013). A magnetic chitosan hydrogel for sustained and prolonged delivery of Bacillus Calmette-Guérin in the treatment of bladder cancer. *Biomaterials*, 34(38), 10258-66.

Zhang H, Zhu X, Ji Y, Jiao X, Chen Q, Hou L, Zhang H and Zhang Z. (2015). Near-infrared-triggered *in situ* hybrid hydrogel system for synergistic cancer therapy. *Journal of Materials Chemistry B*, 3(30), 6310-6326.

Zheng A, Wu D, Fan M, Wang H, Liao Y, Wang Q and Yang Y. (2020). Injectable zwitterionic thermosensitive hydrogels with low-protein adsorption and combined effect of photothermal-chemotherapy. *Journal of Materials Chemistry B*, 8(46), 10637-10649.

Zheng BD, Ye J, Yang YC, Huang YY and Xiao MT. (2022). Self-healing polysaccharide-based injectable hydrogels with antibacterial activity for wound healing. *Carbohydrate Polymers*, 275, 118770.

Zhou M, Liu S, Jiang Y, Ma H, Shi M, Wang Q, Zhong W, Liao W and Xing MMQ. (2015). Doxorubicin-loaded single wall nanotube thermo-sensitive hydrogel for gastric cancer chemo-photothermal therapy. *Advanced Functional Materials*, 25(29), 4730-4739.

Chapter 7

Nanocomposite Hydrogel Wound Dressings

Mohammad Sirousazar*
Yasaman Shaabani
Azin Jahani-Javanmardi
and Sahel Gozalzadeh
Faculty of Chemical Engineering, Urmia University of Technology, Urmia, Iran

Abstract

Wound healing is a complicated dynamic process of cellular and biochemical events in different but overlapping phases. Wound dressings can encourage the phases of the wound healing process and create better healing conditions. The principal function of wound dressings is to provide the optimum healing milieu to accelerate the natural healing operation. In 1962, it was found that a moist environment supplies a more enhanced and accelerated wound healing process compared to a dehydrated environment. Since then, numerous materials have been developed for use as moist wound dressings. Among the moist wound dressings, special attention has been paid to hydrogels because of their unique properties, which can meet the essential requirements of ideal wound dressings. A novel class of hydrogel wound dressings based on nanocomposite hydrogels called nanocomposite hydrogel wound dressings was introduced in 2007, which showed excellent promoted performance both *in vitro* and *in vivo*. In this chapter, the basic concepts of wounds, the wound healing process, and different types of wound dressings will be introduced. Furthermore, we will focus on hydrogels and hydrogel wound dressings. Finally, the nanocomposite hydrogel wound dressings will be discussed in detail.

Keywords: nanocomposite, wound dressing, wound healing, hydrogel, biomaterial

7.1. Introduction

Skin is the largest organ of the body with a broad range of vital roles. The most important function of the skin is to protect the rest of the body from the harmful effects of heat, ultraviolet

* Corresponding Author's Email: m.sirousazar@uut.ac.ir.

In: Nanocomposite Hydrogels and their Emerging Applications
Editor: Mohammad Sirousazar
ISBN: 979-8-88697-675-5

(UV) light, injury, and infection. It also performs various other functions, such as preventing the loss of moisture, synthesizing vitamin D, and regulating body temperature (Shai and Maibach, 2005; Thornton, 2002). The skin consists of two main layers including the epidermis and the dermis. The epidermis, the outermost layer of the skin, is mainly composed of flat and scale-like cells, known as squamous cells. Under the squamous cells, small round cells called basal cells are located. The deepest section of the epidermis contains melanocytes, which produce melanin pigments responsible for providing skin color. The dermis is the second layer of the skin, lying underneath the epidermis, which contains blood and lymph vessels, hair follicles, sweat, and sebaceous glands which release sweat and sebum. Sweat is vital for regulating human body temperature, and sebum is an oily substance that plays an essential role in protecting the skin from dryness. Both sweat and sebum move toward the skin surface through tiny openings in the skin called pores (Breitkreutz et al., 2009).

Being the first line of protective barrier, skin is constantly exposed to injuries and wounds caused by various accidents. Therefore, human beings have always been obsessed with developing wound healing techniques and have accumulated information about wound healing treatments. Appropriately dressing a wound is of great importance for protecting the wound from infection and accelerating the wound healing process. Among various dressings, wet or moist dressings have drawn particular attention. Moist dressings, by increasing the biological activity of the cells, promote the healing of the wound, and also help relieve pain due to their cooling effect (Gonzalez et al., 2014; Singh et al., 2013). The consistency, absorption of fluids and infections, air exchange with the environment, proper mechanical strength, flexibility, transparency, prevention of bacterial penetration, and wound protection against injuries are other advantages of moist dressings. In addition to possessing all mentioned appealing properties, hydrogels and nanocomposite hydrogels have close similarity to natural living tissues as a result of their high water content, porosity, and soft consistency. Consequently, they are the preferable wound dressing agents amongst all other synthetic biomaterials (Singh et al., 2013; Wang et al., 2012). In this chapter, the fundamental concepts of hydrogels and nanocomposite hydrogels will be reviewed and their application in wound healing will be discussed.

7.2. Wound and Wound Healing

7.2.1. Wound

A wound is a breakdown in the protective function of the skin; the loss of continuity of epithelium, with or without loss of underlying connective tissue (i.e., muscle, bone, and nerves) following injury to the skin and underlying tissues/organs caused by surgery, a blow, a cut, chemicals, heat, cold, friction, shear force, pressure, or as a result of diseases, such as leg ulcers or carcinomas (Leaper and Harding, 1998). Wounds could be described by numerous defining characteristics, including their aetiology, location, surrounding skin, dimension and depth, presenting symptoms, clinical appearance, tissue loss, and type of injury. These characteristics are used as a guide to selecting the proper required treatment.

Wounds based on their depth are classified as superficial (loss of epidermis only), partial-thickness wounds (involve the epidermis and dermis), and full-thickness wounds (involve the dermis, subcutaneous fat, and sometimes bone) (Rajendran, 2018). Moreover, based on the process and time of healing, wounds are classified as acute and chronic. Generally, an acute wound occurs suddenly and does not last long. Therefore, it undergoes the normal healing process within the expected and predictable time. In the case of infection-free wounds, it usually takes 8-12 weeks to be cured. This type of wound can happen in any body tissue and vary from superficial scratches to deep injuries. Chronic wounds, in contrast, do not follow the orderly set of healing stages and do not respond to treatments within even three months. In acute wounds, an accurate balance establishes between the production and removal of molecules, such as collagen, whereas in chronic wounds, a higher rate of removal causes the loss of balance (Schönfelder et al., 2005).

7.2.2. Mechanism of Wound Healing

Wound healing is a complex and dynamic biological process of replacing necrotic and damaged tissue with newly produced one. According to the predicted repair process, the wound healing process may be divided into three categories, including primary intention, secondary intention, and delayed primary/tertiary healing (Zahedi et al., 2010). To repair the wounds with minimal tissue and left scar, surgeons close the wound edges together using sutures, staples, or adhesives within hours of its creation with no defect (primary intention). In the healing by secondary intention, the wound is left open, and the healing process happens through epithelialization and contraction, e.g., healing of a deep wound in which wound edges can not be closed together. Ternary healing is formally used for infected wounds. In this case, the wound closure does not happen until several days.

When the tissue is damaged, a set of cellular and biochemical activities start to happen, which finally cause the healing. Generally, wound healing takes place in four phases. These phases do not occur at distinct and different times, and they may happen simultaneously. The first stage of wound healing includes hemostasis and inflammation. Severe tissue damage causes vascular derangement and bleeding. To stop blood flow, the body starts a coagulation cascade process. The vital coagulation factor, fibrinogen, is a glycoprotein complex in the blood plasma which undergoes thrombin cleavage and subsequent polymerization to produce fibrin and then a fibrin-based blood clot. The inflammatory phase occurs soon after the homeostasis phase. The initial cells involved in the inflammation process are platelets, neutrophils, and monocytes. Blood neutrophils are the first immune cells that enter the injured state and penetrate the dead cells (Rajendran, 2018; Shai and Maibach, 2005; Zahedi et al., 2010). This phase is followed by the migratory phase, during which the new and live cells known as the epithelial move toward injured skin to replace damaged or dead cells. The wound is completely covered with epithelial cells about four days after forming granulation tissues in the proliferation stage. Microphages, fibroblasts, and blood vessels migrate to the wound environment and form a single unit within two weeks. As this phase progresses, a gradual decrease happens in the inflammation of the wound. In the remodeling phase, the fibroblast tissues form a new layer of skin and no trace of the wound remains (Zahedi et al., 2010).

Typically, wound healing is an uneventful natural process that promptly restores skin integrity. However, as mentioned before, the healing of chronic wounds is complicated and needs special care. Healing of chronic wounds may be interrupted at any level of healing. However, failure occurs in the inflammatory or proliferative phase in which the extracellular matrix and matrix metalloproteinase, such as collagenase and lactase, accumulate more than usual and cause degradation of collagen and growth factors. An optimum microenvironment in the absence of cytotoxic factors is necessary (Edwards and Harding, 2004; Guo and Dipietro, 2010).

Multiple factors can lead to impaired wound healing. The factors that influence repair can be categorized into local and systemic. Local factors, such as infection and tissue ischemia, directly influence the characteristics of the wound itself. Systemic factors, such as aging and nutritional status, are the overall health or disease state of the individual that affect the ability of the body to heal. Many of these factors are related and the systemic factors act through the local effects affecting wound healing (Gosain and Dipietro, 2004). The healing of wounds could be accelerated using wound dressings. A wound dressing covers the wound surface and, in addition to the mechanical protection, prevents the entry of pathogens into the wound and makes the healing process faster. Although wound dressings do not heal wounds, they promote the healing process and prevent further harm to the wound (Rajendran, 2018). In the following section, the functions and types of wound dressings will be reviewed.

7.3. Wound Dressing

7.3.1. History, Basic Concepts, and Requirements

In ancient times, wound treatment methods were based on experience and traditional knowledge. As an example, in the presence of inflammation and infection, moldy bread was used to be applied as an antibiotic. In Mesopotamia, the wound was washed with water or milk and then dressed with honey. Hippocrates in ancient Greece used vinegar for cleaning wounds. They soaked wool inside the boiled vinegar and used it as a dressing. Through the centuries, the necessity and importance of hygiene and the halting of bleeding became clearer which led to the ongoing development of wound dressing techniques. Eventually, the advent of the germ theory of disease, in the final decades of the 19th century, enormously assisted the improvement of wound care management. Many advances in wound treatment are now available in all forms of health care, from wet to dry dressings. From the time of Hippocrates, it was believed that wound healing in a dry environment is faster than in a wet environment. However, in 1962, the results of research on a pig model confirmed that a moist wound environment through accelerating the healing process and decreasing the infection rate leads to lower overall costs and increased outcomes (Daunton et al., 2012; Winter, 1962; Wodash, 2012).

Dressings as covers for wounds have been used for at least two million years, but the basic knowledge about germs that cause infection is not older than a century. So, using dressings for wound care to control and prevent infection is a modern concept. Gamgee (1880) discovered that cotton wool dressings could physically protect the wound, and gauze absorbent dressings are helpful to absorb exudates. Dressings with phenol were used as antiseptics and influenced wound sepsis rate dramatically (Bishop, 1959). However, the toxicity of phenol is the main shortcoming of the proposed therapy. Recently many clinical trials of topical wound therapies have been done and their main aim was to find suitable dressings having antibacterial activity to prevent infection (Lawrence, 1992). Nowadays, medicated dressings are mostly tulle gras-type materials. Dressings containing chlorhexidine significantly prevent bacterial growth in minor bums and open fractures. However, there is little evidence to indicate that antiseptics influence both the bacterial content of colonized wounds and sepsis rates (Lawrence, 1994). The main requirements for a dressing to be known as an ideal wound dressing are (Rajendran, 2018; Sai K. and Babu, 2000; Watson and Hodgkin, 2005):

- Removing excess exudates but preventing saturation of the dressing to its outer surface
- Providing and sustaining a moist environment on the wound surface (maintaining high humidity at the dressing and wound interface)
- Allowing the gaseous exchange
- Providing thermal insulation
- Protecting the wound from bacterial penetration
- Being non-toxic
- Easily removable without causing trauma to the wound
- Having acceptable handling qualities
- Being sterilizable and comfortable
- Providing mechanical protection
- Controlling local temperature, pH, and wound odor
- Minimizing pain from the wound
- Being cosmetically acceptable
- Being non-allergenic
- Being cost-effective.

7.3.2. Types of Wound Dressings

Wound dressings can be classified in different ways. There are antibacterial and absorbent wound dressings (based on the treatment), collagen and hydrocolloid wound dressings (based on the composition), ointment, film, and hydrogel wound dressings (based on the physical form). They can also be simply classified as dry and moist wound dressings. There are also other classifications for wound dressing. For example, they may be adherent or non-adherent. Moreover, based on their allowance for exchanging water or gases, they may be either occlusive or non-occlusive. In the wound healing process, at first, adherent dressings (such as gauze) are used to debride exudate and remove necrotic tissue. The method of using sterile gauze pads may be dry-to-dry (using dry pads on wounds) or wet-to-dry (the pad is thoroughly wetted by sterile saline). Subsequently, non-adherent dressings are applied to reduce pain and tissue trauma during dressing changes which positively affect the healing process (Sirousazar et al., 2014). Different types of wound dressings are reviewed in Table 7.1. Among the wide variety of wound dressings, special attention has been paid to hydrogels and nanocomposite hydrogels as a new class of moist wound dressings. Owing to their exceptional function in healing wounds and burns, various types of hydrogels have been prepared and used as wound dressings to treat different wounds. Before discussing nanocomposite hydrogel wound dressings, the fundamental concepts of hydrogels and their application as wound dressing will be described in the following section.

Table 7.1. Types of wound dressings

Dressing Type	Characteristics and Compositions	Advantages	Disadvantages	Refs.
Gauze	Made of woven and non-woven fibers of cotton, rayon, polyester, or a combination	Inexpensive, readily available, available in many sizes and forms, can be used on a variety of wounds	It needs to be changed frequently if the wound becomes dry, must be held in place by medical tape or another secondary dressing, fibers of the gauze may stick to the wound, can not prevent the entry of exogenous bacteria into the wound	(Purna and Babu, 2000; Kerihuel, 2009; Sirousazar et al., 2014)
Charcoal	Dressings with a large surface area of the activated charcoal	They are like filters or traps which absorb unpleasant smells from wounds	Their activity decreases in the presence of wound exudate	(Wunderlich and Orfanos, 1991)
Silver-based dressing	They are often used for burns and also for infected and heavily colonized wounds	A broad antibacterial spectrum can slow epithelialization and produce hypernatremia and hypochloremia in addition to hypotonicity	Possible systemic toxicity, effectiveness varies between products	(Rajendran, 2009; Wodash, 2012)
Honey	An ancient treatment useful for acute and chronic wounds	It has antibacterial properties, can be easily used, reduces edema, lowers wound pH, debrides slough and escher creates a moist environment by drawing exudate to the wound surface without the risk of bacterial growth occurring	Honey dilutes by exudate and its components absorb into the bloodstream, its osmotic and antiseptic effects last for only two to three days causing intolerable pain because of drawing the skin	(Molan, 2006; Hofman, 2007; Yusof et al., 2007)
Foam	It consists of hydrophobic polyurethane foam sheets	It has an excellent absorbing ability, it can expand and conforms to wound with an unusual configuration, it is comfortable and can be removed easily for cleaning	It absorbs fluid from the environment and is not suitable for use in the bath	(Watson and Hodgkin, 2005; Rajendran, 2009)
Cadexomer iodine dressing	Contain two forms, as povidone-iodine impregnated into a low-adherent dressing or dressings containing concentrated cadexomer iodine paste	Antibacterial activity which removes the infection, provides a sustained release of iodine, the material is biodegradable	Impedes wound healing through cytotoxic action against fibroblasts, keratinocytes, and leukocytes	(Bell and Hyam, 2007; Daunton et al., 2012)
Spray-on dressing	Liquid polymeric mixtures with a significant capability of absorbing exudate	Accelerates the healing process, does not stick on the wound surface, provides a moist environment on the wound, can be used in all parts of the body	Expensive, may cause bacterial growth, is not transparent, may mix with exudate and delay healing	(Purna and Babu, 2000)
Capillary action dressing	Made from a triple layer of polyester and viscose fibers with a perforated wound contact polymer film	Rapidly absorbs exudates and improves the healing process, maintains a moist wound healing environment, useful for very wet sloughy wounds	Adheres to wounds with low exudate, possible traumatic removal	(Hofman, 2007)

Dressing Type	Characteristics and Compositions	Advantages	Disadvantages	Refs.
Soft silicone	The contact layer consists of a polyamide net coated with soft silicone	It does not stick to the surface of a moist wound, does not cause damage, pain, and epidermal stripping during dressing changes, reduces scar formation, is cosmetically acceptable to the patient	Should not be used for patients with allergy to silicone or on bleeding wounds	(Wheeland, 1992; Cubit et al., 2013)
Alginate	Non-woven and non-adherent pads and ribbons made of natural polymers obtained from brown seaweed	Retain large amounts of water, swelling and diffusion throughout the dressing helps the wound exudates to be absorbed	In wounds with low exudate causes excessive drying of the wound, separating the dressing is painful	(Bell and Hyam, 2007; Kim et al., 2008; Rajendran, 2009)
Hydrocolloid	Made of materials such as gelatin, pectin, and carboxymethylcellulose, forms available in adhesive wafers or as pastes or powders	Absorb light-to-moderate levels of exudate, do not require a secondary dressing, can be used under the shower	Very low oxygen exchange in the dry environment, does not prevent pain in heavily exuding wounds	(Wheeland, 1992; Watson and Hodgkin, 2005; Fujimoto, 2008; Rajendran, 2009)
Hydrogel	Three-dimensional networks of hydrophilic natural or synthetic polymers contain up to 96% water	Transparent and comfortable, able to absorb fluids, provide moisture to dry wounds, makes autolysis of necrotic tissue easy, does not help bacterial growth	The low mechanical properties, can cause pain in heavily exuding wounds	(Jones and Vaughan, 2005; Kokabi et al., 2007; Boateng et al., 2008)

7.4. Hydrogels

Hydrogels are crosslinked three dimensional hydrophilic polymer networks that do not dissolve in aqueous solutions due to the chemical or physical crosslinking of the individual polymer chains (Peppas et al., 2006). They can be swollen by absorbing large amounts of water or aqueous solution without losing their integrity. Hydrogels are produced by polymerizing one or more monomers involving interactions, such as hydrogen bonding and strong van der Waals force between polymer chains. Hydrogels can be prepared in different physical forms, such as slabs, microparticles, nanoparticles, coatings, and films (Ganji and Vasheghani-Farahani, 2009). As soft materials, hydrogels show excellent potential in numerous applications, e.g., superabsorbents, molecular filtration, separation, and bio-separation processes. Having excellent water contact makes them flexible and similar to natural tissue, so they are usable in biomedical fields, such as implants, artificial organs, contact lenses, drug delivery devices, and wound dressings (Ahmed, 2015; Kabiri et al., 2011).

7.4.1. Classification of Hydrogels

Hydrogels can be classified in different ways, listed as follows:

7.4.1.1. Based on Source

Hydrogels can be synthesized using either natural or synthetic polymers. Natural polymers are originally macromolecular structures that exist in nature and can be extracted. Proteins, such as collagen and gelatin, and polysaccharides, such as starch, alginate, and agarose are well-known examples of natural polymers utilized for fabricating hydrogels. In contrast, synthetic polymers are man-made polymers often derived from petroleum oil and traditionally prepared using the chemical polymerization method (Ahmed, 2015; Sirousazar et al., 2014).

7.4.1.2. Based on Polymeric Composition

Three main classes of hydrogels can be described based on polymeric composition:

a) Homopolymer hydrogels: They are hydrogels whose polymeric network consists of a single kind of monomer.
b) Copolymer hydrogels: Their network consists of two or more different monomer species. In their structure, at least one hydrophilic component exists. The hydrophilic component may be arranged in a random, block, or alternating configuration along the chain of the polymer network.
c) Interpenetrating polymer network (IPN) hydrogels: They are an essential class of hydrogels including two or more independent crosslinked synthetic and/or natural polymer components contained in a network form. If one component is a crosslinked polymer and the other is a non-crosslinked polymer, the hydrogel is called semi-IPN.

7.4.1.3. Based on Configuration

In this category, hydrogels are classified based on the configuration of polymeric chains inside the network as listed below (Ahmed, 2015):

a) Non-crystalline (amorphous).
b) Semi-crystalline: This group includes hydrogels that are a complex of amorphous and crystalline phases. The regions with a highly ordered structure (i.e., the crystalline parts) act as crosslinked sites.
c) Crystalline.

7.4.1.4. Based on Type of Crosslinking

Hydrogels are divided into two categories based on the physical or chemical nature of the crosslink junctions. Chemically crosslinked hydrogels have permanent junctions, while physical networks have transient junctions that arise from polymer chain entanglements or physical interactions, such as ionic interactions, hydrogen bonds, or hydrophobic interactions.

7.4.1.5. Based on Network Electrical Charges

Depending on the presence or absence of electrical charge located on the crosslinked polymers, hydrogels may fit the criteria of three groups (Ahmed, 2015):

a) Nonionic (neutral).
b) Ionic (anionic or cationic).
c) Amphoteric electrolytes (ampholytic), which contain both acidic and basic groups.

7.4.1.6. Based on Type of Stimulus

Another classification is based on the sensitivity of hydrogels to external stimuli exerted by the environment. These hydrogels can shrink or expand as a response to changes in external environmental conditions. Their volume may change dramatically in response to various physical and chemical stimuli. Physical stimuli include temperature, electrical or magnetic fields, light, pressure, and sound. Chemical stimuli include pH, solvent composition, ionic strength, and molecular species (Qiu and Park, 2001; Sirousazar et al., 2014).

7.4.2. Application of Hydrogels in Wound Management

Owing to their particular characteristics, including high water absorption and biocompatibility, hydrogels are used in various areas for several purposes like drug delivery, agriculture, sanitary pads, transdermal systems, dental materials, implants, injectable polymeric systems, ophthalmic applications, hybrid-type organs, and wound dressings. However, the major application of hydrogels is in tissue engineering, pharmaceutical, and biomedical fields (Peppas et al., 2000).

Hydrogels, as unique and moist wound dressing materials, can absorb and hold wound exudates in their network structure along with foreign bodies such as bacteria. Also, they reduce the fluid loss from the wound surface and protect the wound from external damage, and as a result, fibroblast proliferation promotes, and the healing process accelerates. Hydrogels establish a proper micro-climate for biosynthetic reactions on the wound surface essential for cellular activities (Diez-Pena et al., 2002). *Keratinocyte migration* is the most crucial cellular activity for the re-epithelialization step to cover the wound *surface*. The epithelialization process of the wound starts from the wound margin and is followed by fibroblast proliferation. The sufficient moisture that hydrogel provides on the wound surface helps keratinocytes migrate into the wound. Transparent hydrogels offer the possibility to monitor wound healing while wound dressing is applied.

The process of angiogenesis, induced by temporary hypoxia, can be initiated using semi-occlusive hydrogel dressings. Angiogenesis maintains an adequate supply of oxygen and nutrients to the wound surface for granulation tissue formation. The back of the fabric or polymeric film of the hydrogel sheet contacts the wound surface, and the hydrogel sheet is secured with adhesives or bandages (Pal et al., 2009). However, due to some of the hydrogel dressing's poor mechanical properties, such as low strength and elasticity or low fracture toughness, especially in a swollen state, hydrogels do not entirely satisfy the requirement for an ideal dressing. For instance, they might be crushed under high stress or might stick to the wound surface.

The main reason for the weakness, low strength, and low elasticity of the conventional hydrogels is constrained molecular motion of the polymer chains, as a result of a large number of randomly arranged crosslinks. Hence an ideal hydrogel dressing is one with good strength and elasticity. To overcome the weakness of conventional hydrogel wound dressings, a novel class of dressings called nanocomposite hydrogel wound dressings was developed. Nanocomposite hydrogels and their application in wound management will be discussed in the next section.

7.5. Nanocomposite Hydrogels in Wound Management

As discussed before, among all types of novel wound dressings, special attention has been paid to hydrogels which can meet some of the essential requirements of proper wound dressing because of their unique and interesting properties, such as immediate pain control, easy replacement, transparency to allow healing follow up, absorb and prevent loss of body fluids, the barrier against bacteria, oxygen permeability, exemplary handling, control of drug dosage, and so on (Kokabi et al., 2007). However, there are some limitations in using conventional polymer hydrogels as wound dressing because of their weak mechanical properties, such as low gel strength, poor stability, and low fracture toughness, especially in a swollen state. The main reason for these disadvantages is a large number of randomly arranged crosslinks in the conventional hydrogels which constrain the molecular motion of the polymer chain (Haraguchi and Takada, 2005; Sirousazar and Kokabi, 2012). Due to the demand for dressing materials with good strength and elasticity, improving the mechanical properties of hydrogels is of paramount importance to extend their applications and improve their performance. Accordingly, to enhance the strength and elasticity of hydrogels, small-scale inorganic particles are used as reinforcing agents to prepare a novel class of hydrogels called nanocomposite hydrogels (Kokabi et al., 2007).

Organic/inorganic nanocomposite hydrogels are reinforced three-dimensional networks, which have been made of at least a crosslinked polymeric matrix and a nanoscale reinforcing agent in which the organic and inorganic components are immiscible (Haraguchi and Takada, 2005; Sirousazar and Kokabi, 2012). Compared to conventional hydrogels, nanocomposite hydrogels exhibit extraordinary mechanical, optical, and swelling/deswelling properties. The improvements include mainly increased strength and heat resistance, decreased gas permeability and flammability, high water content, and good biocompatibility (Haraguchi and Takada, 2005; Kaşgöz et al., 2008). So, nanocomposite hydrogels, with properties that overcome the limitations of conventional chemically crosslinked hydrogels, have good potential to be used as functional materials in biomedical applications, especially in the wound healing process. Jahani-Javanmardi et al., 2016; Sirousazar et al., 2016). Nanocomposite hydrogels are promising candidates for various applications, such as contact lenses, absorption of saline, sterilization by autoclaving, blood compatibility, intramuscular implantation, sensor and actuator systems, oil recovery, biomedical scaffolds, and drug delivery devices (Sirousazar and Kokabi, 2012).

7.5.1. Nanocomposites

The definition of nanocomposite material contains many systems, such as one-dimensional, two-dimensional, three-dimensional, and amorphous materials, made of different components and mixed at the nanometer scale (Kamigaito, 1991; Parmoor et al., 2016). The transition of the particles from micro to nano changes their physical and chemical properties. The changes are mainly attributed to the increased surface area to volume ratio with decreasing the size of the particles. As the surface area to volume ratio increases, the influence of the behavior of atoms located on the surface of particles becomes more dominant compared to that of atoms inside the particles. This affects the properties of the particles when they are reacting with other particles and increases the reactivity. The higher surface area of the nanoparticles leads to higher interaction between them within the mixture (Ajayan et al., 2006).

In general, nanocomposites are made of two phases, including the matrix (or continuous) and the reinforcing nanoscale (dispersed) phases. Based on the types of matrix phase, nanocomposites are classified as ceramic-matrix, metal-matrix, and polymer-matrix nanocomposites. In ceramic-matrix nanocomposites, the major part of the volume is filled with a ceramic, i.e., a chemical compound from the group of oxides, nitrides, borides, and silicides. Nanocomposites with ceramic-matrix structures usually contain metal as the second component. Finely both metallic and ceramic components are dispersed in each other to reach the desired nanoscopic properties (Kruis et al., 1998). Nanocomposites derived from these combinations showed improvement in their optical, electrical and magnetic properties as well as tribological, corrosion-resistance, and other protective properties (Birkholz et al., 2004; Zhang et al., 2003).

In metal-matrix nanocomposites, the metallic matrix is reinforced with a small amount of nanoscale fillers. Carbon nanotubes-based metal-matrix nanocomposite is one of the most important nanocomposites developed to take advantage of the high tensile strength and electrical conductivity of carbon nanotubes. In addition, boron nitride-reinforced metal-matrix composites, and carbon nitride metal-matrix composites are the new research areas on metal-matrix nanocomposites (Bakshi et al., 2010; Lalwani et al., 2013b).

Polymer-matrix nanocomposites are a class of nanocomposites that can substitute for conventional filled polymers. In this class of materials, nano-sized inorganic fillers are dispersed in a polymer matrix, leading to significant improvement in the performance properties of the polymer (Sirousazar et al., 2007). Adding nanoparticles to a polymer matrix can improve its performance by simply capitalizing on the natural properties of the nanoscale filler. This method is particularly effective in producing high-performance composites when good dispersion of the filler is achieved, and the properties of the nanoscale filler are better than those of the matrix (Lalwani et al., 2013a; Manias, 2007; Morgan and Wilkie, 2007; Ryan et al., 2009; Singh et al., 2008). Clays are the most commonly used nanoparticles as fillers in the production of polymer-matrix nanocomposites. According to the bonding conditions between the polymer chain and silicate layers of clays, the new material shows excellent mechanical properties, such as higher tensile strength, heat resistance, and chemical resistance than neat polymer (Sapalidis et al., 2011).

7.5.2. Nanocomposite Hydrogels

Nanocomposite hydrogels are a class of polymer-matrix nanocomposites whose matrix phase is made of three-dimensional polymeric hydrogels, reinforced with at least a nanoscale dispersed phase (Shaabani et al., 2016b). It was shown that the nanometric reinforcing particles act as crosslinkers inside the nanocomposite hydrogels network. The gel fraction values and crosslinking densities of hydrogels increase with the incorporation of the nanoparticles into them (Sirousazar and Kokabi, 2012). Nanocomposite hydrogels are usually prepared by polymerization and crosslinking in place. In this method, monomers in the presence of nanoparticles simultaneously become polymerized and crosslinked. Another technique for preparing nanocomposite hydrogels is crosslinking polymer chains in the presence of nanoparticles. In this method, polymer chains that have been already polymerized and crosslinked in a suspension of nanoparticles, create the three-dimensional structure of the nanocomposite hydrogel. In both methods, chemical agents and physical crosslinking methods (e.g., radiation or freezing-thawing) can be used as helping methods for crosslinking (Schexnailder and Schmidt, 2009).

Most research on nanocomposite hydrogels has been focused on the use of clays, such as montmorillonite (MMT), hectorite, laponite, hydrotalcite, bentonite, kaolinite, and other nanostructured particles, such as gold, silver, etc. (Shaabani et al., 2016a; Sirousazar and Kokabi, 2012). All nanocomposite hydrogel's extraordinary properties and abilities to participate in biomedical applications are because of their unique organic/inorganic structures. As mentioned before, in the structure of clay-based nanocomposite hydrogels, each clay layer acts as a multifunctional crosslinker. This was indicated by the classic rubber-elasticity theory for calculating the number of crosslinking points per clay platelets for nanocomposite hydrogel, which is several tens to more than 100.

Polymer-clay nanocomposite hydrogels in morphology and distribution of silicate layers of clay in their structure are similar to other polymer-clay nanocomposites. Clay layers in the hydrogel matrix, according to the synthesis method and production process, give us three morphology, including the immiscible, intercalated, and exfoliated morphologies. X-ray diffractometry (XRD) and transmission electron microscopy (TEM) are two main methods to determine the morphology of nanocomposite hydrogels. In the case of immiscible (micro-composite) morphology, the silicate layers of clay are not separated in the presence of polymer chains. In this case, XRD peaks of clay and nanocomposite hydrogel are almost coincident. In the case of the intercalated nanocomposite, polymer chains penetrate in spaces between layers of clay and silicate layers by maintaining a parallel arrangement, becoming more open. In this case, the XRD peak of nanocomposite hydrogel is transferred to smaller angles. In the case of the exfoliated nanocomposite, greater integration happens between polymer chains and clay layers, and the parallel arrangement of clay layers disappears. In this case, the XRD curve of nanocomposite hydrogel does not show any peaks (Paul and Robeson, 2008).

Tremendous efforts have been recently made to produce nanocomposite hydrogels based on natural and biocompatible hydrogels, called bionanocomposite hydrogels (BNCHs), for different biomedical applications. Due to unique properties like biocompatibility, biodegradability, and high hydrophilicity, BNCHs are promising materials for biomedical and environmental applications (Sirousazar et al., 2020). An example of natural polymers that have gained attention to produce BNCHs is egg white (EW) which is a natural source of high-quality proteins (Babaei et al., 2019; Shaabani et al., 2016a). Pure EW is completely soluble in aqueous solutions. However, incorporating Na-MMT as a reinforcing and crosslinking agent in the EW matrix leads to the preparation of insoluble, crosslinked, and swellable bionanocomposites. The crosslinked network of bionanocomposites makes them insoluble in aqueous solutions (Shaabani et al., 2016a). In BNCHs prepared based on EW, polyvinyl alcohol (PVA), and MMT clay via a freezing-thawing cyclic method, the MMT platelets act as crosslinkers in the network of BNCH and interconnect the polymer and protein chains together, and produce a three-dimensional network with pore sizes in the nanometric scale (Jahani-Javanmardi et al., 2016).

7.5.3. Application of Nanocomposite Hydrogels in Wound Care

For the first time, the application of nanocomposite hydrogels, as a new class of wound dressings, was introduced by Kokabi et al., (2007). The introduced nanocomposite hydrogel wound dressings, which were based on PVA and MMT, showed intercalated morphology with excellent improvements in physical, mechanical, and chemical properties. One of the critical aspects that must be considered when using nanocomposite hydrogels for wound management, and generally for biomedical applications, is their biocompatibility. The cytotoxicity assays for assessing the general toxicity of PVA/MMT nanocomposite hydrogel wound dressings proved their non-toxic and biocompatible nature. The cytotoxicity activity of the PVA/MMT nanocomposite hydrogel wound dressings on the K562 cell line is represented in Figure 7.1.

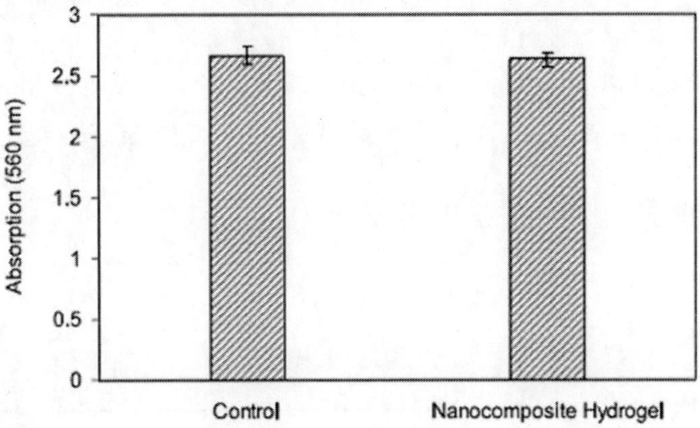

Figure 7.1. The cytotoxic activity of the PVA/MMT nanocomposite hydrogel on K562 cells (Sirousazar et al., 2011).

It can be seen that the absorbance of the medium cultured with nanocomposite hydrogel wound dressing is nearly similar to that of the control medium. The CPII value was calculated as 0.83%, indicating low cytotoxicity of the nanocomposite hydrogel wound dressing. According to Figure 7.2, there is no difference between the morphology of the K562 cells treated with the PVA/MMT nanocomposite hydrogel wound dressing after four days of culture and that of the non-treated samples. It also indicates no considerable dissimilarity in the proliferation index between the treated and non-treated K562 cells with nanocomposite hydrogel wound dressing. These results confirmed that the PVA/MMT nanocomposite hydrogel wound dressings have no significant cytotoxicity effect on cell proliferation (Sirousazar et al., 2011).

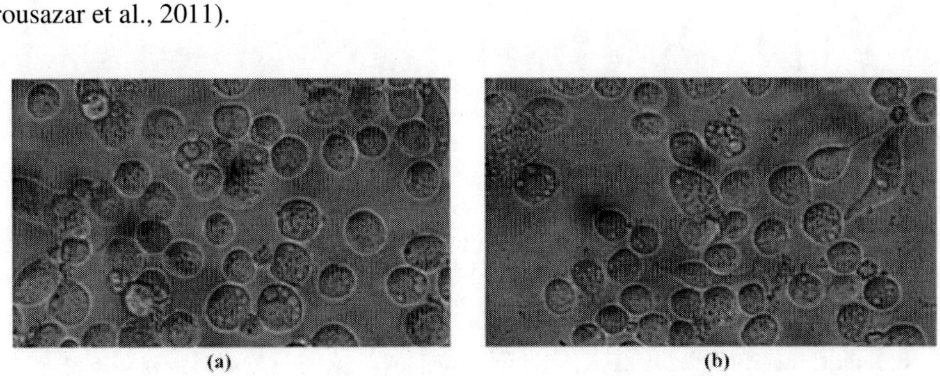

Figure 7.2. The morphology of the K562 cells after four days of culture: (a) not treated with the PVA/MMT nanocomposite hydrogel and (b) treated with the PVA/MMT nanocomposite hydrogel (Sirousazar et al., 2011).

Furthermore, according to *in vitro* investigations, PVA/MMT nanocomposite hydrogel wound dressings had superior performance compared to the conventional hydrogel wound dressings, and due to keeping a moist environment on the interface of the wound for a long time, accelerated the healing process. Figures 7.3 and 7.4 show the swelling characteristics, i.e., equilibrium degree of weight swelling (EDWS), equilibrium degree of volume swelling (EDVS), equilibrium water content (EWC), and the time required to reach the equilibrium state versus the content of clay (MMT) added to the PVA hydrogel matrix.

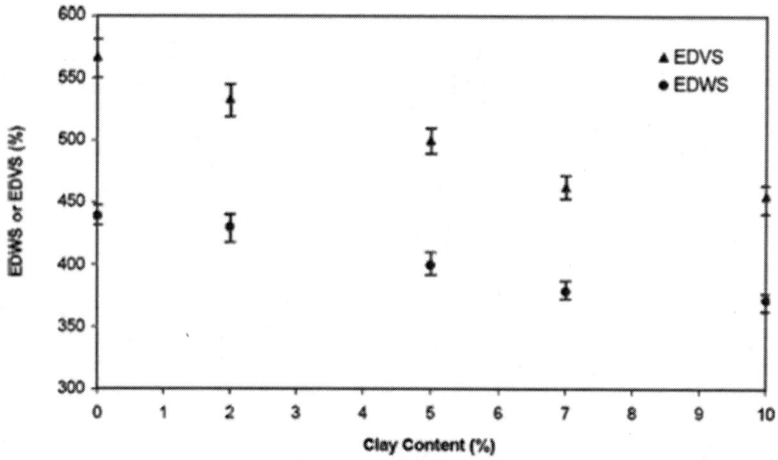

Figure 7.3. Equilibrium degree of weight and volume swelling of PVA/MMT nanocomposite hydrogel wound dressings versus clay content (Sirousazar et al., 2012a).

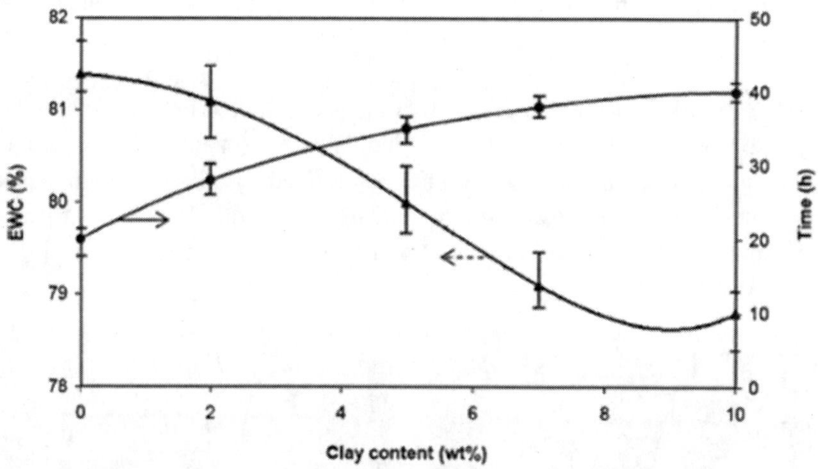

Figure 7.4. Equilibrium water content and the required time to establish equilibrium condition of PVA/MMT nanocomposite hydrogel wound dressings versus clay content (Sirousazar et al., 2012a).

It can be observed that there is an inverse relation between EDWS, EDVS, and EWC and the content of incorporated clay. The lower values of these parameters for PVA/MMT nanocomposite hydrogels compared to pure hydrogel are related to their higher gel fraction values, more entangled structures, and higher values of crosslinking density, which restrict the network relaxation and decrease the water uptake potential. On the other hand, a direct relationship exists between the time required to reach the equilibrium condition and the content of clay in nanocomposite hydrogel wound dressings. The results indicate that the PVA/MMT nanocomposite hydrogel wound dressings, despite having less capability in water uptaking and swelling at equilibrium conditions than conventional hydrogel wound dressings, exhibit a slow swelling process with prolonged duration. The slower swelling of nanocomposite hydrogel wound dressings is attributed to the slower and more prolonged mass transfer process of the medium solution (exudates) in the hydrogel network. So, it can be remarked that nanocomposite hydrogel wound dressings need more time to reach a specified level of swelling, and therefore, they show slower swelling kinetics than clay-free hydrogel wound dressing. This means that the nanocomposite hydrogel wound dressings are suitable candidates for practical applications in wound care where the swelling of hydrogel should be prolonged (Sirousazar et al., 2012a).

As mentioned before, hydrogels used as wound dressing materials would avoid or at least reduce the body's liquid loss by controlling absorption and transmission and keeping the high humidity in the wound area to accelerate the healing process. Water vapor transmission rate (WVTR) is a distinct factor that shows the potential of hydrogel wound dressings in the transmission of body liquid or wound exudates. Higher WVTR means faster wound drying and a tendency to produce eschars. On the other hand, low WVTR shows the accumulation of exudates which may cause the retardation of the healing process and opens up the risk of bacterial growth. Figure 7.5 shows the WVTR for PVA/MMT nanocomposite hydrogel wound dressings containing 0-10 wt.% of MMT. As seen, the WVTR is decreased by increasing the quantity of incorporated clay, and the nanocomposite hydrogel wound dressings have fair values of WVTR, which are nearly around the WVTR of human skin (35 $g/m^2.h$) (Kokabi et al., 2007; Razzak and Darwis, 2001).

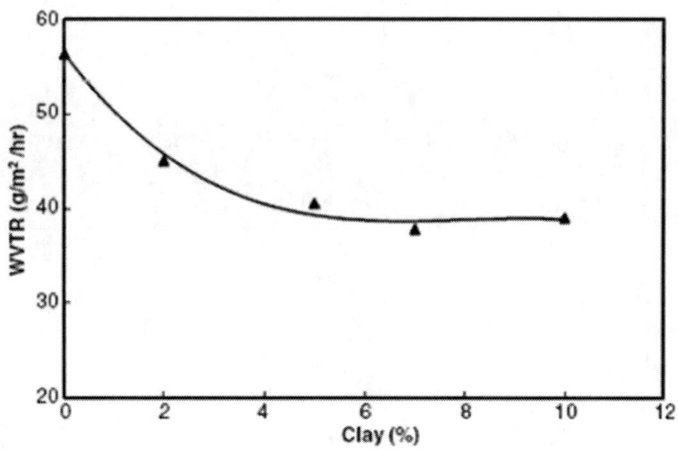

Figure 7.5. Water vapor transmission rates of PVA/MMT nanocomposite hydrogel wound dressings versus clay content (Kokabi et al., 2007).

It has also been shown that nanocomposite hydrogel wound dressings benefit better mechanical properties, i.e., higher elastic modulus and tensile strength and lower elongation at break compared to conventional hydrogel counterparts. The tensile properties of PVA nanocomposite hydrogel wound dressings containing different amounts of MMT (0-15 wt.%) are shown in terms of the tensile modulus, tensile strength, and elongation at break in Figures 7.6-7.8, respectively.

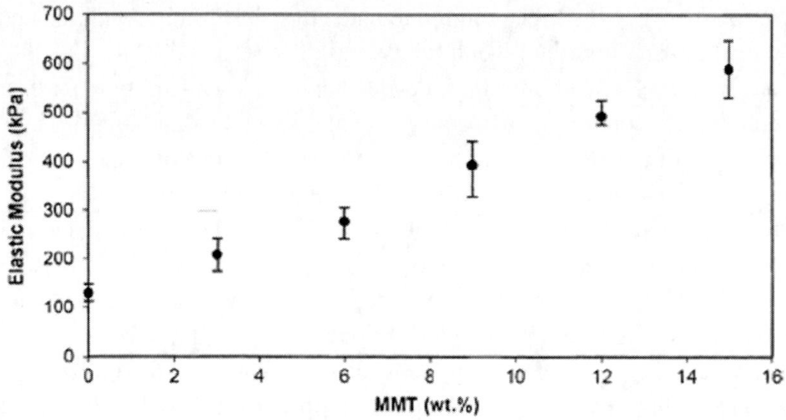

Figure 7.6. Tensile modulus of PVA/MMT nanocomposite hydrogels versus MMT content (Sirousazar et al., 2012b).

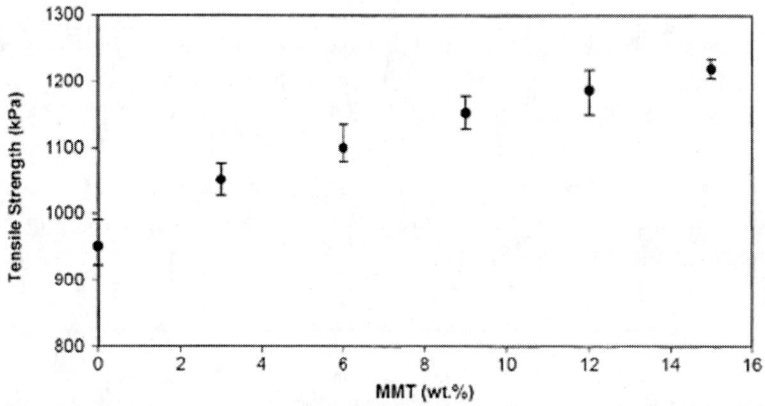

Figure 7.7. Tensile strength of PVA/MMT nanocomposite hydrogels versus MMT content (Sirousazar et al., 2012b).

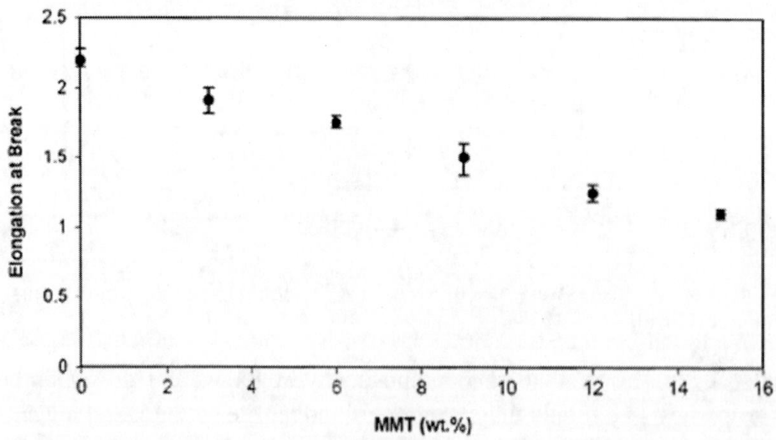

Figure 7.8. Elongation at break of PVA/MMT nanocomposite hydrogels versus MMT content (Sirousazar et al., 2012b).

A marked increase in tensile modulus and tensile strength are observed for the nanocomposite hydrogel wound dressings with increasing loading level of MMT. However, that elongation at break decreased with increasing the MMT content (Sirousazar et al., 2012b). The results confirmed the reinforcing effect of MMT nanoclay in nanocomposite hydrogel wound dressings. It was concluded that the PVA/MMT nanocomposite hydrogel can be considered an excellent choice for wound dressing and its mechanical properties can fulfill the essential requirements for wound dressing to use for wounds, even wounds under high stress (Kokabi et al., 2007).

The hardness of the hydrogel wound dressings can be considered as a factor to evaluate their adhesion to the skin and wound surface. The hardness of a wound dressing should not be so low because, in this case, it may stick to the wound surface and disrupt the healing process. It has been shown that nanocomposite hydrogel wound dressings have a higher level of hardness and, as a result, a lower level of stickiness compared with conventional hydrogel wound dressings. Figure 7.9 shows the hardness values for PVA/MMT nanocomposite hydrogel wound dressings. The results show that incorporating MMT clay into the PVA hydrogel creates nanocomposite hydrogel wound dressings with a high value of hardness and therefore low adhesion to the wound surface, which in practical applications accelerates the wound healing process (Kokabi et al., 2007).

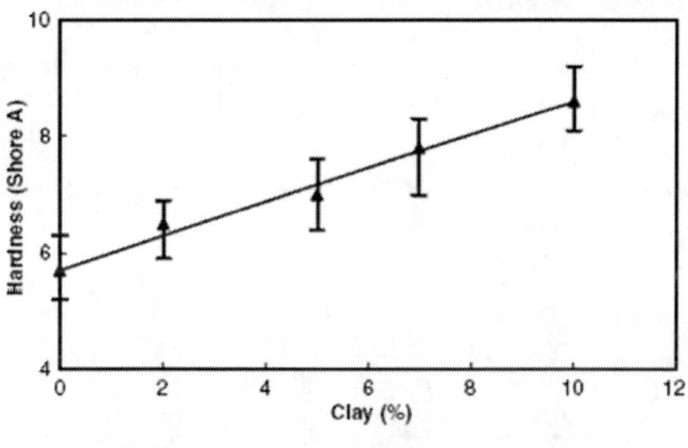

Figure 7.9. The hardness of PVA/MMT nanocomposite hydrogels versus clay content (Kokabi et al., 2007).

Besides the *in vitro* assessments, the nanocomposite hydrogel wound dressings have successfully passed the *in vivo* assays. Figures 7.10 and 7.11 show photographs of the macroscopic appearance of the wound site during the healing process for a control group (wounds healed with sterile gauze) and for a hydrogels group (wounds healed with PVA/MMT nanocomposite hydrogels), respectively. The moist surfaces on the wounds with little scar formation and good closing of wound edges were observed for animals treated with nanocomposite hydrogel wound dressings. The photographic evaluation indicates faster healing of the wounds in the hydrogel group. It can be attributed to the creation of a moist environment on the wound surface in the presence of nanocomposite hydrogel wound dressings and a faster migration rate of the epidermal cells over a moist wound surface compared to the condition that exists under a scab in a dry wound. In other words, the accelerated healing process in the hydrogel group may be due to the presence of growth factors, the increased precipitation of fibrinogen and fibronectin, and the acceleration of debridement, leading to more accessible and faster migration of epidermal cells (Sirousazar et al., 2011).

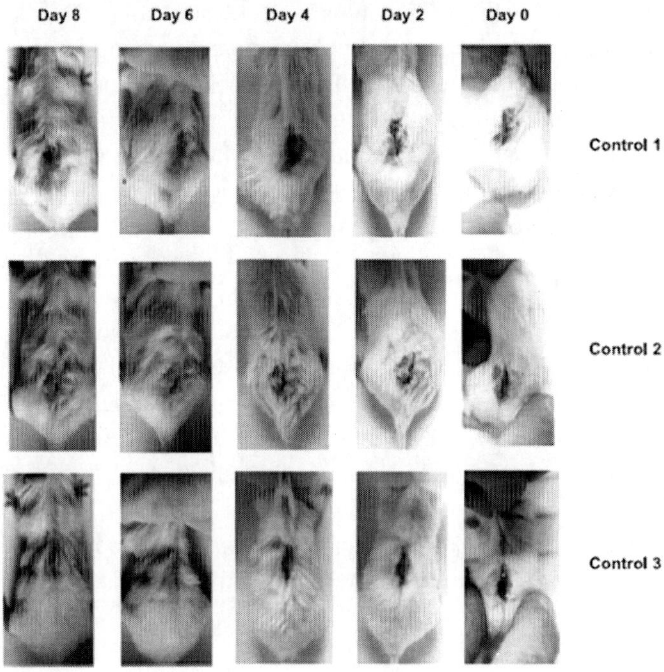

Figure 7.10. Photographs of the macroscopic appearance of the wound site during the healing process of the control animals (not treated with nanocomposite hydrogels) (Sirousazar et al., 2011).

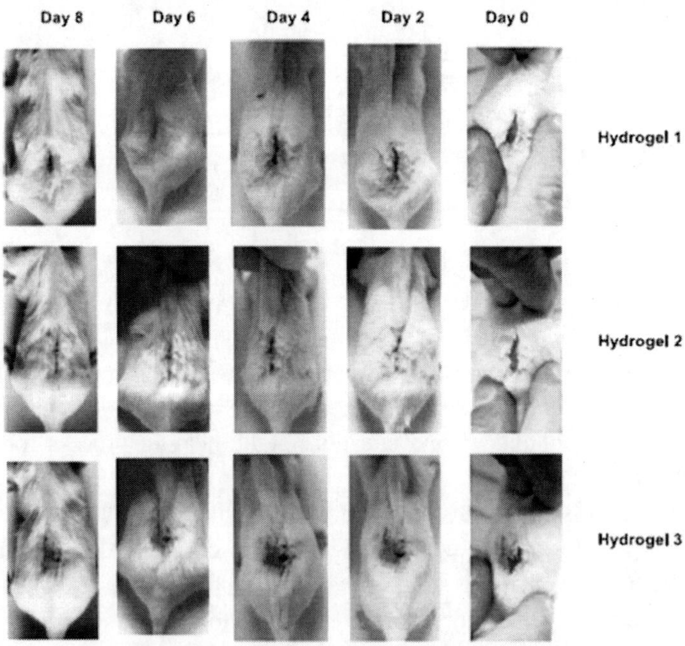

Figure 7.11. Photographs of the macroscopic appearance of the wound site during the healing process of the animals treated with PVA/MMT nanocomposite hydrogel wound dressing (Sirousazar et al., 2011).

Quantitative analysis according to the changes in wound areas at different healing times is shown in Figure 7.12. It can be seen that for both groups, the wound area decreased significantly along with the time duration, and the wound area of the hydrogel group was always smaller than that of the control group at each time interval. The results show the excellent capability of the PVA/MMT nanocomposite hydrogel wound dressings in faster healing of wounds compared to conventional wound dressings (Sirousazar et al., 2011).

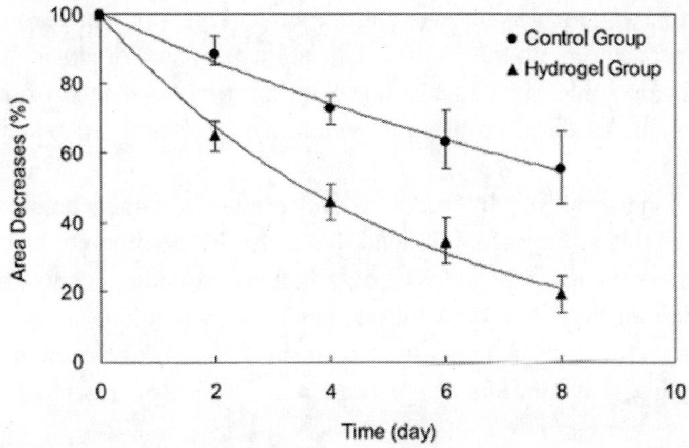

Figure 7.12. Wound area decreases in animals of the control and hydrogel groups during eight days of healing (Sirousazar et al., 2011).

The wounds healed with nanocomposite hydrogel wound dressings also have better qualities, i.e., higher mechanical properties. Figure 7.13 gives the tensile strength and elongation at break of the not-injured skin and excisional wounds of control and hydrogels groups. The tensile strength exhibited by hydrogel groups was 2.7 folds greater than that of the control group. Similarly, the average elongation at break of the healed wounds of the hydrogel group was 1.7 folds higher than that of the control group.

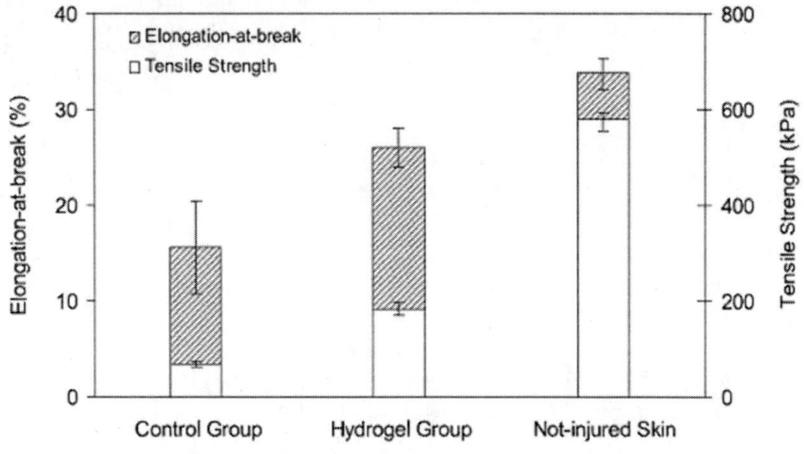

Figure 7.13. Tensile strength and elongation-at-break of the not-injured skins and the healed wounds of control and hydrogel groups (Sirousazar et al., 2011).

Increased tensile strength and elongation at break indicate an increase in collagen matrix due to the creation of a moist environment on the surface of wounds covered by the PVA/MMT nanocomposite hydrogels. In other words, there is rapid biosynthetic activity in hydrogel groups during the initial granulation phase. In the remodeling phase, the maturation of collagen takes place through the formation of intermolecular and intramolecular crosslinks; hence, the tensile properties are increased (Sirousazar et al., 2011). Furthermore, it has been shown that only a 3 mm thickness of the nanocomposite hydrogel dressings could protect wounds from microbe penetration, while it was observed that even 64 layers of gauze could not prevent the entry of exogenous bacteria into the wound. This ability of nanocomposite hydrogel dressings (barrier property against microbe penetration) could protect the wound from further infection and prevent exogenous bacterial infiltration, which may accelerate the wound healing process (Kokabi et al., 2007).

Drug-loaded nanocomposite hydrogel wound dressings, which could release the drugs (e.g., antibiotics) on the surface of the wound, have also been prepared. It was shown that the release of drugs from nanocomposite hydrogel wound dressings is more prolonged with a slower release rate compared with conventional hydrogel wound dressings. Figure 7.14 shows the fractional release curves for gentamicin sulphate in PVA hydrogel wound dressing and PVA nanocomposite hydrogel wound dressings containing 2-10% of organically modified MMT (OMONT).

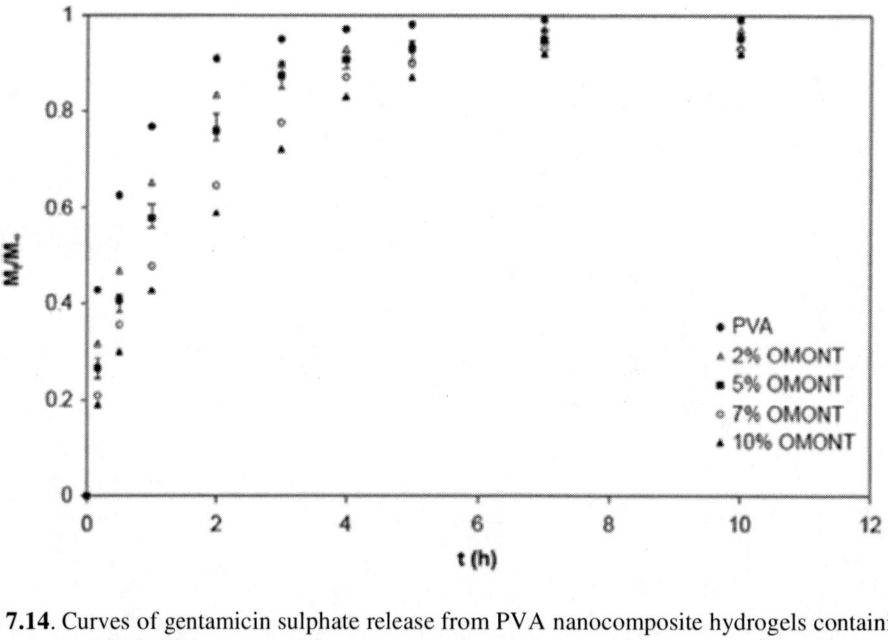

Figure 7.14. Curves of gentamicin sulphate release from PVA nanocomposite hydrogels containing 0-10 wt.% of OMONT (Sirousazar, 2013).

The results exhibit that at any given release time, the fractional gentamicin sulphate release from the system based on pure PVA hydrogel wound dressing was always higher than those based on nanocomposite hydrogels. In addition, the fractional release of the drug from nanocomposite hydrogel drug delivery systems is decreased by increasing the loading level of OMONT. On the other view of point, the results showed that the required time to reach a specified gentamicin sulphate release level was shorter for pure PVA hydrogel wound dressing than the PVA/OMONT nanocomposite hydrogel wound dressings. Furthermore, this time was longer for nanocomposite hydrogel wound dressings with higher loading levels of incorporated OMONT. The results indicate that the gentamicin sulphate release from the nanocomposite hydrogel wound dressings is more prolonged at a slower rate. The antibiotics-loaded nanocomposite hydrogel wound dressings have been recommended for infected wounds, where continuous antibiotic release is required during the healing process of the wound (Sirousazar, 2013).

BNCH wound dressings, as a novel series of nanocomposite hydrogel wound dressings, are also considered to show non-cytotoxic activity and have a satisfactory level of biocompatibility. The result of *in vitro* cytotoxicity assay of a BNCH wound dressing based on PVA and EW, as the matrix phase of BNCHs, and MMT nanoclay, as the reinforcing phase, is shown in Figure 7.15. As can be seen from Figure 7.15, the optical density (OD) absorption of BNCH was higher than the control sample. It is attributed to the presence of nutritive species, for example, albumin, which helps to increase the number of alive cells by feeding them (Sirousazar et al., 2016).

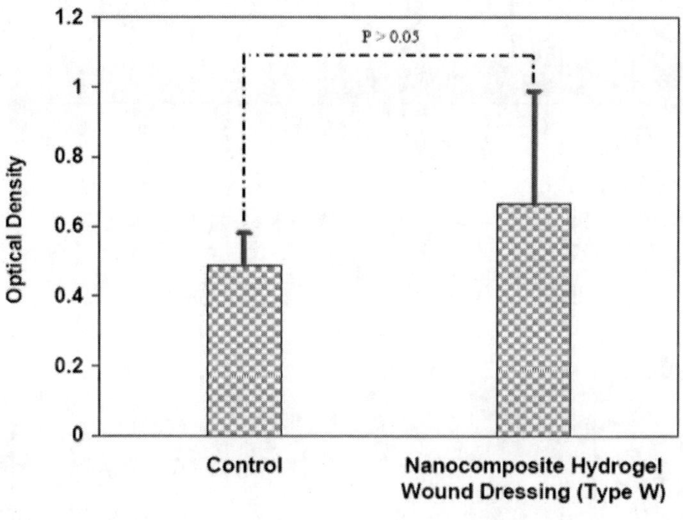

Figure 7.15. MTT assay results of PVA/EW/MMT nanocomposite hydrogel wound dressing on PBMC cells (Sirousazar et al., 2016).

In addition, according to observations, a thin layer of BNCH results in extraordinary protection of the wound against infection, while even 64 layers of sterile gauze are not able to prevent bacterial penetration (Sirousazar et al., 2016). It has also been shown that the EWC and WVTR of nanocomposite hydrogel wound dressings were very close to the properties of human skin. Therefore, they can interact appropriately with the damaged tissue and protect the wound like artificial skin until complete healing. Figures 7.16 and 7.17 show the EWC and WVTR values of PVA/EW/MMT nanocomposite hydrogel wound dressings against MMT content, respectively. Incorporating MMT into the nanocomposite hydrogel wound dressing matrix results in increasing the gel content and decreasing the average pore size of the network. As a result, the water absorption ability of nanocomposite hydrogel wound dressing was lower than that of the corresponding MMT-free sample. Similarly, the value of WVTR is dependent on MMT content. Highly dispersed layers of MMT make more tortuous and indirect paths for water molecules to exit from the wound dressing (Jahani-Javanmardi et al., 2016). The lower EWC and higher gel fraction restrict the motion of drug molecules inside the interconnected network of BNCH due to the creation of labyrinthine paths, which in turn causes a prolonged release of the loaded drugs inside the dressing. On that account, BNCH wound dressings can be helpful for a more extended time without displacement (Delir et al., 2020).

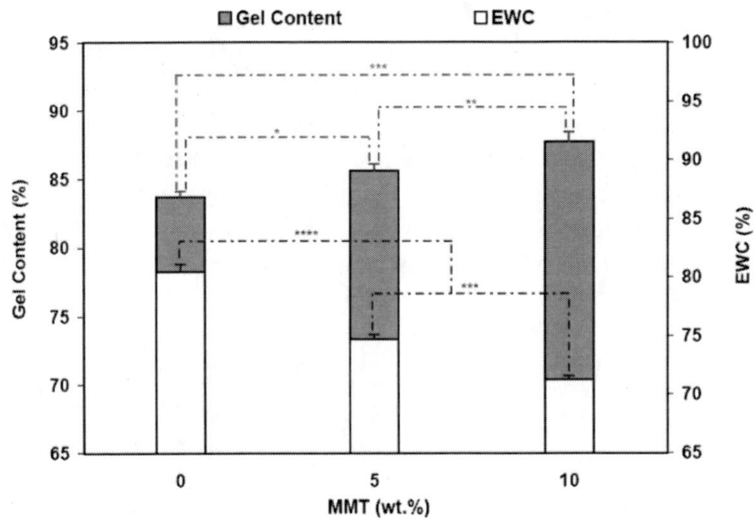

Figure 7.16. Gel content and EWC values of PVA/EW/MMT nanocomposite hydrogels (Jahani-Javanmardi et al., 2016).

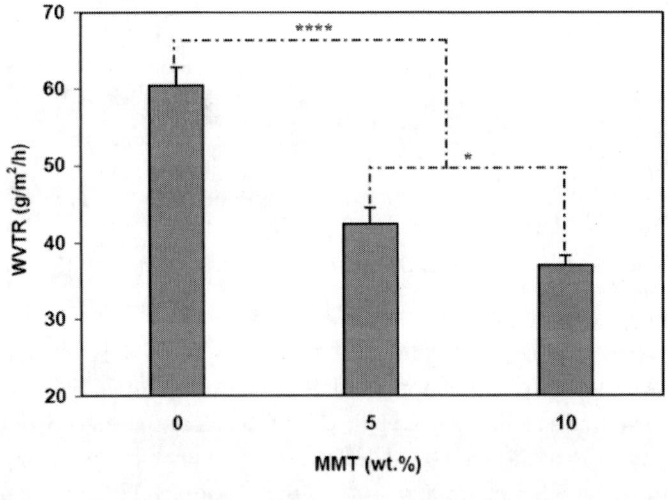

Figure 7.17. WVTR of PVA/EW/MMT nanocomposite hydrogels versus MMT loading level (Jahani-Javanmardi et al., 2016).

Owing to the properties mentioned above, BNCHs can be considered as promising and functional wound dressings for practical applications. As expected, the *in vivo* studies confirmed the capability of these wound dressings in accelerating the healing process of wounds. Figure 7.18 shows the wound healing progress in three representative groups of animals for honey-loaded PVA/EW/MMT nanocomposite hydrogel wound dressings.

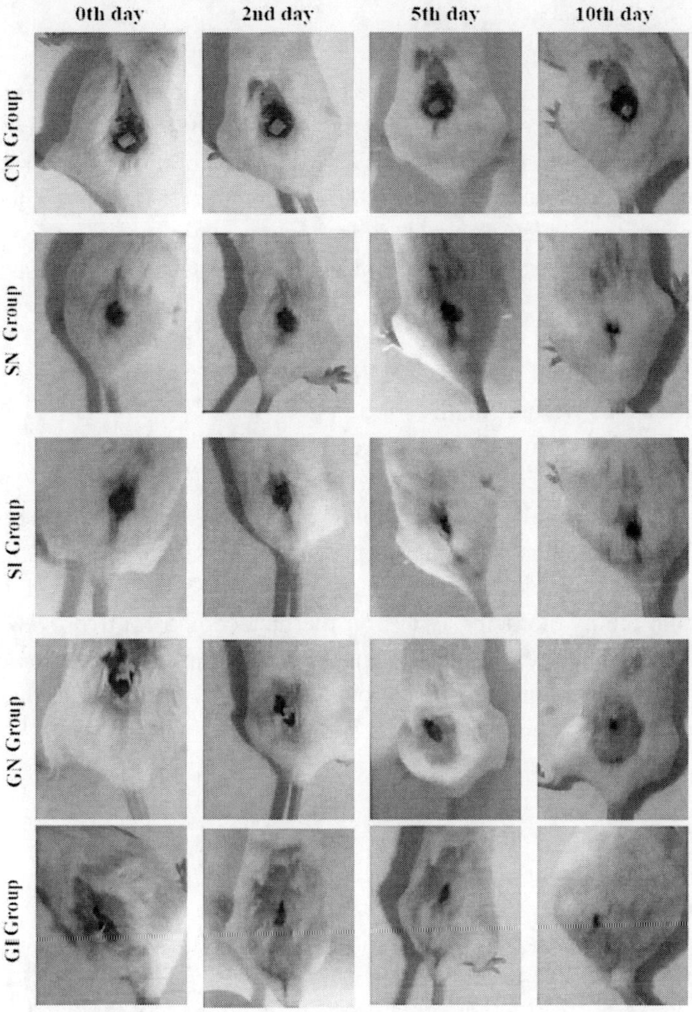

Figure 7.18. Macroscopic appearances of the wounds of animals of CN, SN, and SI groups during the wound healing process (Rafati et al., 2020).

In the CN group (as the control group), the wounds were covered by commercial wound dressing, i.e., Chitopad. In SN and SI groups, drug (honey)-loaded BNCH wound dressings were used for covering the non-infected and infected wounds, respectively. It can be seen that on the second day, there is no significant difference among the groups. However, after the second day, the healing progress in SN and SI groups is more pronounced. The faster healing of SN and S1 groups than the CN group can be attributed to their ability to sustain a wet condition on top of the wound throughout the healing process. It is worth mentioning that proteins of EW play an essential role in collagen building and accelerating the healing process. In addition, the presence of honey, as a natural antibiotic, inside the nanocomposite hydrogel wound dressings can control the infection in wounds and accelerate the wound healing process (Rafati et al., 2020).

Conclusion

Nanocomposite hydrogel wound dressings, as a novel class of materials used in wound care, demonstrated excellent ability in promoting the healing process of wounds. Their high water contents control prolonged swelling and dehydration abilities, and proper water vapor transmission rates donate them the ability to keep a continuous permanent moist environment on the wound's surface, which causes the fast healing of wounds. Nanocomposite hydrogel wound dressings also have excellent mechanical properties, such as increased tensile strength, elastic modulus, and hardness (decreased adhesion), and decreased elongation at break. Therefore, they are recognized as unique choices for wound management and their mechanical properties fulfill the essential requirements for wound dressing to use on wounds even under high stresses such as the wounds on the sole.

The nanocomposite hydrogel wound dressings, which are entirely impenetrable against microbes, have successfully passed the *in vivo* assays on animal models. The *in vivo* assays show their excellent potential in faster healing of animal wounds with better qualities than conventional wound dressings. The nanocomposite hydrogel wound dressings have the drug loading ability and the antibiotic-loaded ones seem to be excellent candidates for infected wounds. Although the cytotoxicity assays on the nanocomposite hydrogel wound dressings proved their non-toxic nature, the design and fabrication of more non-toxic nanocomposite hydrogel wound dressings with acceptable biocompatibility could be an interesting issue for the following researches in this field.

References

Ahmed, E. M. (2015). Hydrogel: preparation, characterization, and applications: a review. *Journal of Advanced Research*, 6(2), 105-121.

Ajayan, P. M., Schadler, L. S., and Braun, P. V. (2006). *Nanocomposite Science and Technology*. John Wiley and Sons.

Babaei, J., Mohammadian, M., and Madadlou, A. (2019). Gelatin as texture modifier and porogen in egg white hydrogel. *Food Chemistry*, 270, 189-195.

Bakshi, S. R., Lahiri, D., and Agarwal, A. (2010). Carbon nanotube reinforced metal matrix composites-a review. *International Materials Reviews*, 55(1), 41-64.

Birkholz, M., Albers, U., and Jung, T. (2004). Nanocomposite layers of ceramic oxides and metals prepared by reactive gas-flow sputtering. *Surface and Coatings Technology*, 179(2-3), 279-285.

Bishop, W. J. (1959). *A History of Surgical Dressings*. Wheat Bridge Mills, Chesterfield.

Breitkreutz, D., Mirancea, N., and Nischt, R. (2009). Basement membranes in skin: unique matrix structures with diverse functions? *Histochemistry and Cell Biology*, 132(1), 1-10.

Daunton, C., Kothari, S., Smith, L., and Steele, D. (2012). A history of materials and practices for wound management. *Wound Practice and Research*, 20(4), 174-186.

Delir, S., Sirousazar, M., and Kheiri, F. (2020). Clindamycin releasing bionanocomposite hydrogels as potential wound dressings for the treatment of infected wounds. *Journal of Biomaterials Science, Polymer Edition*, 31(12), 1489-1514.

Diez-Pena, E., Quijada-Garrido, I., and Barrales-Rienda, J. M. (2002). Hydrogen-bonding effects on the dynamic swelling of P (N-iPAAm-co-MAA) copolymers. A case of autocatalytic swelling kinetics. *Macromolecules*, 35(23), 8882-8888.

Edwards, R., and Harding, K. G. (2004). Bacteria and wound healing. *Current Opinion in Infectious Diseases*, 17(2), 91-96.

Gamgee, S. (1880). Absorbent and medicated surgical dressings. *The Lancet*, 115(2947), 308-309.

Ganji, F., and Vasheghani-Farahani, E. (2009). Hydrogels in controlled drug delivery systems. *Iranian Polymer Journal*, 18(1), 63-88.

Gonzalez, J. S., Ludueña, L. N., Ponce, A., and Alvarez, V. A. (2014). Poly (vinyl alcohol)/cellulose nanowhiskers nanocomposite hydrogels for potential wound dressings. *Materials Science and Engineering: C*, 34, 54-61.

Gosain, A., and DiPietro, L. A. (2004). Aging and wound healing. *World Journal of Surgery*, 28(3), 321-326.

Guo, S., and Dipietro, L. A. (2010). Factors affecting wound healing. *Journal of Dental Research*, 89(3), 219-229.

Haraguchi, K. (2007). Nanocomposite hydrogels. *Current Opinion in Solid State and Materials Science*, 11(3-4), 47-54.

Haraguchi, K., and Takada, T. (2005). Characteristic sliding frictional behavior on the surface of nanocomposite hydrogels consisting of organic-inorganic network structure. *Macromolecular Chemistry and Physics*, 206(15), 1530-1540.

Jahani-Javanmardi, A., Sirousazar, M., Shaabani, Y., and Kheiri, F. (2016). Egg white/poly (vinyl alcohol)/MMT nanocomposite hydrogels for wound dressing. *Journal of Biomaterials Science, Polymer Edition*, 27(12), 1262-1276.

Kabiri, K., Omidian, H., Zohuriaan-Mehr, M. J., and Doroudiani, S. (2011). Superabsorbent hydrogel composites and nanocomposites: a review. *Polymer Composites*, 32(2), 277-289.

Kamigaito, O. (1991). What can be improved by nanometer composites? *Journal of the Japan Society of Powder and Powder Metallurgy*, 38(3), 315-321.

Kaşgöz, H., Durmuş, A., and Kaşgöz, A. (2008). Enhanced swelling and adsorption properties of AAm-AMPSNa/clay hydrogel nanocomposites for heavy metal ion removal. *Polymers for Advanced Technologies*, 19(3), 213-220.

Kokabi, M., Sirousazar, M., and Hassan, Z. M. (2007). PVA-clay nanocomposite hydrogels for wound dressing. *European Polymer Journal*, 43(3), 773-781.

Kruis, F. E., Fissan, H., and Peled, A. (1998). Synthesis of nanoparticles in the gas phase for electronic, optical and magnetic applications-a review. *Journal of Aerosol Science*, 29(5-6), 511-535.

Lalwani, G., Henslee, A. M., Farshid, B., Lin, L., Kasper, F. K., Qin, Y.-X., Mikos, A. G., and Sitharaman, B. (2013a). Two-dimensional nanostructure-reinforced biodegradable polymeric nanocomposites for bone tissue engineering. *Biomacromolecules*, 14(3), 900-909.

Lalwani, G., Henslee, A. M., Farshid, B., Parmar, P., Lin, L., Qin, Y.-X., Kasper, F. K., Mikos, A. G., and Sitharaman, B. (2013b). Tungsten disulfide nanotubes reinforced biodegradable polymers for bone tissue engineering. *Acta Biomaterialia*, 9(9), 8365-8373.

Lawrence, J. C. (1992). Burn bacteriology during the last 50 years. *Burns*, 18, S23-S29.

Lawrence, J. C. (1994). Dressings and wound infection. *The American Journal of Surgery*, 167(1), S21-S24.

Leaper, D. J., and Harding, K. G. (1998). *Wounds: Biology and Management*. Oxford Medical Publications.

Manias, E. (2007). Stiffer by design. *Nature Materials*, 6(1), 9-11.

Morgan, A. B., and Wilkie, C. A. (2007). *Flame Retardant Polymer Nanocomposites*. John Wiley and Sons.

Pal, K., Banthia, A. K., and Majumdar, D. K. (2009). Polymeric hydrogels: characterization and biomedical applications. *Designed Monomers and Polymers*, 12(3), 197-220.

Parmoor, S., Sirousazar, M., Kheiri, F., and Kokabi, M. (2016). Nanoclay and Cu nanoparticles loaded polyethylene nanocomposites for natural gas transfer applications. *Journal of Macromolecular Science, Part B*, 55(4), 331-343.

Paul, D. R., and Robeson, L. M. (2008). Polymer nanotechnology: nanocomposites. *Polymer*, 49(15), 3187-3204.

Peppas, N. A, Bures, P., Leobandung, W. S., and Ichikawa, H. (2000). Hydrogels in pharmaceutical formulations. *European Journal of Pharmaceutics and Biopharmaceutics*, 50(1), 27-46.

Peppas, N. A, Hilt, J. Z., Khademhosseini, A., and Langer, R. (2006). Hydrogels in biology and medicine: from molecular principles to bionanotechnology. *Advanced Materials*, 18(11), 1345-1360.

Qiu, Y., and Park, K. (2001). Environment-sensitive hydrogels for drug delivery. *Advanced Drug Delivery Reviews*, 53(3), 321-339.

Rafati, Z., Sirousazar, M., Hassan, Z. M., and Kheiri, F. (2020). Honey-loaded egg white/poly (vinyl alcohol)/clay bionanocomposite hydrogel wound dressings: *in vitro* and *in vivo* evaluations. *Journal of Polymers and the Environment*, 28(1), 32-46.

Rajendran, S. (2018). *Advanced Textiles for Wound Care*. Woodhead Publishing.

Razzak, M. T., and Darwis, D. (2001). Irradiation of polyvinyl alcohol and polyvinyl pyrrolidone blended hydrogel for wound dressing. *Radiation Physics and Chemistry*, 62(1), 107-113.

Ryan, K. R., Gourley, J. R., and Jones, S. E. (2009). Environmental anomalies at the world trade center: evidence for energetic materials. *The Environmentalist*, 29(1), 56-63.

Sai K., P., and Babu, M. (2000). Collagen based dressings: a review. *Burns*, 26, 54-62.

Sapalidis, A. A., Katsaros, F. K., and Kanellopoulos, N. K. (2011). PVA/montmorillonite nanocomposites: development and properties. *Nanocomposites and Polymers with Analytical Methods*, 29-50.

Schexnailder, P., and Schmidt, G. (2009). Nanocomposite polymer hydrogels. *Colloid and Polymer Science*, 287(1), 1-11.

Schönfelder, U., Abel, M., Wiegand, C., Klemm, D., Elsner, P., and Hipler, U. C. (2005). Influence of selected wound dressings on PMN elastase in chronic wound fluid and their antioxidative potential *in vitro*. *Biomaterials*, 26(33), 6664-6673.

Shaabani, Y, Sirousazar, M., and Kheiri, F. (2016b). Synthetic-natural bionanocomposite hydrogels on the basis of polyvinyl alcohol and egg white. *Journal of Macromolecular Science, Part B*, 55(8), 849-865.

Shaabani, Yasaman, Sirousazar, M., and Kheiri, F. (2016a). Crosslinked swellable clay/egg white bionanocomposites. *Applied Clay Science*, 126, 287-296.

Shai, A., and Maibach, H. I., (2005). Wound Healing and Ulcers of The Skin. Springer.

Singh, B. P., Singh, D., Mathur, R. B., and Dhami, T. L. (2008). Influence of surface modified MWCNTs on the mechanical, electrical and thermal properties of polyimide nanocomposites. *Nanoscale Research Letters*, 3(11), 444-453.

Singh, B., Sharma, S., and Dhiman, A. (2013). Design of antibiotic containing hydrogel wound dressings: biomedical properties and histological study of wound healing. *International Journal of Pharmaceutics*, 457(1), 82-91.

Sirousazar, M. (2013). Mechanism of gentamicin sulphate release in nanocomposite hydrogel drug delivery systems. *Journal of Drug Delivery Science and Technology*, 23(6), 619-621.

Sirousazar, M., and Kokabi, M. (2012). Intelligent nanocomposite hydrogels. *Intelligent Nanomaterials: Processes, Properties, and Applications*, 487-531.

Sirousazar, M., Forough, M., Farhadi, K., Shaabani, Y., and Molaei, R. (2014). Hydrogels: properties, preparation, characterization and biomedical, applications in tissue engineering, drug, delivery and wound care. *Advanced Healthcare Materials*, 295-357.

Sirousazar, M., Jahani-Javanmardi, A., Kheiri, F., and Hassan, Z. M. (2016). *In vitro* and *in vivo* assays on egg white/polyvinyl alcohol/clay nanocomposite hydrogel wound dressings. *Journal of Biomaterials Science, Polymer Edition*, 27(16), 1569-1583.

Sirousazar, M., Khadivi, H., and Delir, S. (2020). Swelling and drying mechanisms of freeze-thawed polyvinyl alcohol/egg white/montmorillonite bionanocomposite hydrogels. *Journal of Macromolecular Science, Part B*, 59(5), 309-330.

Sirousazar, M., Kokabi, M., and Hassan, Z. M. (2011). *In vivo* and cytotoxic assays of a poly (vinyl alcohol)/clay nanocomposite hydrogel wound dressing. *Journal of Biomaterials Science, Polymer Edition*, 22(8), 1023-1033.

Sirousazar, M., Kokabi, M., and Hassan, Z. M. (2012a). Swelling behavior and structural characteristics of polyvinyl alcohol/montmorillonite nanocomposite hydrogels. *Journal of Applied Polymer Science*, 123(1), 50-58.

Sirousazar, M., Kokabi, M., Hassan, Z. M., and Bahramian, A. R. (2012b). Nanoporous nanocomposite hydrogels composed of polyvinyl alcohol and Na-montmorillonite. *Journal of Macromolecular Science, Part B*, 51(8), 1583-1595.

Sirousazar, M., Yari, M., Achachlouei, B. F., Arsalani, J., and Mansoori, Y. (2007). Polypropylene/montmorillonite nanocomposites for food packaging. *E-Polymers*, 7(1), 305-313.

Thornton, M. J. (2002). The biological actions of estrogens on skin. *Experimental Dermatology*, 11(6), 487-502.

Wang, T., Zhu, X. K., Xue, X. T., and Wu, D. Y. (2012). Hydrogel sheets of chitosan, honey and gelatin as burn wound dressings. *Carbohydrate Polymers*, 88(1), 75-83.

Watson, N. F. S., and Hodgkin, W. (2005). Wound dressings. *Surgery*, 23(2), 52-55.

Winter, G. D. (1962). Formation of the scab and the rate of epithelization of superficial wounds in the skin of the young domestic pig. *Nature*, 193(4812), 293-294.

Wodash, A. J. (2012). Wet-to-dry dressings do not provide moist wound healing. *Journal of the American College of Clinical Wound Specialists*, 4(3), 63-66.

Zahedi, P., Rezaeian, I., Ranaei-Siadat, S., Jafari, S., and Supaphol, P. (2010). A review on wound dressings with an emphasis on electrospun nanofibrous polymeric bandages. *Polymers for Advanced Technologies*, 21(2), 77-95.

Zhang, S., Sun, D., Fu, Y., and Du, H. (2003). Recent advances of superhard nanocomposite coatings: a review. *Surface and Coatings Technology*, 167(2-3), 113-119.

Chapter 8

Nanocomposite Hydrogels: A Promising Platform for Chronic Wounds and Drug Delivery Systems

Patel Priya
and Patel Gayatri*
Ramanbhai Patel College of Pharmacy,
Charotar University of Science and Technology (CHARUSAT), Changa, India

Abstract

Biomimetic hydrogels and nanostructures are combined in nanocomposite hydrogels. Hydrogels have been extensively studied for usage as scaffolds or carriers of therapeutic agents, such as small medicines and biomolecules, in the treatment of a wide range of diseases. The biocompatibility, oxygen and nutrient permeability, features that are similar to those of the native extracellular matrix, and porous structures of the nanocomposite hydrogels allow therapeutic medicines to be loaded and released in a controlled and extended-release manner. In this chapter, recent works on different nanocomposite hydrogel matrices useful for therapeutic delivery in the treatment of complicated wounds and the production of drug delivery systems are described. Both patent and non-patent literature are also highlighted. The rationale for selecting nanofillers for hydrogel matrices is also discussed.

Keywords: nanofillers, biocompatibility, infected wound, hydrogel matrix, therapeutic carrier

8.1. Introduction to Hydrogels

Hydrogels are polymeric crosslinked networks that have the ability to swell but not dissolve in water. Hydrogels are excellent biomaterials due to their low interfacial tension, increased permeability to tiny molecules, and soft and rubbery characteristics. Hydrogels are a class of materials that exhibit a three-dimensional (3D) network made of hydrophilic/hydrophobic homopolymers or copolymers that are crosslinked to create either soluble or insoluble polymer matrices. Hydrogels have a high absorption capacity for water or biological fluids (Ahmed,

* Corresponding Author's Email: gayatripatel.ph@charusat.ac.in.

In: Nanocomposite Hydrogels and their Emerging Applications
Editor: Mohammad Sirousazar
ISBN: 979-8-88697-675-5

2015). The term hydrogel was employed long ago, in 1894, for the first time by Bemmelen to explain the formulated colloidal gel. During that period of 1894, colloidal gels of inorganic salts were described as hydrogels, which were completely different from the type of hydrogels we are using nowadays. Presently, hydrogels are described as a hydrophilic natural polymer and gum network that is created either chemically or physically in three dimensions. Wichterle and Lim (1960) published the first report on the use of hydrogels, which application was in the biomedical industry. Large amounts of water are absorbed by hydrogels, causing swelling and giving them their rubbery and supple appearance. Hydrogels are advantageous in biomedical applications because of their rubbery and soft characteristics. Thanks to later advancements in 1980, hydrogels are now usable as diaper products in France and Germany. Numerous publications describing different perspectives have been published to thoroughly examine hydrogel materials. Scientists from DuPont created the first synthetic hydrogel in 1936, known as poly (2-hydroxyethyl methacrylate) (PHEMA). The relevance of PHEMA hydrogels in contact lens applications was first established in 1960 by Wichterle and Lim.

Because changes in hydrogels can be seen in relation to changes in environmental circumstances, some hydrogels are also referred to as "smart hydrogels" which respond to changes in solvent composition, temperature, pH, electrical field, etc. The hydrophilic properties of functional groups like hydroxylic (-OH), carboxylic (-COOH), amidic (-CONH-), primary amidic (-$CONH_2$), and sulphonic (-SO_3H) on the polymeric backbone of hydrogels allow them to absorb water, while crosslinks between network chains prevent them from dissolving (Rafieian et al., 2019). Covalent coupling, ion connections, chain entanglements, and other processes that contribute to hydrogel crosslinks are responsible for a variety of physical structures, chemical combinations, and connections.

Surprisingly, the majority of the physicochemical characteristics of the hydrogel for pharmacological research are defined by the crosslinkage itself. The primary synthetic components of hydrogels are typically an initiator, a monomer, and a crosslinker. By adjusting synthetic parameters, including initiator and monomer concentrations, reaction temperature, vessel, and duration, and the crosslinker to monomer ratio, it is possible to control the properties of hydrogels. Hydrogels are considered to be promising materials because of their exceptional qualities, including flexibility and adaptability in fabrication, variety in composition, high tunability in the physical, chemical, and biological properties, high moldability in shape, especially their excellent biocompatibility, and similarity to native extracellular matrix (ECM), and they have been the subject of the most in-depth academic and industrial research. Hydrogels are interesting materials for a variety of applications because of their wide range of characteristics, including their capacity for absorption, swelling behavior, stability and degradation, bioadhesion and bioactivity, permeability, and mechanical, optical, and surface properties. Today, hydrogels are used in a wide range of biomedical applications, such as antibacterial, tissue engineering, medication delivery, and wound dressing. Multifunctional hydrogels and dynamic interactions with cellular microenvironments are necessary for the majority of these applications.

One of their main drawbacks, particularly when utilized for tissue engineering (for example cartilage tissue) or other applications, is the need for high strength, enhanced compressive and tensile characteristics, strong elasticity, and endurance simultaneously. They are challenging to handle and use for a particular purpose due to their poor mechanical qualities. The focus of hydrogel research has recently shifted to improving its chemical and mechanical characteristics, particularly for biomedical applications (Dalwadi et al., 2013, 2015). New forms of hydrogels, such as double network hydrogels and nanocomposite hydrogels, are being created in an effort to enhance these characteristics. One of these is the development of nanocomposite hydrogels, which has drawn a lot of research attention aimed at enhancing the material properties of hydrogels and extending the scope of their applications in the medical and biotechnological industries.

8.2. Introduction to Wounds and Their Healing Process

A wound is a skin defect or split resulting from thermal or physical trauma, or pathological or physiological conditions. Since healthy skin is necessary for the body's capacity to retain fluids, defend against external infections, and provide thermal insulation, skin injuries constitute a major health risk to people. The way that the healing process is structured makes it easier to distinguish between acute and chronic wounds. Annually, millions of people experience acute and chronic wounds whose treatments lead to serious health issues, financial loss, stress, and a reduction in the patient's quality of life. Complex chronic wounds affect at least 1% of people in developed countries and 5% of people in developing countries during the course of their lifetimes (Patel and Joshi, 2019).

Acute wounds are frequently tissue damage that heal within the anticipated time range, typically 8 to 12 weeks, with little to no scarring. Internal injuries brought on by harmful elements, such as puncture wounds and tears brought on by frictional contact between the skin and rough surface, are among the main causes of acute wounds. Knives, gunfire, and surgical incisions made to remove tumors can all result in wounds from mechanical mishaps. Another category of acute wounds encompasses injuries and chemical reactions brought on by a variety of factors, including heat, energy, corrosive substances, and thermal causes. On the other hand, tissue injuries that result in chronic wounds are painful to heal, take longer than 12 weeks to heal, and occasionally recur. The steps of the wound healing process are described in Figure 8.1.

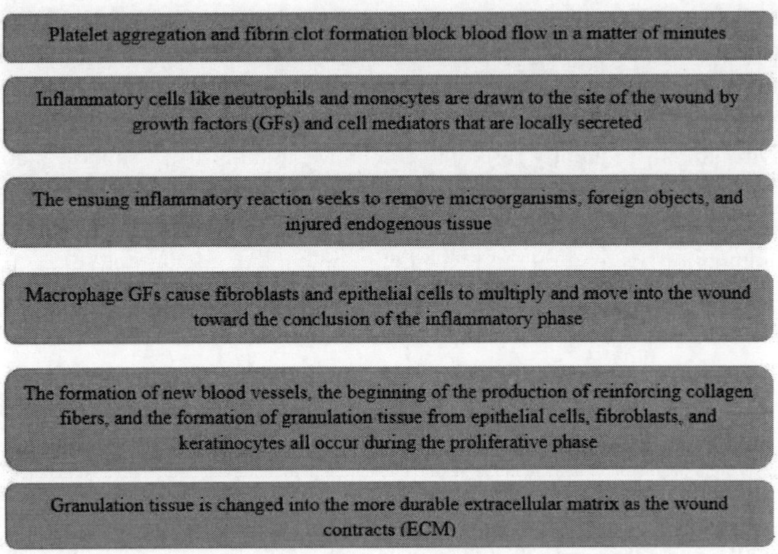

Figure 8.1. Steps of the wound healing process.

8.3. Current Treatments of Chronic Wounds

The environment and location of the wound determine how it is currently treated for chronic wounds. However, non-healing pressure ulcer therapies include dressings that hasten wound healing and release tissue pressure. Venous ulcers and diabetic foot ulcers require dressing that preserves the moist environment conducive to wound healing and compression. Traditional bandages, such as gauze or cotton bandages with a gauze weave, prevent wounds from bacterial contamination and enable gaseous and fluid exchange. They can be used to treat superficial or non-infected wounds, or as secondary dressings to cover complicated dressings. These materials stand out for being affordable and simple to use. However, drawbacks, such as the necessity for frequent replacement, inability to regulate moisture levels, and adhesion to the wound bed have limited their application in wound treatment (Koehler et al., 2018).

New synthetic dressings that can create an appropriate moist environment are readily available nowadays. These dressings can occlude completely or partially. They could take the shape of a hydrocolloid, hydrogel, film, or foam. Polyvinyl alcohol (PVA), poly(lactide-co-glycolide), polyurethanes, polyethylene glycol (PEG), polycaprolactone (PCL), nylon, and silicone are only a few examples of the synthetic materials used to make them. Due to their high-water content, hydrogels are regarded as promising biomaterials appropriate for use as wound dressings. While absorbing excessive exudate, hydrogels can keep the wound interface moist while still allowing gaseous exchange, acting as barriers to bacteria, and not adhering to the wound bed.

In order to acquire the best mechanical qualities, hydrogels are typically comprised of both natural polymers (like collagen and chitosan) and synthetic polymers (like PVA and PEG). Hydrogel dressings are appropriate for treating dry and necrotic wounds due to their high water content, which is around 90% water and 10% natural or synthetic polymers. Due to the sorption properties of hydrogel materials, the moist environment provided by hydrogels promotes the cellular wound resolution process and allows for successful debridement or the elimination of necrotic tissue and foreign material. Hydrogel-based dressings have been demonstrated to absorb up to 1000 g of wound exudate for every gram of dressing (Stefanov et al., 2017). Further enabling unhindered interchange of gaseous CO_2, O_2, and H_2O, permeable hydrogel structures allow the tissue to "breathe."

It is interesting to note that hydrogel structures, which are polymeric networks in an aqueous environment, frequently resemble the ECM, making it possible to incorporate cells and biomolecules into dressings for next-generation biologically enhanced wound treatments (da Silva et al., 2019). By carefully designing the structure of the hydrogel's polymer network using a combination of natural and synthetic precursors, additional bioactivity may be generated, providing wound dressings with brand-new wound healing capabilities. Some rationals for use of hydrogels in wound healing are:

- Help in the rehydration of dead tissues and increase the healing of debridement
- Permeable to metabolites
- Non-irritant
- Non-adherent
- Cool the surface of the wound
- Prevent the hardening of the wound by keeping it moist

- Provide the barrier for external infection and reduce the frequency of dressing
- Provide flexibility according to wound size and shape
- Comfortable delivery of drugs
- Facilitate controlled delivery of drugs to their intended site.

Chronic wounds share the following characteristics despite having no common origin or cause and being clinically and molecularly heterogeneous: bacterial contamination and a high quantity of reactive oxidative species (ROS), myeloperoxidase (MPO), and matrix metalloproteinases (MMPs). ECM and growth factors are excessively degraded in chronic wounds because of an imbalance between the high levels of MMPs and the tissue inhibitors of matrix metalloproteinases (TIMPs). Anyone therapy technique is unlikely to have a satisfactory healing effect since chronic wounds are complex in nature (Koehler et al., 2018). An effective therapeutic strategy for the healing of chronic wounds must include:

- Control the oxidative and proteolytic enzymatic activities that are caused by the infection at the wound site
- Create an environment devoid of microorganisms
- Keep the tissue wet while absorbing the excessive exudates, keep the tissue wet.

Therefore, there is a need for new delivery systems for chronic wounds which satisfy the limitation of conventional as well as standard wound dressings. Due to their properties, nanotechnology platforms have shown new potentials and advantages in the field. The transport of biomolecules, like DNA/RNA or growth factors that can be used in the treatment of chronic wounds, has been made possible by recent advances in nanotechnology. These agents can be delivered intracellularly due to their small size and physicochemical characteristics. Together, they enable topical administration and lengthen these agents' half-lives, reducing the quantity of administrations and expenses. Additionally, diverse drug release patterns that can accommodate the needs of wound healing are made possible by the encapsulation of pharmaceuticals and biomolecules inside nanocarriers. We will outline the research findings of the hydrogel with nanocarriers integrated utilized to treat chronic wounds in the following sections (Rafieian et al., 2019).

Over the past two decades, rapid advancements in nanotechnology have made it possible to include various therapeutic, sensor, and targeting agents into the nanocarrier for the purpose of detection, prevention, and therapy. Hydrogels have undergone much research and study in this area. The topic of nanomedicine, which deals with the medical uses of nanomaterials and biological devices, is still in its early stages. With the help of biological molecules' synergistic interactions with nanomaterials, nanomedicine has made significant strides. Because of the interaction between biological molecules and nanomaterials, novel devices, analytical instruments, physical therapies, and carriers for proteins and enzymes have all been developed (Rafieian et al., 2019). It is important to note that research on the use of hydrogels in nanomedicine is moving extremely quickly all around the world. Today's materials call for the creation of inorganic-organic frameworks with soft hydrogel networks.

Haraguchi created ground-breaking clay nanocomposite hydrogels in 2002. When compared to traditional hydrogels, these nanocomposite hydrogels showed superior mechanical and swelling/deswelling characteristics. A significant development in the field of nanocomposite hydrogels resulted from this study. The crosslinked three-dimensional networks created in the presence of nanostructures are what are known as nanocomposite hydrogels. Crosslinking can be done either physically or chemically to create these structures. In addition to hydrophobic and ionic interactions, hydrogen bonds are also used to create physical links. Strong covalent bonds are formed as a result of chemical crosslinking. These nanocomposite hydrogels have the capacity to alter their characteristics in response to outside stimuli. The crosslinked polymeric network and the numerous nanofillers work together to enhance the distinctive features of the nanocomposite hydrogel systems in a remarkable way (Sun et al., 2017). Designing precisely specified three-dimensional structures that enable customizing their mechanical characteristics can be done by manipulating and controlling hydrogel structures at the nanoscale range. The hydrogel's mechanical and physical properties can be controlled in part by the incorporation of nanometric-sized structures with various morphologies. Because lithographic methods cannot be used due to the high-water content, developing hydrogel nanostructuring technologies is still difficult. On the other hand, the ability of hydrophobic moieties to self-assemble in an aqueous environment can be used to broadly and reasonably approach hydrogel nanostructuring.

Nanocarriers play a crucial role in improving medication formulation by increasing the water-solubility of hydrophobic drugs and stabilizing chemicals that are easily degraded. The augmentation of drug retention in tissues, protection from enzymatic degradation, improvement of cellular uptake, and highly targeted administration are some of the tremendous potentialities of nanosystems, all of which have a significant impact on the effectiveness of treatments. Additionally, nanocarriers are being thoroughly researched as theranostic and diagnostic tools that can be used to simultaneously detect, treat, and track the development of cancers. As shown by the use of lipid nanoparticles (NPs) for the administration of the mRNA-based vaccine against SARS-CoV-19, the delivery of mRNA represents the most current application of nano-based technology. Depending on the substance utilized to generate them, nanosized systems can be categorized as lipidic, polymeric, inorganic, peptide-based, and virus-like nanocarriers (Sun et al., 2017).

8.4. Finding Framework

To increase survey reliability, search systems were initially built. Independent researchers accurately and methodically analyzed the available literature to create a list of potential publications. To search for papers, keywords, such as "nanoparticles," "liposomes," "nanoemulsion," "microemulsion," "nanogel," "nanotube," and "nanomicelles" integrated into hydrogel carrier for the treatment of chronic wounds and combinations of these were used. All watchwords were evaluated using Google Scholar, Web of Science, Scopus, Google Patents, Patseer, Lens, clinicaltrials.org, and ScienceDirect. Different nanocomposite hydrogels have been researched recently for the treatment of chronic wounds. Nanocarriers typically have sizes between 1 and 100 nm, are designed using techniques that significantly affect the physical and chemical properties of molecular-scale structures, and can theoretically be combined to form bigger structures. The research on nanocomposite hydrogels for the treatment of chronic wounds is shown in a flowchart, as presented in Figure 8.2.

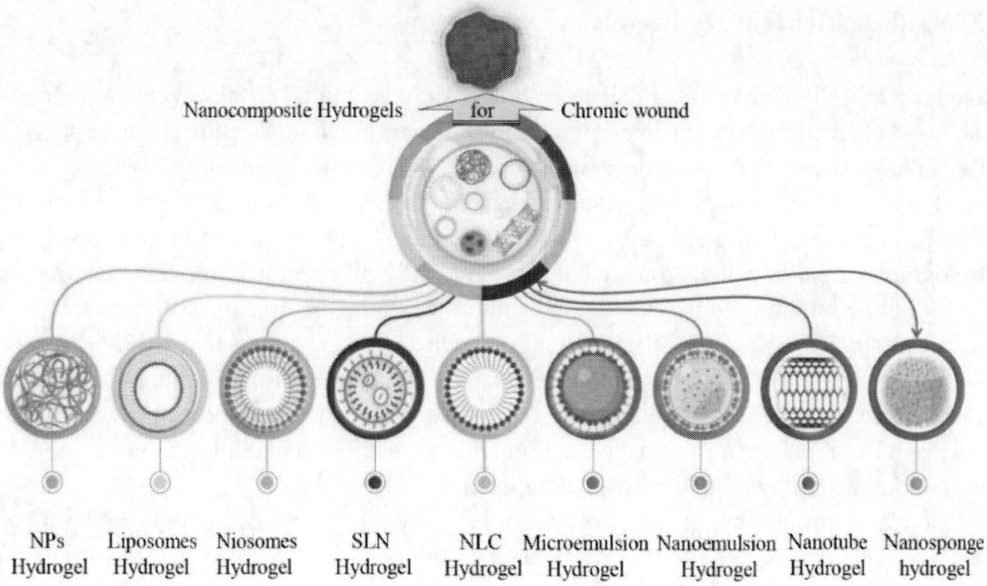

Figure 8.2. Various nanocomposite hydrogels for the treatment of chronic wounds.

More than 870 patent publications on the use of these nanocarriers for various administration routes using hydrogel carriers have been published in the last decade, attesting to the significant research that has been done in the field of wound healing. However, very few clinical trials have been completed or are in progress, and very few products have been approved. Figure 8.3 indicates all the possible documents showing the nanocomposite hydrogel in the field of wound healing.

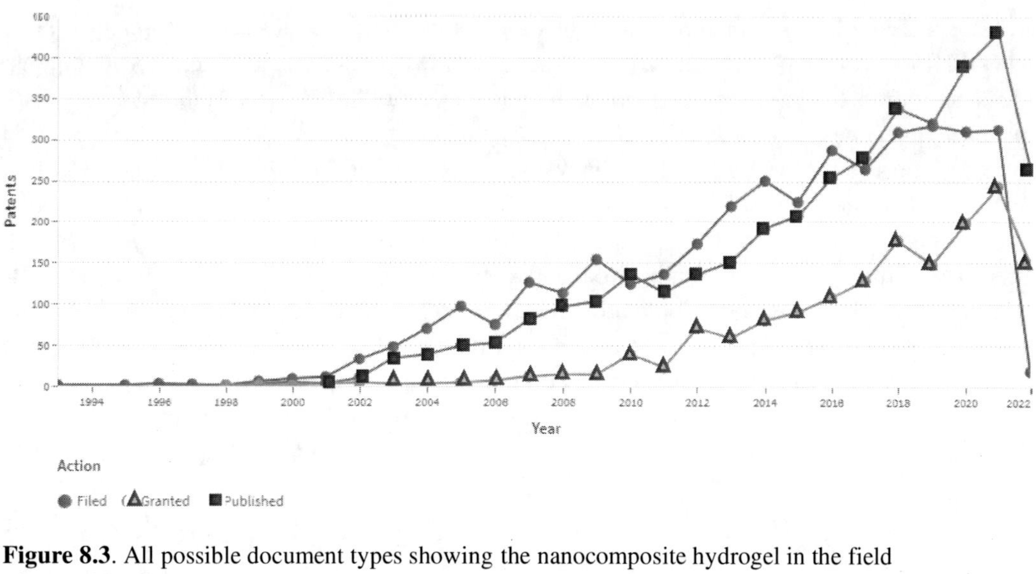

Figure 8.3. All possible document types showing the nanocomposite hydrogel in the field of wound healing.

8.5. Nanoparticulate Hydrogels

Nanoparticles (NPs) are nanoparticulate systems that have unique physicochemical properties, nano-sized characteristics with controlled shape, and versatile modification (Jiang et al., 2020). NPs can enter the body through the skin, ingestion, inhalation, and injection. When NPs enter a biological environment, they invariably come into contact with a wide range of biomolecules, including sugars, proteins, and lipids, which may rapidly coat the NP surfaces and result in the development of the so-called "protein corona." The NPs will eventually reach the target tissues or cells and be internalized by the targeted cells if they are able to avoid being cleared by the reticuloendothelial system (RES) and/or mononuclear phagocytic system (MPS) (De et al., 2008). The physicochemical characteristics of NPs, such as hydrophilic-lipophilic property, surface feature, and particle size shape, are what determine their transport behavior in blood, uptake and clearance by macrophages, subsequent biodistribution, and metabolism, as well as their internalization by targeted cells (Thoniyot et al., 2015).

There are various types of NPs, including silica-, metal-, polymer-, carbon-, and lipid-based NPs, as well as magnetic NPs. Possible uses of NPs incorporated into hydrogels are immunoassay and biomedical diagnosis, and drug loading and delivery for antibacterial, anti-viral, and anti-fungal drugs. The behavior and interactions of NPs with *in vivo* cells are significantly influenced by their form. Because elongated NPs required more time to wrap around the membrane, some studies claimed that spherical NPs of identical size were uptake substantially more efficiently than non-spherical NPs. Non-spherical NPs, on the other hand, outperform their spherical counterparts in terms of their ability to evade phagocytosis, circulate in the blood, and bind to the intended tissues. It should be crucial to show how the physicochemical features of NPs interact with subcellular organelles, cells, tissues, and organisms before NPs are widely used in the biomedical and pharmaceutical fields (Jiang et al., 2020).

Verma et al., (2017) proved wound healing application of sericin/chitosan-capped silver NPs (SNPs) incorporated hydrogel. By capping SNPs with chitosan and sericin, an attempt was made to address the problems of silver staining and stability difficulties. Additionally, because chitosan and silver are known to be antibacterial agents, they decreased the chance of bacterial infection during the healing process. It was discovered that the created hydrogel was non-irritating, had the potential to cure wounds, and had a classy appearance for greater patient compliance. Resmi et al., (2017) prepared gelatin-hydroxypropyl methacrylate hydrogels with silver NPs that have been synthesized and described are used as wound dressings. The hydrogel's mechanical and swelling properties were determined to be sufficient for application in a moist wound environment. After the addition of silver NPs, it exhibited antibacterial properties. Dermal fibroblasts with no adherent nature and cytocompatibility with the hydrogel matrix demonstrated the hydrogel's potential as a temporary wound dressing material.

Xia et al., (2017) experimented the chronic wound treatment with NPs and thermosensitive hydrogel supplemented with chitin whiskers. The hydrogel demonstrated sustained drug delivery capacity, antibacterial action against *Escherichia coli*, *Staphylococcus aureus*, and *Monilia albican* for 7 days, and cell proliferation for up to 5 days.

The hydrogel dressing had a faster wound healing rate, good re-epithelialization, dense collagen deposition, and angiogenesis capabilities, according to an *in vivo* chronic wound healing evaluation in diabetic rats. As a promising bandage for chronic wounds, this composite hydrogel showed a lot of promise. Representative studies based on nanoparticulate hydrogel systems are reported in Table 8.1.

Table 8.1. Representative studies based on nanoparticulate hydrogel systems

Types of NP	Type of Hydrogel Matrix	Molecule Used	Outcomes	Refs.
Polymeric cationic NPs	Silk fibroin hydrogel	Curcumin	The creation of a thin layer of gel (occlusive effect), which prolonged the contact duration of the formulation with the skin and maintained the skin's surface in a well-hydrated state, enabled the translocation of the NPs through the deeper layers of the skin	(Mao et al., 2017)
Reduced polydopamine NPs	Oxidized dextran/ chitosan hybrid hydrogels	Polydopamine	Increase in the wound healing capacity	(Fu et al., 2021)
Copper NPs	Starch hydrogel	-	The formed hydrogel was active against both gram-positive and gram-negative bacteria	(Villanueva et al., 2016)
Solid lipid NPs	Poloxamer-based hydrogels	Resina Draconis	The developed system was useful for the effective transcorneal delivery of lipophilic drugs	(Hao et al., 2014)
Oligopeptide-terminated poly(β-aminoester) (pBAE) NPs	Polyamidoamine	siRNA	siRNA was successfully and continuously released in a breast malignancy	(Segovia et al., 2015)
ZnO NPs	Heparinized PVA/chitosan hydrogels	-	The prepared formulation increased wound healing process	(Khorasani et al., 2018)
Propoxylated glyceryl triacylate NPs	Silicone hydrogel	Timolol	Timolol can be released from contact lenses slowly and continuously, which can have the desired pharmacodynamics result of lowering intraocular pressure	(Jung et al., 2013)
Poly-L-lactic-co glycolic acid (PLGA) NPs	Carbomer hydrogel	Pranoprofen	Reduction in ocular edema	(Abrego et al., 2015)
Solid lipid NPs	-	Ketoconazole	more efficient in the treatment of candidiasis	(Ramasamy et al., 2012)
Lignin NPs	PVAl/chitosan hydrogels	-	Effective against gram-negative bacteria	(Yang et al., 2018)
Silver oxide NPs	Methylcellulose hydrogel	-	The antibacterial activity and burn wound healing of the hydrogel with silver oxide NPs were outstanding	(Kim et al., 2018)
Solid lipid NPs	-	Risperidone	Plastic flow with thixotropy and high adhesiveness was obtained	(Silva et al., 2012)

8.6. Vesicular-Based Hydrogels

Colloidal systems smaller than a micrometer in size are called vesicles. They may be made of lipids, surfactants, or polymers. In these systems, the active principle is trapped in an oil or water cavity surrounded by a membrane. Vesicles can therefore encapsulate larger amounts of medication than matrix systems, allowing for the administration of much smaller doses of vectors. The majority of these systems can vectorize both hydrophilic and lipophilic substances and have flexible transport properties.

8.6.1. Liposomes

The phospholipid structure of liposomes is comparable to the lipid bilayer seen in the cell membranes of living things. Additionally, medicines that are both hydrophilic and hydrophobic can be placed inside liposomes. Liposomes are very biocompatible and have low toxicity. These factors make liposomes a popular drug delivery carrier. Through interactions with the lipid bilayer or compartmentalization inside their aqueous core, liposomes have the capacity to retain both polar and non-polar substances. The liposome bilayer's specific composition enables medication release over time periods of days to weeks. Additionally, liposomes' biocompatibility guarantees that they will be safely broken down and assimilated by the host once their drug supply has run out (Hurler et al., 2013).

To provide sustained drug release throughout their extended stay at the administration site, liposomes-in-hydrogel is used as a delivery mechanism. Different aspects connected to the physicochemical properties of the drug have an impact on how pharmaceuticals are released from drugs-in-liposomes-in-hydrogel systems. The necessary rheological qualities for the included liposomes are provided by the hydrogels used as carriers. Additionally, hydrogels' high viscosity serves as a defense mechanism that can keep liposomes stable. Here are a few examples that support the possibility of liposomes in the hydrogel. Vanic and colleagues (2014) incorporated metronidazole or clotrimazole into deformable propylene glycol-containing liposomes (DPGLs), which were created and tested as an effective medication delivery mechanism to enhance the treatment of vaginal microbial infections.

Results from studies on *in vitro* release, and characterization of the final compositions' rheological and textural properties, support DPGL-in-potential hydrogel as a promising technique for controlled and targeted nano-delivery prolonged vaginal release of antibacterial medications. Park et al., (2013) prepared the quercetin and rutin-loaded ceramide liposomes and studied the drug-releasing effect of liposomes in the hydrogel complex. According to the study's findings, the liposome/hydrogel complex system may function as a potent drug delivery mechanism, facilitating the transdermal absorption of antioxidants that are not water-soluble. Hurler and the team (2013) studied the release characteristics of liposomes in the hydrogel which are influenced by lipid composition and liposome size.

Thus, it was demonstrated that the liposomes-in-chitosan hydrogel formulations could effectively manage both oozing and dry wounds which would have potential in modern wound dressings. Representative studies based on liposomes in hydrogel systems are reported in Table 8.2.

Table 8.2. Representative studies based on liposomes in hydrogel systems

Type of Hydrogel Matrix	Drug Used	Outcomes	Refs.
Pluronic® F127 gel	Paclitaxel	Exhibited the longest drug-release duration when compared to Taxol®, general gel, and liposomes	(Nie et al., 2011)
Chitosan/oxidized hydroxyethyl cellulose (CS/OHEC) hydrogel	Asiaticoside	The hydrogel's asiaticoside release significantly inhibited fibroblast proliferation and collagen secretion, removing scarring for regenerating neurons	(Zheng et al., 2020)
Chitosan-based hydrogel	Resveratrol or epicatechin	Exhibited potent anti-inflammatory activity	(Jøraholmen et al., 2019)
Polymeric hydrogel	Mupirocin	Displayed enhanced bioadhesiveness and prolonged mupirocin release patterns	(Psimadas et al., 2012)
Chitosan-based hydrogel	Doxorubicin	Promising antitumor drug carrier for topical cancer therapy	(Wang et al., 2013)
Silk Fibroin Hydrogel	-	Potential carrier of growth factors for wound healing could be the liposome with the hydrogel core	(Xu et al., 2017)
Chitosan hydrogel	Linseed oil and quercetin	This study has provided a novel and effective strategy for the delivery of quercetin and linseed oil	(Huang et al., 2020)

8.6.2. Niosomes

Niosomes are vesicles made of a nonionic surfactant that are frequently made of cholesterol or products derived from it. Niosomal gel is a brand-new and powerful substance that causes little toxicity or irritation while also having the potential to improve drug absorption via the skin. Niosomes can also improve the effectiveness of drug entrapment, increase the rate at which medications penetrate the skin, and make it easier for drugs to reach their intended tissues. Niosomes can also extend the duration of drug retention time by carefully controlling the release of the active ingredient. More importantly, when compared to other acne treatment options, niosomes have milder adverse effects such as skin irritability. The majority of pharmaceuticals and nutraceuticals, including vitamins and antioxidants, are highly lipophilic molecules that must be encapsulated in various delivery vehicles to get around a number of restrictions. These are generally caused by the molecules' physicochemical instability and/or poor biodisposibility under physiological settings. The formulation has strong mechanical properties, including stiffness and thixotropy. After being enclosed in the hydrogels, the niosomes' structural integrity is also preserved. The total concentration of biopolymers in the hydrogel can be altered in order to control how the encapsulated resveratrol (RSV) is released (Hashim et al., 2014). Additionally, RSV is shielded from photoisomerization by being enclosed in niosome-hydrogel systems, keeping a larger concentration of its physiologically active trans isomer for later delivery.

Here are a few studies of niosomes incorporated into the hydrogel. Wang et al., (2020) formulated and characterized a 3D-printed cryptotanshinone-loaded niosomal hydrogel for topical therapy of acne. It has shown to be a successful and promising topical delivery technology with enormous potential for the use of individualized acne treatments. Machado et al., (2019) encapsulated niosomes in biohydrogels for tunable delivery of phytoalexin RSV. RSV can be delivered in a variety of ways with the help of biohydrogels made of gelatin that contain RSV-niosomes. The formulation has strong mechanical qualities including stiffness and thixotropy. After being enclosed in the hydrogels, the niosomes' structural integrity is also preserved. The total concentration of biopolymers in the hydrogel can be altered in order to control how the encapsulated RSV is released. Hashim and colleagues (2014) studied niosomal hydrogel to deliver atenolol to the eyes. An *in vivo* investigation demonstrated that, when compared to other drug formulations of the same concentration, niosomal hydrogel showed the most substantial persistent decrease in intraocular pressure. Niosomal hydrogel, with its enhanced ocular bioavailability and sustained drug release patterns, may be a potential atenolol administration strategy. Representative studies based on the application of niosomes in hydrogel systems are reported in Table 8.3.

Table 8.3. Representative studies based on the application of niosomes in hydrogel systems

Type of Hydrogel Matrix	Drug Used	Outcomes	Refs.
Carbopol hydrogel	RSV	Niosomal hydrogel may increase the utilization of RSV as an anti-inflammatory agent	(Negi et al., 2017)
Hyaluronic acid hydrogel	Nioplex	Capable of effectively transfecting murine mesenchymal stromal cells (mMSCs) that have excellent cellular viability	(Villate-Beitia et al., 2018)
Chitosan gel	Moxifloxacin	The niosomal formulation of moxifloxacin was the most efficient system against *P. aeruginosa*	(Sohrabi et al., 2016)
Sodium carboxy methylcellulose-based hydrogel	8-Methoxypsoralen	Improved the efficacy and safety of long-term treatment	(Kassem et al., 2017)
Polymeric hydrogel	Curcumin	Provided a controlled delivery system for curcumin	(Bashash et al., 2022)
Polymeric hydrogel	Celastrol	Celastrol's ability to penetrate the skin and become more water-soluble after being encapsulated by niosomes dramatically improved mice's ability to treat their psoriasis	(Meng et al., 2019)
Chitosan-based hydrogel	Triaryl-(Z)-olefin	*In situ* pH-sensitive TZO-loaded niosomes could be an effective treatment for breast cancer	(Salem et al., 2020)

8.7. Nanoemulsion Hydrogels

Nanoemulsions are used to improve the delivery of pharmacologically active compounds, which are manufactured in nanoscale diameters. Nanoemulsion falls within the area of novel drug delivery methods. A drug's effectiveness is increased once it has reached the systemic circulation of the body since it is less water soluble there. A nanoemulsion is produced when two immiscible liquids are mixed and one phase is distributed throughout the other. They typically range in size from 50 to 200 nm. Nanoemulsions frequently contain oil molecules or droplets dispersed in a watery phase or water droplets dispersed in an oily phase, depending on the production method. The separation of two separate layers within nanoemulsion serves as a demonstration of phase separation. Nanoemulsions, which are water-dispersed oil droplets of nanoscale, are typically stabilized by the addition of surfactants and polymers, which also enhance the solubility, stability, and biological activity of the bioactive compounds. Representative studies based on nanoemulsion hydrogel systems are reported in Table 8.4.

Table 8.4. Representative studies based on nanoemulsion hydrogel systems

Type of Hydrogel Matrix	Drug Used	Outcomes	Refs.
Acrylic-acid hydrogels	Soybean isoflavone aglycones	The formulation encouraged better chemical retention and permeation on healthy skin and sufficient retention and permeation on damaged skin	(Back et al., 2020)
Hydroxyethylcellulose hydrogel	β-caryophyllene	Faster healing was achieved and has reduced the skin irritation	(Parisotto-Peterle et al., 2020)
Polyvinylpyrrolidone	Protamine	Significantly more skin permeability was made possible by the low molecular weight proteinuria (LMWP)-conjugated growth factors than the hydrogel with NE-dispersed native growth factors	(Choi et al., 2017)
Carbopol Hydrogel	Tea tree oil	Formulations were able to boost oil's biological activity and shield skin from UV-B radiation-related harm	(Flores et al., 2015)
Polymeric hydrogel	Ketoprofen	In comparison to the marketed formulation, nanoemulgel demonstrated considerably higher cumulative drug flow and permeation, as well as decreased lag time and skin retention	(Arora et al., 2014)
Chitosan hydrogel	Thyme oil	Cell attachment studies revealed that the hydrogel had excellent biocompatibility and may further be used for wound healing purposes	(Moradi et al., 2019)

8.8. Nanotube-Based Hydrogel

Carbon nanotubes (CNTs) are enormous cylindrical big molecules made up of a hexagonal arrangement of carbon atoms that have undergone sp2 hybridization (the C-C distance is approximately 1.4 Å). Since the wall of CNTs is made up of one or more layers of graphene sheets, it is possible to distinguish between single-walled CNTs (SWCNTs), which are produced by rolling up a single graphene sheet, and multi-walled CNTs (MWCNTs), which are produced by rolling up multiple graphene sheets. In a hemispherical configuration of carbon networks known as fullerenes distorted up by the graphene sheet, SWCNTs and MWCNTs are capped at both ends of the tubes (Cirillo et al., 2014). Due to their utility in numerous human endeavors, interest in the production of polymeric hydrogels impregnated with CNTs has recently increased significantly. One-dimensional CNTs have a well-defined structure, a regular pore structure, good conductivity, great electrochemical stability, low mass density, high mechanical strength, and high specific area.

In comparison to typical hydrogels devoid of CNT NPs, the majority of CNT-based nanocomposite hydrogels have been shown to exhibit considerable and beneficial thermal, mechanical, and electrical properties. This is because CNTs are ideal candidates for utilizing the development of a potentially revolutionary material for a variety of applications due to their shape, size, and other characteristics. As with the increase of mechanical, thermal, and electrical aspects, the presence of synergetic capabilities in the matrix of composite hydrogel embedded with CNTs is increased (Adewunmi et al., 2016). The physical and chemical characteristics of polymeric nanocomposite hydrogels impregnated with CNTs can be customized into different multifunctional components by manipulating interactions between the polymer and other polymers or between the polymer and CNTs.

Without a doubt, CNT-based nanocomposite hydrogels have a bright future with a variety of fantastic prospective applications in science and industry. Due to their exceptional qualities, such as their regular pore structure, high conductivity, excellent electrochemical stability, well-defined one-dimensional structure, low mass density, high mechanical strength, and high specific area, CNTs have been widely used in the matrices of numerous polymeric composite hydrogels. Incorporating CNTs into 3D structures is quickly emerging as a creative way to create multicomponent systems with a variety of functionalities within a network of polymeric hydrogel (Cirillo et al., 2014). It is feasible to create hydrogels that are both extremely durable and electrically conductive by adding CNTs to hydrogel formulations.

CNT-based nanocomposites offer a flexible framework for creating hydrogels with exceptional mechanical performance and a range of responsive characteristics. However, there is a question regarding the biocompatibility of hydrogels and the toxic effects of CNTs. According to studies, CNT toxicity appeared to be dose-dependent but might be diminished by functionalizing and integrating CNTs into networks. Due to the difficulties in adequately dispersing CNTs in the matrix, some of the preparation processes are highly difficult and time-consuming. It would be advantageous to have fully physically crosslinked networks, which, in contrast to covalent systems, would allow for easy processability, manipulation, and reshaping features of the hydrogels. This would be in addition to improved mechanical behavior, various responsive properties, and biocompatibility (Adewunmi et al., 2016).

Zhang et al., (2011) studied the programmable materials based on CNT-Hydrogel polymer composites that are optically and thermally responsive. They showed that the SWNT-poly (N-isopropyl acrylamide) hydrogel actuators responded to near-infrared (NIR) laser excitation with a well-defined, ultrafast response, making this design applicable to a variety of optically triggered applications. Kameta et al., (2009) studied the remarkable effect of confined proteins on resistance to denaturants in supramolecular nanotube hydrogels. The pH-dependent nanotube hydrogel was able to fix a protein independently into both one-dimensional hollow nanotubes and independent nanospaces made up of three-dimensional meshworks created between the nanotubes.

For previous supramolecular and polymer hydrogels with three-dimensional meshworks produced in between nanofibers, special property, such as protein stabilization by encapsulation and delayed release of a fixed protein without deforming the nanotube hydrogel has never been seen. As a unique soft material, the nanotube hydrogel is hence widely suited to biological and medical nanotechnologies. Venkatesan et al., (2014) showed that chitosan-CNT hydrogels have antimicrobial properties. The results showed that the created chitosan-CNT hydrogel had a clearly defined pore structure and was capable of absorbing and holding onto water. A high antibacterial action was also demonstrated by the chitosan-CNT hydrogel against *Staphylococcus aureus*, *Escherichia coli*, and *Candida tropicalis*. In light of this, they proposed that the CNT dispersed in a chitosan matrix will be a promising material for biomedical as well as food science applications.

8.9. Polymeric Micelles-Incorporated Hydrogels

Since they were first used for drug delivery, polymeric micelles (PM) have always been at the forefront of the industry. In an aqueous environment, they feature a core-shell structure with an interior core made of hydrophobic segments and an outer corona made of a hydrophilic shell. The hydrophilic outside allows for the retention of PM stability in aqueous media, while the hydrophobic interior serves as a storage space for hydrophobic drugs (Pekar, 2015). Micelle drug delivery systems have a number of benefits that have been demonstrated both *in vitro* and *in vivo*, including simplicity in preparation, small and homogeneous particles, high drug loading, and adjustable drug release profiles. The careful balancing act of relatively weak contacts, such as hydrophobic, hydrogen bonding, and electrostatic, is what drives interactions between drugs and the PM core domain (Anirudhan et al., 2016). PMs are therefore limited in drug delivery systems since they break down after being diluted by bodily fluids, resulting in an initial burst of drug release if they are exposed. Micelles can be added to the hydrogels by two methods. In the first method, the micelles are scattered in hydrogel and a straightforward combination of micelles and hydrogel matrix produces the product. In the second method, micelles are incorporated into the hydrogel. In this method, dispersed micelles are at least partially attached to the hydrogel network and the hydrophobic segments are directly attached to the precursor of the gel to generate micelles (McKenzie et al., 2015).

Micelles can crosslink the hydrogel network physically or chemically. In terms of the polarity of their constituents, hydrogels with hydrophobic, particularly micellar nanodomains, are hybrid materials. Their distinct structural and colloidal features are based on their hybrid nature. When used effectively, they can be used in a wide range of applications, particularly those related to biomedicine, biomaterials, medication delivery, cosmetics, agriculture, and environmental protection (Ghezzi et al., 2021). Amphiphilic di-block copolymers (like polystyrene and PEG) and triblock copolymers (like poloxamers) are the most frequently used polymers for the development of micelles, but graft copolymers (like PEG-poly(caprolactone)-g-polyethyleneimine) are also used. Poly (propylene oxide), polyesters, copolymers of glycolic and lactic acids, as well as poly(vinyl pyrrolidone), poly(acryloylmorpholine), and poly(trimethylene carbonate) can be used to create the hydrophilic and hydrophobic segments. PEG is typically used to create the hydrophilic part (Emoto et al., 2000).

Polymeric micelles are intriguing carriers for various administration routes because of their tiny size, simple synthesis, and good solubilization capabilities. They can increase a medicine's bioavailability and create a controlled and focused drug release, both of which are helpful for minimizing side effects. The most researched method of administering micelles is intravenous injection/infusion (mostly used for chemotherapy), although oral and topical (ocular, nasal, and buccal) delivery have also produced some intriguing outcomes in terms of enhanced drug bioavailability. Micelles are produced through reversible limits and can be destroyed by a number of destabilizing conditions as thermodynamically self-assembled structures (Pekar, 2015).

Conclusion

Nanocomposite hydrogels are cutting-edge biomaterials with potential uses in a variety of biological and pharmacological fields. Nanocomposite hydrogels offer better physical, chemical, mechanical, and electrical properties than traditional polymer hydrogels. Numerous varieties of multifunctional nanocomposite hydrogels, including those with specific stimuli-responsive, catalytic, antibacterial, and wound healing capabilities, have been created and will be researched further in the future. Nanocomposite hydrogels have unique qualities that enable them to be used in a particular biomedical purpose. For instance, hydrogels made of carbon-based nanocomposite materials have excellent mechanical strength and good conductivity, making them suitable for conductive tissue engineering.

For applications involving either hydrophilic or hydrophobic drug administration, polymeric micelles-based nanocomposite hydrogels are employed. Nanocomposite hydrogels made of metal or metal oxide NPs are stimuli-responsive and have been researched for use in antibacterial and remote-controlled medication delivery. Although there is a great deal of research interest in creating nanocomposite hydrogels with specific functions, we must acknowledge that the preparation and characterization of the synthesized nanocomposite hydrogels are still in their early phases. For their ultimate uses in crucial sectors like wound dressing and drug delivery, various fine-tuning is required. The creation of novel materials with enhanced qualities above the individual components is the main goal of hybrid materials. Therefore, the introduction of innovative nanocomposite hydrogels with creative functionalization techniques could best achieve future progress.

References

Abrego, G., Alvarado, H., Souto, E. B., Guevara, B., Bellowa, L. H., Parra, A., Calpena, A., and Garcia, M. L. (2015). Biopharmaceutical profile of pranoprofen-loaded PLGA nanoparticles containing hydrogels for ocular administration. *European Journal of Pharmaceutics and Biopharmaceutics*, 95, 261-270.

Adewunmi, A. A., Ismail, S., and Sultan, A. S. (2016). Carbon nanotubes (CNTs) nanocomposite hydrogels developed for various applications: a critical review. *Journal of Inorganic and Organometallic Polymers and Materials*, 26(4), 717-737.

Ahmed, E. M. (2015). Hydrogel: Preparation, characterization and applications: a review. *Journal of Advanced Research*, 6(2), 105-121.

Anirudhan, T. S., Parvathy, J., and Nair, A. S. (2016). A novel composite matrix based on polymeric micelle and hydrogel as a drug carrier for the controlled release of dual drugs. *Carbohydrate Polymers*, 136, 1118-1127.

Arora, R., Aggarwal, G., Harikumar, S. L., and Kaur, K. (2014). Nanoemulsion based hydrogel for enhanced transdermal delivery of ketoprofen. *Advances in Pharmaceutics*, 2014, 468456.

Back, P. I., Balestrin, L. A., Fachel, F. N. S., Nemitz, M. C., Falkembach, M., Soares, G., Marques, M. da S., Silveira, T., Dal Prá, M., Horn, A. P., Braganhol, E., von Poser, G. L., Dora, C. L., and Teixeira, H. F. (2020). Hydrogels containing soybean isoflavone aglycones-rich fraction-loaded nanoemulsions for wound healing treatment - *in vitro* and *in vivo* studies. *Colloids and Surfaces B: Biointerfaces*, 196, 111301.

Bashash, M., Varidi, M., and Varshosaz, J. (2022). Composite hydrogel-embedded sucrose stearate niosomes: unique curcumin delivery system. *Food and Bioprocess Technology*, 15(9), 2020-2034.

Choi, S. W., Pangeni, R., and Park, J. W. (2017). Nanoemulsion-based hydrogel for topical delivery of highly skin-permeable growth factor combinations: preparation and *in vitro* evaluation. *Journal of Nanoscience and Nanotechnology*, 17(4), 2363-2369.

Cirillo, G., Hampel, S., Spizzirri, U. G., Parisi, O. I., Picci, N., and Iemma, F. (2014). Carbon nanotubes hybrid hydrogels in drug delivery: a perspective review. *BioMed Research International*, 2014, 825017.

da Silva, L. P., Reis, R. L., Correlo, V. M., and Marques, A. P. (2019). Hydrogel-based strategies to advance therapies for chronic skin wounds. *Annual Review of Biomedical Engineering*, 21, 145-169.

Dalwadi C., Patel G. (2013). Recent patents on stimuli responsive hydrogel drug delivery system. *Recent Patents on Drug Delivery and Formulation*, 7(3), 206-215.

Dalwadi C., Patel G. (2015). Application of nanohydrogels in drug delivery systems: recent patents review. *Recent Patents on Nanotechnology*, 9(1), 17-25.

De, M., Ghosh, P. S., and Rotello, V. M. (2008). Applications of nanoparticles in biology. *Advanced Materials*, 20(22), 4225-4241.

Emoto, K., Iijima, M., Nagasaki, Y., and Kataoka, K. (2000). Functionality of polymeric micelle hydrogels with organized three-dimensional architecture on surfaces. *Journal of the American Chemical Society*, 122(11), 2653-2654.

Flores, F. C., De Lima, J. A., da Silva, C. R., Benvegnú, D., Ferreira, J., Burger, M. E., Beck, R. C. R., Rolim, C. M. B., Rocha, M. I. U. M., Da Veiga, M. L., and Da Silva, C. D. B. (2015). Hydrogels containing nanocapsules and nanoemulsions of tea tree oil provide antiedematogenic effect and improved skin wound healing. *Journal of Nanoscience and Nanotechnology*, 15(1), 800-809.

Fu, Y., Zhang, J., Wang, Y., Li, J., Bao, J., Xu, X., Zhang, C., Li, Y., Wu, H., and Gu, Z. (2021). Reduced polydopamine nanoparticles incorporated oxidized dextran/chitosan hybrid hydrogels with enhanced antioxidative and antibacterial properties for accelerated wound healing. *Carbohydrate Polymers*, 257, 117598.

Ghezzi, M., Pescina, S., Padula, C., Santi, P., Del Favero, E., Cantù, L., and Nicoli, S. (2021). Polymeric micelles in drug delivery: an insight of the techniques for their characterization and assessment in biorelevant conditions. *Journal of Controlled Release*, 332, 312-336.

Hao, J., Wang, X., Bi, Y., Teng, Y., Wang, J., Li, F., Li, Q., Zhang, J., Guo, F., and Liu, J. (2014). Fabrication of a composite system combining solid lipid nanoparticles and thermosensitive hydrogel for challenging ophthalmic drug delivery. *Colloids and Surfaces B: Biointerfaces*, 114, 111-120.

Hashim, I. I. A., El-dahan, M. S., Yusif, R. M., Abd-ElGawad, A. E. G. H., and Arima, H. (2014). Potential use of niosomal hydrogel as an ocular delivery system for atenolol. *Biological and Pharmaceutical Bulletin*, 37(4), 541-551.

Huang, J., Wang, Q., Chu, L., and Xia, Q. (2020). Liposome-chitosan hydrogel bead delivery system for the encapsulation of linseed oil and quercetin: preparation and *in vitro* characterization studies. *LWT*, 117, 108615.

Hurler, J., Žakelj, S., Mravljak, J., Pajk, S., Kristl, A., Schubert, R., and Škalko-Basnet, N. (2013). The effect of lipid composition and liposome size on the release properties of liposomes-in-hydrogel. *International Journal of Pharmaceutics*, 456(1), 49-57.

Jiang, Y., Krishnan, N., Heo, J., Fang, R. H., and Zhang, L. (2020). Nanoparticle-hydrogel superstructures for biomedical applications. *Journal of Controlled Release*, 324, 505-521.

Jøraholmen, M. W., Basnet, P., Tostrup, M. J., Moueffaq, S., and Škalko-Basnet, N. (2019). Localized therapy of vaginal infections and inflammation: liposomes-in-hydrogel delivery system for polyphenols. *Pharmaceutics*, 11(2), 53.

Jung, H. J., Abou-Jaoude, M., Carbia, B. E., Plummer, C., and Chauhan, A. (2013). Glaucoma therapy by extended release of timolol from nanoparticle loaded silicone-hydrogel contact lenses. *Journal of Controlled Release*, 165(1), 82-89.

Kameta, N., Yoshida, K., Masuda, M., and Shimizu, T. (2009). Supramolecular nanotube hydrogels: remarkable resistance effect of confined proteins to denaturants. *Chemistry of Materials*, 21(24), 5892-5898.

Kassem, A. A., Abd El-Alim, S. H., and Asfour, M. H. (2017). Enhancement of 8-methoxypsoralen topical delivery via nanosized niosomal vesicles: formulation development, *in vitro* and *in vivo* evaluation of skin deposition. *International Journal of Pharmaceutics*, 517(1-2), 256-268.

Khorasani, M. T., Joorabloo, A., Moghaddam, A., Shamsi, H., and MansooriMoghadam, Z. (2018). Incorporation of ZnO nanoparticles into heparinised polyvinyl alcohol/chitosan hydrogels for wound dressing application. *International Journal of Biological Macromolecules*, 114, 1203-1215.

Kim, M. H., Park, H., Nam, H. C., Park, S. R., Jung, J. Y., and Park, W. H. (2018). Injectable methylcellulose hydrogel containing silver oxide nanoparticles for burn wound healing. *Carbohydrate Polymers*, 181, 579-586.

Koehler, J., Brandl, F. P., and Goepferich, A. M. (2018). Hydrogel wound dressings for bioactive treatment of acute and chronic wounds. *European Polymer Journal*, 100, 1-11.

Machado, N. D., Fernández, M. A., Häring, M., Saldías, C., and Díaz, D. D. (2019). Niosomes encapsulated in biohydrogels for tunable delivery of phytoalexin resveratrol. *RSC Advances*, 9(14), 7601-7609.

Mao, K. L., Fan, Z. L., Yuan, J. D., Chen, P. P., Yang, J. J., Xu, J., ZhuGe, D. L., Jin, B. H., Zhu, Q. Y., Shen, B. X., Sohawon, Y., Zhao, Y. Z., and Xu, H. L. (2017). Skin-penetrating polymeric nanoparticles incorporated in silk fibroin hydrogel for topical delivery of curcumin to improve its therapeutic effect on psoriasis mouse model. *Colloids and Surfaces B: Biointerfaces*, 160, 704-714.

McKenzie, M., Betts, D., Suh, A., Bui, K., Kim, L. D., and Cho, H. (2015). Hydrogel-based drug delivery systems for poorly water-soluble drugs. *Molecules*, 20(11), 20397-20408.

Meng, S., Sun, L., Wang, L., Lin, Z., Liu, Z., Xi, L., Wang, Z., and Zheng, Y. (2019). Loading of water-insoluble celastrol into niosome hydrogels for improved topical permeation and anti-psoriasis activity. *Colloids and Surfaces B: Biointerfaces*, 182, 110352.

Moradi, S., Barati, A., Salehi, E., Tonelli, A. E., and Hamedi, H. (2019). Preparation and characterization of chitosan based hydrogels containing cyclodextrin inclusion compounds or nanoemulsions of thyme oil. *Polymer International*, 68(11), 1891-1902.

Negi, P., Aggarwal, M., Sharma, G., Rathore, C., Sharma, G., Singh, B., and Katare, O. P. (2017). Niosome-based hydrogel of resveratrol for topical applications: an effective therapy for pain related disorder(s). *Biomedicine and Pharmacotherapy*, 88, 480-487.

Nie, S., Hsiao, W. W., Pan, W., and Yang, Z. (2011). Thermoreversible pluronic® F127-based hydrogel containing liposomes for the controlled delivery of paclitaxel: *in vitro* drug release, cell cytotoxicity, and uptake studies. *International Journal of Nanomedicine*, 6(1), 151-166.

Parisotto-Peterle, J., Bidone, J., Lucca, L. G., Araújo, G. de M. S., Falkembach, M. C., da Silva Marques, M., Horn, A. P., dos Santos, M. K., da Veiga, V. F., Limberger, R. P., Teixeira, H. F., Dora, C. L., and Koester, L. S. (2020). Healing activity of hydrogel containing nanoemulsified β-caryophyllene. *European Journal of Pharmaceutical Sciences*, 148, 105318.

Park, S. N., Lee, M. H., Kim, S. J., and Yu, E. R. (2013). Preparation of quercetin and rutin-loaded ceramide liposomes and drug-releasing effect in liposome-in-hydrogel complex system. *Biochemical and Biophysical Research Communications*, 435(3), 361-366.

Patel G. C., Joshi S. A. (2019). Targeting aspects of hydrogels in drug delivery. *Biomedical Applications of Nanoparticles*, 195-231.

Pekar, M. (2015). Hydrogels with micellar hydrophobic (nano)domains. *Frontiers in Materials*, 1, 35.

Psimadas, D., Georgoulias, P., Valotassiou, V., and Loudos, G. (2012). Molecular nanomedicine towards cancer. *Journal of Pharmaceutical Sciences*, 101(7), 2271-2280.

Rafieian, S., Mirzadeh, H., Mahdavi, H., and Masoumi M. E. (2019). A review on nanocomposite hydrogels and their biomedical applications. *Science and Engineering of Composite Materials*, 26(1), 154-174.

Resmi, R., Unnikrishnan, S., Krishnan, L. K., and Kalliyana Krishnan, V. (2017). Synthesis and characterization of silver nanoparticle incorporated gelatin-hydroxypropyl methacrylate hydrogels for wound dressing applications. *Journal of Applied Polymer Science*, 134(10), 44529.

Salem, H. F., Kharshoum, R. M., Gamal F, A., Abo El-Ela, F. I., and Abdellatif, K. R. A. (2020). Treatment of breast cancer with engineered novel pH-sensitive triaryl-(Z)-olefin niosomes containing hydrogel: an *in vitro* and *in vivo* study. *Journal of Liposome Research*, 30(2), 126-135.

Segovia, N., Pont, M., Oliva, N., Ramos, V., Borrós, S., and Artzi, N. (2015). Hydrogel doped with nanoparticles for local sustained release of siRNA in breast cancer. *Advanced Healthcare Materials*, 4(2), 271-280.

Sohrabi, S., Haeri, A., Mahboubi, A., Mortazavi, A., and Dadashzadeh, S. (2016). Chitosan gel-embedded moxifloxacin niosomes: An efficient antimicrobial hybrid system for burn infection. *International Journal of Biological Macromolecules*, 85, 625-633.

Stefanov, I., Pérez-Rafael, S., Hoyo, J., Cailloux, J., Santana Pérez, O. O., Hinojosa-Caballero, D., and Tzanov, T. (2017). Multifunctional enzymatically generated hydrogels for chronic wound application. *Biomacromolecules*, 18(5), 1544-1555.

Sun, H., Zhou, J., Huang, Z., Qu, L., Lin, N., Liang, C., Dai, R., Tang, L., and Tian, F. (2017). Carbon nanotube-incorporated collagen hydrogels improve cell alignment and the performance of cardiac constructs. *International Journal of Nanomedicine*, 12, 3109-3120.

Thoniyot, P., Tan, M. J., Karim, A. A., Young, D. J., and Loh, X. J. (2015). Nanoparticle-hydrogel composites: concept, design, and applications of these promising, multi-functional materials. *Advanced Science*, 2(1-2), 1400010.

Vanic, Ž., Hurler, J., Ferderber, K., Golja Gašparović, P., Škalko-Basnet, N., and Filipović-Grčić, J. (2014). Novel vaginal drug delivery system: deformable propylene glycol liposomes-in-hydrogel. *Journal of Liposome Research*, 24(1), 27-36.

Venkatesan, J., Jayakumar, R., Mohandas, A., Bhatnagar, I., and Kim, S. K. (2014). Antimicrobial activity of chitosan-carbon nanotube hydrogels. *Materials*, 7(5), 3946-3955.

Verma, J., Kanoujia, J., Parashar, P., Tripathi, C. B., and Saraf, S. A. (2017). Wound healing applications of sericin/chitosan-capped silver nanoparticles incorporated hydrogel. *Drug Delivery and Translational Research*, 7(1), 77-88.

Villanueva, M. E., Diez, A. M. D. R., González, J. A., Pérez, C. J., Orrego, M., Piehl, L., Teves, S., and Copello, G. J. (2016). Antimicrobial activity of starch hydrogel incorporated with copper nanoparticles. *ACS Applied Materials and Interfaces*, 8(25), 16280-16288.

Villate-Beitia, I., Truong, N. F., Gallego, I., Zárate, J., Puras, G., Pedraz, J. L., and Segura, T. (2018). Hyaluronic acid hydrogel scaffolds loaded with cationic niosomes for efficient non-viral gene delivery. *RSC Advances*, 8(56), 31934-31942.

Wang, W., Zhang, P., Shan, W., Gao, J., and Liang, W. (2013). A novel chitosan-based thermosensitive hydrogel containing doxorubicin liposomes for topical cancer therapy. *Journal of Biomaterials Science, Polymer Edition*, 24(14), 1649-1659.

Wang, Z., Liu, L., Xiang, S., Jiang, C., Wu, W., Ruan, S., Du, Q., Chen, T., Xue, Y., Chen, H., Weng, L., Zhu, H., Shen, Q., and Liu, Q. (2020). Formulation and characterization of a 3D-Printed cryptotanshinone-loaded niosomal hydrogel for topical therapy of acne. *AAPS PharmSciTech*, 21, 159.

Wichterle, O., and Lim, D. (1960). Hydrophilic gels for biological use. *Nature*, 185(4706), 117-118.

Xia, G., Liu, Y., Tian, M., Gao, P., Bao, Z., Bai, X., Yu, X., Lang, X., Hu, S., and Chen, X. (2017). Nanoparticles/thermosensitive hydrogel reinforced with chitin whiskers as a wound dressing for treating chronic wounds. *Journal of Materials Chemistry B*, 5(17), 3172-3185.

Xu, H. L., Chen, P. P., ZhuGe, D. L., Zhu, Q. Y., Jin, B. H., Shen, B. X., Xiao, J., and Zhao, Y. Z. (2017). Liposomes with silk fibroin hydrogel core to stabilize bFGF and promote the wound healing of mice with deep second-degree scald. *Advanced Healthcare Materials*, 6(19), 1700344.

Yang, W., Fortunati, E., Bertoglio, F., Owczarek, J. S., Bruni, G., Kozanecki, M., Kenny, J. M., Torre, L., Visai, L., and Puglia, D. (2018). Polyvinyl alcohol/chitosan hydrogels with enhanced antioxidant and antibacterial properties induced by lignin nanoparticles. *Carbohydrate Polymers*, 181, 275-284.

Zhang, X., Pint, C. L., Lee, M. H., Schubert, B. E., Jamshidi, A., Takei, K., Ko, H., Gillies, A., Bardhan, R., Urban, J. J., Wu, M., Fearing, R., and Javey, A. (2011). Optically- and thermally-responsive programmable materials based on carbon nanotube-hydrogel polymer composites. *Nano Letters*, 11(8), 3239-3244.

Zheng, F., Li, R., He, Q., Koral, K., Tao, J., Fan, L., Xiang, R., Ma, J., Wang, N., Yin, Y., Huang, Z., Xu, P., and Xu, H. (2020). The electrostimulation and scar inhibition effect of chitosan/oxidized hydroxyethyl cellulose/reduced graphene oxide/asiaticoside liposome based hydrogel on peripheral nerve regeneration *in vitro*. *Materials Science and Engineering C*, 109, 110560.

Chapter 9

Nanocomposite Hydrogels and Their Emerging Applications in Orthopedics

Tung-Yi Lin[1,2]
Ting-Hsiang Chiu[1]
Bo-Yan Li[1]
Chen-Cheng Yu[1]
Hong-Xiang Wei[1]
and Yi-Cheun Yeh[1,*]

[1] Institute of Polymer Science and Engineering, National Taiwan University, Taipei, Taiwan
[2] Department of Orthopaedic Surgery, Chang Gung Memorial Hospital, Keelung Branch, and Chang Gung University College of Medicine, Taoyuan, Taiwan

Abstract

Nanocomposite hydrogels have been developed with versatile properties and served as emerging biomaterials to address the general clinical problems (e.g., bone healing, cartilage defect, and infection control of the musculoskeletal system) encountered by medical physicians in orthopedics. With superior performance (e.g., mechanical strength, injectability, and biocompatibility), nanocomposite hydrogels play a critical role in improving clinical treatment outcomes. In this chapter, we discuss recent applications of nanocomposite hydrogels in the orthopedic field and summarize the potential therapeutic benefits in unsolved problems.

Keywords: nanocomposite hydrogels, bone healing, cartilage defect, infection control

9.1. Introduction of Nanocomposite Hydrogels in Orthopedics

9.1.1. Introduction

Hydrogels are highly hydrated 3D network structures composed of crosslinked polymers with the ability to swell in aqueous solutions. The high water content (70-90%) of hydrogels provides biocompatibility and the ability to mimic natural tissues, showing valuable features

* Corresponding Author's Email: yicheun@ntu.edu.tw.

In: Nanocomposite Hydrogels and their Emerging Applications
Editor: Mohammad Sirousazar
ISBN: 979-8-88697-675-5

for utilizing hydrogels in biomedical applications. Compared to other biomaterials, the superior characteristics of hydrogels (e.g., desirable mechanical strength, biocompatibility, injectability, flexibility, and multifunctionality) allow hydrogels to be designed for specific applications. As an example, hydrogels encapsulated with drugs or growth factors can minimal invasively be filled in the bone or cartilage defects through a simple injection. In particular, hydrogels have been widely demonstrated to mimic the extracellular matrix (ECM) for fundamental studies and practical applications in biomedicine. ECM is a non-cellular component composed of nanoscale collagen fibers that offer structural integrity to all tissues (Frantz, Stewart, and Weaver, 2010; Yue, 2014). ECM has been demonstrated to play several critical roles in tissues (Chan and Leong, 2008), such as mechanical support, a cell-friendly environment, and a reservoir of growth factors. Most importantly, remodeling and neovascularization processes in response to pathological changes (e.g., bone fracture and homeostasis) can be presented in ECM due to a degradable physical environment. Given that cells, scaffolds, and growth factors are generally considered the critical triad and the key components in tissue engineering, a suitable scaffold provides structural support for cells as well as a similar environment to natural ECM-containing materials for tissue development.

Nevertheless, the complex composition and multiple functions of the natural ECM make it challenging to mimic hydrogels precisely with simple polymeric networks. Recently, nanocomposite (NC) hydrogel, a form of nanoparticles (NPs) embedded in a hydrogel matrix, presents variable properties based on its components and preparation conditions (e.g., physical or chemical interactions) to address the limitations of hydrogels (Gaharwar, Peppas, and Khademhosseini, 2014). As an example, one critical disadvantage of hydrogels is their weak mechanical strength, as some natural tissues demand high mechanical strength with good elasticity and resistance to compression force (e.g., cartilage tissues) (Haraguchi, 2007). NC hydrogel can be designed as advanced biomaterials to mimic the structures and functions of natural ECM in a more precise way. The nanoscaled components of NC hydrogels develop a micro-fabricated structure and 3D environment, resulting in desirable mechanical strength for cartilage tissue engineering. Besides, the different choices of nanoscaled structures of NC hydrogels can lead to more exclusive properties than traditional hydrogels (Sasaki and Akiyoshi, 2010; Song, Li, Wang, Liao, and Zhang, 2015). Thus, NC hydrogels with novel behaviors and desirable applications are achieved by the combination of nanotechnology and biomaterials sciences.

9.1.2. Current Challenges in Orthopedics

In recent years, with improved biomaterials and surgical techniques, patients may have a good prognosis even if they suffer from severe trauma, deformity, degeneration disease, tumor, or infection. However, there are still limitations and clinical challenges that physicians need to address. The following section list three main problems in orthopedic practice, including cartilage defect, bone healing, and infection of the musculoskeletal system.

9.1.2.1. Cartilage Defect

Cartilage tissue comprises chondrocytes and ECM that includes water (up to 60-80%), type II collagen, and proteoglycans (Armiento, Stoddart, Alini, and Eglin, 2018). For a healthy joint, cartilage provides a low-friction gliding surface and facilitates the transmission of loads.

Articular cartilage damage caused by accident trauma, sports injury, chronic inflammation (rheumatoid arthritis, RA), or degeneration (osteoarthritis, OA) is one of the significant problems in orthopedics (Ahmed and Hincke, 2014). In particular, the properties of cartilage tissue (e.g., avascular and very few cells inside) result in limited self-healing capacity, especially the damaged area with a more than 4 mm diameter is hard to heal (Asadi et al., 2018).

At present, common surgical treatments (e.g., mosaicplasty, micro-fracture, and autologous chondrocyte injection) lead to similar clinical results. However, the newly-made articular cartilage, unlike a previous native structural organization, presents worse biomechanical properties in comparison to native cartilage (Kock, van Donkelaar, and Ito, 2012). Also, donor-site complications are presented in transplantation procedures, and metal substitutes operated through total or partial joint replacement surgery only replace the defective tissue rather than repair and regenerate natural tissue. Long-term complications (e.g., implant loosening and components wearing) may develop even if the surgery can achieve an acceptable prognosis. It is no doubt that the advance in cartilage tissue engineering is essential to provide cartilage tissue regeneration and recovery of cartilage function.

9.1.2.2. Bone Healing

Bone injury and defects caused by mechanical trauma, pathologic fracture, or infection remain another puzzling problem encountered by medical physicians in orthopedics. Patients suffering from the above complications may lead to deformity, dysfunction, persistent pain, independent life, and high medical cost. Surgical intervention with metallic implants (e.g., plates, screws, and pins) is a standard procedure to realign and stabilize a displaced bone fracture. Bone graft is critical for bone defects and can improve bone healing. Autograft (bone from another site of the same patient) is considered a safe and fast-healing bone graft by providing stem cells, growth factors, and a scaffold (Baldwin et al., 2019). However, the disadvantages of autograft include donor site complications and inadequate volume.

On the other hand, allograft (bone graft from another individual) can avoid donor site morbidity but may provide less osteogenic property. Synthetic bone graft substitutes, such as calcium phosphate and calcium sulfate synthetics, have been commonly used to provide structural support and adequate volume availability. The lack of enzymes or stem cells in synthetic bone grafts leads to less osteogenesis and osteoinductive properties. In addition, growth factors (e.g., bone morphogenetic proteins (BMPs), transforming growth factor-beta (TGF-β), insulin-like growth factors I and II (IGF-I and IGF-II), and platelet-derived growth factor (PDGF)) can be applied to bone defect with a graft to achieve bone healing. The ideal bone graft possesses a suitable scaffold for cell differentiation and adequate enzyme for bone tissue growth.

9.1.2.3. Infection of the Musculoskeletal System

Infections of the musculoskeletal system are critical issues that can involve bone, soft tissue, and implant-related materials. The development of those infections can lead to high medical costs and severe complications in patients, such as function disability, prolonged hospitalization, and even mortality. Adequate surgical debridement and effective antibiotics treatment for weeks are goal standards for infection problems. In clinical, localized drug delivery agents called polymethyl methacrylate (PMMA) beads (antibiotics-PMMA composites) have been used to avoid systemic side effects (Klemm, 2001; Zalavras, Patzakis, and Holtom, 2004). However, there were still several drawbacks to using PMMA as a delivery system. Residual monomers were toxic and may cause tissue necrosis (Stanczyk and van Rietbergen, 2004), and PMMA mixture was hard to control the drug release at a steady rate and effectiveness (Kuechle, Landon, Musher, and Noble, 1991; van de Belt et al., 2000). Furthermore, non-degradable beads become foreign bodies after treatment, and surgical removal may be necessary to avoid another infection.

9.1.3. NC Hydrogels for Orthopedics

NC hydrogels are popular biomaterials designed to mimic the suitable structure and function of natural ECM for bone or cartilage tissue growth. In addition, varied growth factors and antibiotics can also be embedded inside NC hydrogels to regenerate bone/cartilage, presenting as promising scaffolds and carriers for tissue engineering and drug delivery system for infection control, respectively. In the following sections, we will briefly describe the preparation of NC hydrogels and their applications in cartilage tissue engineering, bone healing, and infection control. Several representative examples of NC hydrogels for orthopedics are summarized in Table 9.1.

Table 9.1. Compositions and applications of NC hydrogels in orthopedics

NPs	Polymers	Applications	Refs.
Nano-hydroxyapatite (HA)	Polyvinyl alcohol (PVA) and poly(lactic-*co*-glycolic acid) (PLGA)	Cartilage repair	(Su et al., 2019)
Laponite	Polyethylene glycol (PEG) and poly (amic acid)	Cartilage repair	(Nojoomi, Tamjid, Simchi, Bonakdar, and Stroeve, 2017)
Pluronic F-68	Alginate, chitosan (CS), and PVA	Cartilage repair	(Lim et al., 2010)
Nanodiamonds (NDs)	CS and gelatin	Bone healing	(Pacelli et al., 2017)
Silver NPs	Hyaluronic acid and β-tricalcium phosphate	Bone healing	(Makvandi, Ali, Della Sala, Abdel-Fattah, and Borzacchiello, 2020)
Cerium oxide NPs	Gelatin and alginate	Bone healing	(Purohit et al., 2020)
Poly(γ-glutamic acid) NPs	N-trimethyl CS	Bone healing	(Mi et al., 2008)
Silicon dioxide NPs	Alginate	Bone infection	(Chen et al., 2018)
Vancomycin (VCM) NPs	CS	Bone infection	(Tao et al., 2020)
Calcium phosphate NPs	PVA and κ-Carrageenan	Bone infection	(Alinavaz, Mahdavinia, Jafari, Hazrati, and Akbari, 2021)
Copper crystal	Citric acid and alginate	Bone infection	(Wu et al., 2021)

9.2. Preparation of NC Hydrogels

The structural underpinning of NC hydrogels is the interaction between NPs and polymers. Given the general interactions (e.g., hydrogen bonds, van der Waals contacts, and electrostatic interactions) between NPs and polymers are relatively weak, NPs play a key role in enhancing the structural mechanics of NC hydrogels. The four most common methods for introducing NPs into NC hydrogels include hydrogel formation in an NP suspension, physically embedding the NPs into the hydrogel matrix after gelation, NP formation within a hydrogel matrix, and using

NPs to crosslink polymers for hydrogel formation (He, Shi, Liang, and Guo, 2020; Kumar and Jaiswal, 2016; Tsai et al., 2021) (Figure 9.1).

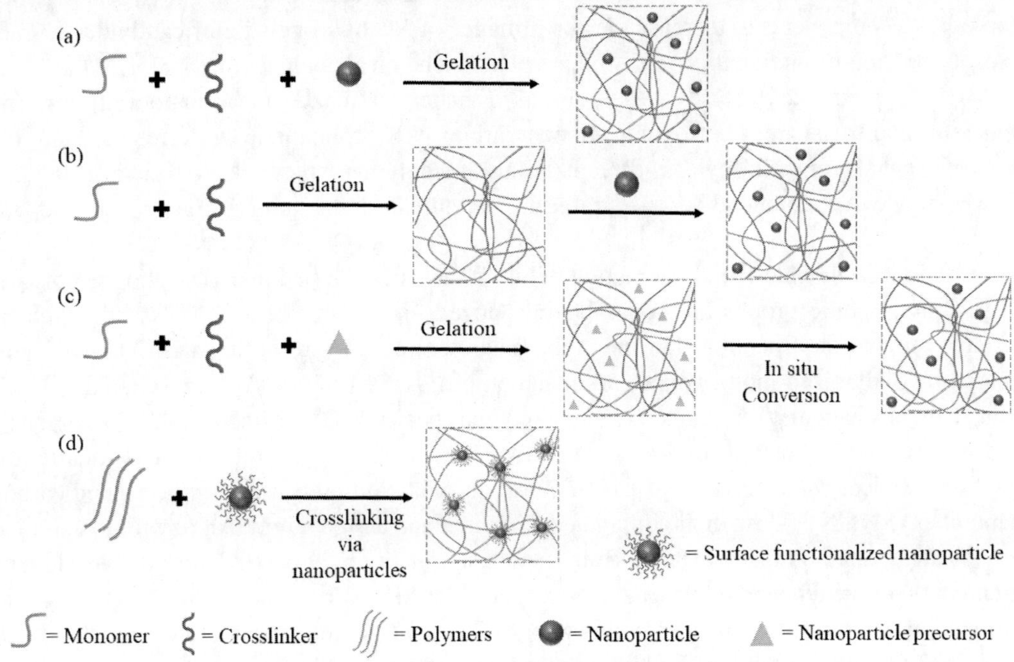

Figure 9.1. Four main approaches used to obtain hydrogel-NP conjugates with uniform distribution: (a) hydrogel formation in an NP suspension, (b) physically embedding the NPs into the hydrogel matrix after gelation, (c) NP formation within a hydrogel matrix, and (d) using NPs to crosslink polymers for hydrogel formation.

In situ polymerization for hydrogel formation is a conventional method used for physically embedding NPs with polar groups on their surfaces (e.g., nanoclay and graphene oxide) as the NPs can establish strong hydrogen bonds and electrostatic interactions with monomers. On the other hand, NPs can be introduced to the hydrogel matrix through the swelling process. Both two methods of physical mixing can be simply achieved by immobilizing NPs directly in the polymeric network while potentially resulting in particle aggregation and a loss of control over the NP-polymer interface. Thus, physical mixing might not be an ideal method for producing NC hydrogels with excellent mechanical characteristics (Kumar and Jaiswal, 2016).

In situ NP growth-based NC hydrogel preparation is carried out to produce NC hydrogels with a stable structure and outstanding mechanical characteristics. NPs having intrinsic functionalities but slight surface modification (e.g., polyaniline NPs, metal, and metal oxide NPs) are usually ideal for *in situ* NP growth hydrogels (He et al., 2020). Using surface-functionalized NPs as crosslinkers in the hydrogel formation can potentially provide additional crosslinks in the hydrogel network as well as present a more homogeneous distribution of NPs (Chen et al., 2018). In this section, we describe the classification of common NPs and concentrate primarily on the dispersion and interactions of NPs in the hydrogels, considering the functional and constructional influences of NPs and the recent advancements in NC hydrogels.

9.2.1. Carbon-Based NPs in NC Hydrogels

Carbon-based nanomaterials (e.g., carbon nanotubes (CNTs) and graphene) have been deemed appropriate components to increase the performance of NC hydrogels significantly due to their exceptional multifunctionality and unique physicochemical and geometrical features. The hydrophobic intrinsic feature of carbon-based materials makes the creation of uniform, coherent, and transparent gels using pristine carbon NPs problematic. As a result, using the poor interfacial compatibility of carbon materials and polymeric networks to form a strong link between the two components for a desirable architecture is challenging (Li, Tang, Bai, and Bai, 2017).

To increase the interfacial interaction between the carbon-based materials and the polymer matrix, hydrophilic groups are generally introduced on the surface of CNTs and graphene through surface modification. For example, Yaghoubi et al., (2021) prepared NC hydrogels consisting of oxidized-multi-walled carbon nanotube (O-MWCNT) and polyacrylamide (PAM) using a mechanism that is reliant on reactive oxygen species (ROS). Mixed acid (H_2SO_4/HNO_3) and strong oxidizing chemicals are used mainly in chemical degradation, which might alter the physicochemical characteristics of CNTs owing to oxidation processes. Single-walled carbon nanotubes (SWCNTs) lose their conjugated structure, and a high number of oxygen-containing groups are created on their surfaces. Simultaneously, *in situ* polymerization of the acrylamide monomers occurred using free radicals to form PAM chains. The O-MWCNT acted as crosslinkers by forming hydrogen bonds with PAM chains to generate a hydrogel network. O-MWCNT dispersed nicely in the hydrogel matrix, and the NC hydrogels presented excellent mechanical properties. The polymer shell was demonstrated to improve the CNTs dispersibility and the interactions between encapsulated-carbon nanotube-polyacrylic acid (E-CNT-PAA) and N-isopropylacrylamide (NIPAM) (Li et al., 2017). However, the polymer wrapping approach may result in a low level of adhesion between the wrapping molecules and the carbon nanomaterials.

Furthermore, the presence of the polymer on the surface may disrupt the carbon-carbon interactions required for electron transmission. To solve this challenge, Emily and coworkers found that the dispersive action of nanocellulose is mediated by oscillations in the counterions on its surface (De France, Hoare, and Cranston, 2017), which aids in the formation of attractive contacts between carbon nanomaterials and nanocellulose. Thus, carbon-based materials can be dispersed in an aqueous medium using the cellulose-assisted technique to provide long-term stable colloidal dispersions. In short, the inclusion of carbon-based materials improves the mechanical and multifunctional properties of NC hydrogels. However, the carbon-based NC hydrogels are still in the early stages of research due to difficult manufacturing methods, low transparency, and concerns about *in vivo* biocompatibility.

9.2.2. Metal and Metal Oxide NPs in NC Hydrogels

The increasing usage of hydrogels in biomedical and biomimetic applications creates the need for the development of multifunctional hydrogels with potential properties of magnetic, electrical, and optical stimuli-responsiveness. Metal and metal oxide NPs possess these distinct characteristics that are not observed in polymeric materials. Metal and metal oxide NPs can create NC hydrogels with antimicrobial activity, magnetic performance, and electrical properties that are appropriate for biomedical applications due to their intrinsic antibacterial, ferromagnetic, and conducting capabilities, respectively. As a result, metal and metal oxide NPs are excellent reinforcing materials with a lot of promise for creating NC hydrogels with distinct properties and customizable qualities.

NC hydrogels have been successfully manufactured using a variety of metallic NPs, including gold (Au), silver (Ag), and metal oxide NPs, such as iron oxide (Fe_3O_4), copper oxide (CuO), titanium oxide (TiO_2), and zirconia (ZrO_2) (Mahmoud, Abouelanwar, Mahmoud, and Salam, 2021; Oun and Rhim, 2017; Wahid, Wang, Lu, Zhong, and Chu, 2017). However, obtaining a stable dispersion in an aqueous solution is difficult due to the high surface energy and strong dipole-dipole interactions among metallic NPs.

Metallic NC hydrogels with good dispersion of metallic NPs can be synthesized using two primary approaches. The first method is the *in situ* growth of metallic NPs in the hydrogel matrix (Zhu et al., 2012). Hydrogels with a 3D network structure often function as both a container for loading metallic NPs and a reactor for NP nucleation and growth. The second method is to utilize reactive metallic NPs as crosslinkers to make hydrogels that polymerize through the free radical reaction (Zhang et al., 2019). Compared to the first embedding procedure, the latter yields metallic NC hydrogels with increased mechanical strength due to strong interfacial contacts between the polymer and surface-modified metallic NPs. As an example, Zang and coworkers used polymer grafted with ligands and metal ions to generate a series of self-assembled metallic-ion NC hydrogels that were stabilized by the *in situ* formed ligand-metal ion NPs (Zhang et al., 2019). Such NC hydrogels exhibit a broad spectrum of mechanical features and remarkable dynamic properties.

9.2.3. Inorganic Clay NPs in NC Hydrogels

Clays, including SiO_2 (0D), halloysite-nanotubes (1D), montmorillonite (MMT) (2D), and synthetic hectorite (2D), are widely used as nanofillers in NC hydrogels. Clay nanoplatelets, such as MMT and synthetic hectorite (laponite), are the most often utilized materials. One octahedral sheet sandwiched between two tetrahedral sheets makes up each layer of clay nanoplatelets. The octahedral sheets usually contain cations, such as magnesium (Mg) and aluminum (Al) ions, that are octahedrally coordinated by oxygen, whereas the tetrahedral sheets are made up of silicon dioxide with negative charges due to partial isomorphous substitution, typically of Al^{3+} by Mg^{2+} in MMT and Mg^{2+} by Li^{+} in laponite (Mousa, Evans, Oreffo, and Dawson, 2018).

Given their interfacial contacts, exfoliated 2D nanoclay (approximately $100\times100\times1$ nm^3 for MMT) with negative charges in polymeric networks endow the resulting hydrogels with outstanding mechanical capabilities (Coleman et al., 2011). As an example, Ma and coworkers successfully prepared a series of NC hydrogels in which the clay was exfoliated and dispersed randomly in the hydrogels, which acted as a multifunctional crosslinker (Wang, Ma, Yang, and Xu, 2011). Notably, clay content can be adjusted to customize the mechanical characteristics and swelling/deswelling kinetics of NC hydrogels. Furthermore, nanoclay NPs possess high biocompatibility to allow the nanoclay-incorporated NC hydrogels to be used as biomaterials for biomedical applications.

9.3. NC Hydrogels for Cartilage Tissue Engineering

Articular cartilage is a highly structured tissue with exceptional load-bearing and low-friction qualities, offering a smooth surface for articulation and a reduced frictional load transfer (Krishnan and Grodzinsky, 2018; Madeira, Santhagunam, Salgueiro, and Cabral, 2015). The sparsely distributed chondrocyte is integrated inside ECM to form articular cartilage. Owing to poor cellularity, low vascularization, non-lymphatic, and limited proliferative potential of resident chondrocytes, articular cartilage has limits in terms of self-regeneration and intrinsic healing (Ahmed and Hincke, 2009; Huey Daniel, Hu Jerry, and Athanasiou Kyriacos, 2012).

Current strategies for repairing cartilage defects have progressed from debridement, microfracture, osteochondral transplantation, and autologous chondrocyte implantation (Brittberg, 2008; Huang, Hu, and Athanasiou, 2016; Matsiko, Levingstone, and O'Brien, 2013). Although these therapies have shown promising results in repairing cartilage defects, these therapeutic treatments have limitations in surgical procedures, immune response danger, and the quality of regenerated cartilage (Killen and Charalambous, 2020; Natoli et al., 2010). In most cases, regenerated cartilage lacks normal mechanical characteristics and zonal structure, which might lead to degeneration. To increase the utilization of biomaterials in cartilage healing, hydrogels have sparked a lot of interest in cartilage tissue engineering. The scaffolds for cartilage tissue engineering produced by natural hydrogels have grown in popularity with their similar structure and biological features to native ECMs to promote cell transplantation, proliferation, and differentiation (Kwon et al., 2013).

9.3.1. Polymeric NP/Alginate NC Hydrogels

The monovalent salt form of alginic acid, alginate, is a linear block copolymer consisting of D-mannuronate (M-block) and L-guluronate (G-block) joined by a 1,4-glycoside bond (Bucke, 1987). The G-block of alginate has correspondingly high affinities for divalent ions, such as calcium (Ca^{2+}), strontium (Sr^{2+}), and barium (Ba^{2+}). The alginate chains are rapidly crosslinked via the staking of G-blocks to form an egg-box structure and then become a gel in the presence of divalent ions. With the properties of biocompatibility and no side effects in humans, alginate hydrogels have been authorized for human use as a wound dressing material and as a food additive by the US Food and Drug Administration (FDA) in the United States (Lee et al., 2015).

Lim et al., (2010) developed an NP/hydrogel system for the regulated distribution of dual growth factors, where bone morphogenetic protein 7 (BMP-7) and transforming growth factor-beta 2 (TGF-2) were delivered via TGF-2-immobilized polyion complex NPs and the BMP-7-immobilized alginate hydrogel system. TGF-2-immobilized polyion complex NPs were created by the complexation of polymers and the growth factor. Adding TGF-2 to the heparin solution created ionic interactions between heparin (the negative charge for the carboxylic acid and sulfate groups in heparin) and TGF-2 (positive charge). Following that, CS solution was added to the solution combination to create NPs via an ionic interaction between heparin (negative charge) and CS (positive charge). BMP-7 and TGF-2 have been shown to be beneficial in promoting chondrogenic differentiation, particularly when combined (typically larger dosages of BMP-7 than TGF-2) (Kim and Im, 2009).

Given that the NP production and subsequent growth factor immobilization do not include any hazardous organic solvents or chemical alteration procedures, the growth factors can retain their bioactivity. The release behavior of dual growth factors (BMP-7/TGF-2) from an NP-containing alginate hydrogel was compared to that of dual growth factors from a non-NP-containing alginate hydrogel. The double growth factor-loaded NP/hydrogel system demonstrated controlled release of both growth factors. A faster release of BMP-7 and a slower release of TGF-2 were observed at approximately 80 and 30% release at the end of an incubation period (21 days), respectively, which may be highly desirable for chondrogenic differentiation of mesenchymal stem cells (MSCs). On the contrary, the release of each growth factor from the dual growth factor-loaded hydrogel (without the NPs) was much slower than that of the NP/hydrogel system, approximately 36% (BMP-7) and 16% (TGF-2) for 21 days. This system can more efficiently improve the chondrogenesis of MSCs and cartilage regeneration (Kim and Im, 2009).

9.3.2. PLGA/Nano-HA/PVA NC Hydrogels

PVA is a commercially available polymer used for hydrogel preparation. However, the biggest drawback of PVA's potential therapeutic use is that PVA hydrogel is not suitable for cell attachment and proliferation (Gonzalez and Alvarez, 2014; Pereira, Silva-Correia, Oliveira, and Reis, 2013). Thus, the composition scaffold using PVA has sparked a lot of curiosity recently (Bailey, Nail, and Grunlan, 2013; Dormer, Singh, Wang, Berkland, and Detamore, 2010; Khandan, Ozada, Saber-Samandari, and Ghadiri Nejad, 2018). A number of research have demonstrated that PLGA and nano-HA can be employed to improve the cellular affinity and biomechanics of PVA hydrogel (Lin, Tsai, and Chang, 2017; Nie et al., 2015; Pan and Xiong, 2009).

Su et al., (2019) created a unique new double-layer hydrogel composed of the core and peripheral components to present strong biocompatibility and great mechanical capabilities to overcome the limitations of PVA hydrogel. The core (nano-HA/PVA hydrogel) was meant to sustain mechanical load, while the peripheral structure (PLGA/nano-HA/PVA) was designed to interface with surrounding cartilage. PLGA/nano-HA/PVA hydrogels co-cultured with chondrocytes were stained histologically and immunohistochemically. There were three experimental groups (E1 contained 5 wt.% PVA/30 wt.% PLGA/5 wt.% nano-HA, E2 contained 10 wt.% PVA/30 wt.% PLGA/5 wt.% nano-HA, and E3 contained 15 wt.% PVA/30 wt.% PLGA/5 wt.% nano-HA) and one control group (CG contained 15 wt.% PVA/5 wt.% HA) in the following. Hematoxylin-eosin (HE) staining revealed a large number of cells in the pores of the hydrogels (Figure 9.2), showing cell proliferation in the experimental groups (E1, E2, and E3) with the matrix brownish yellow and the coarse collagen fibers visible. The time of co-culture had a good correlation with type II collagen expression. The E3 group had the highest level of type II collagen expression among the experimental groups, indicating the double-layer nano-HA/PVA hydrogel covered with PLGA/nano-HA/PVA scaffold can heal cartilage.

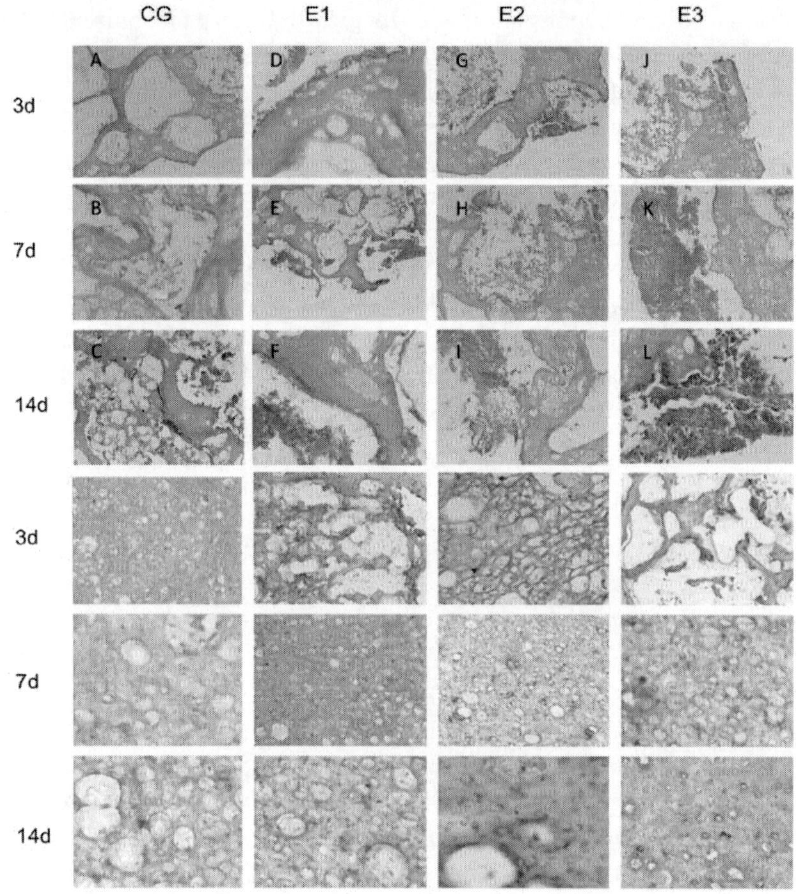

Figure 9.2. PVA mass fractions of 5 wt.% (E1), 10 wt.% (E2), or 15 wt.% (E3), PLGA of 30 wt.% and nano-HA of 5 wt.% were investigated after co-culturing rabbit chondrocytes with scaffolds for 3, 7, and 14 days. The control group (CG) included HA (5 wt.%) and PVA (15 wt.%) (Su et al., 2019).

9.3.3. Laponite/PEG NC Hydrogels

One of the most valuable hydrogels in regenerative medicine and tissue engineering is PEG (Annabi et al., 2010). PEG hydrogels, in particular, have been employed successfully as stem cell delivery vehicles for cartilage formation *in vivo* and as chondrocyte cell carriers to construct cartilaginous tissue for tissue engineering (Bryant and Anseth, 2002, 2003). Nojoomi et al., (2017) created injectable PEG-based hydrogels containing laponite particles, which have mechanical and structural qualities similar to natural articular cartilage. In the research, imide ring-opening reactions with manufactured copolymers incorporating PEG blocks and laponite particles were used to create the nanocomposites in a two-step thermal poly (amic acid) procedure. Butane diamine (BDA) was utilized as a nucleophilic reagent, and hydrogels with interconnected holes ranging in size from 100 to 250 μm were created. Compared to the pristine PEG hydrogels, laponite/PEG hydrogels improved the elastic modulus of hydrogels by 20% and the viscous modulus by 50% at each frequency. These results bring the viscoelastic characteristics of laponite/PEG NC hydrogels closer to those reported for articular cartilage in animal models (380-650 kPa). Cell viability experiments of human bone marrow-derived stromal stem cells (hMSCs) revealed that the PEG hydrogels were biocompatible. Although the inclusion of laponite particles reduced cell viability marginally, the NC hydrogels demonstrated good biocompatibility (>90%), indicating that the scaffold from this hydrogel can be applied in cartilage tissue engineering.

9.4. NC Hydrogels for Bone Healing

Bone defects, which may result from trauma, tumor, and complications after previous surgery, can lead to disability and deformity. Nowadays, even with the advantages of medical biomaterials, bone defect is still an acritical challenge for physicians. For the treatment of bone defects, surgical intervention of implant fixation and the usage of bone grafts were considered the gold standard procedure. Commonly used bone grafts, such as autologous bone grafting (Kofron and Laurencin, 2006; Kondiah et al., 2016), allograft bone grafting (Ito et al., 2005), and synthetic bone grafting alternatives have their limitations that we have discussed in the introduction. Therefore, medical physicians still search for optimal biomaterials and advanced tissue engineering to improve the clinical prognosis after the operation.

Hydrogels can imitate the natural tissue environment and offer structural support for the damaged area in bone tissue engineering. Crosslinked polymer chains provide the basis of hydrogels, and the 3D hydrophilic network structures possess good biocompatibility, biodegradability, drug-carrying capability, and regulated drug-release ability. The traits can be exploited to create a range of therapeutic options for various disorders (Zhang, Li, Guan, Mao, and Zhou, 2021). Although simple hydrogels have the above-mentioned advantages, they can have more potential benefits when hydrogels combine with nanomaterials. In addition to having a complete network, NC hydrogels have enhanced strength, structure to mimic ECM, and more abilities to bind growth factors. Those characteristics of NC hydrogels may be applied to bone healing treatment and address current limitations.

9.4.1. Silver NP/Corn Silk Extract NC Hydrogels

The antibacterial, anti-angiogenic, anti-inflammatory, and anticancer properties of silver NPs (Ag NPs) have been widely documented (Zare et al., 2020). Biocompatible and biodegradable complexes with bactericidal activity and low cytotoxicity properties point to their suitability for the prevention and treatment of chronic wounds, as well as for analgesic therapies, anticancer strategies, and the detection of chemical substances and malignant cells. As an example, spherical Ag NPs were biosynthesized with corn silk extract (CSE) in an organic solvent-free media through a microwave-assisted green method. The Ag NP-containing samples exhibited antibacterial activity against both gram-positive (*Bacillus subtilis* and *Staphylococcus aureus*) and gram-negative (*Pseudomonas aeruginosa* and *Escherichia coli*) bacteria without cytotoxicity, making the CSE-Ag NPs as an excellent scaffold for the regeneration of bone tissue (Makvandi et al., 2020).

9.4.2. Cerium Oxide NP/Gelatin/Alginate NC Hydrogels

Cerium oxide NPs (nanoceria) exhibit excellent free radical scavenging ability to facilitate the bone healing process as the ROS/free radicals are presented at the fracture site (Das et al., 2014). To create NC hydrogels for bone healing, nanoceria was added to gelatin-alginate (GA) hydrogel and then freeze-dried before use (Purohit et al., 2020). The addition of nanoceria improved the mechanical property of hydrogels and reduced the swelling and *in vitro* weight loss of the composite scaffolds. GA hydrogel was used as a control group, and three composite scaffolds with different nanoceria amounts (i.e., 100, 500, and 1000 μg/ml for GA-NC1, GA-NC2, and GA-NC3, respectively) were prepared. Osteogenic differentiation studies (i.e., alkaline phosphatase (ALP) activity, RunX2, and osteocalcin expression) were performed to investigate the potential of GA-NCs scaffolds for the differentiation of MSCs to osteoblast. ALP is a well-defined osteogenic differentiation marker (Li, Kim, Choi, and Han, 2019) and RunX2 and osteocalcin regulate the transcription of essential genes in MSCs to differentiate into osteoblasts by interacting with particular DNA sequences (Detsch, Mayr, and Ziegler, 2008).

The results showed that ALP activity was significantly increased in the GA and GA-NCs scaffolds on day 14. In particular, ALP activity was higher on the GA-NC3 scaffold than that on the GA scaffold due to the higher concentration of nanoceria, suggesting that nanoceria might aid in the process of biomineralization. The relative ALP activity for GA and GA-NCs scaffolds was reduced on day 21, with the GA-NCs scaffolds showing a more significant drop in relative ALP activity than the GA scaffold. This indicates that more MSCs have developed into osteoblasts with the integration of nanoceria in the scaffolds, speeding up the osteogenic differentiation of the MSCs. Immuno-fluorescent markers (RunX2 and osteocalcin) were used to validate the osteogenic differentiation potential of GA and GA-NCs scaffolds through confocal imaging. It was noticed that higher expression of RunX2 was observed in the cell nuclei in the GA-NC2 and GA-NC3 scaffolds than that in the GA and GA-NC1 scaffolds, indicating that MSCs were differentiated into osteoblasts.

On the other hand, qualitative expression of osteocalcin was elicited by MSCs cultured on GA and GA-NCs scaffolds on days 14 and 21 (Figure 9.3a), and osteocalcin expression has risen over time for all of the scaffolds (Figure 9.3b).

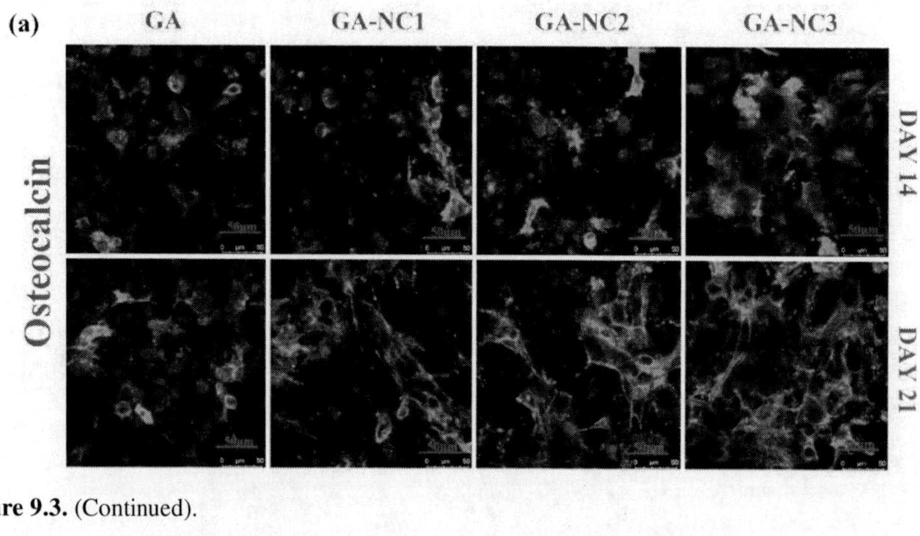

Figure 9.3. (Continued).

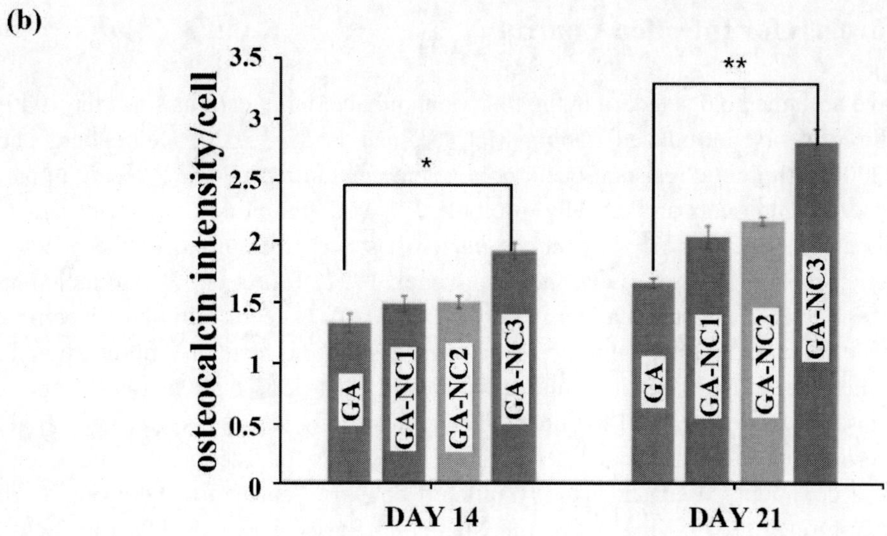

Figure 9.3. (a) Fluorescence pictures of osteocalcin expression on days 14 and 21 and (b) MSCs cultivated on GA and GA-NCs hydrogel expressed osteogenic markers (Purohit et al., 2020).

Notably, MCSs on the GA-NCs scaffolds expressed osteocalcin at a higher level than on the GA scaffold, with MCSs on the GA-NC3 scaffold inducing the most osteocalcin expression. These results might be due to the GA-NC3 scaffold possessing more Ca/P nucleation sites with a higher amount of nanoceria. In sum, the nanoceria-containing scaffolds have been demonstrated to present a beneficial impact on facilitating the differentiation of MSCs into osteoblasts.

9.4.3. NDs/CS/Gelatin NC Hydrogels

NDs (nanodiamonds)-vascular endothelial growth factor (VEGF) complexes were added to the polymeric mixture of gelation to create injectable NC hydrogels (Mochalin, Shenderova, Ho, and Gogotsi, 2020). The prepared NC hydrogels were also formed using VEGF that had not been altered as a control to evaluate the release of VEGF from the NP complexes. An early burst release of VEGF for the first 12 h was followed by a slower release for the remainder of the research in both the control gels containing VEGF and the experimental hydrogels containing ND-VEGF. The release profile plateaued after five days, suggesting that the release of VEGF from the control and experimental gels appeared to have stabilized at a constant rate. The release profiles for both hydrogels were fitted to the Korsmeyer-Peppas model to mimic the release of VEGF (Costa and Lobo, 2001). The release exponent for the control gel containing free VEGF showed n= 0.47 by the Korsmeyer-Peppas model, indicating Fickian diffusion. In contrast, the release exponent for the experimental gel containing the ND-VEGF complex showed n= 0.57, indicating that the release exhibits non-Fickian behavior and the complexation of the protein with NDs resulted in additional transport phenomena (Pacelli et al., 2017).

9.5. NC Hydrogels for Infection Control

Antibiotics have been the gold standard in the treatment of bacterial infections since the 1940s when penicillin was first introduced (Cobb, McCabe, and Priddy, 2020; Deurenberg and Stobberingh, 2008). Antibiotic-resistant bacteria have emerged in recent years, owing in part to the widespread, empirical, and usually prolonged use of antibiotics, rendering many antibiotic regimens ineffective. *Staphylococcus aureus* has been recognized as the primary cause of infection-related diseases (Lew and Waldvogel, 1997; Tao et al., 2020), including endocarditis, dermatitis, and osteomyelitis (Cobb et al., 2020). Osteomyelitis has become a major worldwide problem and is potentially devastating due to the facts of limb-threatening sequelae and mortality. Therefore, new and effective drug carriers, e.g., foams (Cabezas, Gracia, de Lucas, and Rodríguez, 2014), films (Chen et al., 2013), hydrogels (Krishnan et al., 2017), and microparticles (Gallon et al., 2017), are urgent needs for osteomyelitis treatment. Hydrogels are a commonly used and handily tailored delivery vehicle for therapeutics for osteomyelitis (Cobb et al., 2019; Kim et al., 2021; Mahmoud, Sayed, Thabit, and Safwat, 2020). Currently, there are many NC hydrogels used for the treatment of bone infections due to their excellent mechanical performance (e.g., tensile, compression, and rheological properties), encapsulation efficiency, drug loading, and antibacterial activity (Chen et al., 2018; Haraguchi, 2007; Tao et al., 2020).

9.5.1. Silica Nanosphere/Alginate NC Hydrogels

Chen et al., (2018) prepared pH-responsive antibacterial microspheres with green tea polyphenol (TP) for the invasive treatment of osteomyelitis. TP was mixed with porous silica (SiO_2) nanospheres (NSs) to form TP-loaded SiO_2 NSs (TP@SiO_2), and sodium alginate (SA) was used to encapsulate TP@SiO_2 to form TP@SiO_2@SA. TP@SiO_2@SA@$CaCO_3$ MSs (microgel spheres) were further obtained from the cladding of TP@SiO_2@SA with $CaCO_3$.The morphological changes of osteocytes at different pHs (i.e., 7.4 and 5.5) were revealed with confocal laser scanning microscopy (CLSM) (Figure 9.4), where pH=7.4 is the normal physiological environment (Figure 9.4a), while pH=5.5 is a mimic to the weakly acidic medium of osteomyelitis-related conditions (Figure 9.4b). Osteoblasts become flattened in the ossified environment (pH=5.5) (Figure 9.4c), which is attributed to the fact that the acidic environment favors bacterial growth and is detrimental to normal cells.

When H_2O_2 (200 μM) was added for oxidative stimulation in the pH=5.5 condition (Figure 9.4d), the osteoblasts were shortened and even died. As a control, osteoblasts remained flat in the conditions of H_2O_2 and pH=5.5 after adding antioxidant TP. In the presence of TP@SiO_2@SA@$CaCO_3$ MSs (within 200 μM TP), the morphology of osteoblasts returned to normal, and their shape resumed to integral and continuous (Figure 9.4e). This result showed that TP@SiO_2@SA@$CaCO_3$ MSs significantly improved the proliferation and differentiation of osteoblasts.

Furthermore, the release of TP was conducted at varying pH conditions (pH of 5.5 and 7.4). Only a tiny quantum (12.9%) of TP was released in phosphate-buffered saline (PBS) solution at pH=7.4 after dialysis for 6 h. In contrast, TP (42%) was released rapidly at pH=5.5 within 1 h and nearly 100% within 6 hr. This result should be due to the desolvation of $CaCO_3$

at a low pH (pH=5.5), and Ca^{2+} ions released from dissolved $CaCO_3$ cannot form stable calcium phosphate. Thus, the release of TP from TP@SiO_2@SA@$CaCO_3$ MSs was pH-responsive, allowing TP to be released precisely into osteomyelitis cells. The turbidity test was used to quantify the antibacterial effect of TP@SiO_2@SA@$CaCO_3$ MSs, showing that the turbidity of *Staphylococcus aureus* decreased in the presence of TP@SiO_2@SA@$CaCO_3$ MSs. The bacteriostatic rate of these TP@SiO_2@SA@$CaCO_3$ MSs was about 72% of VCM by using VCM as a positive control group.

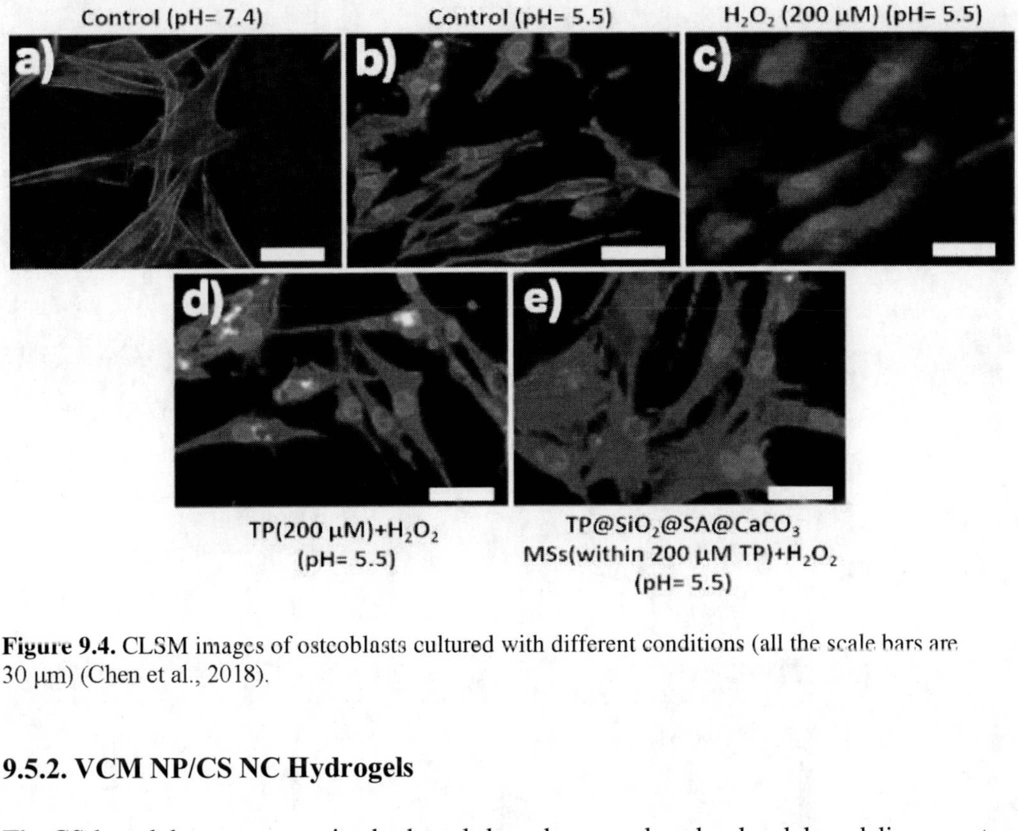

Figure 9.4. CLSM images of osteoblasts cultured with different conditions (all the scale bars are 30 μm) (Chen et al., 2018).

9.5.2. VCM NP/CS NC Hydrogels

The CS-based thermo-responsive hydrogels have been produced as local drug delivery systems of VCM (vancomycin) (VCM-NPs/Gel) (Cabezas et al., 2014; Tao et al., 2020). VCM-NPs were formulated with quaternary ammonium CS and carboxylated CS through the electrostatic interaction-driven assembly to improve the encapsulation and drug loading efficiency of VCM, as well as to prevent infection and repair broken bones. This VCM-NPs/Gel system was evaluated in a rabbit osteomyelitis model. The CS/GP(glycerophosphate disodium salt)/NPs mixture possessed good injectability and were translucent, slightly viscous liquids (sols).

Additionally, due to CS-GP electrostatic interactions between the ammonium and phosphate groups, they consolidated into semisolids (gels) at physiological temperature (Figure 9.5A). The higher concentrations of β-GP(glycerol 2-phosphate) possessed faster gelation periods when examining the gelation times for different ratios of CS/β-GP (Figure 9.5B), and the addition of α-GP (DL-glyceraldehyde-3-phosphate) can reduce the gelation process even more (Figure 9.5C). The gelation time of the VCM-NPs/Gel was slightly increased with NP loading compared with the VCM/Gel (Figure 9.5D).

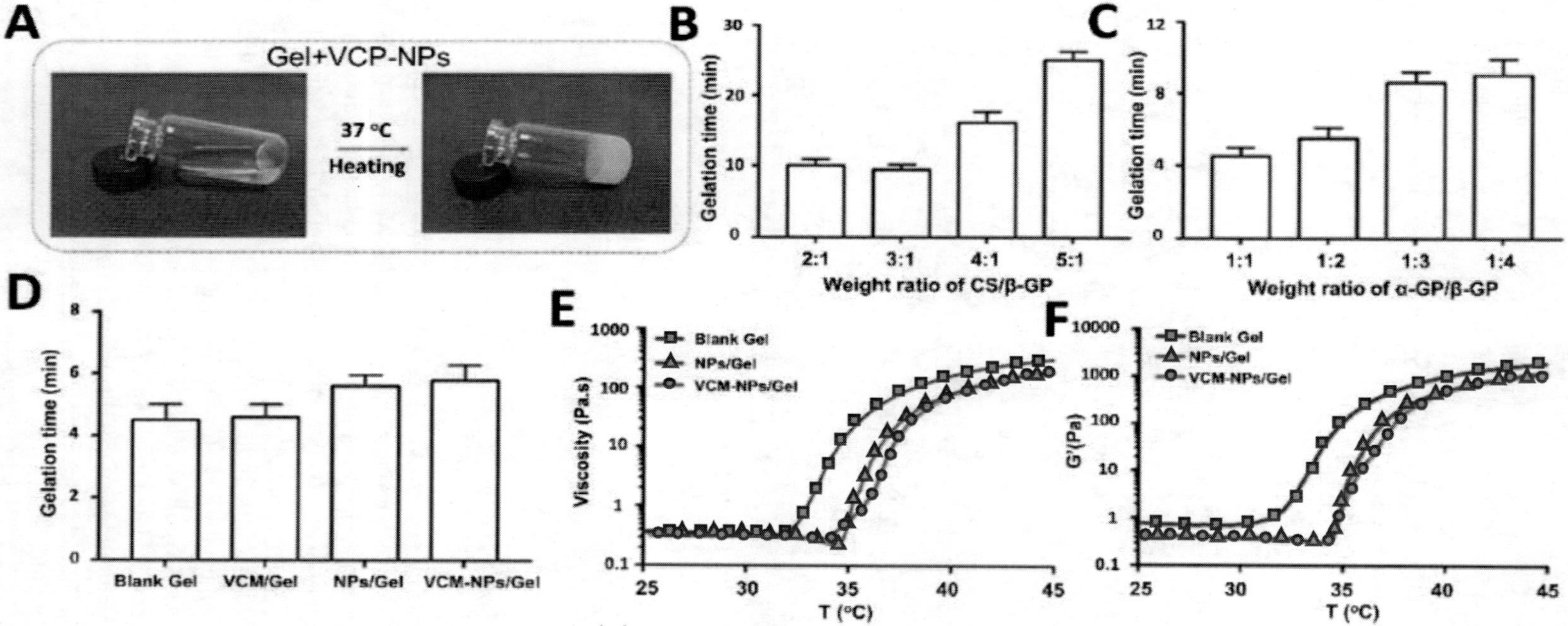

Figure 9.5. (Continued).

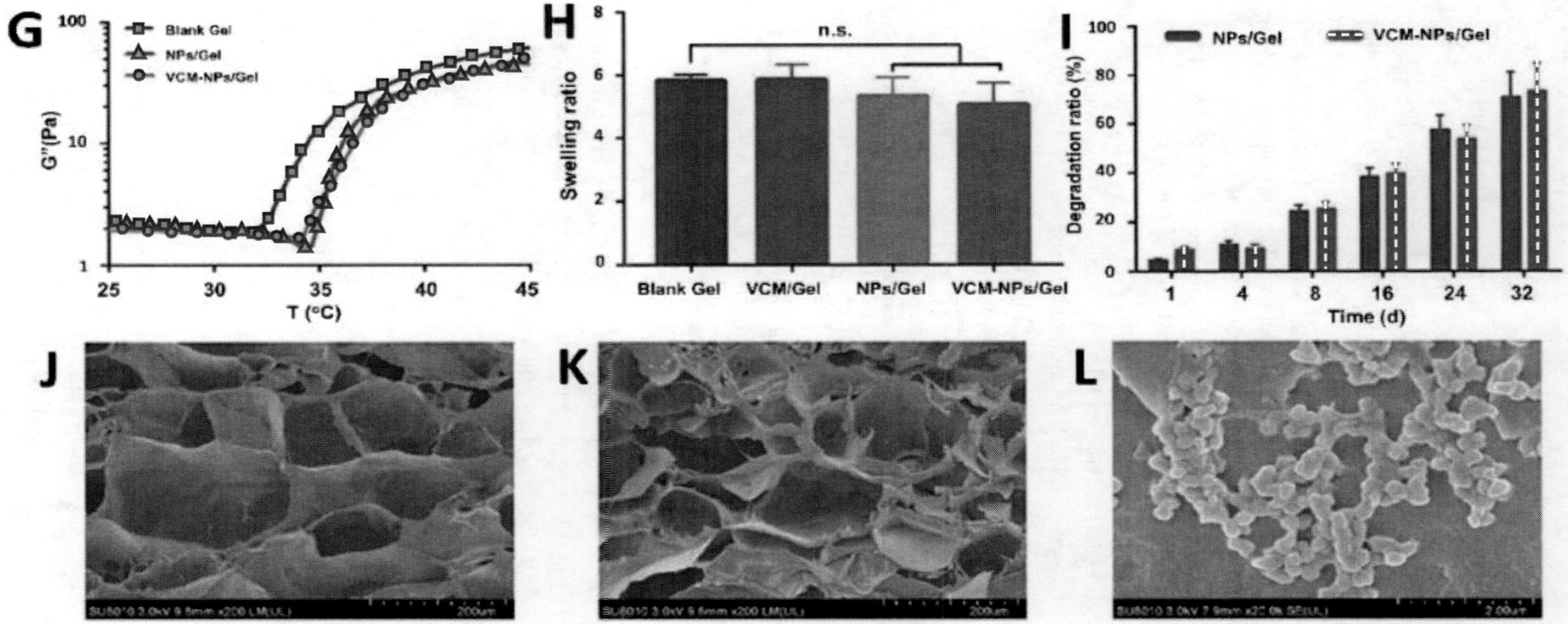

Figure 9.5. Characterizations of VCM-NPs/Gel: (A) Photographs of VCM-NPs/Gel aqueous solutions exhibiting sol and gel states at 20 and 37°C, respectively, (B-D) gelation time of hydrogel with CS/β-GP, α-GP/β-GP, and VCM-NPs, respectively, (E-G) rheological properties exhibiting η, G', and G'' of the aqueous solutions of the blank Gel, NPs/Gel, and VCM-NPs/Gel as a function of temperature, respectively, (H) equilibrium swelling ratio of hydrogels after swelling for 24 h in PBS (pH=7.4), (I) degradation profiles of the hydrogels in PBS with lysozyme, (J) SEM image of VCM/Gel, and (K and L) SEM image and enlarged SEM image of VCM-NPs/Gel (Tao et al., 2020).

The sol-to-gel transition temperature of VCM-NPs/Gel was between physiological and room temperatures, showing a suitable gelation behavior for clinical applications. Below 30 °C, the blank gel and NPs/Gel had low values of viscosity (η), storage modulus (*G'*), and loss modulus (*G"*). In addition, *G"* was larger than *G'*, signifying good flowability (Figure 9.5E-G). As the temperature increased, η, *G',* and *G"* increased rapidly by several orders of magnitude, and *G'* exceeded *G"*, indicating the *in situ* formation of physical hydrogels. Hydrogel swelling results showed that both hydrogels with and without NPs exhibited strong water uptake and that the swelling ratio for the blank gel and the VCM/Gel was approximately equal to 6 (Figure 9.5H). It appears that all hydrogels have enough holes to absorb a lot of water as the water uptake capabilities of the NPs/Gel and VCM-NPs/Gel did not significantly be altered when the NPs were added compared to the blank gel. The degradation behaviors of the VCM/Gel and VCM-NPs/Gel in the lysozyme solution were outstanding, and the presence of NPs didn't affect the hydrogel degradation (Figure 9.5I). The produced hydrogel microstructure was analyzed using a scanning electron microscope (SEM). The architecture of the VCM-NPs/Gel presented a porous network that improved nutrient exchange, cell entrance, and sustained drug release (Figure 9.5J), which was similar to VCM/Gel (Figure 9.5K). High-power field SEM was used to investigate the morphology and particle size of VCM-NPs in more detail (Figure 9.5L).

Additionally, with the results of transmission electron microscopy (TEM) and dynamic light scattering, the VCM-NPs had a particle size of ~200 nm, suggesting that the VCM-NPs were still incorporated into the gel structure. The antibacterial activity of the gel system showed that the hydrogels loaded with VCM or VCM-NPs did not exhibit antibacterial effects against *Staphylococcus aureus*. As a result of the rapid VCM release, the inhibition zone of the VCM/Gel demonstrated a sudden and significant decrease on day 5 and exhibited no effect against *Staphylococcus aureus* on day 10. In contrast, the VCM-NPs/Gel showed sustained antibacterial activity against *Staphylococcus aureus* for more than 20 days, indicating that the VCM-NPs/Gel had better antibacterial activity.

Thus, VCM-NPs/Gel increased osteoblast growth and activity against *Staphylococcus aureus* based on the *in vitro* testing. For the *in vivo* test, rabbits with osteomyelitis were treated with VCM-NPs/Gel (containing 10 mg VCM per rabbit). Under osteomyelitis conditions, the VCM-NPs/Gel presented strong anti-infection characteristics and accelerated bone repair. In conclusion, the VCM-NPs/Gel system has demonstrated its great potential for local antibiotic delivery for bone regeneration promotion and anti-infection, presenting as a promising scaffold for osteomyelitis treatment.

Conclusion

Cartilage defect, bone healing, and infection of the musculoskeletal system remain the most complex challenges for clinical physicians. Successful treatment relays on correct diagnosis, suitable surgical intervention, and effective augmentations by biomedical materials. Hydrogel-based strategies for treating musculoskeletal disease provide novel management in progress with the strength of versatile structures controlled by the preparation conditions (e.g., polymers and crosslinking mechanisms) as well as multiple functions regulated by different components (e.g., growth factors and antibiotics).

In particular, injectable hydrogels provide a minimally invasive procedure to precisely fill the defects of tissues and bones in different sizes, facilitating the design and processing of therapy. NC hydrogels have been recognized as promising next-generation biomaterials to make hydrogels similar to the native scaffold, carry desirable particles into the tissue, and match suitable degradation rates. Nevertheless, there is still a long way to develop and control the structural features and biophysical/biochemical properties of NC hydrogels to improve their utility in orthopedics. Taken together, versatile NC hydrogels with a variety of characteristics and properties can be achieved by regulating numerous components and changing preparation conditions, with some reliable results of *in vivo* trials still needed for specific and practical clinical uses in orthopedics.

References

Ahmed, T. A. E., and Hincke, M. T. (2009). Strategies for articular cartilage lesion repair and functional restoration. *Tissue Engineering Part B: Reviews,* 16(3), 305-329.

Ahmed, T. A. E., and Hincke, M. T. (2014). Mesenchymal stem cell-based tissue engineering strategies for repair of articular cartilage. *Histol Histopathol,* 29(6), 669-689.

Alinavaz, S., Mahdavinia, G. R., Jafari, H., Hazrati, M., and Akbari, A. (2021). Hydroxyapatite (HA)-based hybrid bionanocomposite hydrogels: ciprofloxacin delivery, release kinetics and antibacterial activity. *Journal of Molecular Structure,* 1225, 129095.

Annabi, N., Nichol, J. W., Zhong, X., Ji, C., Koshy, S., Khademhosseini, A., and Dehghani, F. (2010). Controlling the porosity and microarchitecture of hydrogels for tissue engineering. *Tissue Engineering Part B: Reviews*, 16(4), 371-383.

Armiento, A. R., Stoddart, M. J., Alini, M., and Eglin, D. (2018). Biomaterials for articular cartilage tissue engineering: learning from biology. *Acta Biomaterialia,* 65, 1-20.

Asadi, N., Alizadeh, E., Salehi, R., Khalandi, B., Davaran, S., and Akbarzadeh, A. (2018). Nanocomposite hydrogels for cartilage tissue engineering: a review. *Artificial Cells, Nanomedicine, and Biotechnology,* 46(3), 465-471.

Bailey, B. M., Nail, L. N., and Grunlan, M. A. (2013). Continuous gradient scaffolds for rapid screening of cell material interactions and interfacial tissue regeneration. *Acta Biomaterialia,* 9(9), 8254-8261.

Baldwin, P., Li, D. J., Auston, D. A., Mir, H. S., Yoon, R. S., and Koval, K. J. (2019). Autograft, allograft, and bone graft substitutes: clinical evidence and indications for use in the setting of orthopaedic traumasurgery. *Journal of Orthopaedic Trauma,* 33(4), 203-213.

Brittberg, M. (2008). Autologous chondrocyte implantation-technique and long-term follow-up. *Injury,* 39(1), 40-49.

Bryant, S. J., and Anseth, K. S. (2002). Hydrogel properties influence ECM production by chondrocytes photoencapsulated in poly(ethylene glycol) hydrogels. *Journal of Biomedical Materials Research,* 59(1), 63-72.

Bryant, S. J., and Anseth, K. S. (2003). Controlling the spatial distribution of ECM components in degradable PEG hydrogels for tissue engineering cartilage. *Biomedical Materials Research Part A,* 64(1), 70-79.

Bucke, C. (1987). Cell immobilization in calcium alginate. *Methods in Enzymology*, 135, 175-189.

Cabezas, L. I., Gracia, I., de Lucas, A., and Rodríguez, J. F. (2014). Validation of a mathematical model for the description of hydrophilic and hydrophobic drug delivery from biodegradable foams: experimental and comparison using indomethacin as released drug. *Industrial and Engineering Chemistry Research,* 53(21), 8866-8873.

Chan, B. P., and Leong, K. W. (2008). Scaffolding in tissue engineering: general approaches and tissue-specific considerations. *European Spine Journal, 17(4)*, 467-479.

Chen, D., Chen, J., Wu, M., Tian, H., Chen, X., and Sun, J. (2013). Robust and flexible free-standing films for unidirectional drug delivery. *Langmuir,* 29(26), 8328-8334.

Chen, T., Hou, K., Ren, Q., Chen, G., Wei, P., and Zhu, M. (2018). Nanoparticle-polymer synergies in nanocomposite hydrogels: from design to application. *Macromolecular Rapid Communications,* 39(21), 1800337.

Chen, Z., Lv, X., Zhao, M., Zhang, P., Ren, X., and Mei, X. (2018). Encapsulation of green tea polyphenol by pH responsive, antibacterial, alginate microgels used for minimally invasive treatment of bone infection. *Colloids and Surfaces B: Biointerfaces,* 170, 648-655.

Cobb, L. H., McCabe, E. M., and Priddy, L. B. (2020). Therapeutics and delivery vehicles for local treatment of osteomyelitis. *Journal of Orthopaedic Research,* 38(10), 2091-2103.

Cobb, L. H., Park, J., Swanson, E. A., Beard, M. C., McCabe, E. M., Rourke, A. S., Seo, K. S., Olivier, A. K., and Priddy, L. B. (2019). CRISPR-Cas9 modified bacteriophage for treatment of *Staphylococcus aureus* induced osteomyelitis and soft tissue infection. *PloS One,* 14(11), e0220421.

Coleman, J. N., Lotya, M., O'Neill, A., Bergin, S. D., King, P. J., Khan, U., Young, K., Gaucher, A., De, S., Smith, R. J., Shvets, I. V., Arora, S. K., Stanton, G., Kim, H. Y., Lee, K., Kim, G. T., Duesberg, G. S., Hallam, T., Boland, J. J., Wang, J. J., Donegan, J. F., Grunlan, J. C., Moriarty, G., Shmeliov, A., Nicholls, R. J., Perkins, J. M., Grieveson, E. M., Theuwissen, K., McComb, D. W., Nellist, P. D., and Nicolosi, V. (2011). Two-dimensional nanosheets produced by liquid exfoliation of layered materials. *Science,* 331(6017), 568-571.

Costa, P., and Lobo, J. M. S. (2001). Modeling and comparison of dissolution profiles. *European Journal of Pharmaceutical Sciences,* 13(2), 123-133.

Das, S., Chigurupati, S., Dowding, J., Munusamy, P., Baer, D. R., McGinnis, J. F., Mattson, M. P., Self, W., and Seal, S. (2014). Therapeutic potential of nanoceria in regenerative medicine. *MRS Bulletin,* 39(11), 976-983.

De France, K. J., Hoare, T., and Cranston, E. D. (2017). Review of hydrogels and aerogels containing nanocellulose. *Chemistry of Materials,* 29(11), 4609-4631.

Detsch, R., Mayr, H., and Ziegler, G. (2008). Formation of osteoclast-like cells on HA and TCP ceramics. *Acta Biomaterialia,* 4(1), 139-148.

Deurenberg, R. H., and Stobberingh, E. E. (2008). The evolution of Staphylococcus aureus. *Infection, Genetics and Evolution,* 8(6), 747-763.

Dormer, N. H., Singh, M., Wang, L., Berkland, C. J., and Detamore, M. S. (2010). Osteochondral interface tissue engineering using macroscopic gradients of bioactive signals. *Annals of Biomedical Engineering,* 38(6), 2167-2182.

Frantz, C., Stewart, K. M., and Weaver, V. M. (2010). The extracellular matrix at a glance. *Journal of Cell Science,* 123(24), 4195-4200.

Gaharwar, A. K., Peppas, N. A., and Khademhosseini, A. (2014). Nanocomposite hydrogels for biomedical applications. *Biotechnology and Bioengineering,* 111(3), 441-453.

Gallon, G., Lapinte, V., Robin, J. J., Chopineau, J., Devoisselle, J. M., and Aubert-Pouëssel, A. (2017). Cross-linked castor oil-based hybrid microparticles as drug delivery systems. *ACS Sustainable Chemistry and Engineering,* 5(5), 4311-4319.

Gonzalez, J. S., and Alvarez, V. A. (2014). Mechanical properties of polyvinylalcohol/hydroxyapatite cryogel as potential artificial cartilage. *Journal of the Mechanical Behavior of Biomedical Materials,* 34, 47-56.

Haraguchi, K. (2007). Nanocomposite hydrogels. *Current Opinion in Solid State and Materials Science,* 11(3-4,), 47-54.

He, J., Shi, M., Liang, Y., and Guo, B. (2020). Conductive adhesive self-healing nanocomposite hydrogel wound dressing for photothermal therapy of infected full-thickness skin wounds. *Chemical Engineering Journal,* 394, 124888.

Huang, B. J., Hu, J. C., and Athanasiou, K. A. (2016). Cell-based tissue engineering strategies used in the clinical repair of articular cartilage. *Biomaterials,* 98, 1-22.

Huey Daniel, J., Hu Jerry, C., and Athanasiou Kyriacos, A. (2012). Unlike bone, cartilage regeneration remains elusive. *Science,* 338(6109), 917-921.

Ito, H., Koefoed, M., Tiyapatanaputi, P., Gromov, K., Goater, J. J., Carmouche, J., Zhang, X., Rubery, P. T., Rabinowitz, J., Samulski, R. J., Nakamura, T., Soballe, K., O'Keefe, R. J., Boyce, B. F., and Schwarz, E. M. (2005). Remodeling of cortical bone allografts mediated by adherent rAAV-RANKL and VEGF gene therapy. *Nature Medicine,* 11(3), 291-297.

Khandan, A., Ozada, N., Saber-Samandari, S., and Ghadiri Nejad, M. (2018). On the mechanical and biological properties of bredigite-magnetite ($Ca_7MgSi_4O_{16}$-Fe_3O_4) nanocomposite scaffolds. *Ceramics International,* 44(3), 3141-3148.

Killen, M. C., and Charalambous, C. P. (2020). Advances in cartilage restoration techniques. *Advances in Medical and Surgical Engineering*, 71-83.

Kim, H. J., and Im, G. I. (2009). Combination of transforming growth factor-beta2 and bone morphogenetic protein 7 enhances chondrogenesis from adipose tissue-derived mesenchymal stem cells. *Tissue Engineering Part A,* 15(7), 1543-1551.

Kim, J., Hlaing, S. P., Lee, J., Saparbayeva, A., Kim, S., Hwang, D. S., Lee, E. H., Yoon, I. S., Yun, H., Kim, M. S., Moon, H. R., Jung, Y., and Yoo, J. W. (2021). Exfoliated bentonite/alginate nanocomposite hydrogel enhances intestinal delivery of probiotics by resistance to gastric pH and on-demand disintegration. *Carbohydrate Polymers,* 272, 118462.

Klemm, K. (2001). The use of antibiotic-containing bead chains in the treatment of chronic bone infections. *Clinical Microbiology and Infection,* 7(1), 28-31.

Kock, L., van Donkelaar, C. C., and Ito, K. (2012). Tissue engineering of functional articular cartilage: the current status. *Cell and Tissue Research,* 347(3), 613-627.

Kofron, M. D., and Laurencin, C. T. (2006). Bone tissue engineering by gene delivery. *Advanced Drug Delivery Reviews,* 58(4), 555-576.

Kondiah, P. J., Choonara, Y. E., Kondiah, P. P. D., Marimuthu, T., Kumar, P., Du Toit, L. C., and Pillay, V. (2016). A Review of injectable polymeric hydrogel systems for application in bone tissue engineering. *Molecules,* 21(11), 1580.

Krishnan, L., Priddy, L. B., Esancy, C., Klosterhoff, B. S., Stevens, H. Y., Tran, L., and Guldberg, R. E. (2017). Delivery vehicle effects on bone regeneration and heterotopic ossification induced by high dose BMP-2. *Acta Biomaterialia,* 49, 101-112.

Krishnan, Y., and Grodzinsky, A. J. (2018). Cartilage diseases. *Matrix Biology,* 71, 51-69.

Kuechle, D. K., Landon, G. C., Musher, D. M., and Noble, P. C. (1991). Elution of vancomycin, daptomycin, and amikacin from acrylic bone cement. *Clinical Orthopaedics and Related Research*, 264, 302-308.

Kumar, A., and Jaiswal, M. (2016). Design and *in vitro* investigation of nanocomposite hydrogel based *in situ* spray dressing for chronic wounds and synthesis of silver nanoparticles using green chemistry. *Journal of Applied Polymer Science,* 133(14), 43260.

Kwon, J. S., Yoon, S. M., Kwon, D. Y., Kim, D. Y., Tai, G. Z., Jin, L. M., Song, B., Lee, B., Kim, J. H., Han, D. K., Min, B. H. and Kim, M. S. (2013). Injectable *in situ*-forming hydrogel for cartilage tissue engineering. *Journal of Materials Chemistry B,* 1(26), 3314-3321.

Lee, R. J., Hinson, A., Bauernschmitt, R., Matschke, K., Fang, Q., Mann, D. L., Dowling, R., Schiller, N. and Sabbah, H. N. (2015). The feasibility and safety of Algisyl-LVR™ as a method of left ventricular augmentation in patients with dilated cardiomyopathy: initial first in man clinical results. *International Journal of Cardiology,* 199, 18-24.

Lew, D. P., and Waldvogel, F. A. (1997). Osteomyelitis. *New England Journal of Medicine,* 336(14), 999-1007.

Li, Y., Kim, J. H., Choi, E. H., and Han, I. (2019). Promotion of osteogenic differentiation by non-thermal biocompatible plasma treated chitosan scaffold. *Scientific Reports,* 9(1), 3712.

Li, Z., Tang, M., Bai, W., and Bai, R. (2017). Preparation of hydrophilic encapsulated carbon nanotubes with polymer brushes and its application in composite hydrogels. *Langmuir,* 33(24), 6092-6101.

Lim, S. M., Oh, S. H., Lee, H. H., Yuk, S. H., Im, G. I., and Lee, J. H. (2010). Dual growth factor-releasing nanoparticle/hydrogel system for cartilage tissue engineering. *Journal of Materials Science: Materials in Medicine,* 21(9), 2593-2600.

Lin, H. Y., Tsai, W. C., and Chang, S. H. (2017). Collagen-PVA aligned nanofiber on collagen sponge as bi-layered scaffold for surface cartilage repair. *Journal of Biomaterials Science, Polymer Edition,* 28(7), 664-678.

Madeira, C., Santhagunam, A., Salgueiro, J. B., and Cabral, J. M. S. (2015). Advanced cell therapies for articular cartilage regeneration. *Trends in Biotechnology,* 33(1), 35-42.

Mahmoud, G. A., Sayed, A., Thabit, M., and Safwat, G. (2020). Chitosan biopolymer based nanocomposite hydrogels for removal of methylene blue dye. *SN Applied Sciences,* 2(5), 968.

Mahmoud, M. E., Abouelanwar, M. E., Mahmoud, S. E. M. E., and Salam, M. A. (2021). Doping starch-gelatin mixed hydrogels with magnetic spinel ferrite@biochar@molybdenum oxide as a highly efficient nanocomposite for removal of lead (II) ions. *Journal of Environmental Chemical Engineering,* 9(6), 106682.

Makvandi, P., Ali, G. W., Della Sala, F., Abdel-Fattah, W. I., and Borzacchiello, A. (2020). Hyaluronic acid/corn silk extract based injectable nanocomposite: a biomimetic antibacterial scaffold for bone tissue regeneration. *Materials Science and Engineering: C,* 107, 110195.

Matsiko, A., Levingstone, T. J., and O'Brien, F. J. (2013). Advanced strategies for articular cartilage defect repair. *Materials*, 6(2), 637-668.

Mi, F. L., Wu, Y. Y., Lin, Y. H., Sonaje, K., Ho, Y. C., Chen, C. T., Juang, J. H., and Sung, H. W. (2008). Oral delivery of peptide drugs using nanoparticles self-assembled by poly(γ-glutamic acid) and a chitosan derivative functionalized by trimethylation. *Bioconjugate Chemistry,* 19(6), 1248-1255.

Mochalin, V., Shenderova, O., Ho, D., and Gogotsi, Y. (2020). The properties and applications of nanodiamonds. *Nano-Enabled Medical Applications*, 313-350.

Mousa, M., Evans, N. D., Oreffo, R. O. C., and Dawson, J. I. (2018). Clay nanoparticles for regenerative medicine and biomaterial design: a review of clay bioactivity. *Biomaterials,* 159, 204-214.

Natoli, R. M., Skaalure, S., Bijlani, S., Chen, K. X., Hu, J., and Athanasiou, K. A. (2010). Intracellular Na^+ and Ca^{2+} modulation increases the tensile properties of developing engineered articular cartilage. *Arthritis & Rheumatism*, 62(4), 1097-1107.

Nie, L., Zhang, G., Hou, R., Xu, H., Li, Y., and Fu, J. (2015). Controllable promotion of chondrocyte adhesion and growth on PVA hydrogels by controlled release of TGF-β1 from porous PLGA microspheres. *Colloids and Surfaces B: Biointerfaces,* 125, 51-57.

Nojoomi, A., Tamjid, E., Simchi, A., Bonakdar, S., and Stroeve, P. (2017). Injectable polyethylene glycol-laponite composite hydrogels as articular cartilage scaffolds with superior mechanical and rheological properties. *International Journal of Polymeric Materials and Polymeric Biomaterials,* 66(3), 105-114.

Oun, A. A., and Rhim, J. W. (2017). Carrageenan-based hydrogels and films: effect of ZnO and CuO nanoparticles on the physical, mechanical, and antimicrobial properties. *Food Hydrocolloids,* 67, 45-53.

Pacelli, S., Acosta, F., Chakravarti, A. R., Samanta, S. G., Whitlow, J., Modaresi, S., Ahmed, R. P., Rajasingh, J., and Paul, A. (2017). Nanodiamond-based injectable hydrogel for sustained growth factor release: preparation, characterization and *in vitro* analysis. *Acta Biomaterialia,* 58, 479-491.

Pan, Y., and Xiong, D. (2009). Study on compressive mechanical properties of nanohydroxyapatite reinforced poly(vinyl alcohol) gel composites as biomaterial. *Journal of Materials Science: Materials in Medicine,* 20(6), 1291-1297.

Pereira, D. R., Silva-Correia, J., Oliveira, J. M., and Reis, R. L. (2013). Hydrogels in acellular and cellular strategies for intervertebral disc regeneration. *Journal of Tissue Engineering and Regenerative Medicine*, 7(2), 85-98.

Purohit, S. D., Singh, H., Bhaskar, R., Yadav, I., Chou, C. F., Gupta, M. K., and Mishra, N. C. (2020). Gelatin-alginate-cerium oxide nanocomposite scaffold for bone regeneration. *Materials Science and Engineering: C,* 116, 111111.

Sasaki, Y., and Akiyoshi, K. (2010). Nanogel engineering for new nanobiomaterials: from chaperoning engineering to biomedical applications. *The Chemical Record,* 10(6), 366-376.

Song, F., Li, X., Wang, Q., Liao, L., and Zhang, C. (2015). Nanocomposite hydrogels and their applications in drug delivery and tissue engineering. *Journal of Biomedical Nanotechnology,* 11(1), 40-52.

Stanczyk, M., and van Rietbergen, B. (2004). Thermal analysis of bone cement polymerisation at the cement-bone interface. *Journal of Biomechanics,* 37(12), 1803-1810.

Su, W., Hu, Y., Zeng, M., Li, M., Lin, S., Zhou, Y., and Xie, J. (2019). Design and evaluation of nano-hydroxyapatite/poly(vinyl alcohol) hydrogels coated with poly(lactic-co-glycolic acid)/nano-hydroxyapatite/poly(vinyl alcohol) scaffolds for cartilage repair. *Journal of Orthopaedic Surgery and Research,* 14(1), 446.

Tao, J., Zhang, Y., Shen, A., Yang, Y., Diao, L., Wang, L., Cai, D., and Hu, Y. (2020). Injectable chitosan-based thermosensitive hydrogel/nanoparticle-loaded system for local delivery of vancomycin in the treatment of osteomyelitis. *International Journal of Nanomedicine,* 15, 5855-5871.

Tsai, T. Y., Shen, K. H., Chang, C. W., Jovanska, L., Wang, R., and Yeh, Y. C. (2021). *In situ* formation of nanocomposite double-network hydrogels with shear-thinning and self-healing properties. *Biomaterials Science,* 9(3), 985-999.

van de Belt, H., Neut, D., Schenk, W., van Horn, J. R., van der Mei, H. C., and Busscher, H. J. (2000). Gentamicin release from polymethylmethacrylate bone cements and Staphylococcus aureus biofilm formation. *Acta Orthopaedica Scandinavica,* 71(6), 625-629.

Wahid, F., Wang, H. S., Lu, Y. S., Zhong, C., and Chu, L. Q. (2017). Preparation, characterization and antibacterial applications of carboxymethyl chitosan/CuO nanocomposite hydrogels. *International Journal of Biological Macromolecules,* 101, 690-695.

Wang, Y., Ma, J., Yang, S., and Xu, J. (2011). PDMAA/Clay nanocomposite hydrogels based on two different initiations. *Colloids and Surfaces A: Physicochemical and Engineering Aspects,* 390(1), 20-24.

Wu, X. X., Zhang, Y., Hu, T., Li, W. X., Li, Z. L., Hu, H. J., Zhu, S. R., Chen, W. Z., Zhou, C. S., and Jiang, G. B. (2021). Long-term antibacterial composite via alginate aerogel sustained release of antibiotics and Cu used for bone tissue bacteria infection. *International Journal of Biological Macromolecules,* 167, 1211-1220.

Yaghoubi, A., Ramazani, A., and Ghasemzadeh, H. (2021). Synthesis of physically crosslinked PAM/CNT flakes nanocomposite hydrogel films via a destructive approach. *RSC Advances,* 11(62), 39095-39107.

Yue, B. (2014). Biology of the extracellular matrix: an overview. *Journal of Glaucoma,* 23(8 Suppl 1), S20-23.

Zalavras, C. G., Patzakis, M. J., and Holtom, P. (2004). Local antibiotic therapy in the treatment of open fractures and osteomyelitis. *Clinical Orthopaedics and Related Research®,* 427, 86-93.

Zare, E. N., Jamaledin, R., Naserzadeh, P., Afjeh-Dana, E., Ashtari, B., Hosseinzadeh, M., Vecchione, R., Wu, A., Tay, F. R., Borzacchiello, A., and Makvandi, P. (2020). Metal-based nanostructures/PLGA nanocomposites: antimicrobial activity, cytotoxicity, and their biomedical applications. *ACS Applied Materials &* Interfaces, 12(3), 3279-3300.

Zhang, K., Yuan, W., Wei, K., Yang, B., Chen, X., Li, Z., Zhang, Z., and Bian, L. (2019). Highly dynamic nanocomposite hydrogels self-assembled by metal ion-ligand coordination. *Small*, 15(15), 1900242.

Zhang, Y., Li, Z., Guan, J., Mao, Y., and Zhou, P. (2021). Hydrogel: a potential therapeutic material for bone tissue engineering. *AIP Advances,* 11(1), 010701.

Zhu, C. H., Hai, Z. B., Cui, C. H., Li, H. H., Chen, J. F., and Yu, S. H. (2012). *In situ* controlled synthesis of thermosensitive poly(N-isopropylacrylamide)/Au nanocomposite hydrogels by gamma radiation for catalytic application. *Small,* 8(6), 930-936.

Chapter 10

Nanocomposite Hydrogels for Food Packaging

Ahmad Etemadi*
and Atiyeh Heydari-Zefre
Faculty of Chemical Engineering, Urmia University of Technology, Urmia, Iran

Abstract

As the need for new, high-performance, and environment-friendly packaging materials is growing, researchers have recently been highly interested in bionanocomposite materials. Bionanocomposite, with appropriate biocompatibility and biodegradability, is recognized as a novel platform for food packaging. These materials are used to reduce the consumption of conventional plastic materials and to attain a high recycling efficiency while keeping the favorable barrier and mechanical characteristics. The research works over the past decade have focused on nanocomposite hydrogels, as novel bionanocomposites, which can be appropriately used as food packaging materials due to their suitable properties, including high thermal, mechanical, barrier, antioxidant, and antimicrobial characteristics. The purpose of the present chapter is to review the recent innovations in bionanocomposites and nanocomposite hydrogels for food packaging usage. An overview of the basic concepts of hydrogels and nanocomposite hydrogels is also presented and different approaches for the incorporation of nanomaterials into the hydrogel matrix and surface treatments are briefly discussed with a focus on their applications as packaging materials in the food industry.

Keywords: food packaging, nanocomposite, hydrogel, food industry

10.1. Hydrogels

10.1.1. Introduction

Hydrogels are 3D networks of polymeric chains with random crosslinking by chemical or physical bonds (Pourjavadi et al., 2004; Zhang et al., 2010; Ahmed, 2015; Kalia et al., 2017). Hydrogels may have a branched or linear structure and can absorb high amounts of biological

* Corresponding Author's Email: a.etemadi@che.uut.ac.ir.

In: Nanocomposite Hydrogels and their Emerging Applications
Editor: Mohammad Sirousazar
ISBN: 979-8-88697-675-5

fluids or water. Therefore, they have the ability to balance swelling and forming insoluble structures because of their 3D networks (Hebeish et al., 2013). Some authors also have defined hydrogels as colloidal gels because of their dispersion in water and presenting structural and viscoelastic characteristics because of the polymeric chain bonds' behavior and the system-incorporated solvents (Laftah et al., 2011; Ahmed et al., 2013). There exist two phases in the 3D structure of hydrogels, including a solid phase, based on a polymeric blend or a neat polymer, and a liquid phase, typically biological fluids or water. When immersed in an aqueous solution, the target fluids are absorbed in hydrogel without structural alteration, as a result of its 3D structure, (Bao et al., 2011; Croisier and Jérôme, 2013).

Several factors affect the hydrogel's water sorption capacity, including the density and nature of the used polymer and crosslinker (Buwalda et al., 2014). Swelling potential, moisture retention capacity, and mechanical resistance are some of the major properties of hydrogels (Gulrez et al., 2011; Guilherme et al., 2015; Chen et al., 2016; Ng, et al., 2016). These features are important for some industrial sectors and have had a role in the progress of some significant applications, including wastewater treatment, cosmetics, tissue engineering, biosensing, drug release, biomedicine, agriculture, and food industry (Chen, et al., 2016; Alves and Mano, 2008; Hoare and Kohane, 2008; Deligkaris et al., 2010; Chang and Zhang, 2011; Kabiri et al., 2011; Isana et al., 2013; Feng et al., 2014; Caló and Khutoryanskiy, 2015; Rodríguez-Vázquez et al., 2015; Neto et al., 2017). Concerning the production of hydrogel matrices, hydrogels are mostly formed from synthetic compounds, and their matrices are chiefly bound by chemical crosslinking. Petroleum derivatives are the most commonly used compounds, which include polyvinylpyrrolidone, polysodium acrylate, polyacrylamide, polyacrylic acid, etc. (Kabiri et al., 2011, Ullah et al., 2015).

In general, an ideal hydrogel material must have the following functional properties (Zohuriaan-Mehr and Kabiri, 2008):

- Desired absorption rate (preferred porosity and particle size) depending on the application conditions.
- The maximum absorption capacity (highest equilibrium swelling) in saline.
- The maximum absorbency under load.
- The lowest residual monomer and soluble content.
- The cheapest price.
- The maximum stability and durability during storage and in the swelling environment.
- The maximum biodegradability without forming toxic species after the degradation.
- pH-neutrality after swelling.
- Photostability, odorlessness, non-toxicity, and colorlessness.
- Re-wetting capability (if needed), the hydrogel should have the ability of giving back the imbibed solution or maintaining it, which depends on the application conditions (for example, in hygienic or agricultural applications).

It is not possible for a hydrogel sample to concurrently meet all the mentioned properties. Hence, in reality, it is needed to optimize the production reaction variables in such a way that a suitable balance is achieved between the features. For instance, a hygienic product of hydrogels should have the maximum rate of absorption, the lowest residual monomer, and the lowest re-wetting.

10.1.2. Classification of Hydrogels

It is possible to use various classification criteria for distinguishing different hydrogels. The hydrogels are mainly classified based on the source, ionic charge, crosslinking method, physical properties, degradability, stimulus responsivity, and preparation method, as shown in Figure 10.1 (Bhattarai et al., 2010). Four classification criteria for hydrogels are described in the following section, including the classification of hydrogels based on source, polymer composition, degradability, and type of crosslinking.

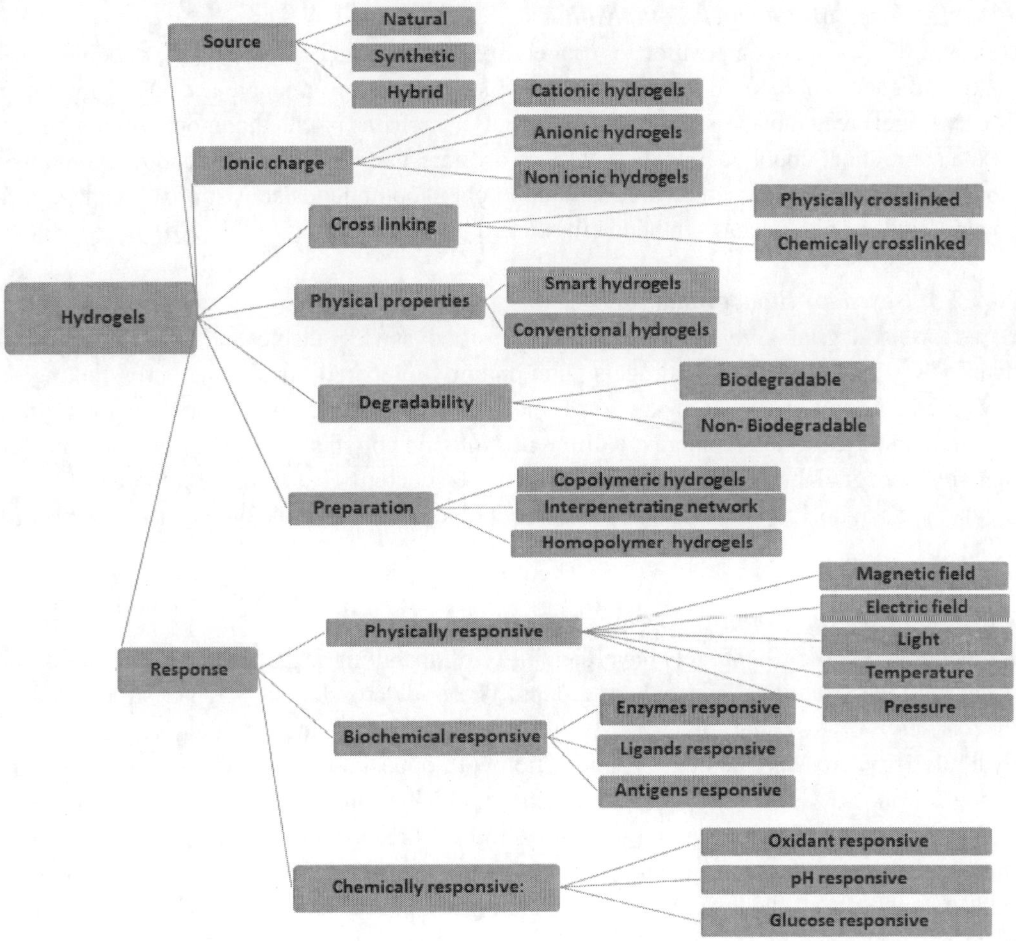

Figure 10.1. Classification of hydrogels based on different criteria.

10.1.2.1. Based on Source

It is possible to prepare a hydrogel by the use of synthetic polymers, natural macromolecules, or by their combination which is called hybrid hydrogel. Most of the existing hydrogels are made of artificial or synthetic sources because of their chemical and mechanical characteristics, offering a variety of applications. Besides, hydrogels made of natural polymers provide less toxicity and greater biodegradability and biocompatibility compared to synthetic ones. As a result, a significant increase has been witnessed in hybrid hydrogel production due to the favorable properties of both groups. Using multicomponent mixtures and copolymers is an approach to designing hydrogels with desired properties (Batista et al., 2019). There are two primary categories of natural polymers utilized for the production of hydrogels, including proteins, e.g., gelatin, soy, collagen, and fish proteins, and carbohydrates, such as starch, sodium alginate, cellulose, chitosan, guar gum, and xanthan gum. For achieving a hydrogel with desired properties, especially for application in food packaging, a mixture of the mentioned natural and synthetic polymers is required.

10.1.2.2. Based on Polymer Composition

Based on the polymer composition, hydrogels are classified as homopolymers, copolymers, or multicomponent compounds. A homopolymer is derived from a single species of monomer. The hydrogel created by crosslinking of homopolymer chains is called homopolymer hydrogel. At the same time, copolymers consist of two or more varying types of monomers organized along a chain of a polymer network. Multicomponent compounds consist of two independent combinations of natural and synthetic crosslinked polymers (Batista et al., 2019).

10.1.2.3. Based on Biodegradability

Based on biodegradability, hydrogels are classified as degradable and non-biodegradable hydrogels. Biodegradable hydrogels are mainly prepared from the crosslinking of biodegradable polymers that are generally derived from natural sources (e.g., carbohydrates and proteins). Nevertheless, there are different synthetic polymers utilized in the food industry that are biodegradable, including polyaldehyde gluconate, polyanhydrides, polyacrylic acid, polylactic acid, polyglycolic acid, and methoxy polyethylene glycol (Byun et al., 2008; Batista et al., 2019).

10.1.2.4. Based on Stimulus Responsivity

As mentioned earlier, hydrogels have the ability of absorbing high amounts of biofluids and water via their structures' hydrophilic groups. These 3D crosslinked polymer structures have the capacity of swelling hundreds or thousands of times their initial weight. Charge, hydrophilicity, crosslinking density, interaction with opposite ions in the environment, degree of ionization, pH, and temperature are among the main parameters affecting the swelling of hydrogels. Some hydrogels, known as smart or stimulus-responsive hydrogels, show a response in a short time to changes in environmental stimuli, such as temperature, pH, ionic strength, light, magnetic field, and the presence of biomolecules. This causes the reaction of the stimulus-responsive hydrogel by shrinkage, discoloration, or swelling. There is a relationship between these changes and the balance of hydrophilicity to hydrophobicity in the network structure of smart hydrogels (Mahinroosta et al., 2018). The use of stimulus-responsive hydrogels in creating intelligent or smart food packaging is very important (Ullah et al., 2015; Batista et al., 2019). Based on the type of stimulus applied to the hydrogels, they are classified as, pH, temperature, light, electrical and magnetic field responsive hydrogels (Byun et al., 2008; Qiu and Park, 2001; Mahinroosta et al., 2018).

pH-sensitive hydrogels are able to cause abrupt changes in shape or volume because of changes in the pH of the environment. pH-sensitive hydrogels usually consist of a network of acidic and ionizable basic groups linked to a polymer chain, with the ability to become protonated or release protons by pH changes. Temperature-responsive hydrogels have hydrophilic and hydrophobic groups in their network which are grouped into three categories of negative- and positive- temperature-sensitive and heat-reversible (or thermally reversible) hydrogels. Light-responsive hydrogels can respond to a change in the exerted light and are divided into two groups, including visible light- and ultraviolet (UV) light-sensitive hydrogels. The response mechanism of electrical field-responsive hydrogels is similar to the mechanism of pH-sensitive hydrogels.

10.1.2.5. Based on the Type of Crosslinking

It is possible to classify hydrogels according to the type of crosslinking created among the polymeric chains of the network. In this method of classification, the hydrogels are divided into two types, including physical or reversible hydrogels and chemical or permanent hydrogels, as shown in Figure 10.2 (Caccavo et al., 2018).

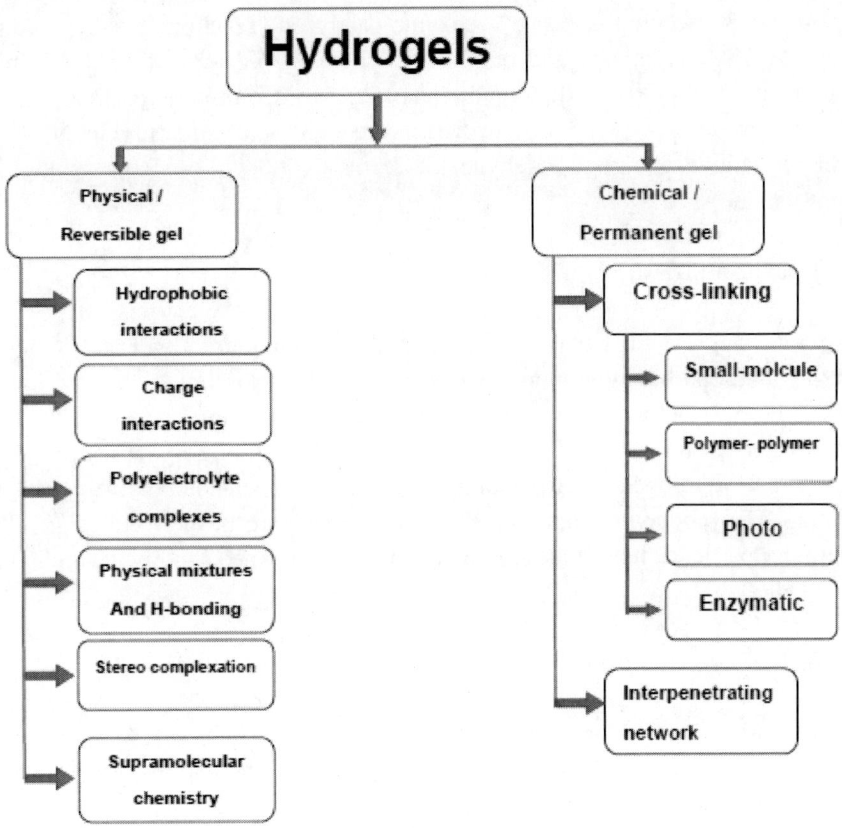

Figure 10.2. Classification of hydrogels based on the type of crosslinking.

- *Physical or reversible hydrogels*: Electrostatic, hydrophobic, as well as hydrogen bonding created among the polymer chains, are the forces involved in the formation of the physical or reversible hydrogel. Gel formation can be reversed since the network formation by all of these interactions is totally physical (Berger et al., 2004; Boucard et al., 2005). Many researchers have been attracted by physical or reversible hydrogels due to their quite easy formation. The most important benefit of this hydrogel type is not using crosslinkers, which prevent them from mixing with trapped materials (e.g., drugs) and thus the need to remove them before use. Various methods for the production of hydrogels with physical crosslinking have been reported in the literature, including heating-cooling of the polymer solution by hydrophobic interactions, freezing-thawing, and creating ionic interactions, hydrogen bonding, and spatial complexation (Mahinroosta et al., 2018). A critical issue in preparing physical hydrogels is the polymer selection, which is dependent on two main criteria: there should be an adequately strong interaction among the chains so that a semi-permanent junction is formed in the molecular network, and the network must embrace a large amount of water molecules inside it.

- *Chemical or permanent hydrogels*: Hydrogels with physical crosslinking are produced without chemical modification or using crosslinkers. In chemical hydrogels, covalent bonding is used between polymer chains for producing a permanent hydrogel. It is done by light or an enzyme-catalyzed reaction (Caccavo et al., 2018). Crosslinking is sometimes accomplished by enzyme-catalyzed reaction or by adding small crosslinking molecules and polymer-polymer conjugation of light-sensitive agents (Caccavo et al., 2018; Hoare and Kohane, 2008). Common crosslinking agents used for the production of chemical hydrogels are malic acid, fumaric acid, N, N-methylene bisacrylamide, and glutaraldehyde.

10.1.3. Preparation Methods of Hydrogels

There are two well-established schemes to produce polymeric hydrogels, including the polymerizing hydrophilic monomers in the presence of crosslinker or functionalizing or crosslinking the existing polymers. The main materials for making hydrogels are synthetic or natural monomers or polymers (Zhou and Wu, 2012). Different polymerization methods are used to prepare the hydrogels, such as bulk, solution, and suspension polymerization as well as base bonding. The products obtained from these methods are in the form of small particles, powders, micro-particles, membranes, fibers, and even deformed liquids (Hamidi et al., 2008; Ahmed, 2015).

10.1.3.1. Bulk Polymerization

Bulk polymerization is one of the simplest methods for the production of hydrogels, only involving monomers and water-soluble primers. Many vinyl monomers are used to produce hydrogels with physical properties intended for specific applications. Bulk hydrogels are prepared with one or more types of monomers. In general, a small amount of crosslinking agent is introduced to the hydrogel formulation. Polymerization begins with UV radiation or a chemical catalyst. The choice of the appropriate primer is dependent on the monomer type and the used solvent. A high rate and degree of polymerization occur due to high concentrations of monomers. Nevertheless, an increase in conversion rate makes a significant increase in the reaction viscosity, generating heat in polymerization. It is possible to avoid these issues by controlling the reaction at low conversion rates. The monomer bulk polymerization to produce a homogeneous hydrogel creates a glassy, clear, rigid polymer substrate. The advantage of bulk polymerization is the production of polymers with high purity and molecular weight in different forms, including membranes, films, emulsions, and particles (Ahmed, 2015).

10.1.3.2. Solution Polymerization

In solution polymerization, the ionic or neutral monomers are combined with the multifunctional crosslinking agent in solution copolymerization/crosslinking reactions. A redox initiator system or UV-irradiation thermally initiates the polymerization. The existence of a solvent that serves as a heat sink is the main benefit of solution polymerization compared to bulk polymerization. It is required to wash the developed hydrogels with distilled water for the removal of the oligomers, monomers, crosslinking agents, extractable and soluble polymer chains, initiators, and other impurities. After the occurrence of phase separation, the heterogeneous hydrogel is produced when the water has a higher volume during polymerization compared to the content of water equivalent to the equilibrium swelling. Ethanol, water-ethanol mixes, benzyl alcohol, and water are the solvents that are typically employed for the solution polymerization of hydrogels. Then, the synthesis solvent can be eliminated following the gel formation by swelling the hydrogels in water (Ahmed, 2015).

10.1.3.3. Suspension Polymerization

Suspension polymerization is an effective approach for preparing microparticles or spherical hydrogels with sizes of 1 μm to 1 mm. In this method, with the dispersion of the monomer solution in the anti-solvent, fine monomer particles are formed and stabilized by adding a stabilizer. Polymerization begins with radicals resulting from the initiator's thermal decomposition. Then, the lattice agent and the newly prepared microparticles are rinsed so that unreacted monomers are removed. The viscosity of the monomer phase affects the shape of the prepared particles while the particle size is controlled by modifying the suspension hydrophobicity-hydrophilicity. Microparticles of poly (hydroxyethyl methacrylate) hydrogel can be prepared by this method. Reverse suspension polymerization is also extensively employed to prepare polyacrylamide hydrogels because of the ease of removal of toxic acrylamide monomers left over from the polymer (Hamidi et al., 2008; Ahmed, 2015).

10.1.3.4. Bonding to the Base

Hydrogels that are generally formed by bulk polymerization are intrinsically weakly structured. For improving the hydrogel's mechanical characteristics, the hydrogel can be bonded to a coated surface with a strong base. It includes the formation of free radicals on a strong base surface and the direct polymerization of the monomers on it, which results in a chain of monomers being attached to the surface by covalent bonding (Hamidi et al., 2008; Ahmed, 2015).

10.1.4. Application of Hydrogels in Food Packaging

Hydrogels have a variety of applications, including in cosmetics, water retainers, moisturizers, sludge coagulants, thickeners, biosensors, and drug delivery systems. They also are applicable as contact lenses in ophthalmology, as a basis for artificial muscles, and as absorbent materials in sanitary clothing. Hydrogels are utilized in agriculture and horticulture to improve the capacity and conservation of water in the soil. They also have applications as a water carrier in fire-retardant compounds (Hamidi et al., 2008). One interesting and potential application of hydrogels is in food packaging systems. Packing is important in the food industry, which is dominated by synthetic packaging substances. Modern plastic packing imposes a negative effect on the environment owing to the non-degradable packaging waste representing further expensive problems for communication. Hence, several research works have been performed for developing environment-friendly food packaging systems (Hanani et al., 2014).

In recent years, researchers have considered dehumidifiers for food products based on absorbent pads with water removal features, particularly because they have the potential of reducing the microbial contamination risk and at the same time, they can preserve the sensory properties of packaged foods. It can have many applications, including absorbing food liquids, modifying empty packaging space, and antimicrobial effects (He et al., 2015; Otoni et al., 2016). Four basic principles are followed in using hydrogel adsorbents in food packaging systems. First, after adsorption, the adsorbent must hold the discharge in its 3D structure. The second is that the good appearance of the sorbent and the sensory properties of the packaged food must be preserved at a low cost. In the third principle, the adsorbent must provide specific functional properties that integrate and guarantee the packaging system's structure when it is stored. The last principle is that the adsorbent should extend the shelf life of stored foods and prevent microbial growth on their surfaces.

According to the above requirements, hydrogels are an appealing option for utilization in food packaging systems and can provide a good opportunity to efficiently design these packaging systems with favorable characteristics (e.g., biodegradability, mechanical strength, and durability). Most of the hydrogels used in food packaging are made in the form of films (Roy et al., 2012; Batista et al., 2019). Other applications of hydrogels include usage as part of an intelligent packaging system or as a carrier system that is directly embedded in food matrices. The application of hydrogels in intelligent packaging falls into three general categories, temporal-temperature, freshness, and emitted indices. In other words, in these systems, the main purpose can be monitoring temperature and tracking temporal-temperature distribution, or providing information on the freshness of existing food products, or as part of a simple diagnostic approach for determining the existence of contaminants, e.g., aflatoxins (Batista et al., 2019).

10.2. Nanocomposite Hydrogels

As mentioned in the previous section, hydrogels can be potentially used in various areas due to their tunable chemical, biological, and physical characteristics, as well as their great versatility in production and biocompatibility. Despite these important properties, there are yet some challenges with hydrogels, such as low strain, poor mechanical strength, and low thermal stability, restricting their applicability. Due to the mentioned shortcomings, researchers have made trials to design novel nanocomposite hydrogels with special enhanced features (Wei et al., 2010; Datta et al., 2013; Annabi et al., 2014; Vickers, 2017). Nanocomposite hydrogels are extremely hydrated polymeric networks also called hybrid hydrogels and have shown greater biological, chemical, physical, and mechanical features (Barrile et al., 2018).

Nanocomposite hydrogels, as hybrid polymeric networks, have a special hydration property. As a result of the existence of hydrophilic moieties, such as carboxylic acids, amides, and alcohols in nanocomposite hydrogels, their water-absorbing capacity is enhanced. The stability of these materials under extreme pH, pressure, and temperature conditions is increased by adding a crosslinker in the hydrogel synthesis. Biodegradability, high hydrophilicity, and great mechanical strength are among the characteristics of natural polymer-based nanocomposite hydrogels. As a result of the very porous structure of nanocomposite hydrogels, they swell in the water presence. Due to the superior swelling behavior of nanocomposite hydrogels, they are distinctive from hydrogels. Nanocomposite hydrogels are mechanically tough, showing bending and stretching. Nanocomposite hydrogels may include nanoparticles, e.g., clays with great mechanical features. These mechanically tough nanocomposite hydrogels can be potentially used in electronic devices and biomedical engineering. In general, it is possible to stretch nanocomposite hydrogels above 1000% of their original length. Elongation of these hydrogels is dependent on the used nanoparticles and polymers in the synthesis, and crosslinker type (Haraguchi, 2011).

The network of nanocomposite hydrogels is composed using various synthetic and natural polymers. As a result of combining the inorganic and organic parts for the fabrication of nanocomposite hydrogels, biological, chemical, physical, and swelling/ deswelling features are improved (Song et al., 2015). Nanocomposite hydrogels present different superior features, such as good modulus, high heat resistance, and high strength (Shin et al., 2009). For obtaining nanocomposite hydrogels, various nanoparticles are mixed with the polymeric networks, such as carbon-based nanomaterials (e.g., nanodiamonds, graphene, and carbon nanotubes), inorganic nanoparticles (e.g., clays, hydroxyapatite, and calcium phosphate), and metal/metal oxide nanoparticles (e.g., gold, iron oxide, and silver) (Adhikari et al., 2012; Goenka et al., 2014).

Carbon-based nanomaterials, such as graphene, nanodiamonds, and carbon nanotubes are under investigation for the production of nanocomposite hydrogels (Iijima, 1991, Elias et al., 2009; Huang et al., 2009; Bai et al., 2010; Liu et al., 2010; Ma et al., 2010; Zhang et al., 2011; Sant et al., 2012; Siengchin and Karger-Kocsis, 2012; Bae et al., 2014; Bhattacharya et al., 2017; Li et al., 2017; Sharma et al., 2017; Vickers 2017). Carbon nanotube-based nanocomposite hydrogels show great electrical conductivity and mechanical toughness. They are appropriate to be applied in biomedicine, drug delivery, environmental detoxification, tissue engineering, and biosensing (Kuilla et al., 2010). Excellent functionality can be provided to the nanocomposite hydrogels by the combination of metal/metal oxide nanoparticles and polymeric hydrogels. The preparation of metal/metal oxide nanocomposite hydrogels has attracted interest due to their electrical, catalytic, and optical features (Xiang et al., 2006). There are numerous inorganic nanoparticles utilized for the fabrication of nanocomposite hydrogels. The properties and structures of nanocomposite hydrogels can be improved by incorporating inorganic clays (Haraguchi, 2011).

By incorporating clay into the polymer chains, nanocomposite hydrogels can be achieved with enhanced mechanical, chemical, physical, electrical, biological, magnetic, and swelling/deswelling features in comparison with their constituents (Kumar et al., 2015). In recent years, the focus of the studies on nanocomposite hydrogels has been on using inorganic silicate clays as nanoparticles. Clay is a material including layered silicates with organic matter and metal oxide traces. They are aluminum phyllosilicates containing a varying amount of alkali, magnesium, iron, and alkaline earth metals as well as other cations like carbonates and quartz (Biswas et al., 2019). The alternative sheets of AlO_6 and SiO_2 in a 2:1 ratio of (e.g., montmorillonite), 1:1 (e.g., kaolinite), and 2:2 (e.g., chlorite) make the natural clays' basic composition. They are environmentally and economically benign materials for different applications as a result of their least toxicity, copiousness, and natural occurrence. Ion-exchange capacity, specific structure, surface area, surface reactivity, mechano-chemical stability, and water retention capacity are included in the surface chemistry affecting the chemical and physical features of clays (Cherian et al., 2018; Feng et al., 2021).

To modify the clay materials, several approaches were run, including acid activation (for developing surface acidity, porosity, and surface area) (Barakan and Aghazadeh, 2019), heat treatment (enhancing adsorptive properties and water-stability characteristics) (Ouyang et al., 2014), and surfactant amendment (making clay minerals well-matched with hydrophobic polymers and non-polar molecules) (Abdel Ghafar et al., 2020). Research studies have shown various types of clay and their impacts on the mechanical features of nanocomposite hydrogels (Haraguchi, 2004). Nanocomposite hydrogels are not produced by non-water-swellable clay minerals (e.g., sepiolite). However, montmorillonite or modified hectorite can be employed for producing nanocomposite hydrogels (Liu et al., 2006; Zhu et al., 2006; Aalaie and Rahmatpour, 2007; Bignotti et al., 2007). Regarding the clay particle size, a diameter of almost 30 nm (for synthetic hectorite) is adequate to make nanocomposite hydrogels. Thus, clay particles with higher sizes are not essential (more than 300 nm for natural montmorillonite). Larger clay particles are sometimes even inefficient, possibly because of the moderate aggregation and inadequate exfoliation in the aqueous solution (Kaneko et al., 2006, 2007; Kim and Chujo, 2003; Mu and Zheng, 2007).

10.3. Polymeric Nanocomposites and Nanocomposite Hydrogels as Packaging Materials

Numerous polymeric nanocomposites and nanocomposite hydrogels are recently introduced as novel materials for food packaging applications, most of them are based on natural biopolymers and a few on synthetic polymers. The most studied nanocomposite hydrogels for food packaging applications are those prepared by polyvinyl alcohol (PVA), carboxymethyl cellulose (CMC), polyvinyl pyrrolidone (PVP), gelatin, alginate, starch, and chitosan. In the rest of the chapter, we will focus on the food packaging materials prepared by the aforementioned materials, and recent research in this regard will be briefly reviewed.

PVA is a biodegradable synthetic polymer with outstanding film-forming features and excellent biocompatibility. PVA should be crosslinked to form a hydrogel for use in different applications, particularly in the pharmaceutical formulations and food industry. The films made from PVA are the most convenient that are used especially for powdered drinks, spices, and chocolate for retaining the smell and reducing the polluted container waste, fish food packing bags, and resealable pouches with zippers. PVA films have also been proven for food contact uses (Cooper, 2013). Butnaru et al., (2016) prepared PVA/chitosan/montmorillonite nanocomposite hydrogels for food packaging applications using a noninvasive and low-cost freezing-thawing cyclic method. Montmorillonite clay content in the PVA/chitosan hydrogels was varied to establish the optimum amount to obtain improved properties. The increased thermal stability and mechanical features were displayed by the PVA/chitosan/montmorillonite nanocomposite hydrogels by insertion of 5 wt.% of montmorillonite clay in the crosslinked polymeric matrices comprising chitosan of more than 20 wt.%. The thermal features were not affected by incrementing the clay content. Incrementing the clay content considerably prevented the growth of the gram-negative and gram-positive bacteria. By examining the investigated materials against *Listeria monocytogenes* and *Escherichia coli*, antimicrobial activities were obtained for the PVA/chitosan/montmorillonite specimens with the highest efficiency. Based on the obtained results, it was found that the montmorillonite-containing PVA/chitosan nanocomposite hydrogels are auspicious materials for developing food packaging substances.

Gregorova et al., (2015) prepared biodegradable, transparent, flexible, hygroscopic, and breathable PVP/CMC hydrogel and introduced it as a novel material for food packaging applications. In the prepared hydrogel packaging, the hygroscopic feature of PVP and CMC contributes to altering their mechanical features where load caring capacity is a criterion for packaging materials. Their study aimed to assess precisely the moisture effect on the viscoelastic characteristics of straight PVP/CMC hydrogel and CMC neat hydrogel. It was indicated that the stiffness of the PVP/CMC hydrogel film was reduced by changing the humidity from 0 to 10%. Hence, it could be a favorable food packaging material for keeping vegetables and fruits fresh longer.

The morphology, physical-chemical behavior, and roughness of CMC-based hydrogel film were highly affected by the polymeric hydrogel film composition, relative humidity, and temperature. By incorporating PVP into the PVP/CMC hydrogel film, it would be possible to inhibit the plasticizing moisture effects by more than 50%. The existence of PVP in the PVP/CMC hydrogel film is helpful to gain greater stability in hydrothermal treatment. As a result of the increased temperature, an interaction can be triggered between PVP and CMC in PVP/CMC hydrogel film leading to the physical crosslinking and thus increased stiffness. PVP/CMC hydrogel film represented greater strain recovery in the recovery phase and lower creep compliance during creep. It was also ruptured during moisture conditioning.

The prepared PVP/CMC hydrogel film was flexible and transparent and represents acceptable mechanical features along with biodegradability. It was proved that these hydrogel films have biodegradability. The released moisture could be absorbed by the hydrogel film keeping the environment dry. Thus, the fast decay of the food materials is prevented, and oxygen diffusion is also maintained. It is observed that the PVP/CMC hydrogel films did not change until eight weeks (from the surface outline). The degradation happened while studying biodegradation in a liquid state internally. However, the same hydrogel film was diminished mostly or totally within 4 weeks while assessing the biodegradation in a solid-state compost bed. The PVP/CMC hydrogel films can be appropriately used as principal food packaging materials. As hydrogel films, they have more advantages as food packaging materials, particularly for fresh fruits, food, and vegetables, and those releasing moisture.

Recently, Haghighi et al., (2021) synthesized bionanocomposites based on bacterial cellulose nanowhiskers (BCNW) obtained from *Komagataeibacter xylinus* (strain K2G30; UMCC 2756) using sulfuric acid hydrolysis and incorporated into a gelatin-PVA (GL/PVA) blend film matrix. They studied the effect of BCNW content (1-10 wt.% of biopolymer) on the microstructural, mechanical, optical, and water barrier properties of the prepared bionanocomposites. As mentioned before, PVA is a water-soluble, non-toxic, synthetic polymer that is obtained commercially from hydrolyzing polyvinyl acetate. This polymer is biodegradable and represents higher flexibility, gas barrier features, and tensile strength, despite its synthetic characteristics. The PVA and gelatin have higher miscibility and compatibility owing to the existence of multifunctional groups in their polymeric structure. Several studies have been performed on GL/PVA blend-based films as potential biodegradable packaging substances. Nevertheless, films are made with higher sensitivity to water and poor water vapor permeability, because PVA and gelatin have a hydrophilic nature. Therefore, their industrial applications are highly limited when applied to higher-moisture food products since these films may be swollen, dissolved, or disintegrated when contacting with water (Aloui et al., 2016).

According to the physical and microstructural characterizations reported by Haghighi et al., (2021), the film properties were considerably improved by incorporating BCNW up to 7.5 wt.% for food packaging applications. Hence, it is possible to propose this concentration as a threshold level to incorporate BCNW into the GL/PVA film matrix. It is because mechanical, optical, and water barrier features are significantly deteriorated over this level. No differences were found in the films' total color by the naked eye and higher lightness and transparency values were represented by the established films compared to the commercially available synthetic plastic films. Generally, by studying the reinforcement effects of GL/PVA on BCNW films, such bionanocomposite was proposed as a sustainable and effective route for improving water barrier and mechanical features of GL/PVA films that are particularly vital for the preservation of food products. Finaly, they concluded that the characterization of functional properties revealed that GL/PVA blend film reinforced with BCNW could be used as an environmentally friendly packaging material to partially replace or reduce the use of current petroleum-based packaging materials.

In another research, Li et al., (2019) provided a novel and facile strategy for the design of integrated gelatin-based nanocomposites with promising functional properties for application as active packaging materials. They designed tough, hydrophobic, and antibacterial gelatin-based nanocomposites by constructing synergistic multiple coordination bonds between dopamine-grafted cellulose nanofibrils (DCNF) and montmorillonite in a covalent polymer network. Silver nanoparticles were prepared and anchored to the gelatin chains and montmorillonite-DCNF nanohybrids using *in situ* reduction and the strong catechol-metal interactions of polydopamine. An improvement was observed in the toughness and tensile stress of the hybrid films at once by 991.5 and 400.9%, respectively due to the efficient energy dissipation associated with the multiple crosslinked network structure. The strong antibacterial activity was represented by the hybrid films against *Staphylococcus aureus* (gram-positive) and *Escherichia coli* (gram-negative) pathogenic bacteria by the silver nanoparticles incorporation into the gelatin matrix. The nanocomposite films with a gelatin basis were biodegradable and renewable in comparison to traditional active packaging. They showed greater mechanical features, UV-shielding capacity, water resistance, antibacterial activity, and thermal stability as the key features for use in commercial and complex food packaging applications.

Alginate is a crucial algal biopolymer that is chiefly obtained from bacteria and brown seaweed. It is highly acceptable because of its biodegradability, biocompatibility, low cost, and abundance. In terms of structure, alginate is a linear anionic polysaccharide with the blocks of β-D-mannuronic acid (M) and α-L-guluronic acid (G) residues, attached by 1,4 glycosidic linkages. Nevertheless, the distribution and ratio of these blocks are different in different sources, species, and seasonal conditions (Ramaseshan et al., 2006). Presently, different applications use gelatin and alginate-derived materials as supported scaffolds or scaffolds in wound dressing, protein encapsulation, bone and soft tissue engineering, geland edible films in the food industries, and three-dimensional electrochemical printing and bioprinting (Saberianpour et al., 2019). Recently, Lavrič et al., (2021) produced alginate-based biopolymer films made of cellulose nanocrystals (CNC), nano-fibrils (CNF), and bacterial nanocellulose (BNC) in combination with chitosan and alginate in terms of applicability for potential food packaging applications. They prepared 25 different formulations and evaluated them, and selected seven biopolymer films with the best mechanical performances, as listed in Table 10.1, and investigated the morphology, mechanical properties, density, contact angle, surface energy, water absorption, and oxygen and water barrier properties of the selected samples. Since tensile properties are one of the basic criteria for packaging materials, the uniaxial tensile test was performed on the selected smples and the obtained tensile strength versus strain curves were reported (Figure 10.3). The obtained results showed that the tensile strength of the films are increased by incorporating the CNC, CNF and BCN nanoparticles into the alginate and chitosan matrices. Based on the obtained results, they concluded that the prepared biopolymer films would be a good alternative to conventional plastic packaging used for ready-to-eat foods with short storage time.

Table 10.1. The composition and thickness of the selected alginate-based biocomposite films (Lavrič et al., 2021)

Sample Designation	Composition	Thickness (μm)
Film 1	Alginate	55 ± 3
Film 2	Alginate + 5% CNC	50 ± 2
Film 3	Chitosan	123 ± 3
Film 4	Chitosan + 5% CNC	70 ± 1
Film 5	BNC	66.4 ± 0.8
Film 6	BNC+30% glycerol	56 ± 2
Film 7	CNF+30% glycerol	31 ± 3

Starch, as a natural biopolymer, has diverse applications, especially in the food industry. Some limitations have been reported for starch-based materials, including poor long-term stability, high water sensitivity, weak mechanical properties, and poor processability. Nevertheless, these issues can be considered by investigating the thermal and rheological properties of starch (Evžen and Václav, 2017). Some studies have presented various starch-based blends and bionanocomposites with properties and improved performances. In recent years, with advances in polymer nanocomposites, many considerations have been focused on using nanosized fillers so that new functionalities can be added for enhancing the performance of starch-based materials (Xu and Blum, 2008). The starch-based bionanocomposites are mostly used in edible films and food packaging usages in the food industry. Valid packaging materials and approaches are used for reducing food losses and presenting safe and healthy nourishment items, which are the pivotal point of food packaging (Youssef and El-Sayed, 2018). Moreover, consumer trends for higher quality, fresh-like, and advantageous food items have recently increased. Thus, with developing a variety of dynamic packaging developments, healthy, better quality, and safe supplements can be offered, besides the consideration of packaging-related disposal and environmental pollution issues (Chan et al., 2017).

This issue can also be solved using bionanocomposites. Inferable from the nanometer measure particles grown by scattering, there are many interesting points regarding these bionanocomposites, including improved organoleptic properties of food, e.g., odor, taste, and appearance, biodegradability, decreased packaging volume, waste, and weight, control of intercomponents, prolonged shelf life, individual packaging of small particulate foods, enhanced quality of usually nonpackaged items, functionalization as carriers for antioxidant and antimicrobial agents, and controlled release of active ingredients (Chaunier et al., 2012).

Using the technology o nanocomposite materials, Goudarzi and Shahabi-Ghahfarrokhi (2018) fabricated starch/TiO_2 bionanocomposites by photochemical reactions as a biodegradable food packaging material. The physical, mechanical, thermal, and water vapor permeability properties of the prepared starch/TiO_2 bionanocomposites were investigated and the photo-degradation properties of nanocomposite films were studied. Film-forming solutions were exposed to UV-A irradiation at different times. The obtained results showed that UV-A irradiation increased the hydrophobicity of starch films. It was shown that with increasing UV-A exposure time, tensile strength and Young's modulus of the films were decreased while elongation at break was increased. Finally, they concluded that the modification of starch-based biopolymers by UV-A and TiO_2 nanoparticles is an easy and accessible process to improve the packaging properties and photo-degradability of packaging films.

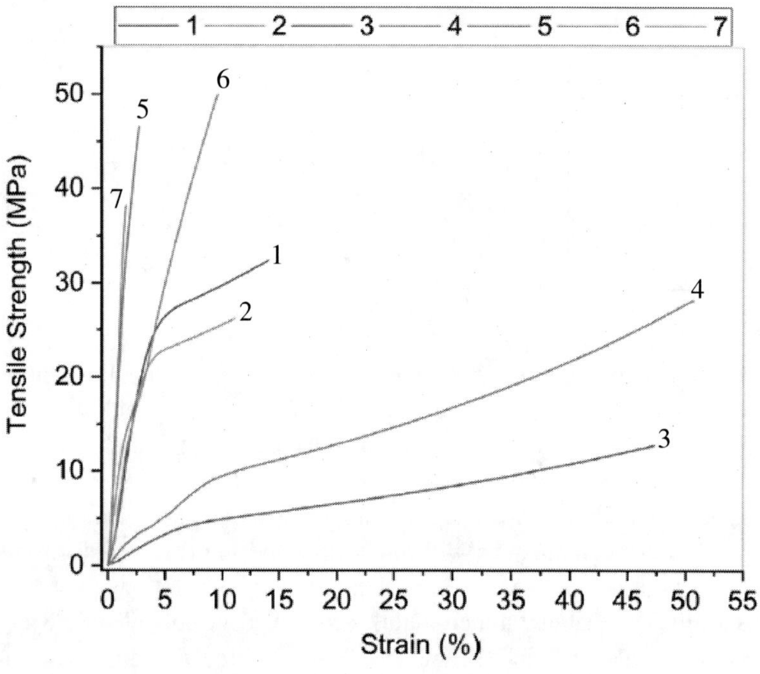

Figure 10.3. Tensile strength versus strain curves obtained for alginate-based biocomposite film (the compositions of films 1-7 have been listed in Table 10.1) (Lavrič et al., 2021).

More recently, a novel chitosan-based nanocomposite hydrogel was prepared by Patil and coworkers (2022) for food packaging application. The prepared nanocomposite hydrogel was a poly (acrylamide-co-2-hydroxyethyl methacrylate)/chitosan semi-interpenetrating polymer network copolymeric membrane embedded with ZnO nanoparticles. The nanocomposite hydrogel packaging was synthesized using one-pot polymerization techniques with ammonium persulfate as an initiator. The oxygen barrier property of the membrane was tested and found to be oxygen barrier and no permeation has been recorded up to 1.5 kg/cm^2 feed pressure. The antimicrobial test demonstrates that the ZnO-loaded membrane is a suitable material for the production of antimicrobial fruit and vegetable packaging materials. The fabricated copolymeric nanocomposite hydrogel exhibited flexibility, oxygen barrier property, and biodegradability and it has proven as a substrate or as a new-generation packaging material in vegetable packaging as well as processed food.

Conclusion

The nanocomposite concept has provided an inspiring path for developing innovative and novel food packaging materials with enhanced performances. Improved barrier characteristics of the nanocomposite packaging against CO_2, O_2, water vapor, and flavor compounds considerably influence the prolonging of the shelf-life of different processed and fresh foods. Moreover, if nanoparticles and polymer matrix are properly selected, it is possible to fine-tune the biodegradability of the bionanocomposite, which can be recognized as an advantage for food packaging. Nanocomposites have a lower weight than traditional composite materials, which makes them competitive with other materials for such applications as food packaging. Besides, the packaging waste related to processed food can be reduced and the protection of packaged foods is supported by prolonged shelf-life by using bionanocomposite packaging.

Using bionanocomposites and nanocomposite hydrogels with these extra functional features, different innovative food packaging technologies can be potentially developed, e.g., intelligent and active packaging, nanosensors, high barrier packaging, nanocoating, freshness indicator, and self-cleaning packagings. The use of nanocomposite hydrogels as food packaging materials creates some concerns regarding safety. Limited scientific data exists regarding the migration of nanostructures into food from packaging materials. Still, considerable research is needed for assessing the possible toxicity of nanocomposite hydrogels and their environmental safety. Overall, it seems that nanocomposite hydrogel packagings offer a very auspicious future for a variety of applications in food industries and active, intelligent, and innovative food packaging with bio-functional characteristics.

References

Aalaie, J., and Rahmatpour, A. (2007). Preparation and swelling behavior of partially hydrolyzed polyacrylamide nanocomposite hydrogels in electrolyte solutions. *Journal of Macromolecular Science, Part B*, 47(1), 98-108.

Abdel Ghafar, H. H., Radwan, E. K., and El-Wakeel, S. T. (2020). Removal of hazardous contaminants from water by natural and zwitterionic surfactant-modified clay. *ACS Omega*, 5(12), 6834-6845.

Adhikari, B., Biswas, A., and Banerjee, A. (2012). Graphene oxide-based hydrogels to make metal nanoparticle-containing reduced graphene oxide-based functional hybrid hydrogels. *ACS Applied Materials & Interfaces,* 4(10), 5472-5482.

Ahmed, E. M. (2015). Hydrogel, preparation, characterization, and applications: a review. *Journal Of Advanced Research,* 6(2), 105-121.

Ahmed, E. M., Aggor, F. S., Awad, A. M., and El-Aref, A. T. (2013). An innovative method for preparation of nanometal hydroxide superabsorbent hydrogel. *Carbohydrate Polymers,* 91(2), 693-698.

Aloui, H., Khwaldia, K., Hamdi, M., Fortunati, E., Kenny, J. M., Buonocore G. G., and Lavorgna M. (2016). Synergistic effect of halloysite and cellulose nanocrystals on the functional properties of PVA based nanocomposites. *ACS Sustainable Chemistry & Engineering,* 4(3), 794-800.

Alves, N. M., and Mano, J. F. (2008). Chitosan derivatives obtained by chemical modifications for biomedical and environmental applications. *International Journal of Biological Macromolecules,* 43(5), 401-414.

Annabi, N., Tamayol, A., Uquillas, J. A., Akbari, M., Bertassoni, L. E., Cha, C., Camci-Unal, G., Dokmeci, M. R., Peppas, N. A., and Khademhosseini, A. (2014). 25th anniversary article: rational design and applications of hydrogels in regenerative medicine. *Advanced Materials*, 26(1), 85-124.

Bae, H., Chu, H., Edalat, F., Cha, J. M., Sant, S., Kashyap, A., Ahari, A. F., Kwon, C. H., Nichol, J. W., and Manoucheri, S. (2014). Development of functional biomaterials with micro-and nanoscale technologies for tissue engineering and drug delivery applications. *Journal of Tissue Engineering and Regenerative Medicine,* 8(1), 1-14.

Bai, H., Li, C., Wang, X., and Shi, G. (2010). A pH-sensitive graphene oxide composite hydrogel. *Chemical Communications,* 46(14), 2376-2378.

Bao, Y., Ma, J., and Li, N. (2011). Synthesis and swelling behaviors of sodium carboxymethyl cellulose-g-poly (AA-co-AM-co-AMPS)/MMT superabsorbent hydrogel. *Carbohydrate Polymers,* 84(1), 76-82.

Barakan, S., and Aghazadeh, V. (2019). Structural modification of nano bentonite by aluminum, iron pillarization and 3D growth of silica mesoporous framework for arsenic removal from gold mine wastewater. *Journal of Hazardous Materials,* 378, 120779.

Barrile, R. A., van der Meer, D., Park, H., Fraser, J. P., Simic, D., Teng, F., Conegliano, D., Nguyen, J., Jain, A., and Zhou, M. (2018). Organ-on-chip recapitulates thrombosis induced by an anti-CD154 monoclonal antibody: translational potential of advanced microengineered systems. *Clinical Pharmacology & Therapeutics,* 104(6), 1240-1248.

Batista, R. A., Espitia, P. J. P., Quintans, J. d. S. S., Freitas, M. M., Cerqueira, M. Â., Teixeira, J. A., and Cardoso, J. C. (2019). Hydrogel as an alternative structure for food packaging systems. *Carbohydrate Polymers,* 205, 106-116.

Berger, J., Reist, M., Mayer, J. M., Felt, O., and Gurny, R. (2004). Structure and interactions in chitosan hydrogels formed by complexation or aggregation for biomedical applications. *European Journal of Pharmaceutics and Biopharmaceutics,* 57(1), 35-52.

Bhattacharya, S., Nandi S., and Jelinek, R. (2017). Carbon-dot-hydrogel for enzyme-mediated bacterial detection. *RSC Advances,* 7(2), 588-594.

Bhattarai, N., Gunn, J., and Zhang, M. (2010). Chitosan-based hydrogels for controlled, localized drug delivery. *Advanced Drug Delivery Reviews,* 62(1), 83-99.

Bignotti, F., Lebon, F., and Peroni, I. (2007). Effect of filler networking on the response of thermosensitive composite hydrogels. *European Polymer Journal,* 43(5), 1996-2006.

Biswas, B., Warr, L. N., Hilder, E. F., Goswami, N., Rahman, M. M., Churchman, J. G., Vasilev, K., Pan, G., and Naidu, R., (2019). Biocompatible functionalisation of nanoclays for improved environmental remediation. *Chemical Society Reviews,* 48(14), 3740-3770.

Boucard, N., Viton, C., and Domard, A. (2005). New aspects of the formation of physical hydrogels of chitosan in a hydroalcoholic medium. *Biomacromolecules,* 6(6), 3227-3237.

Butnaru, E., Cheaburu, C. N., Yilmaz, O., Pricope, G. M., and Vasile, C. (2016). Poly (vinyl alcohol)/chitosan/montmorillonite nanocomposites for food packaging applications, Influence of montmorillonite content. *High Performance Polymers,* 28(10), 1124-1138.

Buwalda, S. J., Boere, K. W., Dijkstra, P. J., Feijen, J., Vermonden, T., and Hennink, W. E. (2014). Hydrogels in a historical perspective: from simple networks to smart materials. *Journal of Controlled Release,* 190, 254-273.

Byun, H., Hong, B., Nam, S. Y., Jung, S. Y., Rhim, J. W., Lee, S. B., and Moon, G. Y. (2008). Swelling behavior and drug release of poly (vinyl alcohol) hydrogel cross-linked with poly (acrylic acid). *Macromolecular Research,* 16(3), 189-193.

Caccavo, D., Cascone, S., Lamberti, G., and Barba, A. (2018). Hydrogels, experimental characterization and mathematical modelling of their mechanical and diffusive behaviour. *Chemical Society Reviews,* 47(7), 2357-2373.

Caló, E., and Khutoryanskiy, V. V. (2015). Biomedical applications of hydrogels: a review of patents and commercial products. *European Polymer Journal,* 65, 252-267.

Chan, G. G., Koch, C. M., and Connors, L. H. (2017). Blood proteomic profiling in inherited (ATTRm) and acquired (ATTRwt) forms of transthyretin-associated cardiac amyloidosis. *Journal of Proteome Research,* 16(4), 1659-1668.

Chang, C., and Zhang, L. (2011). Cellulose-based hydrogels: present status and application prospects. *Carbohydrate Polymers,* 84(1), 40-53.

Chaunier, L., Véchambre, C., and Lourdin, D. (2012). Starch-based foods presenting shape memory capabilities. *Food Research International,* 47(2), 194-196.

Chen, Y., Tang, H., Liu, Y., and Tan, H. (2016). Preparation and study on the volume phase transition properties of novel carboxymethyl chitosan grafted polyampholyte superabsorbent polymers. *Journal of the Taiwan Institute of Chemical Engineers,* 59, 569-577.

Cherian, C., Kollannur, N. J., Bandipally, S., and Arnepalli, D. N. (2018). Calcium adsorption on clays: effects of mineralogy, pore fluid chemistry and temperature. *Applied Clay Science,* 160, 282-289.

Cooper, T. A. (2013). Developments in plastic materials and recycling systems for packaging food, beverages and other fast-moving consumer goods. *Trends in Packaging of Food, Beverages and Other Fast-Moving Consumer Goods (FMCG),* 58-107.

Croisier, F., and Jérôme, C. (2013). Chitosan-based biomaterials for tissue engineering. *European Polymer Journal,* 49(4), 780-792.

Datta, K., Achari, A., and Eswaramoorthy, M. (2013). Aminoclay: a functional layered material with multifaceted applications. J*ournal of Materials Chemistry* A, 1(23), 6707-6718.

Deligkaris, K., Tadele, T. S., Olthuis, W., and van den Berg, A. (2010). Hydrogel-based devices for biomedical applications. *Sensors and Actuators B: Chemical,* 147(2), 765-774.

Elias, D. C., Nair, R. R., Mohiuddin, T. M. G., Morozov, S. V., Blake, P., Halsall, M. P., Ferrari, A. C., Boukhvalov, D. W., Katsnelson, M. I., Geim, A. K., and Novoselov, K. S. (2009). Control of graphene's properties by reversible hydrogenation: evidence for graphane. *Science*, 323(5914), 610-613.

Evžen, Š., and Václav, D. (2017). Biosynthesis of waxy starch-a review. *Plant, Soil and Environment,* 63(8), 335-341.

Feng, E., Ma, G., Wu, Y., Wang, H., and Lei, Z. (2014). Preparation and properties of organic-inorganic composite superabsorbent based on xanthan gum and loess. *Carbohydrate Polymers,* 111, 463-468.

Feng, X., Onel, O., Council-Troche, M., Noble, A., Yoon, R. H., and Morris, J. R. (2021). A study of rare earth ion-adsorption clays: the speciation of rare earth elements on kaolinite at basic pH. *Applied Clay Science,* 201, 105920.

Goenka, S., Sant, V., and Sant, S. (2014). Graphene-based nanomaterials for drug delivery and tissue engineering. *Journal of Controlled Release,* 173, 75-88.

Goudarzi, V., and Shahabi-Ghahfarrokhi, I. (2018). Photo-producible and photo-degradable starch/TiO_2 bionanocomposite as a food packaging material: development and characterization. *International Journal of Biological Macromolecules*, 106, 661-669.

Gregorova, A., Saha, N., Kitano, T., and Saha, P. (2015). Hydrothermal effect and mechanical stress properties of carboxymethylcellulose based hydrogel food packaging. *Carbohydrate Polymers,* 117, 559-568.

Guilherme, M. R., Aouada, F. A., Fajardo, A. R., Martins, A. F., Paulino, A. T., Davi, M. F., Rubira, A. F., and Muniz, E. C. (2015). Superabsorbent hydrogels based on polysaccharides for application in agriculture as soil conditioner and nutrient carrier: a review. *European Polymer Journal,* 72, 365-385.

Gulrez, S. K., Al-Assaf, S., and Phillips, G. O. (2011). Hydrogels, methods of preparation, characterisation and applications. *Progress in Molecular and Environmental Bioengineering-From Analysis and Modeling to Technology Applications*, 117150.

Haghighi, H., Gullo, M., La China, S., Pfeifer, F., Siesler, H. W., Licciardello, F., and Pulvirenti, A. (2021). Characterization of bio-nanocomposite films based on gelatin/polyvinyl alcohol blend reinforced with bacterial cellulose nanowhiskers for food packaging applications. *Food Hydrocolloids,* 113, 106454.

Hamidi, M., Azadi, A., and Rafiei, P. (2008). Hydrogel nanoparticles in drug delivery. *Advanced Drug Delivery Reviews,* 60(15), 1638-1649.

Hanani, Z. N., Roos, Y. H., and Kerry, J. P. (2014). Use and application of gelatin as potential biodegradable packaging materials for food products. *International Journal of Biological Macromolecules,* 71, 94-102.

Haraguchi, K. (2004). Mechanical properties of nanocomposite hydrogels consisting of organic/inorganic networks and the effects of clay modification thereto. *Journal of Network Polymer, Japan,* 25, 2-12.

Haraguchi, K. (2011). Synthesis and properties of soft nanocomposite materials with novel organic/inorganic network structures. *Polymer Journal,* 43(3), 223-241.

He, J., Yap, R. C. C., Wong, S. Y., and Li, X. (2015). Polymer composites for intelligent food packaging. *Journal of Molecular and Engineering Materials,* 3(01n02), 1540005.

Hebeish, A., Hashem, M., Abd El-Hady, M., and Sharaf, S. (2013). Development of CMC hydrogels loaded with silver nano-particles for medical applications. *Carbohydrate Polymers,* 92(1), 407-413.

Hoare, T. R., and Kohane, D. S. (2008). Hydrogels in drug delivery: progress and challenges. *polymer,* 49(8), 1993-2007.

Huang, J. Y., F. Ding, B. I. Yakobson, P. Lu, L. Qi and J. Li (2009). *In situ* observation of graphene sublimation and multi-layer edge reconstructions. *Proceedings of the National Academy of Sciences* 106(25), 10103-10108.

Iijima, S. (1991). Helical microtubules of graphitic carbon. *Nature,* 354(6348), 56-58.

Isana, C. L. S., Marismar, F. d. N., Raquel, G. D. S. N., Jos, C. D. S., Luiz, P. C., Cardoso, J. C.,, Juliana, C. C., and Ricardo, L. C. D. A. J. (2013). Effect of the maltodextrin-induced chemical reticulation on the physical properties and healing potential of collagen-based membranes containing Brazilian red propolis extract. *International Journal of Medicine and Medical Sciences,* 5(12), 514-524.

Jayakumar, R., Nwe, N., Tokura, S., and Tamura, H. (2007). Sulfated chitin and chitosan as novel biomaterials. *International Journal of Biological Macromolecules*, 40(3), 175-181.

Kabiri, K., Omidian, H., Zohuriaan-Mehr, M., and Doroudiani, S. (2011). Superabsorbent hydrogel composites and nanocomposites: a review. *Polymer Composites,* 32(2), 277-289.

Kalia, S. S., Adelman, K., Bale, S. J., Chung, W. K., Eng, C., Evans, J. P., Herman, G. E., Hufnagel, S. B., Klein, T. E., Korf, B. R., and McKelvey, K. D. (2017). Recommendations for reporting of secondary findings in clinical exome and genome sequencing, 2016 update (ACMG SF v2. 0): a policy statement of the American College of Medical Genetics and Genomics. *Genetics in medicine*, 19(2), 249-255.

Kaneko, Y., Noguchi, K., and Kadokawa, J. I. (2007). Synthesis of temperature-responsive organic-inorganic hybrid hydrogel by free-radical polymerization of methacrylamide using water-soluble rigid polysiloxane having acryamido side-chains as a cross-linking agent. *Polymer Journal,* 39(10), 1078-1081.

Kaneko, Y., Sato, S., Kadokawa, J. I., and Iyi, N. (2006). Synthesis of organic-inorganic hybrid hydrogels using rodlike polysiloxane having acrylamido groups as a new cross-linking agent. *Journal of Materials Chemistry,* 16(18), 1746-1750.

Kim, K. M., and Chujo, Y. (2003). Organic-inorganic hybrid gels having functionalized silsesquioxanes. *Journal of Materials Chemistry,* 13(6), 1384-1391.

Kuilla, T., Bhadra, S., Yao, D., Kim, N. H., Bose, S., and Lee, J. H. (2010). Recent advances in graphene based polymer composites. *Progress in Polymer Science*, 35(11), 1350-1375.

Kumar, A., Sharma, G., Naushad, M., and Thakur, S. (2015). SPION/β-cyclodextrin core-shell nanostructures for oil spill remediation and organic pollutant removal from waste water. *Chemical Engineering Journal,* 280, 175-187.

Laftah, W. A., Hashim, S., and Ibrahim, A. N. (2011). Polymer hydrogels: a review. *Polymer-Plastics Technology and Engineering,* 50(14), 1475-1486.

Lavrič, G., Oberlintner, A., Filipova, I., Novak, U., Likozar, B., and Vrabič-Brodnjak, U. (2021). Functional nanocellulose, alginate and chitosan nanocomposites designed as active film packaging materials. *Polymers,* 13(15), 2523.

Li, H., Hou, J., Duan, L., Ji, C., Zhang, Y., and Chen, V. (2017). Graphene oxide-enzyme hybrid nanoflowers for efficient water soluble dye removal. *Journal of Hazardous Materials,* 338, 93-101.

Li, K., Jin, S., Chen, H., and Li, J. (2019). Bioinspired interface engineering of gelatin/cellulose nanofibrils nanocomposites with high mechanical performance and antibacterial properties for active packaging. *Composites Part B, Engineering,* 171, 222-234.

Liu, H., Liu, M., Zhang, L., Ma, L., Chen, J., and Wang, Y. (2010). Dual-stimuli sensitive composites based on multi-walled carbon nanotubes and poly (N, N-diethylacrylamide-co-acrylic acid) hydrogels. *Reactive and Functional Polymers,* 70(5), 294-300.

Liu, Y., Zhu, M., Liu, X., Zhang, W., Sun, B., Chen, Y., and Adler, H. J. P. (2006). High clay content nanocomposite hydrogels with surprising mechanical strength and interesting deswelling kinetics. *Polymer,* 47(1), 1-5.

Ma, P. C., Siddiqui, N. A., Marom, G., and Kim, J. K. (2010). Dispersion and functionalization of carbon nanotubes for polymer-based nanocomposites: a review. *Composites Part A, Applied Science and Manufacturing,* 41(10), 1345-1367.

Mahinroosta, M., Farsangi, Z. J., Allahverdi, A., and Shakoori, Z. (2018). Hydrogels as intelligent materials: A brief review of synthesis, properties and applications. *Materials Today Chemistry,* 8, 42-55.

Mu, J., and Zheng, S. (2007). Poly (N-isopropylacrylamide) nanocrosslinked by polyhedral oligomeric silsesquioxane, temperature-responsive behavior of hydrogels. *Journal of Colloid and Interface Science,* 307(2), 377-385.

Neto, A. D. N. P., dos Santos Cruz, C. F., Serafini, M. R., dos Passos Menezes, P., de Carvalho, Y. M. B. G., Matos, C. R. S., and Nunes, P. S. (2017). Usnic acid-incorporated alginate and gelatin sponges prepared by freeze-drying for biomedical applications. *Journal of Thermal Analysis and Calorimetry,* 127(2), 1707-1714.

Ng, W. L., Yeong, W. Y., and Naing, M. W. (2016). Development of polyelectrolyte chitosan-gelatin hydrogels for skin bioprinting. *Procedia CIRP,* 49, 105-112.

Otoni, C. G., Espitia, P. J., Avena-Bustillos, R. J., and McHugh, T. H. (2016). Trends in antimicrobial food packaging systems: emitting sachets and absorbent pads. *Food Research International,* 83, 60-73.

Ouyang, J., Zhou, Z., Zhang, Y., and Yang, H. (2014). High morphological stability and structural transition of halloysite (Hunan, China) in heat treatment. *Applied Clay Science,* 101, 16-22.

Patil, M. B., Rajamani, S. B., Mathad, S. N., Patil, A. Y., Hussain, M. A., Alorfii, H. S., Khan, A., Asiri, A. M., Khan, I., and Puttegowda, M. (2022) Microwave-assisted synthesis of poly (acrylamide-co-2-hydroxyethyl methacrylate)/chitosan semi-IPN ZnO nanocomposite membranes for food packaging applications. *Journal of Materials Research and Technology*, 20, 3537-3548.

Pourjavadi, A., Harzandi, A., and Hosseinzadeh, H. (2004). Modified carrageenan 3. Synthesis of a novel polysaccharide-based superabsorbent hydrogel via graft copolymerization of acrylic acid onto kappa-carrageenan in air. *European Polymer Journal,* 40(7), 1363-1370.

Qiu, Y., and Park, K. (2001). Environment-sensitive hydrogels for drug delivery. *Advanced Drug Delivery Reviews,* 53(3), 321-339.

Ramaseshan, R., Sundarrajan, S., Liu, Y., Barhate, R., Lala, N. L., and Ramakrishna, S. (2006). Functionalized polymer nanofibre membranes for protection from chemical warfare stimulants. *Nanotechnology,* 17(12), 2947.

Rodríguez-Vázquez, M., Vega-Ruiz, B., Ramos-Zúñiga, R., Saldaña-Koppel, D. A., and Quiñones-Olvera, L. F. (2015). Chitosan and its potential use as a scaffold for tissue engineering in regenerative medicine. *BioMed Research International,* 2015, 821279.

Roy, N., Saha, N., Kitano, T., and Saha, P. (2012). Biodegradation of PVP-CMC hydrogel film: a useful food packaging material. *Carbohydrate Polymers,* 89(2), 346-353.

Saberianpour, S., Karimi, A., Nemati, S., Amini, H., Sardroud, H. A., Khaksar, M., Mamipour, M., Nouri, M., and Rahbarghazi, R. (2019). Encapsulation of rat cardiomyoblasts with alginate-gelatin microspheres preserves stemness feature *in vitro*. *Biomedicine & Pharmacotherapy,* 109, 402-407.

Sant, S., Tao, S. L., Fisher, O. Z., Xu, Q., Peppas, N. A., and Khademhosseini, A. (2012). Microfabrication technologies for oral drug delivery. *Advanced Drug Delivery Reviews,* 64(6), 496-507.

Sharma, G., Bhogal, S., Naushad, M., Kumar, A., and Stadler, F. J. (2017). Microwave assisted fabrication of La/Cu/Zr/carbon dots trimetallic nanocomposites with their adsorptional vs photocatalytic efficiency for remediation of persistent organic pollutants. *Journal of Photochemistry and Photobiology A: Chemistry,* 347, 235-243.

Shin, M. K., Spinks, G. M., Shin, S. R., Kim, S. I., and Kim, S. J. (2009). Nanocomposite hydrogel with high toughness for bioactuators. *Advanced Materials,* 21(17), 1712-1715.

Siengchin, S., and Karger-Kocsis, J. (2012). Polystyrene nanocomposites produced by melt-compounding with polymer-coated magnesium carbonate nanoparticles. *Journal of Reinforced Plastics and Composites,* 31(3), 145-152.

Song, F., Li, X., Wang, Q., Liao, L., and Zhang, C. (2015). Nanocomposite hydrogels and their applications in drug delivery and tissue engineering. *Journal of Biomedical Nanotechnology,* 11(1), 40-52.

Ullah, F., Othman, M. B. H., Javed, F., Ahmad, Z., and Akil, H. M. (2015). Classification, processing and application of hydrogels: a review. *Materials Science and Engineering: C*, 57, 414-433.

Vickers, N. J. (2017). Animal communication: when i'm calling you, will you answer too? *Current Biology,* 27(14), R713-R715.

Wei, L., Hu, N., and Zhang, Y. (2010). Synthesis of polymer-mesoporous silica nanocomposites. *Materials,* 3(7), 4066-4079.

Xiang, Y., Peng, Z., and Chen, D. (2006). A new polymer/clay nano-composite hydrogel with improved response rate and tensile mechanical properties. *European Polymer Journal,* 42(9), 2125-2132.

Xu, G., and Blum, F. D. (2008). Surfactant-enhanced free radical polymerization of styrene in emulsion gels. *Polymer,* 49(15), 3233-3238.

Youssef, A. M., and El-Sayed, S. M. (2018). Bionanocomposites materials for food packaging applications: concepts and future outlook. *Carbohydrate Polymers,* 193, 19-27.

Zhang, P., Zhang, X., Brown, J., Vistisen, D., Sicree, R., Shaw, J., and Nichols, G. (2010). Global healthcare expenditure on diabetes for 2010 and 2030. *Diabetes Research and Clinical Practice,* 87(3), 293-301.

Zhang, X., Pint, C. L., Lee, M. H., Schubert, B. E., Jamshidi, A., Takei, K., Ko, H., Gillies, A., Bardhan, R., and Urban, J. J. (2011). Optically-and thermally-responsive programmable materials based on carbon nanotube-hydrogel polymer composites. *Nano Letters,* 11(8), 3239-3244.

Zhou, C., and Wu, Q. (2012). Recent development in applications of cellulose nanocrystals for advanced polymer-based nanocomposites by novel fabrication strategies. *Nanocrystals: Synthesis, Characterization and Applications*, 103-120.

Zhu, M., Liu, Y., Sun, B., Zhang, W., Liu, X., Yu, H., Zhang, Y., Kuckling, D., and Adler, H. J. P. (2006). A novel highly resilient nanocomposite hydrogel with low hysteresis and ultrahigh elongation. *Macromolecular Rapid Communications,* 27(13), 1023-1028.

Zohuriaan-Mehr, M., and Kabiri, K. (2008). Superabsorbent polymer materials: a review. *Iranian Polymer Journal*, 17(6), 451-477.

Chapter 11

Material Design, Synthesis, and Mechanical Properties of Nanocomposite Hydrogels for Wastewater Contaminant Adsorption

Zhiying Wu[1] and Canhui Yang[2,*]

[1] Department of Chemistry, City University of Hong Kong,
Hong Kong, People's Republic of China
[2] Shenzhen Key Laboratory of Soft Mechanics & Smart Manufacturing,
Department of Mechanics and Aerospace Engineering,
Southern University of Science and Technology, Shenzhen, People's Republic of China

Abstract

In recent years, adsorption has been viewed as one of the most effective and promising wastewater treatment technologies, whereby the development of adsorbents plays a pivotal role. Among the variety of adsorbents developed by far, nanocomposite hydrogels stand out owing to their excellent adsorption performances encompassing large adsorption capacity, superb water affinity, easy preparation, wide tunability, recyclability, and robust mechanical properties.

In this chapter, we discuss the recent progress of nanocomposite hydrogels for water contaminant adsorption, by elaborating the basic materials design principles and synthesis strategies and summarizing the mechanical properties along with a brief introduction regarding the mechanical characterization methods and the associated mechanics. The main advances related to the use of nanocomposite hydrogels containing different fillers for adsorbing various pollutants including heavy metals, toxic dyes, antibiotics, and others, as well as the adsorption kinetic, isotherm, regeneration abilities, and related mechanisms are briefly reviewed.

Keywords: nanocomposite hydrogel, water treatment, synthesis, mechanical properties

---

[*] Corresponding Author's Email: yangch@sustech.edu.cn.

In: Nanocomposite Hydrogels and their Emerging Applications
Editor: Mohammad Sirousazar
ISBN: 979-8-88697-675-5

11.1. Introduction

Water is the most essential substance to humans and other living organisms on the earth. However, the development of human society since the industrial evolution has produced tremendous harmful chemical compounds, plenty of which have been discharged into water sources, detrimentally affecting not only the species living in water but also the broader biological community. As the menace of the global water crisis keeps escalating, decontamination and reutilization of wastewater are becoming increasingly important. To mitigate the water crisis, many water treatment techniques have been developed hitherto, including adsorption (Wang et al., 2020), flocculation (Moghaddam et al., 2010), photodegradation (Gupta et al., 2011), membrane filtration (Getachew et al., 2017), electrochemical treatment (Martínez-Huitle and Brillas, 2009), and so forth.

Among the many water treatment techniques, the adsorption method stands out with the merits of wide applicability, simple operation, low secondary pollution impact, and relatively low cost (Pandey et al., 2020; Pandey and Ramontja, 2016). In general, the adsorption process can be divided into two groups: physisorption and chemisorption. Physisorption is a reversible process that is controlled by physical interactions like hydrogen bonds, van der Waals forces, hydrophobic interactions, etc. Chemisorption is an irreversible process owing to the formation of chemical bonds between the adsorbent and the adsorbate. The important factors that determine the adsorption performance of adsorbents include pH, ionic strength, adsorbent dose, initial concentration, contact time, and temperature. Optimization of these factors should be considered in designing the industrial-scale of the adsorption process (Ahmad et al., 2015). The technical point of the adsorption method lies in the development of adsorbents. Various adsorbents, such as natural materials, agricultural wastes, industrial byproducts, biomass, and nanomaterials have been widely used for wastewater purification (Singh et al., 2018). The fascinating properties of nanomaterials and their combination with existing technologies have brought a bright future for the revolution in wastewater treatment. However, due to the high surface energy, most nanomaterials are prone to irreversible aggregation in solution and thus losing their original advantages.

Hydrogel is a bulky material composed of a three-dimensional polymer network infiltrated with a large amount of water. Hydrogels have been intensively studied for treating wastewater in recent years due to their low cost, recyclability, and ease of fabrication (Weerasundara et al., 2020). However, conventional hydrogels are constrained by their poor mechanical performances and low adsorption capacity (Yang et al., 2013; Kalcioglu et al., 2012). To overcome these issues, conferring nanoparticles (NPs) into a hydrogel matrix to form nanocomposite (NC) hydrogels is an effective means to improve the mechanical property of hydrogel while simultaneously inhibiting the agglomeration of NPs. Since Haraguchi et al., (2002) first reported using synthetic clay composed of silicate nanoplatelets as multifunctional crosslinking agents to synthesize NC hydrogels from poly (*N*-isopropyl acrylamide), NC hydrogels have quickly become a hot research spot (Zhang et al., 2022). Various types of polymer matrices loaded with different types of nanofillers have been used to fabricate NC hydrogels for pollutants adsorption with enhanced structural, physical, chemical, and biological properties (Zhao et al., 2020). The field of synthesizing NC hydrogels of marked adsorbing and mechanical performances is in the ascendant and evolving fast.

In this chapter, we discuss the recent progress of NC hydrogels for water contaminant adsorption, by elaborating the basic materials design principles and synthesis strategies, and summarizing the mechanical properties along with a brief introduction regarding the mechanical characterization methods and the associated mechanics. The main advances related to the use of NC hydrogels containing clay minerals, carbonaceous materials, metallic NPs, metal-organic frameworks (MOFs), transition metal carbides or nitrides (MXenes), and other fillers in the applications for pollutants adsorption including heavy metals, toxic dyes, antibiotics, and others are reviewed. The adsorption kinetic, isotherm, regeneration abilities, and related mechanisms are also briefly mentioned. This chapter concludes with an over-the-horizon outlook discussing current challenges and emerging research opportunities.

11.2. Materials Design and Synthesis

11.2.1. NC Hydrogel Design

NC hydrogels are prepared from the combination of polymers and NPs by physical or chemical interactions (Zhao et al., 2020). By controlling the interactions between polymer chains and NPs, various physical, chemical, optical, and biological properties can be engineered. A wide range of natural and synthetic polymers could be used to form the NC hydrogel network. Natural polymers include gelatin, alginate, chitosan, cellulose, fibrin, starch, etc. The use of natural polymers has advantages, such as biocompatibility, biodegradability, and non-toxicity. During the last two decades, natural hydrogels have been gradually substituted by synthetic hydrogels which have a long service life, high strength, and high capacity for water adsorption (Chamkouri et al., 2021).

Examples of synthetic polymers include polyacrylamide, polyvinyl alcohol (PVA), polyethylene, polyvinylidene fluoride, polypropylene, polyaniline, polyvinyl chloride, polyamide, polyethylene terephthalate, polystyrene, etc. NPs that could be combined with the polymeric networks include clay minerals (montmorillonite, attapulgite, laponite, etc.), carbon-based nanomaterials (carbon nanotubes (CNTs), graphene, biochar, etc.), metal/metal oxide NPs (Fe_3O_4, MoS_2, TiO_2, etc.), MOFs, polymeric NPs, dendrimers, and some others. Figure 11.1 summarizes the three main routes for incorporating NPs into hydrogels.

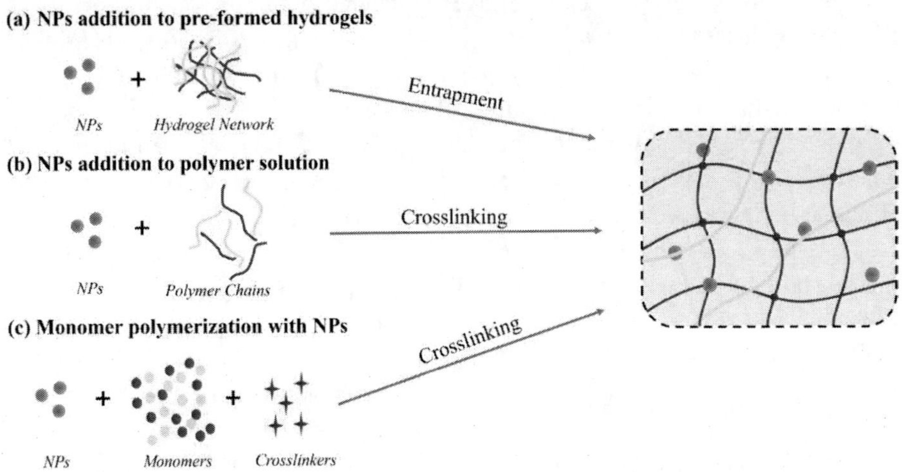

Figure 11.1. Schematics illustrating the strategies to synthesize NC hydrogels: (a) NPs are added to pre-formed hydrogels, (b) NPs are added to polymer solutions, and (c) monomer polymerization in the presence of NPs.

NPs can be added to pre-formed hydrogel matrices, polymers solutions that are subsequently gelled, with the NPs used as the crosslinkers or simply as the additives, or monomer solutions prior to co-crosslinking polymerization. It is also possible to grow the NPs *in situ* from the precursors that are incorporated in the polymer network using strategies (a) and (b) (Dannert et al., 2019). Of course, these routes are applicable to prepare NC hydrogels with single or multiple polymer networks.

11.2.2. Synthetic Strategies

NC hydrogels are typically prepared using two major strategies based on whether NPs are physically embedded in the matrix or chemically bonded to the polymeric network. To physically incorporate NPs in the hydrogel matrix, NPs can either act as fillers in hydrogels without any chemical interaction with the polymeric network or be introduced into the network by blending or *in situ* formation. Besides, surface-functionalized NPs can behave as gelators to chemically crosslink the polymeric network of hydrogels (Lu et al., 2021). The characteristics of the obtained NC hydrogels are determined by the composition and proportional concentration of the nanofillers and polymers, together with the size and the degree of the nanofillers dispersed in the hydrogel (Han et al., 2019). Therefore, controlling the dispersion degree of nanomaterials in the hydrogel and its mutual combination with the molecular chains of the hydrogel is the key to the preparation of NC hydrogels. In accord with the dispersion method of nanomaterials in hydrogels, we briefly discussed several methods employed for the preparation of NC hydrogels.

11.2.2.1. Blending Method

The blending method is the simplest route to prepare NC hydrogels. Monomers, crosslinking agents, and nanomaterials are mixed and directly initiated to form NC hydrogels (Lü and Yang, 2009). For example, blending has been used to synthesize sulfonated graphene (SG)/PVA hydrogel by mixing SG and PVA solution, which showed better mechanical properties and adsorption performance towards cationic dyes compared with pure PVA hydrogel (Li et al., 2015). This method has been also utilized to generate magnetized NC hydrogels from iron oxide NPs and polyethylene glycol hydrogel polymers by co-blending, which showed high elastic modulus and good biocompatibility (Filippi et al., 2019).

A supramolecular NC hydrogel is obtained by directly blending well-designed dendritic macromolecules with laponite NPs (Wang et al., 2010). The interactions between functional groups on the peripheries of the dendritic macromolecules and on the surface of the laponite NPs led to a novel supramolecular assembly. Surprisingly, a high-water-content moldable hydrogel can be achieved without any catalysts or initiators. When the NC hydrogel is prepared by the blending method, the NPs are prone to agglomeration, resulting in weak interactions between the NPs and the polymer matrix (Ponnamma et al., 2019). To address the problem, mechanical methods, such as ultrasonication or mechanical stirring could help to disperse nanomaterials in most solvents. Besides, chemical approaches, such as surface functionalization and surfactant stabilization are often adopted to minimize the interface energies between NPs and polymer matrix, so as to disperse inorganic materials more uniform within hydrogel precursor solution (Du and Huang, 2022).

11.2.2.2. **In Situ** *Polymerization Method*

In situ free radical polymerization is to homogeneously disperse pre-prepared NPs in a monomer solution and then initiate the polymerization by external stimuli, such as thermal, photochemical, or chemical activation, through initiators and/or catalysts (Zhao et al., 2015). In this case, the NPs are nucleated and grown inside the polymer matrices, which could effectively prevent particle agglomeration and maintain uniform distribution. The elasticity of the polymer network and the interactions between the NPs and the polymer chains prevent individual particles from meeting each other by diffusion (Lü and Yang, 2009). Haraguchi et al., (2002) develop NC hydrogels consist of acrylamide derivative, poly (N-alkyl acrylamides), in the presence of inorganic clay. By changing the composition of the initial reaction solution, the properties of the final hybrid gels can be tailored. The increased modulus and mechanical strength, improved deformability, and fracture energy were some of the important improvements obtained compared with the conventional hydrogels. *In situ* polymerization can ensure that the NPs are evenly distributed in the hydrogel and preserve the characteristics of the NPs, so the mechanical strength of the NC hydrogel can be enhanced. However, unreacted educts or byproducts of the *in situ* reaction might scratch the properties of the final material.

11.2.2.3. Grafting onto Method

The grafting method is to modify the surface of the nanomaterials to build a covalent bond between the inorganic nanomaterials and the hydrogel network, which could effectively ameliorate the agglomeration problems of inorganic nanomaterials (Prasad et al., 2020). Various techniques may be used to create free radicals on the surfaces of NPs, e.g., ionizing radiation, peroxide formation by ceric ions, and radical transfer. Other examples of grafting techniques are chemical grafting, enzymatic grafting, free radical grafting, photochemical grafting, ionic grafting, and radiation grafting (Haider et al., 2015; Ju et al., 2018). Hu et al., (2019) prepared a magnetic NC hydrogel by grafting method, with polyacrylamide hydrogel as the matrix and 3-(trimethoxysilyl) propyl methacrylate coated Fe_3O_4 as the fillers. Magnetic hydrogel exhibits a relatively high modulus and toughness compared to pristine hydrogel. Another NC hydrogel was synthesized from the grafting of acrylamide onto hydroxypropyl methylcellulose using methylene bis-acrylamide crosslinker and sodium montmorillonite nanoclay (Mahdavinia et al., 2013). The investigation of the crystal violet dye adsorption onto the gel was performed and both adsorption capacity and kinetic performance were enhanced with the increase of nanoclay content. Therefore, the advantage of the grafting method is that after modifying the surface of the nanomaterial, the NPs and the hydrogel polymer are integrated covalently, which markedly enhances the mechanical strength and stability of the NC hydrogel.

11.2.2.4. Freeze-Thaw Method

The freeze-thaw method is a physical crosslinking method for preparing NC hydrogels through repeatedly freezing and thawing the hydrogel mixed solution at low temperature and room temperature (Stoyneva et al., 2014). In this process, polymer aggregation and non-aggregation areas are formed internally when the hydrogel mixture solidifies. During the thawing process, the polymer aggregation areas will undergo physical crosslinking to form a crosslinked network (Li et al., 2012). Rezazadeh et al., (2020) prepared NC hydrogel pipes upon PVA and montmorillonite clay through a cyclic freezing-thawing method, which showed improved mechanical and thermal properties compared to the pure hydrogel pipes. Other studies have also reported that the freeze-thaw technique could help to increase the pore size of NC hydrogels. Widening of pore size occurred by the formation of a crystalline matrix and immediate cooling of the homogeneous aqueous solution of monomers (Sharma et al., 2018). The disadvantage of the freeze-thaw method is that the physical crosslinking is reversible, so the crosslinking network structure of the synthesized hydrogel is not very strong and thus might be easily decomposed by external influences.

11.2.2.5. Other Methods

Some other technologies could also be used for the preparation of NC hydrogels. For example, the self-assembly method is commonly used for the preparation of 3D graphene-based hydrogels, which self-assembles via π-π interactions of graphene oxide (GO) sheets (Xu et al., 2010). Radiation-induced polymerization is a method to fabricate NC hydrogels using radiation energy, such as γ-ray, ultraviolet (UV) light, electron beam, and microwave (Singhal and Gupta, 2016). Advantages of this method include mild processing conditions, absence of poisonous crosslinking reagents, relatively low cost, and operational convenience. Examples of the polymers used for preparing NC hydrogels by radiation-induced polymerization include polyethylene glycol, PVA, polyacrylamide, polyvinyl pyrrolidone, and some others. Eid (2011) successfully synthesized starch-based polyelectrolyte hydrogel using the gamma radiation polymerization technique at room temperature. In another study, an alginate-based hydrogel was prepared using UV light radiation-induced polymerization, which resulted in the crosslinking of N-methylolacrylamide and acrylic acid to form a three-dimensional matrix (Wu et al., 2019). Moreover, poly (acrylamide co-acrylic acid) hydrogels were obtained under electron beam irradiation and the hydrogels were used for removing heavy metals from wastewater (Manaila et al., 2017).

11.2.3. Mechanical Properties and Characterizations

Robust mechanical properties are the key prerequisites for the practical deployments of bulk hydrogels in many applications, especially for industrial-scale usages such as water treatment. Relevant parameters include but not limited to modulus, fracture strain, strength, work of fracture, hysteresis, toughness, and fatigue threshold. Depending on the specific application scenarios, the mechanical properties of NC hydrogels are assessed from different aspects and characterized via different measurements.

11.2.3.1. Tension

Tension is perhaps the most pervasive test used to probe the mechanical properties of hydrogels. Two types of tension are often used. The first one is uniaxial tension, whereby a slender sample is pulled along the lengthwise direction (Figure 11.2a). The recorded force-displacement curve is converted into the engineering stress-engineering strain curve, by dividing the force and the displacement with the cross-sectional area and the length of the sample, both measured in the undeformed state, respectively. The engineering stress-strain curve readily gives the modulus, the fracture strain, the strength, and the work of fracture of the tested sample (Yang et al., 2019).

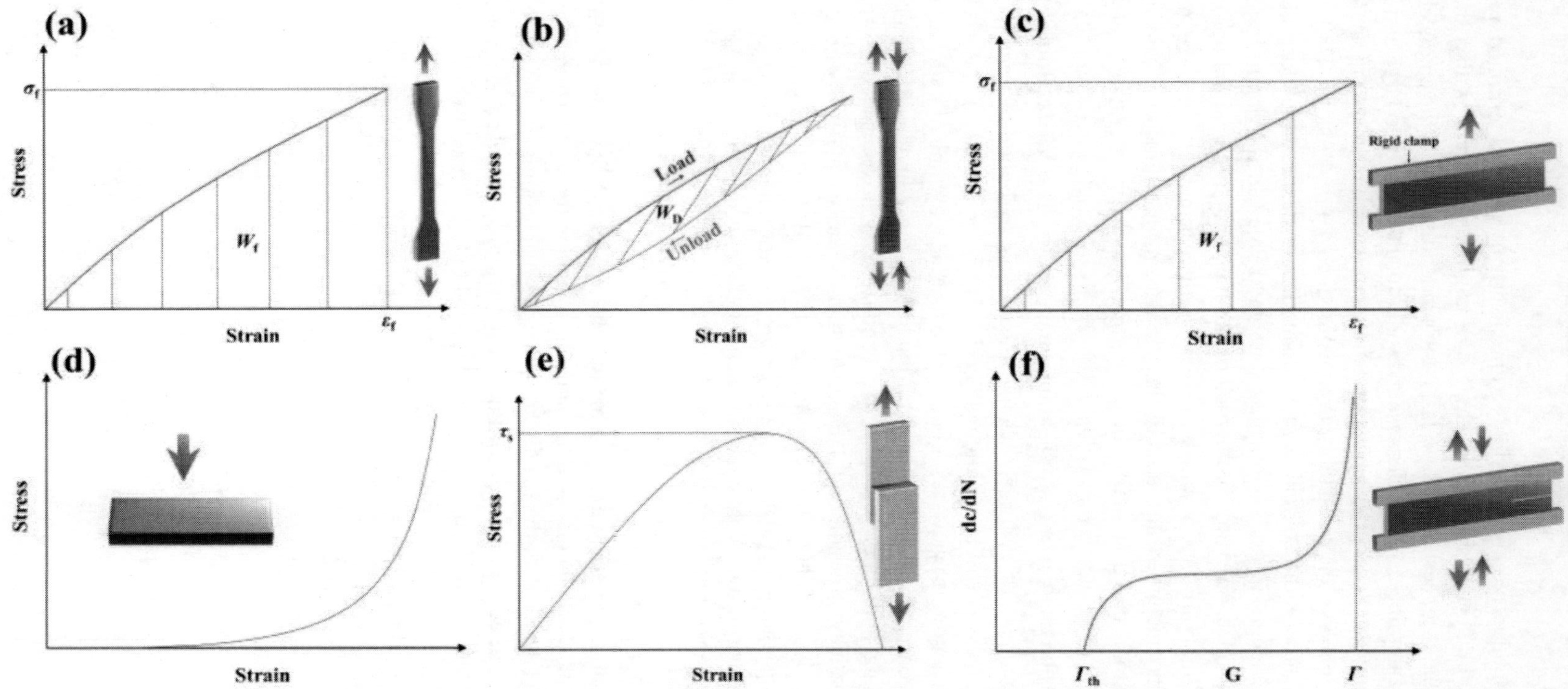

Figure 11.2. Mechanical characterizations: (a) uniaxial tension, (b) cyclic loading and unloading, (c) pure shear, (d) compression, (e) lap-shear, and (f) fatigue fracture.

The Young's modulus, or elastic modulus, is frequently used and determined by fitting the data within the small deformation regime, say 5% of engineering strain, with a straight line, $\sigma = E\varepsilon$, where σ is the engineering stress, E is Young's modulus, and ε is the engineering strain. The slope of the straight line gives E. Because hydrogels usually undergo large and non-linear deformation, alternatively, the engineering stress-strain curve can be fitted to the neo-Hookean model. For uniaxial tension, the neo-Hookean model reduces to $\sigma = \mu(\lambda^2 - \lambda^{-1})$, where μ is the shear modulus and λ is the stretch, defined as the length of the sample in the deformed state divided by the length in the undeformed state and related to the engineering strain as $\lambda = 1 + \varepsilon$. The slop of the curve of the neo-Hookean model at infinitesimal deformation is $d\sigma/d\lambda\,|_{\lambda\to1} = 3\mu$, implying that Young's modulus is three times of the shear modulus in uniaxial tension. The fracture strain, ε_f, and the strength, σ_f, are directly obtained from the engineering stress-strain curve at the onset of rupture. The work of fracture, of unit J/m^3, represents the amount of work per unit volume done by the external force and is calculated as the area underneath the engineering stress-strain curve, $W_f = \int_0^{\varepsilon_f} \sigma(\varepsilon)\, d\varepsilon$. Sometimes, W_f is referred to as the toughness of hydrogel, whereas the rational definition of fracture toughness is different in mechanics, as will be mentioned later.

From the practical point of view, it is conceivable that the recycling of hydrogel is of great importance for the concerns of low-cost and minimal interference on the environment. In this regard, cyclic loading and unloading tests are conducted to investigate the reliability of materials. Due to the inevitable internal damages within the matrix of NC hydrogels, such as the dissociation of polymer chains from the surfaces of nanofillers or the cleavage of covalent bonds along polymer chains (Peak et al., 2013), the loading and unloading curves do not coincide with each other in most cases. The area enclosed between the loading and unloading curves represents the energy dissipated during that load cycle, W_D (Figure 11.2b) (Liu et al., 2019a). The ratio of W_D/W_f determines the hysteresis, a parameter frequently used to quantify the energy dissipation capacity.

In the experiments of uniaxial tension, note that the transverse dimensions, i.e., the width and the thickness for a rectangular sample and the diameter for a cylindrical sample, are required to be much smaller than the length, such that the state of uniaxial tension is fulfilled. For example, the length can be at least 10 times of the width and thickness, for a dumbbell-shaped sample. In addition, two ends of the samples need to be clamped by the fixtures of the testing machine and thus their dimensions are often purposely enlarged to avoid the premature failure of the sample caused by stress concentration at the places of clamping. For a fair comparison between different measurements, the strain rate needs to be specified, which is calculated as the ratio of the loading velocity over the gauge length of the sample, since many NC hydrogels are viscoelastic and exhibit rate-dependent mechanical properties.

The second type of tension, known as the pure shear test, was first proposed by Rivlin and Thomas (1953) for measuring the fracture toughness of rubbers. This method has been widely adopted for hydrogels since its implementation by Suo and co-workers (2012). In the test, a rectangular sample is stretched along the longitudinal direction, while its width is completely constrained (Figure 11.2c). Note that the sample needs to possess a width much larger than the length, which in turn is much larger than the thickness. For the pure shear test, the neo-Hookean model reduces to $\sigma = \mu(\lambda - \lambda^{-3})$ (Xiao et al., 2022). The slop of the curve at infinitesimal deformation is $d\sigma/d\lambda\,|_{\lambda\to1} = 4\mu$, implying that Young's modulus is four times of the shear modulus in the pure shear test. Likewise, the fracture strain, the strength, and the work of fracture are readily obtained from the engineering stress-strain curve at the onset of rupture. Fracture toughness is defined as the energy needed to propagate a crack by unit area and has a unit of J/m^2.

To measure the fracture toughness, two sets of samples are prepared. One set of samples are intact, pulled to a decent strain (may or may not be ruptured), and their engineering stress-strain curves are recorded. Then the other set of samples are pre-cut with an edge notch along the middle line, pulled to rupture, and the critical strain, ε_c, is recorded. The fracture toughness is calculated as $\Gamma = HW(\varepsilon_c)$, where H is the original height of the sample (along the loading direction) and $W(\varepsilon_c) = \int_0^{\varepsilon_c} \sigma(\varepsilon)\, d\varepsilon$ is the strain energy density of the materials far ahead of the crack tip, with $\sigma(\varepsilon)$ being the engineering stress-strain relationship obtained from the intact samples. In the experiments of pure shear tests, particular attentions should be paid to the preparation of the samples, especially the constrains along the width direction. In the literature, polymethyl methacrylate sheets are often used and glued to the surfaces of hydrogel to provide the constraints. In addition, the length of the pre-cut should be much larger than the original height of the sample. Typical dimensions of the sample in undeformed state are 10 cm for width, 1 cm for the height, 1 mm for the thickness, and 2 cm for pre-cut.

11.2.3.2. Compression

Compression is another widely adopted measurement to characterize the mechanical properties of hydrogels. Similar to tension, the engineering stress-engineering strain curve is also constructed in a compressive test, yet the shape of the curve is quite different from that of tension. Generally, the curve of tension increases steeply at the beginning, then enters a relatively flat region whereby the strain increases significantly while the stress increases negligibly, and finally reaches the strain hardening stage, if there is one, before catastrophic rupture. By contrast, the curve of compression increases tardily at the beginning and then goes up steeply before failure (Figure 11.2d). The curve gives the modulus, the fracture strain, the strength, and the work of fracture. In certain situations, incompressibility is assumed and compression is adopted as the equi-biaxial tension (Webber et al., 2007).

In a compressive test, typically used samples are cylinders whose diameter is larger than the height (Haider et al., 2015). The cylindrical shape ensures homogeneous uniaxial compression and the large ratio of diameter over height alleviates buckling instability during the compressing procedure. Note that the top and the bottom surfaces between the sample and the working stage/load cell need to be appropriately lubricated such that the hydrogel sample can deform freely in the lateral directions upon compression. Because the allowable displacement is limited by the height of the sample, careful attention should be paid to avoid jamming the testing machine.

11.2.3.3. Shear

Shear is another commonly encountered loading mode. Usually, two types of shear tests are used in the characterization of hydrogels. The first one is the lap-shear test, which consists of a layer of hydrogel sandwiched between two layers of stiff or flexible backing (Figure 11.2e) (Wang et al., 2021b). Forces are applied through the backings to exert shear deformation of the hydrogel. The force-displacement curve is converted into engineering shear stress-shear strain curve by dividing the force and the displacement with the area and the thickness of the sample, respectively. The peak shear stress is defined as the shear strength, τ_s, a parameter frequently used to characterize the shear performance of the hydrogel. Sometimes the lap-shear test is used to measure the adhesion energy between the hydrogel and other materials, and in this case, the area underneath the shear stress-strain curve before failure gives the adhesion energy. Note that adhesive failure or cohesive failure may occur during the lap-shear test. When the latter case happens, the test actually measures the toughness of the hydrogel under the mode II fracture.

The second one is the rotational shear measured using a rheometer, wherein a cylindrical sample is fixed on the working stage, pre-compressed by the loading head, and sheared as the loading head rotates at a given temperature. In the test, the storage modulus, the loss modulus, and the viscosity of the hydrogel are often plotted against the frequency. Since NC hydrogels are already crosslinked, the storage modulus should be higher than the loss modulus throughout the tested frequency. Depending on the specific constituents, the viscosity of NC hydrogels typically decreases with frequency. Note that, the applied shear strain must be small, typically within 1%, since the instrument and the software calculate the parameters based on linear viscoelasticity assumptions. The rheology test, sometimes, is used to determine the gelling point of NC hydrogels by tracking the variation of the storage modulus and the loss modulus with time. Before gelling, the precursor of NC hydrogels is more liquid-like and the storage modulus is smaller than the loss modulus. After gelling, the sample of NC hydrogels becomes more solid-like and the storage modulus is larger than the loss modulus. The time when the storage modulus equates to the loss modulus is the gelation time.

11.2.3.4. Fatigue

The field of the fatigue of hydrogels is nascent. It has been recently revealed that most hydrogels are prone to fatigue (Bai et al., 2019), a consequence of molecular diseases which was proposed by Lake and Thomas (1967). However, as mentioned before, NC hydrogels are inevitably subjected to cyclic loading and unloading in practical usage. Hence, the fatigue performances of NC hydrogels are of great significance. The fatigue of hydrogel is usually characterized in two ways, namely fatigue damage and fatigue fracture, using the pure shear test configurations (Bai et al., 2018).

In a fatigue damage test, an intact sample is subjected to cyclic tension and the cyclic engineering stress-strain curves are recorded. For analysis, the variation of the mechanical properties, such as the modulus or the stress at the maximum strain before unloading, with the number of cycles is plotted. A hydrogel is said to be fatigue-damage-free or fatigue-damage-resistant if its mechanical properties do not keep degrading with the number of cycles. In a fatigue fracture test, a pre-cut sample is pulled to a certain stretch, corresponding to a certain energy release rate, $G = HW(\lambda)$. If G is equal to the fracture toughness of the hydrogel, the sample fractures in the first cycle. If G is lower than the toughness, the sample may survive a few cycles before rupture. As G continues to decrease, the pre-cut propagates slower and slower, and the number of cycles before rupture increases in proportion. It is sensible that at some point the pre-cut no longer propagates and the corresponding energy release rate is termed as the fatigue threshold, $\Gamma_{th} = HW(\lambda_{th})$ (Figure 11.2f). Note that the strain energy density needs to be determined by using the intact samples subjected to the same number of loading cycles or using the stabilized engineering stress-strain curve.

In experiments, one has to subjectively determine a critical number of cycles for the fatigue threshold. In addition, very often the fatigue threshold is obtained by linearly extrapolating the data points close to zero propagation of pre-cut. Furthermore, since hydrogels are susceptible to desiccation, a humid environment of controlled humidity is required. Besides the fatigue of the bulk materials, occasionally, NC hydrogels may need to adhere to other materials. Whereas the adhesion of hydrogel by itself deserves a comprehensive review (Yang and Suo, 2018; Yang et al., 2020) and is beyond the scope of the current chapter, it is suggested that the fatigue of hydrogel adhesion should be taken into account when both adhesion and fatigue are important (Li et al., 2021; Liu et al., 2020; Zhang et al., 2021a).

11.3. Adsorption Applications

11.3.1. Performances Assessment

To date, various NC hydrogels have been employed for pollutant adsorption. In the following text, we review the main advances related to the use of NC hydrogels containing clay minerals, carbonaceous materials, metallic NPs, MOFs, and other fillers.

11.3.1.1. NC Hydrogels Containing Clay Minerals

Numerous clays, including kaolinite, montmorillonite, attapulgite, diatomite, smectite, vermiculite, biotite, laponite, sepiolite, bentonite, and halloysite have been added to the polymeric matrix to synthesize NC hydrogels for pollutants removal from water/wastewater (Van Tran et al., 2018). In many cases, NC hydrogels embedded with clays exhibited extraordinary ion conductivity, mechanical properties, and faster swelling-deswelling properties than conventional hydrogels (Zhao et al., 2015). Besides, its current market price is cheaper than that of active carbon and some other materials.

Wang et al., (2021a) used montmorillonite nanosheets, acrylic acid, and acrylamide as raw materials to synthesize a 3D reticular-structured hydrogel. The as-synthesized gel achieves the effective removal (99%) for highly concentrated Pb(II) solution via a small hydrogel dosage within a short time. Clay minerals also have a remarkable adsorption capacity to remove antibiotics from wastewater. Peng et al., (2016) reported a cellulose-montmorillonite NC hydrogel which showed high adsorption performance for methylene blue dye through a spontaneous physisorption reaction. The maximum removal efficiencies for methylene blue solution (100 mg/L) could be up to 98%. These performances of clay minerals make their NC hydrogels highly promising as superadsorbents to remove pollutants from sewage water.

11.3.1.2. NC Hydrogels Containing Carbonaceous Materials

Carbonaceous materials (e.g., GO, CNTs, and biochar) have also been associated with hydrogels, which display excellent features, such as high density, high strength, and high hardness (Anjali et al., 2019). In recent years, GO, one of the representative carbon nanomaterials by virtue of its large specific surface and abundant oxygen-containing groups, has received considerable attention for effective pollution treatment. GO possesses a high adsorption capacity for different dye molecules through π-π stacking and electrostatic interactions (Singh et al., 2011). GO is negatively charged and thus usually combined with hydrogels formulated with anionic polymers to be applied as adsorbents with strong anionic behavior for removing cationic dyes. GO was also proved to be a prospective material for the adsorption of heavy metal ions (Liu et al., 2019b) and antibiotics (Yang et al., 2022).

CNTs are also relevant carbon-based nanomaterials often used to prepare NC hydrogels for adsorption. Their well-defined one-dimensional structure, and exceptional mechanical and thermal properties, together with a high specific area, enable the adsorption or conjugation of a wide variety of pollutant compounds (Adewunmi et al., 2016). More recently, embedding biochar in the hydrogel matrix has shown promising applications for the adsorptive removal of pollutants due to its highly-porous structure, large surface area, various oxygen functional groups, and low cost (Wu et al., 2022; Akl et al., 2021; Lu et al., 2020). Recent studies have demonstrated that polymeric biochar adsorbents offer better mechanical strength, enhanced durability, high adsorption capacity, and improved recyclability (Afzal et al., 2018).

11.3.1.3. NC Hydrogels Containing Metallic NPs

In the last decade, the preparation of NC hydrogels containing metallic NPs has received considerably increased attention. Metallic NPs have been used not only to improve the physical, chemical, optical, magnetic, and electronic properties of hydrogels but also to provide active sites in the adsorption of pollutants, allowing convenient adsorption/desorption recycles by employing the external magnetic field.

Konwar et al., (2015) synthesized a magnetic hydrogel based on an alginate polymer matrix embedded with Fe_3O_4 NPs, which acted as an effective adsorbent of ciprofloxacin hydrochloride. The hydrogels could be quickly recovered using a magnet, allowing their reuse in cyclic adsorption-desorption runs. In recent years, molybdenum disulfide (MoS_2) has been reported as an efficient adsorbent for pollutant removal from wastewater (Yang et al., 2021b). Two-dimensional MoS_2-based adsorbents possess much higher adsorption capacity for Pb^{2+}, Hg^{2+}, and Cr^{6+} in comparison with clay-based and activated carbon-based adsorbents, due to their unique layered structures and abundant active sites (Wang et al., 2018). Ma et al., (2022) incorporated MoS_2 into PVA hydrogel, which shows enhanced Pb^{2+} removal performance with a high adsorption capacity of 257 mg/g. The use of other metallic compounds (TiO_2, Cu_2O, CuS, ZnO, etc.) to prepare NC hydrogels for pollutant removal has also been reported (Ponnamma et al., 2019).

11.3.1.4. NC Hydrogels Containing MOFs

MOFs have emerged as promising adsorbents ascribing to their large surface area, ultrahigh porosity as well as modifiable structure and functionality (Yao et al., 2021). They are formed by the linkage between organic ligands and metal-containing nodes through coordination bonds. Until now, a wide range of MOFs with various crystal structures, molecular compositions, topologies, pore structures, and surface morphologies have been discovered based on diverse metal nodes and ligands (Wang et al., 2019). The recently developed MOFs-based hydrogels can overcome the weakness of powdered MOFs, such as fragile, unstable, unprocessable, and unrecoverable, showing increasing potential towards water treatment.

Zhuang et al., (2019) reported a novel one-step method for preparing ZIF-67-alginate hydrogel and used it for tetracycline removal. The maximum adsorption capacity of tetracycline adsorption on this material was 364.89 mg/g and the material exhibited excellent reusability after 10 cycles. Li and coworkers (2019) proposed an *in situ* method to prepare UiO-66-NH_2 NPs on cellulose gels with higher adsorption performance of Pb^{2+} and Cu^{2+} compared to UiO-66 powders. Moreover, it can be easily reused for more than 5 cycles without significant performance degradation.

11.3.1.5. NC Hydrogels Containing MXenes

Among the existing 2D nanomaterials, the thriving family of MXenes show excellent sorption capability and efficient reduction performance for various contaminants of wastewater owing to their unique layered structure, large specific surface area, and excellent hydrophilic properties. Zhang et al., (2021b) synthesized $Ti_3C_2T_x$/sodium alginate gel by ionic crosslinking, which exhibits efficient adsorption capacity on methylene blue (maximum 92.17 mg/g) and the adsorption efficiency could maintain 81.36% after three cycles. Chen et al., (2021) incorporated $Ti_3C_2T_x$ into polyacrylamide via the sol-gel procedure and the synthesized hydrogel showed increased equilibrium adsorption capacity (282.29 mg/g) on methylene blue removal, nearly three times higher than that of batch adsorption method (102.16 mg/g).

11.3.1.6. Others

The increasing interest in cellulose nanocrystals (CNCs) as emerging nanomaterials has made CNC-based hydrogels promising for water purification. Over the last decade, CNC-based hydrogels used for the adsorption of methylene blue have been extensively studied based on various polymers (Rahman and Rimu, 2020). The data of adsorption capacity obtained from different CNC-based hydrogels showed remarkable adsorption compared with hydrogel prepared with other biopolymers. Yang et al., (2021a) provide a detailed and comparative review of chitosan-based hydrogels for the adsorption of pollutants including heavy metal ions, industrial dyes, microplastics, and organics. As a whole, chitosan-based hydrogels have great potential for water treatment applications due to their cost effectiveness, excellent bioactivity, high pollutant adsorption capacity, high water absorption, water holding capacity, and water affinity.

11.3.2. Adsorption Mechanism

It is of vital importance to interpret the adsorption behaviors by NC hydrogels so as to provide a theoretical basis for optimization and further research. To determine the adsorption mechanism, numerous characterizations methods have been used to experimentally explore the functional groups and structure of the adsorbent before and after pollutants adsorption, including Fourier Transform Infrared Spectroscopy, X-ray photoelectron spectroscopy, X-ray diffraction, Zeta potential analyzer, and some others.

The kinetics and isotherms of adsorption are widely investigated to theoretically reveal the mechanisms of adsorption. The sorption kinetics data describe the rate of the solute uptake at the solid-solution interface and provide valuable insights into the reaction pathways and mechanisms (Simonin, 2016). Among all the models, pseudo-first-order and pseudo-second-order adsorption models are frequently applied to fit kinetic data and the equations are as follows:

Pseudo-first-order model:

$$Q_t = Q_e(1 - e^{-k_1 t}) \quad (11\text{-}1)$$

Pseudo-second-order model:

$$Q_t = \frac{K_2 Q_e^2 t}{K_2 Q_e t + 1} \tag{11-2}$$

where Q_e (mg/g) and Q_t (mg/g) are the adsorbed adsorbate per mass of adsorbent in equilibrium and at time *t*, respectively. K_1 (1/h) and K_2 (mg/g.h) are the pseudo-first-order constant and pseudo-second-order constant, respectively. Generally, the pseudo-first-order model suggests that the sorption process is physisorption (Azizian, 2004). The pseudo-second-order kinetic model assumes that the rate-controlling step may be chemisorption involving valency forces through sharing or exchanging of electrons between adsorbent and adsorbate (Ali et al., 2016).

Equilibrium adsorption isotherms are usually employed to determine the capacity, surface properties, and affinity of an adsorbent (Tafakori et al., 2017). The Langmuir and Freundlich equilibrium models are the most widely used sorption isotherms. The equations are expressed according to equations (11-3) and (11-4):

Langmuir model:

$$Q_e = Q_{max} K_L C_e / (1 + K_L C_e) \tag{11-3}$$

Freundlich model:

$$Q_e = K_F C_e^{1/n} \tag{11-4}$$

where C_e (mg/L) is the residual adsorbate concentration in solution at the equilibrium state and Q_e (mg/g) is the adsorption amount per unit mass of adsorbent. Q_{max} (mg/g) and K_L (L/mg) are Langmuir constants related to the maximum monolayer capacity of the adsorbent and affinity of the binding sites, respectively. K_F (mg/g $(L/mg)^{1/n}$) and *n* are the Freundlich constant and heterogeneity factor, respectively. Langmuir isotherm is a monolayer adsorption model for ideal interfaces while Freundlich isotherm indicates the predominance of multilayer adsorption (Wang and Guo, 2020).

Combing experimentally and theoretically analysis, various metal adsorption mechanisms have been reported, with electrostatic interaction being considered to be the main mechanism in the adsorption of pollutants by NC hydrogels (Figure 11.3) (Khan and Lo, 2016). The functional groups on the surfaces of NC hydrogels will protonate to form anions attracting cationic pollutants (metal ions and cationic dyes) or deprotonate to form cations interacting with anionic pollutants. Apart from the electrostatic interaction, ion exchange is also an important mechanism for the adsorption of pollutants from aqueous solutions. Several studies have indicated that some metal ions can be effectively removed from solutions by the ion exchange mechanism. Other studies have proved that the adsorption of pollutants by NC hydrogels also can take place by several other mechanisms, including hydrogen bonding, coordination bonds, chelating interaction, hydrophobic interactions, or a combination thereof (Van Tran et al., 2018).

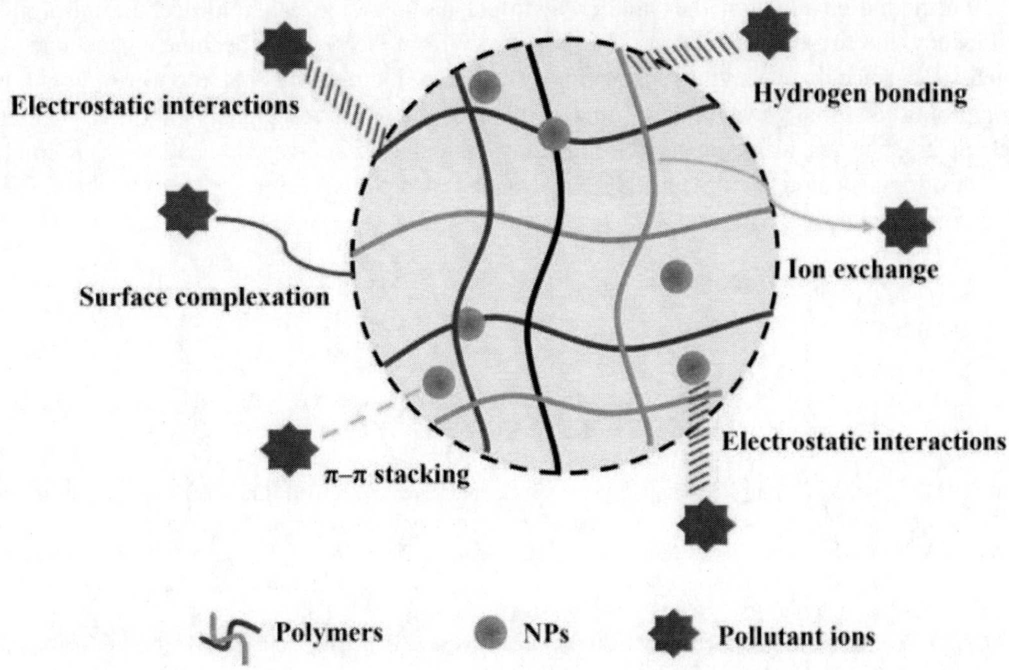

Figure 11.3. Possible adsorption mechanisms of NC hydrogels.

Conclusion

This chapter provides a general overview of the recent progress of NC hydrogels for water contaminant adsorption. Emphases have been placed on the basic materials design principles, the synthesis strategies, and the mechanical properties of NC hydrogels, along with a brief introduction regarding the mechanical characterization methods and the associated mechanics. In general, NC hydrogels have better mechanical properties and adsorption performances compared to their pure parent hydrogels. Despite the significant progresses made by far, further investigations are needed for this emerging and exciting field. For instance, although NC hydrogels have exhibited superior performance in the adsorptive removal of a wide range of aqueous pollutants, their selective adsorption is rarely explored; thus, research efforts are needed to develop hydrogels with desired structures and functionalities possessing higher selectivity toward specific pollutants.

In terms of mechanical performances, the fatigue behaviors of NC hydrogels are rarely probed. Immediate investigations are highly demanded and eventually, fatigue-resistant NC hydrogels for recyclable adsorption should be delivered for the sake of industrial-scale production and implementation. Studies focusing on understanding how the filler materials affect the resulting composite hydrogel properties (intrinsic and adsorption properties) are required. In addition, many NC hydrogels have acquired excellent strength and adsorption properties, but their chemical and biological stability also need to be taken into consideration in wastewater treatment for economic viability and sustainability, which are largely overlooked in current research. These composites' general safety must also be assessed since the leaching of filler materials can increase potential environmental health risks.

Furthermore, although the smaller hydrogel particle size can improve the adsorption efficiency, the subsequent separation and recovery from the solution become a challenge. As such, a balance between the adsorption effect and their facility to recovery should be compromised. Finally, we believe that a dialogue between researchers of different areas (chemistry, physics, mechanics, environmental and materials engineering, and so on) is crucial to developing more efficient NC hydrogel adsorbents to remediate contaminated water/wastewater.

References

Adewunmi, A. A., Ismail, S., and Sultan, A. S. (2016). Carbon nanotubes (CNTs) nanocomposite hydrogels developed for various applications: a critical review. *Journal of Inorganic and Organometallic Polymers and Materials*, 26(4), 717-737.

Afzal, M. Z., Sun, X. F., Liu, J., Song, C., Wang, S. G., and Javed, A. (2018). Enhancement of ciprofloxacin sorption on chitosan/biochar hydrogel beads. *Science of Total Environment*, 639, 560-569.

Ahmad, A., Mohd-Setapar, S. H., Chuong, C. S., Khatoon, A., Wani, W. A., Kumar, R., and Rafatullah, M. (2015). Recent advances in new generation dye removal technologies: novel search for approaches to reprocess wastewater. *RSC Advances*, 5(39), 30801-30818.

Akl, Z. F., Zaki, E. G., and Elsaeed, S. M. (2021). Green hydrogel-biochar composite for enhanced adsorption of uranium. *ACS Omega*, 6(50), 34193-34205.

Ali, R. M., Hamad, H. A., Hussein, M. M., and Malash, G. F. (2016). Potential of using green adsorbent of heavy metal removal from aqueous solutions: adsorption kinetics, isotherm, thermodynamic, mechanism and economic analysis. *Ecological Engineering*, 91, 317-332.

Anjali, J., Jose, V. K., and Lee, J. M. (2019). Carbon-based hydrogels: synthesis and their recent energy applications. *Journal of Materials Chemistry A*, 7(26), 15491-15518.

Azizian, S. (2004). Kinetic models of sorption: a theoretical analysis. *Journal of Colloid Interface Science*, 276(1), 47-52.

Bai, R., Yang, J., Morelle, X. P., Yang, C., and Suo, Z. (2018). Fatigue fracture of self-recovery hydrogels. *ACS Macro Letters*, 7(3), 312-317.

Bai, R., Yang, J., and Suo, Z. (2019). Fatigue of hydrogels. *European Journal of Mechanics-A/Solids*, 74, 337-370.

Chamkouri, H. (2021). A review of hydrogels, their properties and applications in medicine. *American Journal of Biomedical Science & Research*, 11(6), 485-493.

Chen, F., Wang, R., Chen, H., and Lu, H. (2021). Preparation of polyacrylamide/MXene hydrogels as highly-efficient electro-adsorbents for methylene blue removal. *Polymer-Plastics Technology and Materials*, 60(14), 1-17.

Dannert, C., Stokke, B. T., and Dias, R. S. (2019). Nanoparticle-hydrogel composites: from molecular interactions to macroscopic behavior. *Polymers*, 11(2), 275.

Du, C., and Huang, W. (2022). Progress and prospects of nanocomposite hydrogels in bone tissue engineering. *Nanocomposites*, 8(1), 102-124.

Eid, M. (2011). Gamma radiation synthesis and characterization of starch based polyelectrolyte hydrogels loaded silver nanoparticles. *Journal of Inorganic and Organometallic Polymers and Materials*, 21, 297-305.

Filippi, M., Dasen, B., Guerrero, J., Garello, F., Isu, G., Born, G., Ehrbar, M., Martin, I., and Scherberich, A. (2019). Magnetic nanocomposite hydrogels and static magnetic field stimulate the osteoblastic and vasculogenic profile of adipose-derived cells. *Biomaterials*, 223, 119468.

Getachew, B. A., Kim, S. R., and Kim, J. H. (2017). Self-healing hydrogel pore-filled water filtration membranes. *Environmental Science & Technology*, 51, 905-913.

Gupta, V. K., Jain, R., Nayak, A., Agarwal, S., and Shrivastava, M. (2011). Removal of the hazardous dye-tartrazine by photodegradation on titanium dioxide surface. *Materials Science and Engineering: C*, 31, 1062-1067.

Haider, H., Yang, C. H., Zheng, W. J., Yang, J. H., Wang, M. X., Yang, S., Zrínyi, M., Osada, Y., Suo, Z., and Zhang, Q. (2015). Exceptionally tough and notch-insensitive magnetic hydrogels. *Soft Matter*, 11(42), 8253-8261.

Han, I. K., Chung, T., Han, J., and Kim, Y. S. (2019). Nanocomposite hydrogel actuators hybridized with various dimensional nanomaterials for stimuli responsiveness enhancement. *Nano Convergence*, 6(1), 1-21.

Haraguchi, K., Takehisa, T., and Fan, S. (2002). Effects of clay content on the properties of nanocomposite hydrogels composed of Poly(*N*-isopropylacrylamide) and clay. *Macromolecules*, 35(27), 10162-10171.

Hu, X., Nian, G., Liang, X., Wu, L., Yin, T., Lu, H., Qu, S., and Yang, W. (2019). Adhesive tough magnetic hydrogels with high Fe_3O_4 content. *ACS Applied Materials & Interfaces*, 11(10), 10292-10300.

Ju, Y., Zhang, Y., and Zhao, H. (2018). Fabrication of polymer-protein hybrids. *Macromolecular Rapid Communications*, 39(7), 1700737.

Kalcioglu, Z. I., Mahmoodian, R., Hu, Y., Suo, Z., and Vliet, K. J. Van. (2012). From macro- to microscale poroelastic characterization of polymeric hydrogels via indentation. *Soft Matter*, 8, 3393-3398.

Khan, M., and Lo, I. M. C. (2016). A holistic review of hydrogel applications in the adsorptive removal of aqueous pollutants: recent progress, challenges, and perspectives. *Water Research*, 106, 259-271.

Konwar, A., Gogoi, A., and Chowdhury, D. (2015). Magnetic alginate-Fe_3O_4 hydrogel fiber capable of ciprofloxacin hydrochloride adsorption/separation in aqueous solution. *RSC Advances*, 5(99), 81573-81582.

Lake, G., and Thomas, A. (1967). The strength of highly elastic materials. *Proceedings of the Royal Society of London. Series A. Mathematical and Physical Sciences*, 300(1460), 108-119.

Lei, C., Gao, J., Ren, W., Xie, Y., Abdalkarim, S. Y. H., Wang, S., Ni, Q., and Yao, J. (2019). Fabrication of metal-organic frameworks@cellulose aerogels composite materials for removal of heavy metal ions in water. *Carbohydrate Polymers*, 205, 35-41.

Li, H., Fan, J., Shi, Z., Lian, M., Tian, M., and Yin, J. (2015). Preparation and characterization of sulfonated graphene-enhanced poly (vinyl alcohol) composite hydrogel and its application as dye absorbent. *Polymer*, 60, 96-106.

Li, Q., Wang, L., Liu, Q., Hong, W., and Yang, C. (2021). Fatigue damage-resistant physical hydrogel adhesion. *Frontiers in Robotics and AI*. 8, 666343.

Li, X., Dong, S., Yan, H., and Wu, C. (2012). Fabrication and properties of PVA-TiO_2 hydrogel composites. *Procedia Engineering*, 27, 1488-1491.

Liu, J., Lin, S., Liu, X., Qin, Z., Yang, Y., Zang, J., and Zhao, X. (2020). Fatigue-resistant adhesion of hydrogels. *Nature Communications*, 11(1), 1-9.

Liu, J., Yang, C., Yin, T., Wang, Z., Qu, S., and Suo, Z. (2019a). Polyacrylamide hydrogels. II. elastic dissipater. *Journal of the Mechanics and Physics of Solids*, 133, 103737.

Liu, X., Ma, R., Wang, X., Ma, Y., Yang, Y., Zhuang, L., Zhang, S., Jehan, R., Chen, J., and Wang, X. (2019b). Graphene oxide-based materials for efficient removal of heavy metal ions from aqueous solution: a review. *Environmental Pollution*, 252, 62-73.

Lü, C., and Yang, B. (2009). High refractive index organic-inorganic nanocomposites: design, synthesis and application. *Journal of Materials Chemistry*, 19, 2884-2901.

Lu, C. H., Yu, C. H., and Yeh, Y. C. (2021). Engineering nanocomposite hydrogels using dynamic bonds. *Acta Biomaterialia*, 130, 66-79.

Lu, L., Yu, W., Wang, Y., Zhang, K., Zhu, X., Zhang, Y., Wu, Y., Ullah, H., Xiao, X., and Chen, B. (2020). Application of biochar-based materials in environmental remediation: from multi-level structures to specific devices. *Biochar*, 2(1), 1-31.

Ma, Q., Ge, W., Wu, C., Xia, L., Wang, W., and Song, S. (2022). MoS_2 composite hydrogel supported by two-dimensional montmorillonite nanosheets for Pb^{2+} removal from water. *Chemical Physics*, 556, 111477.

Mahdavinia, G. R., Hasanpour, J., Rahmani, Z., Karami, S., and Etemadi, H. (2013). Nanocomposite hydrogel from grafting of acrylamide onto HPMC using sodium montmorillonite nanoclay and removal of crystal violet dye. *Cellulose*, 20(5), 2591-2604.

Manaila, E., Craciun, G., Ighigeanu, D., Cimpeanu, C., Barna, C., and Fugaru, V. (2017). hydrogels synthesized by electron beam irradiation for heavy metal adsorption. *Materials*, 10(5), 540.

Martínez-Huitle, C. A., and Brillas, E. (2009). Decontamination of wastewaters containing synthetic organic dyes by electrochemical methods: a general review. *Applied Catalysis B: Environmental*, 87, 105-145.

Moghaddam, S. S., Moghaddam, M. R., and Arami, M. (2010). Coagulation/flocculation process for dye removal using sludge from water treatment plant: optimization through response surface methodology. *Journal of Hazard Materials*, 175, 651-657.

Pandey, S., Do, J. Y., Kim, J., and Kang, M. (2020). Fast and highly efficient removal of dye from aqueous solution using natural locust bean gum based hydrogels as adsorbent. *International Journal of Biological Macromolecules*, 143, 60-75.

Pandey, S., and Ramontja, J. (2016). Guar gum-grafted poly(acrylonitrile)-templated silica xerogel: nanoengineered material for lead ion removal. *Journal of Analytical Science and Technology*, 7(1), 1-15.

Peak, C. W., Wilker, J. J., and Schmidt, G. (2013). A review on tough and sticky hydrogels. *Colloid and Polymer Science*, 291(9), 2031-2047.

Peng, N., Hu, D., Zeng, J., Li, Y., Liang, L., and Chang, C. (2016). Superabsorbent cellulose-clay nanocomposite hydrogels for highly efficient removal of dye in water. *ACS Sustainable Chemistry & Engineering*, 4(12), 7217-7224.

Ponnamma, D., Cabibihan, J. J., Rajan, M., Pethaiah, S. S., Deshmukh, K., Gogoi, J. P., Pasha, S. K. K., Ahamed, M. B., Krishnegowda, J., Chandrashekar, B. N., Polu, A. R., and Cheng, C. (2019). Synthesis, optimization and applications of ZnO/polymer nanocomposites. *Materials Science & Engineering C-Materials for Biological Applications*, 98, 1210-1240.

Prasad, A. S., Wang, Y., Li, X., Iyer, A., Chen, W., Brinson, L. C., and Schadler, L. S. (2020). Investigating the effect of surface modification on the dispersion process of polymer nanocomposites. *Nanocomposites*, 6(3), 111-124.

Rahman, M. M., and Rimu, S. H. (2020). Recent development in cellulose nanocrystal-based hydrogel for decolouration of methylene blue from aqueous solution: a review. *International Journal of Environmental Analytical Chemistry*, In press, doi: 10.1080/03067319.2020.1817424.

Rezazadeh, B., Sirousazar, M., Abbasi-Chianeh, V., and Kheiri, F. (2020). Polymer-clay nanocomposite hydrogels for molecular irrigation application. *Journal of Applied Polymer Science*,137(18), 48631.

Rivlin, R., and Thomas, A. G. (1953). Rupture of rubber. I. characteristic energy for tearing. *Journal of Polymer Science*, 10(3), 291-318.

Sharma, G., Thakur, B., Naushad, M., Kumar, A., Stadler, F. J., Alfadul, S. M., and Mola, G. T. (2018). Applications of nanocomposite hydrogels for biomedical engineering and environmental protection. *Environmental Chemistry Letters*, 16(1), 113-146.

Simonin, J. P. (2016). On the comparison of pseudo-first order and pseudo-second order rate laws in the modeling of adsorption kinetics. *Chemical Engineering Journal*, 300, 254-263.

Singh, N. B., Nagpal, G., Agrawal, S., and Rachna. (2018). Water purification by using adsorbents: a review. *Environmental Technology & Innovation*, 11, 187-240.

Singh, V., Joung, D., Zhai, L., Das, S., Khondaker, S. I., and Seal, S. (2011). Graphene based materials: past, present and future. *Progress in Materials Science*, 56(8), 1178-1271.

Singhal, R., and Gupta, K. (2016). A review: tailor-made hydrogel structures (classifications and synthesis parameters). *Polymer-Plastics Technology and Engineering*, 55(1), 54-70.

Stoyneva, V., Momekova, D., Kostova, B., and Petrov, P. (2014). Stimuli sensitive super-macroporous cryogels based on photo-crosslinked 2-hydroxyethylcellulose and chitosan. *Carbohydrate Polymers*, 99, 825-830.

Sun, J. Y., Zhao, X., Illeperuma, W. R., Chaudhuri, O., Oh, K. H., Mooney, D. J., Vlassak, J. J., and Suo, Z. (2012). Highly stretchable and tough hydrogels. *Nature*, 489(7414), 133-136.

Tafakori, V., Zadmard, R., Tabandeh, F., Amoozegar, M. A., and Ahmadian, G. (2017). Equilibrium isotherm, kinetic modeling, optimization, and characterization studies of cadmium adsorption by surface-engineered Escherichia coli. *Iranian Biomedical Journal*, 21(6), 380-391.

Van Tran, V., Park, D., and Lee, Y. C. (2018). Hydrogel applications for adsorption of contaminants in water and wastewater treatment. *Environmental Science and Pollution Research*, 25(25), 24569-24599.

Wang, G., Li, G., Huan, Y., Hao, C., and Chen, W. (2020). Acrylic acid functionalized graphene oxide: high-efficient removal of cationic dyes from wastewater and exploration on adsorption mechanism. *Chemosphere*, 261, 127736.

Wang, J., and Guo, X. (2020). Adsorption isotherm models: classification, physical meaning, application and solving method. *Chemosphere*, 258, 127279.

Wang, J., Wang, W., Ai, Z., Li, M., Li, H., Peng, W., Zhao, Y., and Song, S. (2021a). Adsorption toward Pb(II) occurring on three-dimensional reticular-structured montmorillonite hydrogel surface. *Applied Clay Science*, 210, 106153.

Wang, J., Wang, X., Zhao, G., Song, G., Chen, D., Chen, H., Xie, J., Hayat, T., Alsaedi, A., and Wang, X. (2018). Polyvinylpyrrolidone and polyacrylamide intercalated molybdenum disulfide as adsorbents for enhanced removal of chromium(VI) from aqueous solutions. *Chemical Engineering Journal*, 334, 569-578.

Wang, L., Xu, H., Gao, J., Yao, J., and Zhang, Q. (2019). Recent progress in metal-organic frameworks-based hydrogels and aerogels and their applications. *Coordination Chemistry Reviews*, 398, 213016.

Wang, Q., Mynar, J. L., Yoshida, M., Lee, E., Lee, M., Okuro, K., Kinbara, K., and Aida, T. (2010). High-water-content mouldable hydrogels by mixing clay and a dendritic molecular binder. *Nature*, 463(7279), 339-343.

Wang, Y., Yin, T., and Suo, Z. (2021b). Polyacrylamide hydrogels. III. lap shear and peel. *Journal of the Mechanics and Physics of Solids*, 150, 104348.

Webber, R. E., Creton, C., Brown, H. R., and Gong, J. P. (2007). Large strain hysteresis and mullins effect of tough double-network hydrogels. *Macromolecules*, 40(8), 2919-2927.

Weerasundara, L., Gabriele, B., Figoli, A., Ok, Y. S., and Bundschuh, J. (2020). Hydrogels: novel materials for contaminant removal in water-a review. *Critical Reviews in Environmental Science and Technology*, 51(1), 1-45.

Wu, C., Li, C., Zhang, X., Cheng, C., and Wang, J. (2019). An alginate-based hydrogel composite obtained by UV radiation and its release of 5-fluorouracil. *Polymer Bulletin*, 76, 1167-1182.

Wu, Z., Zhang, P., Zhang, H., Li, X., He, Y., Qin, P., and Yang, C. (2022). Tough porous nanocomposite hydrogel for water treatment. *Journal of Hazard Materials*, 421, 126754.

Xiao, Y., Li, Q., Yao, X., Bai, R., Hong, W., and Yang, C. (2022). Fatigue of amorphous hydrogels with dynamic covalent bonds. *Extreme Mechanics Letters*, 53, 101679.

Xu, Y., Sheng, K., Li, C., and Shi, G. (2010). Self-assembled graphene hydrogel via a one-step hydrothermal process. *ACS Nano*, 4(7), 4324-4330.

Yang, C., and Suo, Z. (2018). Hydrogel ionotronics. *Nature Reviews Materials*, 3(6), 125-142.

Yang, C., Yin, T., and Suo, Z. (2019). Polyacrylamide hydrogels. I. network imperfection. *Journal of the Mechanics and Physics of Solids*, 131, 43-55.

Yang, C. H., Wang, M. X., Haider, H., Yang, J. H., Sun, J. Y., Chen, Y. M., Zhou, J., and Suo, Z. (2013). Strengthening alginate/polyacrylamide hydrogels using various multivalent cations. *ACS Applied Materials & Interfaces*, 5(21), 10418-10422.

Yang, J., Bai, R., Chen, B., and Suo, Z. (2020). Hydrogel adhesion: a supramolecular synergy of chemistry, topology, and mechanics. *Advanced Functional Materials*, 30(2), 1901693.

Yang, J., Chen, X., Zhang, J., Wang, Y., Wen, H., and Xie, J. (2021a). Role of chitosan-based hydrogels in pollutants adsorption and freshwater harvesting: a critical review. *International Journal of Biological Macromolecules*, 189, 53-64.

Yang, J., Wang, K., Lv, Z., Li, W., Luo, K., and Cao, Z. (2021b). Facile preparation and dye adsorption performance of poly(N-isopropylacrylamide-co-acrylic acid)/molybdenum disulfide composite hydrogels. *ACS Omega*, 6(42), 28285-28296.

Yang, P., Yu, F., Yang, Z., Zhang, X., and Ma, J. (2022). Graphene oxide modified kappa-carrageenan/sodium alginate double-network hydrogel for effective adsorption of antibiotics in a batch and fixed-bed column system. *Science of Total Environmental*, 837, 155662.

Yao, Y., Wang, C., Na, J., Hossain, M. S. A., Yan, X., Zhang, H., Amin, M. A., Qi, J., Yamauchi, Y., and Li, J. (2021). Macroscopic MOF architectures: effective strategies for practical application in water treatment. *Small*, 18(8), 2104387.

Zhang, W., Hu, J., Yang, H., Suo, Z., and Lu, T. (2021a). Fatigue-resistant adhesion II: swell tolerance. *Extreme Mechanics Letters*, 43, 101182.

Zhang, X., Xiang, J., Hong, Y., and Shen, L. (2022). Recent advances in design strategies of tough hydrogels. *Macromolecular Rapid Communications*, 43(15), 220075.

Zhang, Z. H., Xu, J. Y., and Yang, X. L. (2021b). MXene/sodium alginate gel beads for adsorption of methylene blue. *Materials Chemistry and Physics*, 260, 124123.

Zhao, H., Liu, M., Zhang, Y., Yin, J., and Pei, R. (2020). Nanocomposite hydrogels for tissue engineering applications. *Nanoscale*, 12(28), 14976-14995.

Zhao, L. Z., Zhou, C. H., Wang, J., Tong, D. S., Yu, W. H., and Wang, H. (2015). Recent advances in clay mineral-containing nanocomposite hydrogels. *Soft Matter*, 11(48), 9229-9246.

Zhuang, Y., Kong, Y., Wang, X., and Shi, B. (2019). Novel one step preparation of a 3D alginate based MOF hydrogel for water treatment. *New Journal of Chemistry*, 43(19), 7202-7208.

Chapter 12

Iron-Based Nanocomposite Hydrogels for Dye Removal from Wastewater

Godfred Ohemeng-Boahen[1,*]
Patrick Boakye[1]
Divine Damertey Sewu[2,3]
Yen Adams Sokama-Neuyam[4]
Gati Godwin[1]
Barnabas Asamoah Osei[1]
Osei Akoto Fofie[1]
and Seung Han Woo[2,3]

[1] Department of Chemical Engineering,
Kwame Nkrumah University of Science and Technology, UPO, Kumasi, Ghana
[2] Department of Chemical and Biological Engineering,
Hanbat National University, Yuseong-gu, Daejeon, Republic of Korea
[3] Life Green Technology Co. Ltd., Yuseong-Gu, Daejeon, Republic of Korea
[4] Department of Petroleum Engineering,
Kwame Nkrumah University of Science and Technology, UPO, Kumasi, Ghana

Abstract

Iron-based nanoparticles possessing unique properties are being utilized to enhance the properties of hydrogels used in stripping wastewater off of various dye compounds. In this chapter, several representative iron-based nanocomposite hydrogels employed for dye degradation or removal in the treatment of wastewater have been discussed. The current methods for synthesizing these materials have been highlighted here along with the properties that make them interesting for use in wastewater treatment. The chapter concludes by providing a guide for next-generation research by outlining the limitations related to the preparations and applications of iron-based nanocomposite hydrogels in dye removal from wastewater.

* Corresponding Author's Email: godfred.ohemeng-boahen@knust.edu.gh.

In: Nanocomposite Hydrogels and their Emerging Applications
Editor: Mohammad Sirousazar
ISBN: 979-8-88697-675-5

Keywords: nanocomposites, hydrogels, adsorption, nanoparticles, dye removal

12.1. Introduction

Rapid urbanization and industrialization advancements have led to an increase in the concentration of harmful materials, such as heavy metals and dyes in water bodies beyond acceptable levels (Marcelo et al., 2021). The proliferation of industries like paper, plastics, leather, and textiles over the past few decades has adversely impacted the environment and water bodies due to the quantity of dye-laden wastewater they produce (Belachew and Bekele, 2019). Dyes and pigments make up a significant class of contaminants that affect water quality due to their extensive use in these industries, which obviously changes the aesthetic appeal of waterbodies via coloration. These dyes mostly contain the azo group (-N=N-), which pose serious health dangers to aquatic life and humans when exposed to them. They are also extremely poisonous, mutagenic, and carcinogenic substances (Nizamuddin et al., 2019). Therefore, improper disposal of these hazardous dyes into acquatic and terrestrial habitats can result in serious problems including diminishing the oxygen levels in the water, rise in health problems including hemolysis, hypertension, respiratory diseases, jaundice, organ damage, and tissue necrosis in humans (Sarkar et al., 2020). Due to the potential toxicity and high visibility of small amounts of dyes and pigments in effluents, the effluents containing these dangerous compounds have received much attention in the wake of a rise in their concentration and or quantity in effluents into the environment directly or indirectly (Shi et al., 2017).

Conventional methods for treating water and wastewater include coagulation and flocculation, reverse osmosis, activated sludge, bacterial action, chemical oxidation, ozonation, and physical methods, like membrane filtration, ion exchange, and electrochemical techniques (Mittal et al., 2016). These methods have some disadvantages, including high operational and maintenance costs, residual toxicity, the inability to recover materials, etc. (Ayalew, 2022). Adsorption is one of the most promising and often employed methods for removing contaminants from wastewater because of its simplicity, high efficiency, convenience of use, and accessibility to a variety of adsorbents. Various adsorbents have been employed in the past to remove contaminants from wastewater, but most of them are relatively costly, have poor selectivity, and also produce some residual waste (Fan et al., 2012).

The development of nanomaterials for water purification has been the subject of several investigations, which have been driven by the evident recognition that the world's water sources pose serious threats to public health in the form of pollution (Kumar et al., 2014). Recent years have seen the emergence of nanomaterial-based technologies as viable alternatives to the present water treatment methods, offering cheaper prices, higher efficiency, and the ability to simultaneously fulfill the ever-stringent water quality regulations (Gutierrez et al., 2017). Nanoparticles (NPs) have drawn interest because of their distinctive morphological and physicochemical characteristics, such as their extremely tiny size, shape, and size distribution (Bolade et al., 2020). These useful nanomaterials include bioactive NPs, nanostructured membranes, and redox and catalytically active NPs (Kumar et al., 2014). These nanomaterials, particularly iron-based NPs, are of great interest. They are also referred to as "smart nanomaterials" because they not only share with other NPs the basic benefits of good mechanical, electrical, optical, catalytic, and photonic capabilities, but they also have the extra advantageous property of magnetism. They will quickly cluster in the presence of an external magnetic field, which makes them easy to manipulate, separate, and regenerate (Zhou et al., 2016). However, some challenges must be overcome before they can be utilized. Bare iron-based magnetic NPs are rapidly oxidized, unstable in air or moisture, and prone to aggregating due to their substantial surface area and high surface energy (Wu et al., 2020). To overcome this drawback, research has been done to create high-performance nanomaterials made of iron-based NPs that are contained in micro- or macroscale supports, like silica, carbon, clay, or polymer matrix to produce nanocomposite materials. These supports allow the NPs to retain their reactivity while enhancing the adsorbent properties of the materials they are combined with (Gutierrez et al., 2017). Nanocomposites have the potential to be used in numerous processes, including water and wastewater treatment, by combining the properties of two or more materials with different physicochemical characteristics (Marcelo et al., 2021). Commonly used iron-based NPs include goethite (α-FeOOH), hematite (α-Fe_2O_3), magnetite, maghemite (γ-Fe_2O_3), and magnetite (Fe_3O_4).

Hydrogels are 3D polymeric networks that may be extensively employed in sensing, actuation, tissue engineering, and drug administration due to their high swelling ratio, biocompatibility, and stimulus-responsive qualities (Zhang et al., 2017). In addition to causing the hydrogel network to absorb a lot of water, the inclusion of different ionizable functional groups, like AOH, ACOOH, $ACONH_2$, and ASO_3H affects the hydrogel's permeability, biocompatibility, and mechanical stability (Mohammadi et al., 2021). Despite their numerous benefits, their application is typically restricted by their weak mechanical and response properties. The mechanical characteristics of hydrogels can be greatly enhanced by converting them into composite or nanocomposite forms (Rezazadeh et al., 2020). Nanocomposite hydrogels made from polymers and NPs have additional chemical, physical, biological swelling/de-swelling, and optical capabilities that are not possible for any one material to achieve alone. Consequently, these multifunctional nanocomposite hydrogels may be particularly beneficial in environmental detoxification due to their improved properties, increased functionalities, and stimulus sensitivity (Naushad et al., 2019). Nanocomposite hydrogels have a considerable impact on the parent materials' improvement in mechanical performance. They have a high aspect ratio, which results in a huge surface area. Nanocomposite hydrogels can withstand broadening, twisting, tying, crushing, and other modifications due to their strength (Su and Chen, 2018).

12.2. Techniques for Fabrication of Nanocomposite Hydrogels

The main components of the nanocomposite hydrogel are the polymeric material and the NPs. A hydrogel is a 3D crosslinked polymer with hydrophilic characteristics. The polymer network can be formed by linking existing polymer chains or by first creating the polymer chains from monomers followed by interconnecting the synthesized polymeric materials (Omidian et al., 2005). Hydrophilic monomers are the main constituent of hydrogels prepared by the latter approach. However, hydrophobic monomers are sometimes used in the synthesis to regulate some specific properties of the hydrogel. It is reported that monomers, initiators, and crosslinkers are integral parts of hydrogel synthesis (Ahmed, 2015; Omidian et al., 2005; Peppas and Hoffman, 2020).

Song et al., (2015) reported the disadvantages of hydrogel in terms of its mechanical property and multifunctionality. NPs represent a bridge between bulk materials and structures at the atomic level (Gubin et al., 2005). Incorporating NPs in the hydrogel matrix to fabricate nanocomposite hydrogels is one strategy employed to enhance the physicochemical properties as well as the effectiveness of hydrogels during application (Haraguchi, 2007; Song et al., 2015).

The improved performance of nanocomposite hydrogels is a result of the combined advantages of nanofillers and hydrogel matrices (Song et al., 2015). For instance, physical properties, such as quick response to mechanical, optical, thermal, and electrical stimuli can be conferred uniquely to nanocomposite hydrogels thereby enhancing the applicability of the resulting composite (Odaci et al., 2010). For instance, Ohemeng-Boahen et al., (2020) reported the significantly differing physicochemical properties of the chitosan-goethite multi-membrane composite hydrogel relative to the pristine chitosan hydrogel. When tested for its applicability in wastewater treatment, specifically for anionic dye removal, the chitosan-goethite multi-membrane composite hydrogel showed significant improvement in the adsorptive capacity due to the synergistic effect of the impregnated goethite NPs.

12.3. Precursors for Nanocomposite Hydrogels

Inorganic and polymeric NPs have been used as fillers to reinforce the hydrogel matrix with the aim of improving functionalities. Clay, graphene, metals, and metal oxides as well as compounds such as hydroxyapatite are some of the common inorganic precursors for the synthesis of NPs which are used in the synthesis of nanocomposite hydrogels (Haraguchi, 2007; Song et al., 2015).

12.3.1. Clay-Based Nanocomposite Hydrogels

Attapulgite, laponite, and montmorillonite clays have been utilized for the synthesis of clay-based nanocomposite hydrogels. Song et al., (2015) reported that clay-based nanocomposite hydrogel with characteristics such as enhanced mechanical properties and tunable cell adhesion among a few other properties could be prepared via physical or chemical crosslinking with water-soluble monomers containing amide groups. Monomers, such as N-isopropyl acrylamide, N, N-dimethyl acrylamide, and acrylamide interact adequately with a clay surface. Clay pellets are allowed to act as multifunctional crosslinkers in the formation of polymer/clay networks (Haraguchi, 2007). Clay in the smectite group such as those mentioned above, and their modification can be used as a precursor if it can be swollen and exfoliated in water. The physical properties of clay nanocomposite hydrogels can be controlled by varying the content of clay.

12.3.2. Carbon-Based Nanocomposite Hydrogels

Carbon-based NPs, such as carbon nanotubes, graphene, and nanodiamonds have potential applications. The carbon nanotube is hydrophobic. Therefore, modifying the surface of the carbon nanotube with polar groups, such as hydroxyl and carboxyl enhances its incorporation with the hydrophilic polymers and nanocomposite hydrogels thereby widening its scope of applicability (Rafieian et al., 2019).

12.3.3. Metal-Based Nanocomposite Hydrogels

Several NPs of metals, such as Ag, Au, and especially transition metals, such as Ti, Mn, Fe, and Zn, among others have been used in the production of nanocomposite hydrogels. *In situ* polymerization and *in situ* generation are two common methods employed in the synthesis of nanocomposite hydrogels (Haraguchi, 2007). Gold-based nanocomposite hydrogel is synthesized based on the *in-situ* generation method. This method involves the reduction of the metal salt to NPs and the simultaneous polymerization of difunctional monomers to crosslink the polymer matrix (Jiang et al., 2007; Lee and Tsao, 2010; Odaci et al., 2010). Lee and Tsao (2010) reported the synthesis of silver nanocomposite using the polymerization-reduction method with 1-vinyl-2-pyrrolidone as the precursor solution. The monomer solution was exposed to UV radiation.

12.4. Synthesis of Common Iron-Based NPs

Iron-based NPs have received considerable attention in several areas of human endeavor due to their reported low toxicity, biocompatibility, and chemical inertness. Specifically, in wastewater treatment, iron-based NPs are one of the most studied metal oxides due to their high adsorptive capacity, stability in the aqueous environment, and ease of post-treatment separation via magnetization among others. Additionally, iron oxide NPs generally show shorter equilibrating time and hence have increased adsorption kinetics and increased adsorption of contaminants with an added potential to recover the adsorbed contaminants (Shipley et al., 2013).

The common precursors for the production of iron-based NPs are iron oxides. These iron oxides are grouped into oxide forms, hydroxides, and oxy-hydroxides. The oxide forms consist of $Fe(OH)_3$, $Fe(OH)_2$, and $Fe_5HO_8.4H_2O$ (ferrihydrite). The hydroxide forms consist of hematite (α-Fe_2O_3), magnetite (Fe_3O_4), maghemite (γ-Fe_2O_3), and ε-Fe_2O_3. The oxy-hydroxides on the other hand consist of goethite (α-FeOOH), akaganéite (β-FeOOH), lepidocrocite (γ-FeOOH), δ-FeOOH, and $FeOOH{\cdot}0.4H_2O$ (Aragaw et al., 2021). Studies have shown that advancement in technology has led to increased plausibility of synthesizing the majority of these iron oxide NPs in the laboratory from various iron precursors at relatively lower costs further buttresses this rise in interest.

Several methods used in the synthesis of some iron-based nanocomposites specifically those with magnetic properties have been reported (Aragaw et al., 2021; Biftu et al., 2020; Fatimah et al., 2021). Methods such as co-precipitation, chemical vapor deposition, electrochemical synthesis, solvothermal method, microemulsion, and microwave-assisted methods were among the methods reported. Some of the common iron-based NPs and the method of synthesizing them are discussed hereafter.

12.4.1. Ferrite NPs

Cobalt ferrite ($CoFe_2O_4$) (Karthikeyan et al., 2020) and zinc oxide ferrite ($ZnFe_2O_4$) (Chandrika et al., 2019; Gubin et al., 2005; Kerroum et al., 2019) are synthesized using the co-precipitation method. The co-precipitation method has been reported as the most efficient and straightforward method used in the synthesis of magnetic NPs. The method involves the simultaneous precipitation of aqueous solutions of the magnetic precursor by the addition of a base (Chandrika et al., 2019; Karthikeyan et al., 2020; Kerroum et al., 2019). Dhand et al., (2017) reported synthesizing ferrite NPs employing the chemical vapor disposition (CVD) method. Solutions of the ferrite composite NPs were prepared using a sol-gel process. This was done by stoichiometrically mixing respective metal nitrite with citric acid.

12.4.2. Hematite NPs

Nguyen et al., (2020) reported synthesizing orthoferrite ($HoFeO_3$) employing a co-precipitation method using ethanol as a solvent. A green pathway for the synthesis of hematite NPs has been reported in the literature. The magnetic source, Fe_2O_3, was synthesized *in situ* by dropping the iron solution into cellulose solution. The cellulose acts as a template and a nucleation site leading to the formation of magnetic composites at low temperatures (Yu et al., 2013). Lin et al., (2014) also reported synthesizing hematite NPs using the hydrothermal method. Nano-hematite can also be synthesized by reacting HCl with $FeCl_3$ solution (Chen and Li, 2010).

12.4.3. Magnetite NPs

Radwan et al., (2019) prepared Fe_3O_4 NPs employing a co-precipitation technique using $FeCl_3.6H_2O$ and $FeCl_2.4H_2O$ as precursors. Magnetite NPs can also be synthesized by employing the glycol-assisted co-precipitation method. The glycol acts as a cheap, common, and high boiling solvent (Radoń et al., 2020).

12.4.4. Goethite NPs

Goethite is an antiferromagnetic iron oxyhydroxide mostly synthesized using the co-precipitation method (Burleson and Penn, 2006; Chen and Li, 2010). Chen and Li., (2010) synthesized goethite by adding $Fe_2(SO_4)_3$ to NaOH. Some authors also reported synthesizing goethite NPs using a microwave-assisted method (Nguyen et al., 2017). In this method, $Fe(NO_3)_3.9H_2O$ is dissolved in distilled water, and then NaOH is added to the solution to adjust the pH. The obtained suspension is microwaved at 100°C for about 30 mins.

12.4.5. Iron (III) Oxide-Hydroxide NPs

Zelmanov and Semiat (2013) reported a sol-gel method for synthesizing iron (III) oxide-hydroxide NPs. In their report, iron (III) oxide-hydroxide sol was obtained by hydrolysis and pH changes (Aragaw et al., 2021). Notwithstanding, the use of NPs especially in wastewater treatment plants is still not prevalent to say the least mostly due to the relatively higher cost of post-treatment separation necessary to ensure the sustainability via reusability of the nanosorbents and also to prevent the NPs of the metal oxides from getting into the aquatic habitat and eventually into the food chain. Some common precursors and methods used in the synthesis of hematite and magnetite NPs are summarized in Table 12.1.

Table 12.1. Common precursors and methods used in the synthesis of hematite and magnetite NPs

Magnetic Source	Method of Synthesis	Refs.
Hematite (α-Fe_2O_3)		
$Ni_{0.5}Zn_{0.5}Fe_2O_4$	CVD method	(Mustaffa et al., 2019)
Iron rods	Electrochemical method	(Lucas-Granados et al., 2018)
$FeCl_3.4H_2O$ and $FeCl_3.6H_2O$	Solvothermal method	(Rebekah et al., 2020)
$FeCl_3.6H_2O$	Microemulsion	(Nikolic et al., 2017)
Magnetite (Fe_3O_4)		
$FeCl_3$ and $Fe_3(CO)_{12}$	CVD method	(Vangelista et al., 2012; Wang et al., 2015)
$FeSO_4$ and $(NH_4)_2Fe$	Electrochemical method	(Ansari et al., 2020; Sadeghfar et al., 2018)
$FeCl_3.6H_2O$ and $FeSO_4.7H_2O$	Solvothermal method	(Kaewmanee et al., 2020; Wang et al., 2020)
$FeCl_3.6H_2O$ and $FeCl_2$, $NiCl_2.6H_2O$ and $FeCl_3$	Microemulsion	(Dung et al., 2016; Nikolic et al., 2017; Yu et al., 2018)
$FeCl_3.6H_2O$, $Fe(C_5H_5)_2$, and $FeCl_3$	Microwave-assisted method	(Kumar et al., 2018; Li et al., 2019; Mahmoud et al., 2016)

12.5. Classification and Fabrication of Hydrogels

The physicochemical properties and performance of hydrogels are dependent on the nature and the arrangement of the polymeric material and the method of synthesis as well as the crosslinking agent. As such, it is convenient to generalize the behavior of hydrogels by classifying them based on these parameters (Madduma-Bandarage and Madihally, 2021). Thus, hydrogels can be categorized according to the type of crosslinking, the origin of the constituent polymers, and the charge on polymer chains.

12.5.1. Classification Based on Crosslinking

The peculiar properties of hydrogels, such as sorption capacity, physical integrity, mechanical strength, and unique chemical conformation different from their raw materials result mainly from the crosslinking that exists between the polymer chains (Varaprasad et al., 2017). As such, an important basis for categorizing hydrogels is the type of crosslink junctions in the 3D network. Depending on the type of crosslink junction, hydrogels can be grouped into physically and chemically crosslinked hydrogels. Moreover, since the type of crosslinking depends on the method of synthesis, the synthesis approach, chemical or physical synthesis route, can also be used to categorize hydrogels based on crosslinks.

12.5.1.1. Chemically Crosslinked Hydrogels

In chemically crosslinked hydrogels, permanent linkages exist between different polymer chains via covalent bonding. Chemically crosslinked hydrogels exhibit relatively high mechanical strength and depending on the chemical synthesis route, also demonstrate higher stability compared to physically crosslinked hydrogels (Lu et al., 2018). The different methods for obtaining chemical crosslinks in hydrogels are discussed in the following sections.

- *Synthesis via the reaction of functional groups*: The functional groups, such as amine, carboxylic acid, and hydroxyl groups present in various natural and synthetic polymers can be utilized in the formation of hydrogels by reacting with crosslinkers, such as epichlorohydrin, glutaraldehyde, polyaldehydes, and adipic acid dihydrazide. Glutaraldehyde, for example, has been used as a crosslinking agent in the preparation of chitosan hydrogel beads studied for the adsorption of Cr(VI) ions. The results indicated that the chitosan beads could remove up to 100% of the initial concentration of Cr(VI) ions after only 90 mins (Torres-Badajoz et al., 2022). In another study, the adsorption of Congo red dye by novel chitosan hydrogel prepared by crosslinking with trimellitic anhydride isothiocyanate was investigated. The authors reported an enhanced affinity for the Congo red dye (maximum adsorption capacity of 63.05 mg/g) due to the incorporation of the functional groups of the crosslinker into the same structure (Mohamed et al., 2020). In addition, the complementary reactions involving isocyanate-OH/NH_2 or amine-carboxylic acid functional groups, the click chemistry or Schiff base formation, are usually employed to establish covalent linkages between different polymeric chains (Gennen et al., 2016).
- *Synthesis via grafting*: Polymerization of a monomer on the backbone of a base polymeric unit underlies the preparation of hydrogels by grafting. Grafting introduces new functional entities on the pre-formed polymer which increases the number of sorption sites, hydrophilicity, surface polarity, and in some cases the selectivity toward a specific dye or heavy metal (Aly and El-Bisi, 2018). Using ammonium persulfate as the initiator, Han et al., (2021) grafted phytic acid (PA) groups on carboxymethyl chitosan (CMCS) backbone to derive a composite hydrogel for the adsorption of methyl orange and Congo red dyes from aqueous solution. The researchers identified that the composite hydrogel possessed more surface wrinkles and porous structure due to the interaction between CMCS and PA. Salama (2018) synthesized an anionic hydrogel by grafting poly (3-sulfopropyl methacrylate) onto carboxymethyl cellulose (CMC). Besides having good regenerability, the hydrogel displayed excellent adsorption of methylene blue with a maximum capacity of 1675 mg/g.
- *Synthesis via radical polymerization*: Radical polymerization employs a free radical generating agent, such as tetramethylene-diamine (TEMED), ammonium persulphate (APS), and potassium persulphate (KPS) to obtain chemically crosslinked gels from low molecular weight monomers in the presence of a crosslinker (Varaprasad et al., 2017). Redox reaction, light, temperature, or ultraviolet (UV) or gamma radiation decomposes the radical-generating chemical which initiates the formation and propagation of the polymeric chains which subsequently form the 3D gel network. Fabrication of hydrogels by this method is rapid even under mild conditions (Laftah et al., 2011). Hu et al., (2018) used APS and TEMED as initiators to prepare polyacrylic

acid-based super-adsorbent Nanocomposite hydrogel via free radical *in situ* polymerization which showed excellent removal of methylene blue dye with an adsorption capacity up to 2100 mg/g. In another study, Varaprasad et al., (2018) synthesized CMC/acrylamide/nano-hydroxyapatite composite hydrogels via free radical polymerization with APS as an initiator and N, N′-methylene-bis-acrylamide as a crosslinking agent. The composite hydrogel showed a high potential for the removal of acid blue dye with an adsorption capacity of up to 29.52 mg/g.

- *Synthesis via high energy irradiation*: Chemically crosslinked hydrogels can also be synthesized by polymerization of unsaturated compounds using ionizing high-energy radiation, such as gamma, UV, and electron beam radiations. Exposure to high-energy radiation leads to the formation of radicals on polymer chains by homolytic scission. Hydroxyl groups from cleaved water molecules also attack polymer chains hence creating microradicals. The radicals and microradicals recombine on several chains, forming covalent bonds that result in a crosslinked structure. The dosage and duration of the irradiation determine the density of linkages (Sinha and Chakma, 2019). This method is efficient even under mild conditions and no waste results from the process since crosslinkers or chemical initiators are not used (Laftah et al., 2011). Moghaddam et al., (2019) synthesized a dopamine-pectin hydrogel using electron beam irradiation at a dose of 15-55 kGy. They examined the ability of the hydrogel to remove methylene blue from aqueous solutions and found that the hydrogel had a maximum sorption capacity of 111.1 mg/g for the gel material. The authors reported an even greater sorption capacity of 186.7 mg/g for the same hydrogel modified with tannic acid, prepared by the same method.

12.5.1.2. Physically Crosslinked Hydrogels

Hydrogels prepared by physical crosslinking exhibit non-covalent bonding among polymeric chains. Physical interactions, such as hydrogen bonding, ionic interaction, stereo-complex formation, or hydrophobic interaction are responsible for the mechanical strength of these materials (Ullah et al., 2015). Some of the synthesis techniques reported in the literature for obtaining physically crosslinked hydrogels include ionic interaction and crystallization.

- *Synthesis via ionic interaction*: Physical linkages can be achieved between charge-bearing polymer chains by the addition of multivalent ions of opposite charge. Different researchers have prepared hydrogels this way from polymers, including alginate, chitosan, and poly[di(carboxylatophenoxy)phosphazene] via interaction with various ionic solutions, including calcium (Varaprasad et al., 2017), glycerol-phosphate (Zhao et al., 2009), and sodium dodecyl sulfate (SDS). For example, Chatterjee et al., (2010) prepared chitosan hydrogel beads by dropwise addition of polyelectrolyte chitosan solution into aqueous SDS. They ensured consistency in the concentration of SDS in each gel material by keeping the SDS solution in excess of the chitosan drops throughout each gelation process.
- *Synthesis via crystallization*: Hydrogel synthesis by repeated freezing and thawing is another method of forming physical crosslinks as a result of the formation of crystallite in the gel structure. In particular, polyvinyl alcohol (PVA) hydrogels have commonly been prepared by this method due to the peculiar property of crystallite formation in

PVA solutions after repeated freezing and thawing cycles (Hassan and Peppas, 2000). These hydrogels have found extensive use in biomedical and pharmaceutical applications owing to their high porosity, high degree of swelling in water, high mechanical strength, rubbery, spongy, higher elastic, and non-toxic nature compared to PVA hydrogels synthesized by other methods (Abdel-Mohsen et al., 2011; Hassan and Peppas, 2000; Yang et al., 2008). The blending of PVA and other polymeric materials has extended the application of hydrogels prepared by the freeze-thaw method to other fields. For example, Wang and Wang (2016) demonstrated the potential of composite PVA/CMC hydrogels prepared by the freeze-thaw method for the removal of heavy metal ions (Ag^+, Ni^{2+}, Cu^{2+}, and Zn^{2+}). They found that the synthesized composite hydrogel material was particularly selective toward the Ag^+ ion and could adsorb up to 8.4 mg of the ion per gram of the hydrogel.

12.5.2. Classification Based on Material Resources

Based on the source of the constituent polymeric materials, hydrogels can be grouped into natural or synthetic hydrogels. Polysaccharides, such as starch, alginate, chitosan, and agarose as well as proteins, such as gelatin and collagen are examples of natural polymers that are used to prepare hydrogels (Ahmed, 2015). Chemical polymerization techniques are typically used to prepare synthetic polymers that are used for developing hydrogels. Synthetic polymers used in the synthesis of hydrogels include PVA, polyethylene oxide (PEO), and polyvinyl pyrrolidone (PVP) (Sionkowska, 2011). Natural hydrogels have the advantage of biodegradability, low cytotoxicity, and biocompatibility, and can be prepared at a lower cost, especially for hydrogels prepared from relatively abundant polymers (Catoira et al., 2019). However, natural hydrogels usually possess poor mechanical strength, and their properties are difficult to control due to potential batch-to-batch variations. It is therefore common to combine synthetic polymers, which confer unique mechanical and structural properties, with natural polymers to form composite hydrogels (Jabbari et al., 2016; Tang et al., 2011).

12.5.3. Classification Based on Charge

According to the type of electrical charge present on the crosslinked chains, hydrogels are categorized into four groups, including nonionic (neutral), ionic (cationic or anionic), ampholytic (containing both acidic and basic groups), and zwitterionic (containing both anionic and cationic groups on constituent monomers) (Ahmed, 2015). In the context of wastewater treatment, the type of hydrogel based on the charge indicates the type of pollutants to which the hydrogel can be applied. For instance, an ampholytic pullulan-graft-polyacrylamide porous hydrogel has been developed by Saber-Samandari et al., (2014) that could interact with both cationic and anionic dyes. The authors reported respective maximum adsorption capacities of 2728 and 1822 mg/g for methylene blue and reactive blue 2 as the representative cationic and anionic dyes, respectively.

12.6. Modification of Hydrogels to Enhance Adsorption Characteristics

Hydrogels have found wide applicability in several areas from food and beverages, to cosmetics, pharmaceuticals, and biomedicine. Particularly, in wastewater treatment, hydrogels and their composite materials have proven to have superior adsorption capacity compared to other adsorbents due to the presence of several functional groups, such as $CONH_2$, OH-, NH_2, and COOH just to mention a few in and on both the internal and external surface areas ensuring enhanced interaction between the adsorbate and the hydrogel (Ozay et al., 2009). Traditional hydrogels synthesized from synthetic and natural sources, such as chitosan, alginate, cellulose, pectin, starch, gum acacia, and guar gum have seen prodigious interest in the purification of wastewater due to their remarkable sorption capacity, hydrophilicity, swelling ratio, and versatility in fabrication (Naushad et al., 2019; Shalla et al., 2018). Even so, their poor mechanical strength, narrow effective pH range, and low thermal stability limit their efficient application, especially in the industrialization of these technologies (Adewunmi et al., 2016). As such, different researchers have explored various modification techniques to synthesize hydrogels with improved mechanical properties, wide thermal and pH stability, less crystallinity, improved hydrophilic properties, large surface area and porosity, and increased specificity to enhance their adsorption efficiency for specific contaminants. The different modification techniques that have been employed in the literature to synthesize hydrogels with enhanced properties include functionalization, surface oxidation, impregnation of metal/metal oxides, and inorganic and organic nanocomposites (Sinha and Chakma, 2019).

In recent times, incorporating NPs into hydrogel networks has been extensively studied, especially for wastewater purification, since these materials possess a large surface area to volume ratio, large adsorption capacity, and great potential for removing specific target pollutants (Xu et al., 2012). The types of nanomaterials that have been explored for this purpose are nanosized polymeric, inorganic/ceramic e.g., silica and silicates (Haraguchi, 2007), carbon-based e.g., carbon nanotubes (Adewunmi et al., 2016), and metal/metal oxide e.g., silver and titanium oxide (Madduma-Bandarage and Madihally, 2021; Monerris et al., 2019) materials. Immobilization of these particles onto gel matrices is also advantageous because it allows for easy regeneration of the NPs which without the gel platforms are rather difficult to recover (Tesh and Scott, 2014). Consequently, nanocomposite hydrogels combine the important properties of constituents for improved utility in wastewater treatment (Wang et al., 2018). In particular, iron oxide NPs, such as maghemite, goethite, and hematite have drawn a lot of interest because of their large specific surface area, high catalytic activity, low toxicity, chemical inertness, biocompatibility, high surface hydroxyl content, wide operating pH range, and the prospect of being synthesized from a variety of iron precursors at a reasonably low cost (Afkhami and Moosavi, 2010; Y. Wang et al., 2015; Xu et al., 2012). Iron oxide NPs also possess magnetic properties which allow for easy separation from aqueous solutions by the application of an external magnetic field (Alvarez-Torrellas et al., 2019).

Even though iron oxide NPs possess the aforementioned characteristics, their effective impregnation and dispersion in polymeric gel networks are crucial in realizing the proposed benefits in prepared nanocomposite hydrogels. Various approaches have been employed by different researchers to synthesize hydrogels impregnated with iron oxide NPs a summary of which is presented in Table 12.2.

Table 12.2. Summary of some synthesis methods used for the preparation of iron oxide-based nanocomposite hydrogels

Hydrogel	Preparation Method	Adsorbate	Refs.
N-vinylpyrrolidon/chitosan/Fe_3O_4	In situ co-precipitation of iron salts within CS/NVP hydrogel network or gel matrix	Methyl orange	(Abou Taleb et al., 2021)
Kappa-iota-carrageenan/maghemite Nanocomposite hydrogel	Crosslinking of carrageenan and maghemite mixture with KCl	Cd(II) and Pb(II)	(Maciel et al., 2019)
Fe_3O_4@SiO_2 starch-graft-polyacrylic acid nanocomposite hydrogel	Grafting of acrylic acid and methylenebisacrylamide (MBA) on starch in solution with silica-coated magnetite (Fe_3O_4@SiO_2) initiated by APS	Crystal violet	(Pourjavadi et al., 2013)
Magnetic kappa-carrageenan/ PVA nanocomposite hydrogel	In situ co-precipitation of iron salts within a binary mixture of kappa-carrageenan and PVA followed by crosslinking via freeze-thawing and subsequent treatment with KCl	Crystal violet	(Mahdavinia et al., 2014)
Chitosan/hematite nanocomposite hydrogel capsule	Crosslinking of chitosan/hematite mixture with SDS	Congo red	(Ohemeng-Boahen et al., 2021)
Gum arabic-crosslinked-polyacrylamide/Ni(OH)$_2$/FeOOH nanocomposite hydrogel	Copolymerization of gum-arabic and acryl amide suspended with Ni(OH)$_2$/FeOOH particles initiated by APS and crosslinked with MBA	Methylene blue	(Naushad et al., 2019)
Goethite impregnated chitosan Nanocomposite hydrogel	Anionic surfactant gelation of goethite-mixed-chitosan solution with SDS	Congo red	(Chatterjee et al., 2022)
Goethite-poly(acrylamide) nanocomposite hydrogel	In situ free radical polymerization of goethite/acrylamide mixture with N,N'-methylenebis(acrylamide)	Arsenic	(Ramirez-Muñiz et al., 2018)
Chitosan/polypropenoic acid/ethylenediamine/magnetite nanocomposite hydrogel	Gamma irradiation of chitosan/polypropenoic acid/ ethylenediamine/magnetite-NPs mixture	Astrazon blue and Lerui acid brilliant blue	(Ali et al., 2022)
κ-carrageenan/poly(acrylamide)/ Fe_3O_4 nanocomposite hydrogel	Grafting of acrylamide monomer onto κ-carrageenan in the presence of magnetic laponite RD and MBA crosslinker	Crystal violet	(Mahdavinia et al., 2016)
Pectin-graft-poly(N-hydroxyethyl acrylamide)/magnetite (Pec-g-PHEAA/Fe_3O_4)	In situ co-precipitation of iron salts within gel network	Rhodamine 6G	(Kulal and Badalamoole, 2020)
Magnetic PVA/laponite RD nanocomposite hydrogel	Dispersion of magnetic laponite in PVA solution, followed by crosslinking via the freezing-thawing method	Methylene blue	(Mahdavinia et al., 2017)
Magnetic alginate functionalized multiwalled carbon nanotubes beads (A-FMWCNT-Fe_2O_3)	Gelation of dispersion of citrate-coated maghemite, alginate, and FMWCNT with CaCl2 as a crosslinker	Methylene blue	(Boukhalfa et al., 2019)
Poly(vinyl acetate-co-maleic anhydride)/magnetite/melamine nanocomposite hydrogel (PVAcMAn/Fe_3O_4/Me)	Copolymerization of maleic anhydride and vinyl acetate followed by addition of Fe_3O_4 NPs dispersion and subsequent addition of melamine solution	Congo red	(Rezazadeh et al., 2020)

Generally, iron oxide NPs are integrated into hydrogels by three approaches, including the addition of NPs to pre-formed hydrogels, the addition of NPs to polymer solutions, and the polymerization of monomers in the presence of NPs (Dannert et al., 2019). The addition of NPs to pre-formed hydrogels can be achieved either by *in situ* formation of NPs in a pre-formed polymeric network or by blending pre-synthesized NPs into a hydrogel network. Similarly, the method of adding NPs to polymer solutions may involve mixing pre-synthesized NPs with polymeric chains before polymer gelation or blending iron salts with polymeric chains for the simultaneous formation of NPs and crosslinks within the hydrogel matrix (Dannert et al., 2019).

In a study by Abou Taleb et al., (2021), N-vinylpyrrolidone/chitosan/Fe_3O_4 (NVP/CS/Fe_3O_4) nano-adsorbent was synthesized by the *in situ* co-precipitation of iron salts in the NVP/CS hydrogel matrix. With methyl orange as a model dye, the authors observed a higher adsorption capacity by NVP/CS/Fe_3O_4 nanocomposite compared to NVP/CS. They attributed this improved performance to the increased surface area in NVP/CS/Fe_3O_4 hydrogel with the impregnation of Fe_3O_4 NPs. They also noticed that even after trapping the Fe_3O_4 NPs in the hydrogel network, it still maintained its magnetic properties and contributed to the excellent reusability of the nanocomposite hydrogel, which was reported to be 85% of the original capacity after the third cycle.

In another study, Ohemeng-Boahen et al., (2021) prepared chitosan/hematite nanocomposite hydrogel capsules by the anionic surfactant gelation of a homogenized mixture of chitosan and hematite (α-Fe_2O_3) NPs. The researchers applied the developed nanocomposite hydrogel to remove Congo red from an aqueous solution. They reported a massively enhanced maximum adsorption capacity of 4705.6 mg/g compared to the 176.5 mg/g by the pristine chitosan hydrogel capsule. Pu et al., (2017) developed highly porous magnetic chitosan hydrogel beads via *in situ* co-precipitation and simultaneous crosslinking by the addition of Fe^{3+}/Fe^{2+}/chitosan mixture into a sodium hydroxide/sodium citrate. The high maximum sorption capacity of 84.02 mg/g for Pb^{2+} was linked to the amino, hydroxyl, and carboxyl groups in the magnetic chitosan hydrogel beads.

The thermogravimetric analysis performed by the researchers also indicated that the incorporation of the Fe_3O_4 NPs improved the thermal stability of the nano-adsorbent. Ramirez-Muñiz et al., (2018) synthesized goethite/ polyacrylamide hydrogel via *in situ* free radical polymerization of acrylamide solution impregnated with goethite (FeOOH) particles. The goethite/polyacrylamide composite hydrogel examined for the removal of arsenic had an adsorption capacity of 1.22 mg/g. Goethite was found to be immobilized in polyacrylamide through hydrogen bonding with the hydroxyl and amine groups of polyacrylamide, as well as by acid and basic Lewis interactions.

12.7. Characterization of Iron-Based Nanocomposite Hydrogels

Following production or synthesis, the characterization of the iron-based nanocomposite hydrogels is paramount to properly assess their attributes or properties and compare with a set performance/market standard. A thorough characterization of the iron-based nanocomposite hydrogels can enable researchers confidently elucidate the governing adsorption mechanisms for dye removal or otherwise from aqueous solution. This inadvertently informs the researcher on what properties to tweak for an intended end-use. Properties relevant to its performance in the field of water/wastewater treatment may be physical, chemical, mechanical, thermal, and magnetic. These may include properties such as particle and pore size distributions, surface area, mechanical properties (e.g., tensile strength), magnetism, functional groups, surface charges, morphology, and both bulk and surface elemental makeup (Sarkar et al., 2020; Sharma et al., 2019). The aforementioned properties are often studied with both rudimentary gravimetric and/or state-of-the-art approaches using diverse equipment.

12.7.1. Physical Characterization

Morphological and topographical studies of iron-based nanocomposite hydrogels reveal structural features normally within a scale inaccessible to the human eye. As such, microscopic analysis with equipment like SEM, transmission electron microscopy (TEM), and atomic force microscopy (AFM) have been predominantly utilized. For more clarity in image capture, the higher version of SEM, i.e., field emission SEM (FESEM) and higher resolution TEM (HRTEM) have been employed (Sarkar et al., 2020). Consequently, morphologies, such as degree of surface roughness, blocked and/or accessible pores, position, shape, and size of the nanocomposite additives within the hydrogel as well as the uniform distribution or otherwise of the nanocomposite can be assessed (Sharma et al., 2017).

It is noteworthy that occasionally, it is difficult to analyze the surface of nanocomposite hydrogels as it is and, therefore, requires coating with either carbon (Fosso-Kankeu et al., 2016) or gold (Pourjavadi et al., 2013) to obtain clearer microscopy images. Nevertheless, Saberi et al., (2019) analyzed the surface of magnetite (Fe_3O_4)-based starch-polyacrylic acid nanocomposite hydrogel with SEM and noticed cluster porosity than the pristine starch-polyacrylic acid bioadsorbent. With TEM analysis, the authors clearly showed a uniform distribution of ca. 70 nm spherical Fe_3O_4 NPs within the hydrogel matrix. By undertaking AFM analysis, 2D and 3D images of the topography of the sample's surface were generated, which undoubtedly showed and confirmed a rougher surface with Fe_3O_4 NPs incorporation into the hydrogel and vice versa without Fe_3O_4 NPs incorporation, suggesting a successful synthesis pathway. Fosso-Kankeu et al., (2016), reported ellipsoidal and spherical Fe_3O_4 NPs with mean diameters of 60 nm (coated sample) and 35 nm (uncoated sample) distributed within the polymer matrix by employing TEM on both coated and uncoated Fe_3O_4-based nanocomposite hydrogel.

The textural properties, particularly surface area and porosity (e.g., pore size distribution) are amongst the key determiners for higher adsorption performance of iron-based nanocomposite hydrogels. Thus, predominantly, the Brunauer-Emmett-Teller (BET) and Barrett-Joyner-Halenda (BJH) analyses (which determines BET surface area (ABET), pore volume, and pore size via adsorption-desorption of mostly nitrogen gas) and mercury porosimetry measurements are employed to analyze these properties (Fosso-Kankeu et al., 2016). For ABET, however, prior to the analysis, thorough water desorption is undertaken via degassing at a temperature below the decomposition temperature of the iron-based nanocomposite hydrogels for a set time. Following this, the sample can then be exposed to varying pressures of nitrogen to yield adsorption-desorption hysteresis curves for the ABET determination. Hysteresis curves of a particular shape are characteristic of adsorbents with certain properties as compiled and demonstrated by the IUPAC nomenclature (Thommes et al., 2015). As a case in point, Sarkar et al., (2020) performed a BET adsorption isotherm measurement on Fe_3O_4-based PVA/CMC-boric acid@graphene oxide nanocomposite hydrogel. They obtained a type V isotherm, but with a type II pattern, when the low-pressure region was uplifted. These indicated the co-existence of macro and mesopores within the nanocomposite hydrogel. Similarly, Fosso-Kankeu et al., (2016) performed nitrogen adsorption-desorption measurements on polymer samples with and without Fe_3O_4 NP incorporation. The authors reported an increase in all textural parameters of ABET, pore volume, and pore diameter upon Fe_3O_4 NP incorporation from 0.4658 m^2/g, 0.002458 cm^3/g, and 183.6915 Å to 0.9845 m^2/g, 0.008473 cm^3/g, and 216.7254 Å, respectively.

12.7.2. Chemical Characterization

Without a thorough chemical characterization, knowledge of the surface chemistry due to the appearance or disappearance of unique chemical structures arising from the synthesis process may only remain in the realm of guesses. In addition, essential elucidation of the adsorption mechanism which is helpful in fine-tuning the synthesis process to enhance target specificity and selectivity for targeted adsorption applications may elude researchers. Since hydrogels have organic polymeric compounds in the gel matrix (Ohemeng-Boahen et al., 2020), researchers usually perform organic functional group analysis employing Fourier transform and infrared spectroscopy (FTIR) instrument. Before the analysis, the hydrogel sample is thoroughly mixed with KBr and hydraulically pressed into thin disks before insertion into the FTIR instrument. A scanning resolution of 4 cm^{-1} within a range of 400 to 4000 cm^{-1} wavelengths has often been practiced by many researchers (Sharma et al., 2019). That notwithstanding, a range of 500 to 4000 cm^{-1} has also been reported (Sarkar et al., 2020). It is worth mentioning that, by analyzing the shifts in wavenumbers and peak intensity of the spectra for the iron-based nanocomposite hydrogel adsorbent before and after adsorption, one can satisfactorily narrow down the functional groups involved in the adsorption process (Sewu et al., 2021). In addition, the success of a pretreatment approach for nanocomposite hydrogel synthesis can be evaluated using FTIR. For instance, after performing FTIR on both leached and pristine nanocomposite hydrogel samples, Sarkar et al., (2020) demonstrated that the former sample was successfully leached of polyethylene glycol (PEG) using water medium for 5 h since the distinct band corresponding to PEG was absent in the latter sample. In addition, the presence of the Fe_3O_4 NP in both samples was confirmed by the detection of spectral bands at 530 cm^{-1} assigned to the stretching vibration of the Fe-O bond.

Elemental analysis can be performed qualitatively and quantitatively for the bulk sample and/or specific region of interest on the sample's surface. In the case of the former, elemental analyzers that focus on the total carbon (C), hydrogen (H), nitrogen (N), sulphur (S), and oxygen (O) have been the norm. On the latter, however, equipment, such as energy dispersive X-ray spectroscopy (EDS/EDX) and X-ray photoelectron spectroscopy (XPS) have been employed. These are normally used complementarily. EDS is found coupled with SEM equipment for simultaneous image capture and surface elemental analysis. EDS can even be used for mapping specific elements to ascertain their distribution on the surface. For example, via EDS, Sharma et al., (2019) reported surface detection of C, O, and Fe on zerovalent iron-based nanocomposite hydrogel. The authors went a step further to color-map the aforementioned elements on the surface to determine their distribution. Evidently, it was apparent how well or not the various elements were distributed across that specific portion of the surface of the sample. Since the bulk sample elemental analyzer is handicapped in measuring other elements besides C, H, N, S, and O, other elements, particularly, metals can be accurately determined in the bulk sample of iron-based nanocomposite hydrogels via using inductively coupled plasma-absorption/optical emission spectroscopy (ICP-AES/ICP-OES) or its variance. For instance, Ohemeng-Boahen et al., (2021) employed ICP-OES to qualitatively and quantitatively measure the Fe and Na amounts in the nanocomposite hydrogel since the synthesis required the usage of varying amounts of hematite (α-Fe_2O_3) NP and a fixed amount of sodium dodecyl sulphate (SDS). The results clearly showed an increment in Fe with an increase in α-Fe_2O_3 in the nanocomposite hydrogels, which proved the successful inculcation of α-Fe_2O_3 into the hydrogel.

Also, despite the utility of FTIR in assessing the functional groups present in nanocomposite hydrogels, it is handicapped in determining or analyzing the bond strengths or energies of the chemical bonds. As such, equipment such as XPS is predominantly used for such a purpose. Generally, following a general survey of the sample's surface to ascertain the elemental composition, deconvolution of the elemental peak of interest can reveal the plethora of chemical bond energies that the element is involved in. With this characteristic bond energy, specificity regarding the exact chemical bond can be gleaned. Sharma et al., (2019) used this technique in analyzing the elemental composition as well as the bond energies on the surface of a zerovalent iron-based nanocomposite hydrogel. The researchers performed a general survey and deconvolutions of the selected elements (Fe, C, N, P, and O). Atomic concentrations of 80.11% (C1s), 2.08% (N1s), 16.52% (O1s), 1.18% (P2p), and 0.12% (Fe2p) were obtained. Following deconvolution of the N1s spectrum, Fe bonded to nitrogen oxide was observed for the 404.58-408.32 eV peak. In addition, the deconvoluted Fe2p spectrum showed peaks for ZVI at 705.17 and 708.16 eV.

The determination of inorganic compounds and/or their crystalline structure or phases could be conducted with X-ray diffractometry (XRD) (Nakhjiri et al., 2019). Thermodynamically, some other iron polymorphs are sometimes bound to precipitate alongside the polymorph of interest during the synthesis process (Schwertmann and Cornell, 2008). Consequently, it is always advisable to perform XRD analysis to verify which polymorphs are present and confirm the success of the synthesis route. Singh and coworkers (2021) detected the amorphous nature of their hydrogel (typical of polymers) and also concluded the absence of other magnetic phases like α-Fe_2O_3 or maghemite (γ-Fe_2O_3) within their nanocomposite hydrogel. Thus, demonstrating the success of their synthesis route, the hybridization process not having precipitated any phase changes of Fe_3O_4 and the preservation of the intended iron polymorph (Fe_3O_4). Note that the authors compared the diffraction peaks of Fe_3O_4 to standard reference data for cubic Fe_3O_4. In another research, Nakhjiri et al., (2019) performed the XRD analysis of Fe_3O_4, SiO_2, and Fe_3O_4@SiO_2/collagen nanocomposite hydrogel (MNH) using Cu-Kα radiation. The authors detected and matched peaks characteristic of both Fe_3O_4 (2θ=30, 34, 43, 52, and 61) and SiO_2 (2θ=24) within MNH. By virtue of the characteristic peaks, they also determined that the Fe_3O_4 had a face-centered cubic crystalline structure.

Raman spectroscopy can be used to analyze the chemical configuration of iron-based nanocomposite hydrogels and to ascertain the degree of crystallinity of a sample (Singh et al., 2021). As such, it can somewhat complement the FTIR and XRD techniques. To cite an instance, Singh and coworkers (2021) obtained Raman data after irradiating samples with a 633 nm laser. The results showed characteristic peaks at 350 cm^{-1} (E_g), 500 cm^{-1} (T_2g), and 700 cm^{-1} (A_1g), which were indicative of a Fe_3O_4 phase. These peaks were also observed in the spectra of the nanocomposite hydrogels thus confirming the XRD results. Interestingly, via the Raman spectral analysis of the Fe_3O_4-based nanocomposite hydrogels, the authors observed the appearance of peaks at 1431 and 1597 cm^{-1} assigned to the formation of COO^- and Fe complex; which was a clear indication of Fe_3O_4 binding to the polymeric network of the hydrogel and thus, a successful synthesis route.

The surface charge or electrical state determines which class of pollutants the synthesized iron-based nanocomposite hydrogel will be suited. For example, a net positive surface charge of the hydrogel will be well suited for anionic water pollutants and vice versa for cationic pollutants. The predominant approach to determining the electrical state has been through determining either the point of zero charge (PZC) or the isoelectric point (IEP). Unfortunately, many researchers have applied these terms loosely, wrongly or interchangeably; refer to (Tran et al., 2017) for more details. Simply, PZC measures the solution conditions under which the surface charge density equals zero and thus varies in response to the net total (external and internal) surface charge of the adsorbent (Tran et al., 2017); and is determined via potentiometric titrations and the pH drift method. Conversely, IEP occurs when the electrokinetic potential at the shear plane equals zero and thus the obtained value clearly represents only the external surface charge of particles in solution (Tran et al., 2017); and is determined via zeta potential. That notwithstanding, Sharma et al., (2019) determined the PZC of ZVI-based nanocomposite hydrogel by the pH drift method at pH=8. Thus, below pH=8, the adsorbent is positive and vice versa when above it. Consequently, the authors applied the hydrogel for methyl violet removal from the aqueous solution, and expectedly a methyl violet (pKa = 0.8) solution with a pH of 7 yielded the highest removal performance. This is because in a solution with a pH of 7, the ZVI-based nanocomposite hydrogel surface will be positive whereas that of methyl violet will be negative. Other researchers have also successfully employed the pH drift method to determine PZC (Malatji et al., 2020).

12.7.3. Magnetic Characterization

Many research employing iron-based nanocomposite hydrogels for water treatment applications utilized Fe_3O_4 NPs. Others also employed iron (nitrate salts) (Sharma et al., 2017), α-Fe_2O_3 (Ohemeng-Boahen et al., 2021), and ZVI (Sharma et al., 2019) NPs in the synthesis process. Except for α-Fe_2O_3, inculcation of the aforementioned iron-based NPs generally imparted magnetic attributes to the as-produced nanocomposite hydrogel for easy post-adsorption separation. Hence the need to determine if the saturation magnetization of the hydrogel would facilitate its practical adoption in wastewater treatment. To this end, the vibrating sample magnetometer (VSM) has been the go-to instrument. VSM exposes the sample to different magnetic field intensities/strengths and records the magnetization at a particular temperature thus yielding a hysteresis loop from which the saturation magnetization (M_s) can be determined. Figure 12.1 depicts the typical shape of a hysteresis loop for Ms determination.

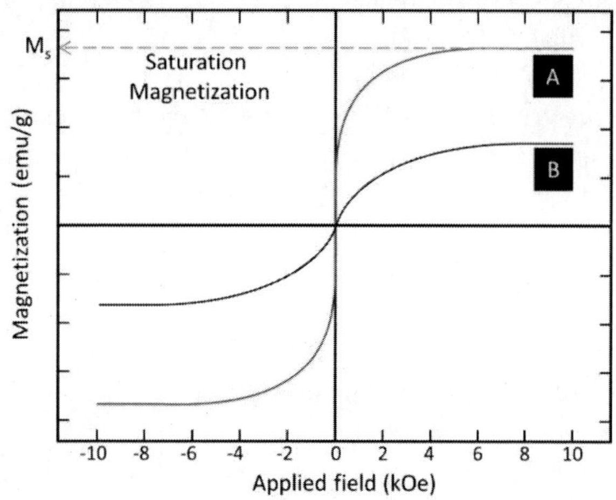

Figure 12.1. A typical shape of the magnetization vs applied magnetic field output of VSM analysis for iron-based nanocomposite hydrogels.

To illustrate, an applied magnetic field ranging from -10,000 to +10,000 Oe was applied to the Fe_3O_4-based starch-polyacrylic acid nanocomposite hydrogel sample at room temperature conditions by Saberi et al., (2019) and reported the Ms of 22.14 emu/g. For nanocomposite hydrogels with Fe_3O_4, the Ms is lower than the standalone M_s for Fe_3O_4, albeit still applicable to external magnetic separation. For instance, Sarkar et al., (2020) reported a drastic decline in M_s from 45.80 emu/g (nano-Fe_3O_4) to 20.55 emu/g, 15.50 emu/g and 11.91 emu/g following incorporation into different gel networks for nanocomposite hydrogel synthesis. The authors eventually proved and thus concluded that the resulting M_s were adequate for magnetic separation post-adsorption of the nanocomposite hydrogel adsorbent.

12.7.4. Mechanical Characterization

One attribute that hampers the large-scale application of hydrogels as adsorbents for water treatment is its poor mechanical properties (Malatji et al., 2020; Peng et al., 2016). Consequently, researchers such as Malatji et al., (2020) analyzed the mechanical stability of their samples employing a dynamic mechanical analysis instrument with a bending mode of a single-cantilever. At a fixed temperature of 30°C, a strain amplitude of 0.05%, a ramp rate of 5°C /min, and a frequency range set at 0.01 to 100 Hz, the dependences of loss modulus and storage modulus on frequency were investigated. Another research showed an increase in the mechanical strength of nanocomposite hydrogels with Fe_3O_4 NPs loading (Saberi et al., 2019). An increment of Fe_3O_4 NPs from 0.1 to 0.5 g resulted in a significant rise in the hardness and Young's modulus of the nanocomposite hydrogels, from 6.7 and 3.1 MPa to 7.8 and 3.7 MPa, respectively. They also reported an optimum value of 7.2 and 3.4 MPa for hardness and Young's modulus, respectively, whereas that of the pristine hydrogels was 5.1 and 2.5 MPa, respectively.

12.7.5. Thermal Characterization

Some authors utilize a cocktail of nanomaterials of iron and thermal property enhancement species to impart thermal stability to the resultant nanocomposite hydrogel. Typical characterizations were performed to ascertain the above-mentioned property, including but not limited to thermogravimetric analysis (TGA) and differential thermogravimetric analysis (DTG) (Saberi et al., 2019). Many researchers have satisfactorily concluded on the effect that the inculcation of iron-based NPs into polymeric networks of hydrogels has led to an enhancement in thermal stability (Malatji et al., 2020; Nakhjiri et al., 2019; Singh et al., 2021).

12.8. Application of Iron-Based Nanocomposite to Water and Wastewater Treatment: Adsorption Studies

Iron-based NPs have a wide range of uses in several fields. They can be utilized as pigments in the industrial sector, catalyst materials, and adsorbents in the treatment of water and wastewater (Fayazi and Ghanbarian, 2019; Hao et al., 2020; Mufti et al., 2014). Additionally, they serve as crucial raw materials for applications such as ion exchangers, coatings, gas sensors, magnetic recording and data storage devices, magnetic resonance imaging, bio-separation, and pharmaceuticals (Aragaw et al., 2021; Bardajee et al., 2014; Chen et al., 2009; Ma et al., 2012; Mahdavinia and Etemadi, 2014; Park et al., 2007; Suri et al., 2002; Teng and Yang, 2005; Vilardi et al., 2018).

Magnetic iron-based nanocomposites have been widely employed in the water treatment industry to remove contaminants from aqueous solutions by chemical, photochemical, or adsorption methods (Fadillah et al., 2020). There have been reports of the removal and decrease of impurities from water, including chemical elements, like dye, heavy metals, and other organic pollutants (Fadillah et al., 2020). The textile and pulp industries are the principal users of dyes, which are mostly organic compounds. These substances cause large volumes of poisonous, colored, and cancer-causing effluent to be produced, which poses a risk to marine life like fish and is extremely hazardous to both humans and animals. These are as a result of the non-biodegradability of the dyes and increased resistance to aerobic digestion. Nanocomposite hydrogels made of iron-based oxide have shown a wide variety of sorption characteristics that are effective in concurrently removing dyes, organic pollutants, and inorganic pollutants (Nizamuddin et al., 2019). This section highlights the enormous potential of iron-based nanocomposite hydrogel adsorbents for wastewater treatment processes.

12.8.1. Effectiveness of Nanocomposite Hydrogels for Wastewater Treatment

Magnetic hydrogel absorbents have proved to be easier to collect after the adsorption process by an external magnetic field compared to conventional water pollution absorbents. The mechanical and chemical characteristics can also be improved by magnetic NPs (MNPs) incorporated into the hydrogel. As a result, magnetic hydrogels have demonstrated very promising capabilities for the treatment of water. The adsorption of ions and dyes has been studied using several magnetic hydrogels (Zhang et al., 2016). Shirsath et al., (2011) conducted a study to determine the viability of using polyacrylic acid-bentonite-FeCo hydrogel to remove crystal violet dye from aqueous solution. In contrast to pure polyacrylic acid hydrogel, which disintegrated when added to the dye solution, the addition of bentonite and Fe-Co to the hydrogel matrix enhanced its mechanical strength and prevented its collapse. The inclusion of bentonite-Fe had the twin effects of giving the hydrogel mechanical strength and enhancing the electrostatic interaction, which improved the adsorption capacity of the hydrogel.

In a different study, Mittal et al., (2016) investigated how the addition of Fe_3O_4 MNPs altered the thermal stability of gum xanthan-crosslinked-polyacrylic acid using TGA. The thermal degradations of the nanocomposite hydrogels were shown to occur at greater temperatures than those of the parent polymeric matrix. Additionally, reduced weight percent degradation was seen in nanocomposite hydrogel, demonstrating that the addition of Fe_3O_4 MNPs improved the thermal stability of the original polymer matrix. In a similar study, the thermal stability of synthesized materials with Fe_3O_4 particle loading of 0, 3, 6, and 9% was investigated. Less weight loss was observed in nanocomposite hydrogel with 9% particle loading, demonstrating how the addition of Fe_3O_4 NPs to the polymeric matrix enhances the thermal stability of the nanocomposite (Mohammadi et al., 2021).

Furthermore, Mittal et al., (2016) found that the nanocomposite hydrogel adsorption effectiveness was improved due to the increased surface area. Furthermore, the mass transfer rate, which depends on the surface area of the adsorbent, had a significant impact on adsorption efficiency. In addition, it was discovered that the presence of NPs in the polymer matrix contributed to an increase in the binding sites for the adsorption of dye molecules. Wang et al., (2020) created hydrogel from cellulose solution and incorporated β-FeOOH NPs to form β-FeOOH/cellulose composite hydrogels. Different β-FeOOH NP loadings were examined and the results showed a rise in the nanocomposites' emerging decomposition temperature. This observation was ascribed to the hydrogen bond interactions within the matrix and the homogeneous distribution of β-FeOOH throughout the cellulose matrix. With increasing amounts of β-FeOOH, there was a corresponding rise in the efficiency of methyl blue degradation.

In another work conducted by Konwar et al., (2016), iron oxide coated graphene oxide (GIO) NPs were incorporated in varying amounts into a chitosan matrix system, which increased the mechanical strength of the hydrogel. This enhancement was due to the robust secondary contacts, such as hydrogen bonding and polar-polar interactions between various functional groups present in the matrix and the reinforcing nanomaterial phase. However, increasing the nanomaterial loading over the ideal level led to a decline in the tensile strength and modulus value of the nanocomposite films. Zheng et al., (2014) showed that mechanical strength is also influenced by the concentration of Fe_3O_4 NPs. Young's modulus and compressive strength were somewhat improved with an increase in Fe_3O_4 content from 0.02 to 0.06%. Because the Fe_3O_4 NP has a significant capacity for light absorption, the hydrogelation could not develop entirely when the concentration of the Fe_3O_4 NP was raised to 0.08%.

Magnetic Fe_3O_4@SiO_2 starch-graft-polyacrylic acid nanocomposite hydrogels were employed as absorbents in a work by Pourjavadi et al., (2013) to remove crystal violet from aqueous solutions. However, Fe_3O_4@SiO_2 served as a crosslinker for copolymer chains and limited their expansion. It is evident from comparing the swelling kinetics of the samples that the swelling of the nonmagnetic nanocomposite is larger than that of the other samples. Higher concentrations of Fe_3O_4@SiO_2 increase the crosslinking density and reduce the distance between the copolymer chains, resulting in a stiff, highly crosslinked structure with narrow holes that cannot be expanded to contain a significant amount of water. Furthermore, according to Li et al., (2011), the adsorption capabilities of magnetic nanocomposite hydrogels are somewhat lower than those of non-magnetic hydrogels. The decrease in swelling for magnetic hydrogels with less macroporous structure is ascribed to magnetic Fe_3O_4 NPs acting as crosslinkers, tying up the polymer chains, and limiting their expansions. This reduces the diffusion of the dye molecules into the hydrogels to reach adsorbing and interacting sites.

In contrast, a study by Zhang et al., (2017) showed that Fe^{3+} and quaternized tunicate cellulose nanocrystals functioned as crosslinkers in the hydrogel network via coordination bonds and electrostatic interactions between $-COO^-$ and Fe^{3+}, and $-N(CH_3)^{3+}$ and $-COO^-$, respectively. The dual crosslinked hydrogels displayed high tensile strength, ductility, and toughness because these physical links acted as reversible sacrificial bonds that may rupture to efficiently dissipate energy. A summary of the method of preparation of iron-based nanocomposites and their performances in dye, heavy metal, and other contaminants removal from wastewater have been provided in Table 12.3.

Table 12.3. Method of preparation of iron-based nanocomposites and their performances in dye and heavy metal adsorption

Nanocomposite	Method of Preparation	Dye Used for Studies	qe (mg/g)	Refs.
Gum xanthan/Fe_3O_4	Free radical graft copolymerization	Methyl violet	642	(Mittal et al., 2016)
Poly(vinyl acetate-co-maleic anhydride)/magnetite/melamine nanocomposite hydrogel	Polymerization	Congo red	48.6	(Rezazadeh et al., 2020)
Chitosan/hematite	Anionic surfactant gelation	Congo red	4706	(Ohemeng-Boahen et al., 2021)
Polyacrylic acid-bentonite-FeCo	Ultrasound-assisted in situ emulsion polymerization	Crystal violet	31.25	(Shirsath et al., 2011)
Collagen-g-poly(acrylic acid-co-N-vinylpyrrolidone)/ Fe_3O_4@SiO_2	Graft copolymerization	Methylene blue Brilliant Green Rhodamine B	207.73 212.68 221.99	(Nakhjiri et al., 2019)
Collagen-(2-acrylamido-2-methylpropane sulfonic acid - methacrylic acid/ acrylamide)-Fe_3O_4@SiO_2	Ultrasound-assisted in situ free radical copolymerization	Phenol 4-nitrophenol	21.193 22.103	(Nakhjiri et al., 2021)
Magnetic poly(acrylic acid-acrylamide-butyl methacrylate)	In situ free radical polymerization	Crystal violet basic magenta	10.787 12.642	(Li et al., 2011)
Starch-graft-polyacrylic acid/ Fe_3O_4@SiO_2	Crosslinking graft copolymerization	Crystal violet	80.64	(Pourjavadi et al., 2013)
Graphene Oxide/Mn Fe_2O_4	Co-precipitation	Pb(II) As(III) As(V)	673 146 207	(Kumar et al., 2014)
Graphene Oxide/Zerovalent Iron	Chemical deposition	Rhodamine B	87.72	(Shi et al., 2017)
Magnetite/graphene/Layered double hydroxides	Two-step reaction	Arsenate	73.14	(Wu et al., 2011)
Graphene/Fe_3O_4@carbon	One-pot solvothermal process	Methylene blue	73.26	(Fan et al., 2012)
Reduced graphene oxide/ zerovalent Iron	Chemical deposition	Cd(II)	47.84	(Fan et al., 2017)
Magnetic chitosan/graphene oxide	Copolymerization	Cu(II)	217.4	(Hosseinzadeh and Ramin, 2018)
Magnetite/kaolinite	Chemical co-precipitation	Pb(II) Cu(II) Cr(VI) Ni(II) Cd(II)	106 98 100 95.2 97	(Lasheen et al., 2015)
Reduced graphene oxide/Fe_2O_3	Co-precipitation	Pb(II) 1-naphthol 1-naphthylamine	454.55 243.16 303.03	(Yang et al., 2012)
Graphene manganese ferrite	Solvothermal process	Pb(II) Cd(II)	100 76.90	(Chella et al., 2015)
Magnetic hydroxyapatite	Co-precipitation	Cd(II) Zn(II)	220.77 140.63	(Feng et al., 2010)
Hydroxyapatite/magnetite	In situ precipitation	Pb(II)	598.8	(Wan et al., 2016)
Nano-magnetic cellulose hybrid	Co-precipitation	Hg(II) Cu(II) Ag(I)	401.18 101.67 140.23	(Donia et al., 2012)

Table 12.3. (Continued)

Nanocomposite	Method of Preparation	Dye Used for Studies	qe (mg/g)	Refs.
$Fe_3O_4@SiO_2/NH_2$	Sol-gel	Pb(II)	243.9	(Zhang et al., 2013)
Amino-modified $Fe_3O_4@SiO_2@$ cellulose	Co-precipitation	Cr(VI)	171.5	(Sun et al., 2014)
EDTA functionalized magnetic graphene oxide	Mechanical rabbling	Pb(II) Hg(II) Cu(II)	508.4 268.4 301.2	(Cui et al., 2015)
Polypyrrole/maghemite	In situ chemical polymerization	Cr(VI) Cu(II)	208.8 170.7	(Chávez-Guajardo et al., 2015)
Fe_3O_4- cetyltrimethylammonium chloride/sodium eicosenoate - montmorillonite	Blending	Methylene blue	246	(Rahmani et al., 2020)
Hydroxyapatite/Fe_3O_4	Hydrothermal	Pb(II)	434.8	(Zhuang et al., 2015)
Nano ferrite bentonite clay	Sol-gel method	Cr(VI)	161.29	(Kaur et al., 2015)
Chitosan magnetic graphene oxide	Co-precipitation	As(III)	45	(Sherlala et al., 2019)
Fe_3O_4/activated carbon/chitosan	Co-precipitation	Ciprofloxacin Erythromycin Amoxicillin	90.10 178.57 526.31	(Danalıoğlu et al., 2017)
Fe_3O_4/bentonite	Solvothermal method	Pb(II) Cd(II) Cu(II)	81.5 21.7 19.6	(Yan et al., 2016)
Graphene/magnetite/montmorillonite	Simple green hydrothermal	Methylene blue	225.0	(Zhang et al., 2019)
Polyethylenimine enhanced magnetic carboxymethyl Chitosan	Blending	Pb(II)	124.0	(Wang et al., 2017)
Montmorillonite/Fe_3O_4/ polyethylenimine	One-Pot Hydrothermal	Cr(VI)	62.89	(Fayazi and Ghanbarian, 2019)

12.8.2. Adsorption Mechanism

Several processes, such as adsorption, chemical oxidation, and photo-oxidation processes have been studied for the effective removal of contaminants from aqueous solutions (Ayalew, 2022). Adsorption is the most widely used method for the removal of most organic and inorganic pollutants from water and wastewater (Gutierrez et al., 2017). Active carbon, zeolites, chitosan, and, more recently, magnetic NPs have all been employed as water treatment adsorbents throughout the years (Alver and Metin, 2012; Hussain et al., 2021; Maneerung et al., 2016). Iron-based nanocomposites have gained considerably high attention in adsorption studies of inorganic and organic contaminants from water and wastewater. Several factors affect the adsorption process and the adsorbate-adsorbent interaction (Marcelo et al., 2021).

Some concepts or principles used in the adsorption of pollutants from contaminated areas or wastewater streams onto an adsorbent include electrostatic interaction, ligand-ion combination, and surface binding (Aragaw et al., 2021). According to Marcelo et al., (2021), the adsorption mechanism also involves hydrogen bonds, π-π interactions, complexation, electrostatic interactions, chemisorption, and ion exchange. It is worth mentioning that it is very rare for the adsorption of a contaminant from an aqueous solution onto the internal and external surfaces of an adsorbent could be defined by only one adsorption mechanism. Rather, a cocktail of these mechanisms may be involved. In iron oxide NPs, the interactions are driven by the surface hydroxyl groups and surface charges of iron oxides. Chatterjee et al., (2022) reported that the adsorption of Congo red from an aqueous solution can be attributed to the combination of hydrophobic interaction such as Ruderman-Kittel-Kasuya-Yoshida interaction, between the hydroxyl group of chitosan and the aromatic residue in Congo red, coordination between Fe and amine groups ($-NH_2$) of Congo red, hydrogen bonding between -OH group of chitosan and electronegative residues in the dye molecule, as well as electrostatic interaction between the amine in chitosan and the hydroxyl groups in Congo red (Chatterjee et al., 2007; Ohemeng-Boahen et al., 2021; Wei et al., 2013). Additionally, other important factors in the adsorption mechanism include the distribution of NPs, hydrophobicity, pore distribution, and pore radius on the surface. Higher specific surface areas theoretically offer greater room for adsorption. The mechanism of iron oxide-based nanocomposites for the adsorption of different pollutants has been the subject of several investigations.

According to a study by Shirsath et al., (2011), exposed magnetic particles in water-based systems are prone to oxidation and aggregation, which decreases the effectiveness of separation and adsorption. As a result, the presence of COOH groups in the hydrogel can increase the effectiveness of Fe-Co NPs. Higher dye adsorption in basic pH is caused by the COOH functional groups of acrylic acid that are present in the hydrogel matrix dissociating at higher pH to produce COO^- ions. Positively charged dye and negatively charged COO^- ions are electrostatically attracted to one another, creating an ionic complex that accelerates dye removal. However, research conducted by Ohemeng-Boahen et al., (2021) on the impact of initial solution pH on the adsorption of Congo red onto various iron-based hydrogels revealed that the adsorptive abilities of the adsorbents continuously decreased as initial solution pH increased. This is explained by the rise in H^+ ions on the adsorbent's surface at $pH<7$, which enhances the electrostatic interaction between the surface of the adsorbent and the negatively charged Congo red molecules. This shows that the initial solution pH is a crucial factor in the adsorption process as it affects the chemistry of both the dye and the adsorbent.

12.9. Future Perspectives

The procedure for the fabrication of iron-based nanocomposites can be very dynamic in an attempt to confer specific properties to the as-obtained nanocomposite hydrogel. Notwithstanding, key parameters, such as concentration of precursor, polymer, and crosslinker, type of crosslinker, temperature, electrolyte, salts, and pH among several other factors influence the stability of the precursor material and effectively play a vital role in the quality of nanocomposite hydrogel produced. As such, optimizing these parameters will be highly beneficial. Again, the synthesis procedures should be modeled to optimize factors that influence the performance of nanocomposite hydrogels in order to hasten commercial-scale production and application.

Furthermore, studies on the applicability of iron-based nanocomposite hydrogels for wastewater treatment should focus more on column application rather than conventional bench-scale adsorption studies. Modeling of these column adsorption studies will also go a long way to answer most of the budding questions surrounding the use of iron-based nanocomposite hydrogels in wastewater treatment plants. Additionally, although NPs are synthesized from various means, synthesizing NPs and nanocomposite hydrogels from biogenic sources using an eco-friendly method needs to be considered. More so, the overall effectiveness of nanocomposite hydrogels especially for water and wastewater treatment should not just be limited to adsorptive capacity without recourse to a wholistic assessment of the synthesis approach, cost of chemicals, and the chemical waste generated during synthesis. Thus, a life cycle analysis of the iron-based nanocomposite hydrogels should be studied.

Conclusion

Although studies on the application of NPs, especially iron-based NPs, in wastewater treatment are on the ascendency, the incorporation of these unique materials into polymer hydrogel matrices and using the same as nanocomposite adsorbent seems a massive game-changer as it has a high potential of saving engineers and investors not only the high cost of post-treatment separation that will otherwise be required to recover and reuse these NPs but also helps in curtailing the long term ecotoxicological effects of NPs freely dispersed in the environment especially unbeknown to humans, aquatic lives and other living organisms. The electrochemical, solvothermal, microwave-assisted, micro-emulsion, and CVD methods are some of the means by which iron-based NPs can be synthesized.

In summary, iron oxide NPs are integrated into hydrogels by three approaches, including the addition of NPs to pre-formed hydrogels, the addition of NPs to polymer solutions, and the polymerization of monomers in the presence of NPs. Thus, the formation of iron-based nanocomposite is very dynamic. By employing characterization techniques, such as SEM, TEM, AFM, EDS/EDX, ICP-AES/ICP-OES, XPS, FTIR, XRD, BET, and Raman spectroscopy, the physicochemical properties of the iron-based nanocomposites hydrogels could be thoroughly analyzed providing engineers and scientists with relevant information about the chemistry of these materials and helping expatiate the adsorption mechanism.

For iron-based nanocomposite hydrogels which exhibit magnetic properties, specialized techniques, such as VSM, are also required to estimate the saturation magnetization. Evidently, iron-based nanocomposite hydrogels compared to pristine polymer hydrogels have improved characteristics, such as higher surface area, porosity, mechanical and acid stability, as well as multifunctionality and more importantly, tremendously higher adsorptive capacities in wastewater treatment. As such, iron-based nanocomposite hydrogels exhibit a high affinity for contaminants especially dyes in water and wastewater and they can be considered suitable adsorbents.

References

Abdel-Mohsen, A., Aly, A., Hrdina, R., Montaser, A., and Hebeish, A. (2011). Eco-synthesis of PVA/chitosan hydrogels for biomedical application. *Journal of Polymers and the Environment*, 19(4), 1005-1012.

Abou Taleb, M. F., Abou El Fadl, F. I., and Albalwi, H. (2021). Adsorption of toxic dye in wastewater onto magnetic NVP/CS nanocomposite hydrogels synthesized using gamma radiation. *Separation and Purification Technology*, 266, 118551.

Adewunmi, A. A., Ismail, S., and Sultan, A. S. (2016). Carbon nanotubes (CNTs) nanocomposite hydrogels developed for various applications: a critical review. *Journal of Inorganic and Organometallic Polymers and Materials*, 26(4), 717-737.

Afkhami, A., and Moosavi, R. (2010). Adsorptive removal of Congo red, a carcinogenic textile dye, from aqueous solutions by maghemite nanoparticles. *Journal of Hazardous Materials*, 174(1-3), 398-403.

Ahmed, E. M. (2015). Hydrogel: preparation, characterization, and applications: a review. *Journal of Advanced Research*, 6(2), 105-121.

Ali, H. E., Nasef, S. M., and Gad, Y. H. (2022). Remediation of Astrazon blue and Lerui acid brilliant blue dyes from waste solutions using amphoteric superparamagnetic nanocomposite hydrogels based on chitosan prepared by gamma rays. *Carbohydrate Polymers*, 283, 119149.

Alvarez-Torrellas, S., Boutahala, M., Boukhalfa, N., and Munoz, M. (2019). Effective adsorption of methylene blue dye onto magnetic nanocomposites. Modeling and reuse studies. *Applied Sciences*, 9(21), 4563.

Alver, E., and Metin, A. Ü. (2012). Anionic dye removal from aqueous solutions using modified zeolite: adsorption kinetics and isotherm studies. *Chemical Engineering Journal*, 200-202, 59-67.

Aly, A. A., and El-Bisi, M. K. (2018). Grafting of polysaccharides: recent advances. *Biopolymer Grafting*, 469-519.

Ansari, N., Payami, Z., and Feghhi, F. (2020). Synthesis of iron/graphene composites with controlled magnetization by electrochemical exfoliation/deposition using sodium dodecyl sulfate as surfactant. *Journal of Magnetism and Magnetic Materials*, 500, 166398.

Aragaw, T. A., Bogale, F. M., and Aragaw, B. A. (2021). Iron-based nanoparticles in wastewater treatment: a review on synthesis methods, applications, and removal mechanisms. *Journal of Saudi Chemical Society*, 25(8), 101280.

Ayalew, A. A. (2022). A critical review on clay-based nanocomposite particles for application of wastewater treatment. *Water Science and Technology*, 85(10), 3002-3022.

Bardajee, G. R., Hooshyar, Z., Asli, M. J., Shahidi, F. E., and Dianatnejad, N. (2014). Synthesis of a novel supermagnetic iron oxide nanocomposite hydrogel based on graft copolymerization of poly((2-dimethylamino)ethyl methacrylate) onto salep for controlled release of drug. *Materials Science and Engineering: C*, 36, 277-286.

Belachew, N., and Bekele, G. (2019). Synergy of magnetite intercalated bentonite for enhanced adsorption of Congo red dye. *Silicon*, 12(3), 603-612.

Biftu, W. K., Ravindhranath, K., and Ramamoorty, M. (2020). New research trends in the processing and applications of iron-based nanoparticles as adsorbents in water remediation methods. *Nanotechnology for Environmental Engineering*, 5, 12.

Bolade, O. P., Williams, A. B., and Benson, N. U. (2020). Green synthesis of iron-based nanomaterials for environmental remediation: a review. *Environmental Nanotechnology, Monitoring & Management*, 13, 100279.

Boukhalfa, N., Boutahala, M., Djebri, N., and Idris, A. (2019). Maghemite/alginate/functionalized multiwalled carbon nanotubes beads for methylene blue removal: adsorption and desorption studies. *Journal of Molecular Liquids*, 275, 431-440.

Burleson, D. J., and Penn, R. L. (2006). Two-step growth of goethite from ferrihydrite. *Langmuir*, 22(1), 402-409.

Catoira, M. C., Fusaro, L., Di Francesco, D., Ramella, M., and Boccafoschi, F. (2019). Overview of natural hydrogels for regenerative medicine applications. *Journal of Materials Science: Materials in Medicine*, 30, 115.

Chandrika, M., Ravindra, A., Rajesh, C., Ramarao, S., and Ju, S. (2019). Studies on structural and optical properties of nano $ZnFe_2O_4$ and $ZnFe_2O_4$-TiO_2 composite synthesized by co-precipitation route. *Materials Chemistry and Physics*, 230, 107-113.

Chatterjee, S., Chatterjee, S., Chatterjee, B. P., and Guha, A. K. (2007). Adsorptive removal of Congo red, a carcinogenic textile dye by chitosan hydrobeads: binding mechanism, equilibrium and kinetics. *Colloids and Surfaces A: Physicochemical and Engineering Aspects*, 299(1-3), 146-152.

Chatterjee, S., Chatterjee, T., and Woo, S. H. (2010). A new type of chitosan hydrogel sorbent generated by anionic surfactant gelation. *Bioresource Technology*, 101(11), 3853-3858.

Chatterjee, S., Godfred, O. B., Sewu, D. D., Osei, B. A., and Woo, S. H. (2022). Improved adsorption of Congo red from aqueous solution using alkali-treated goethite impregnated chitosan hydrogel capsule. *Journal of Environmental Chemical Engineering*, 10(5), 108244.

Chávez-Guajardo, A. E., Medina-Llamas, J. C., Maqueira, L., Andrade, C. A. S., Alves, K. G. B., and de Melo, C. P. (2015). Efficient removal of Cr (VI) and Cu (II) ions from aqueous media by use of polypyrrole/maghemite and polyaniline/maghemite magnetic nanocomposites. *Chemical Engineering Journal*, 281, 826-836.

Chella, S., Kollu, P., Komarala, E. V. P. R., Doshi, S., Saranya, M., Felix, S., Ramachandran, R., Saravanan, P., Koneru, V. L., Venugopal, V., Jeong, S. K., and Nirmala Grace, A. (2015). Solvothermal synthesis of $MnFe_2O_4$-graphene composite - investigation of its adsorption and antimicrobial properties. *Applied Surface Science*, 327, 27-36.

Chen, J. S., Lim, B. C., Ding, Y. F., Hu, J. F., Chow, G. M., and Ju, G. (2009). Granular L10 FePt-X (X=C, TiO_2, Ta_2O_5) (001) nanocomposite films with small grain size for high density magnetic recording. *Journal of Applied Physics*, 105(7), 093704.

Chen, Y. H., and Li, F. A. (2010). Kinetic study on removal of copper (II) using goethite and hematite nano-photocatalysts. *Journal of Colloid and Interface Science*, 347(2), 277-281.

Cui, L., Wang, Y., Gao, L., Hu, L., Yan, L., Wei, Q., and Du, B. (2015). EDTA functionalized magnetic graphene oxide for removal of Pb(II), Hg(II) and Cu(II) in water treatment: adsorption mechanism and separation property. *Chemical Engineering Journal*, 281, 1-10.

Danalıoğlu, S. T., Bayazit, Ş. S., Kerkez Kuyumcu, Ö., and Salam, M. A. (2017). Efficient removal of antibiotics by a novel magnetic adsorbent: magnetic activated carbon/chitosan (MACC) nanocomposite. *Journal of Molecular Liquids*, 240, 589-596.

Dannert, C., Stokke, B. T., and Dias, R. S. (2019). Nanoparticle-hydrogel composites: from molecular interactions to macroscopic behavior. *Polymers*, 11(2), 275.

Dhand, V., Mittal, G., Rhee, K. Y., and Park, S. J. (2017). Synthesis and comparison of different spinel ferrites and their catalytic activity during chemical vapor deposition of polymorphic nanocarbons. *International Journal of Precision Engineering and Manufacturing-Green Technology*, 4(4), 441-451.

Donia, A. M., Atia, A. A., and Abouzayed, F. I. (2012). Preparation and characterization of nano-magnetic cellulose with fast kinetic properties towards the adsorption of some metal ions. *Chemical Engineering Journal*, 191, 22-30.

Dung, C. T., Quynh, L. M., Linh, N. P., Nam, N. H., and Luong, N. H. (2016). Synthesis of ZnS: Mn-Fe_3O_4 bifunctional nanoparticles by inverse microemulsion method. *Journal of Science: Advanced Materials and Devices*, 1(2), 200-203.

Fadillah, G., Yudha, S. P., Sagadevan, S., Fatimah, I., and Muraza, O. (2020). Magnetic iron oxide/clay nanocomposites for adsorption and catalytic oxidation in water treatment applications. *Open Chemistry*, 18(1), 1148-1166.

Fan, M., Li, T., Hu, J., Cao, R., Wei, X., Shi, X., and Ruan, W. (2017). Artificial neural network modeling and genetic algorithm optimization for cadmium removal from aqueous solutions by reduced graphene oxide-supported nanoscale zero-valent iron (nZVI/rGO) composites. *Materials*, 10(5), 544.

Fan, W., Gao, W., Zhang, C., Tjiu, W. W., Pan, J., and Liu, T. (2012). Hybridization of graphene sheets and carbon-coated Fe_3O_4 nanoparticles as a synergistic adsorbent of organic dyes. *Journal of Materials Chemistry*, 22(48), 25108-25115.

Fatimah, I., Fadillah, G., and Yudha, S. P. (2021). Synthesis of iron-based magnetic nanocomposites: a review. *Arabian Journal of Chemistry*, 14(8), 103301.

Fayazi, M., and Ghanbarian, M. (2019). One-pot hydrothermal synthesis of polyethylenimine functionalized magnetic clay for efficient removal of noxious Cr(VI) from aqueous solutions. *Silicon*, 12(1), 125-134

Feng, Y., Gong, J. L., Zeng, G. M., Niu, Q. Y., Zhang, H. Y., Niu, C. G., Deng, J. H., and Yan, M. (2010). Adsorption of Cd (II) and Zn (II) from aqueous solutions using magnetic hydroxyapatite nanoparticles as adsorbents. *Chemical Engineering Journal*, 162(2), 487-494.

Fosso-Kankeu, E., Mittal, H., Waanders, F., Ntwampe, I., and Ray, S. S. (2016). Preparation and characterization of gum karaya hydrogel nanocomposite flocculant for metal ions removal from mine effluents. *International Journal of Environmental Science and Technology*, 13(2), 711-724.

Gennen, S., Grignard, B., Thomassin, J. M., Gilbert, B., Vertruyen, B., Jerome, C., and Detrembleur, C. (2016). Polyhydroxyurethane hydrogels: synthesis and characterizations. *European Polymer Journal*, 84, 849-862.

Gubin, S. P., Koksharov, Y. A., Khomutov, G., and Yurkov, G. Y. (2005). Magnetic nanoparticles: preparation, structure and properties. *Russian Chemical Reviews*, 74(6), 489.

Gutierrez, A. M., Dziubla, T. D., and Hilt, J. Z. (2017). Recent advances on iron oxide magnetic nanoparticles as sorbents of organic pollutants in water and wastewater treatment. *Reviews on Environmental Health*, 32(1-2), 111-117.

Han, D., Zhao, H., Gao, L., Qin, Z., Ma, J., Han, Y., and Jiao, T. (2021). Preparation of carboxymethyl chitosan/phytic acid composite hydrogels for rapid dye adsorption in wastewater treatment. *Colloids and Surfaces A: Physicochemical and Engineering Aspects*, 628, 127355.

Hao, R., Ren, J. T., Lv, X. W., Li, W., Liu, Y. P., and Yuan, Z. Y. (2020). N-doped porous carbon hollow microspheres encapsulated with iron-based nanocomposites as advanced bifunctional catalysts for rechargeable Zn-air battery. *Journal of Energy Chemistry*, 49, 14-21.

Haraguchi, K. (2007). Nanocomposite hydrogels. *Current Opinion in Solid State and Materials Science*, 11(3-4), 47-54.

Hassan, C. M., and Peppas, N. A. (2000). Structure and applications of poly (vinyl alcohol) hydrogels produced by conventional crosslinking or by freezing/thawing methods. *Biopolymers. PVA Hydrogels Anionic Polymerisation Nanocomposites*, 37-65.

Hosseinzadeh, H., and Ramin, S. (2018). Effective removal of copper from aqueous solutions by modified magnetic chitosan/graphene oxide nanocomposites. *International Journal of Biological Macromolecules*, 113, 859-868.

Hu, X. S., Liang, R., and Sun, G. (2018). Super-adsorbent hydrogel for removal of methylene blue dye from aqueous solution. *Journal of Materials Chemistry A*, 6(36), 17612-17624.

Hussain, S., Kamran, M., Khan, S. A., Shaheen, K., Shah, Z., Suo, H., Khan, Q., Shah, A. B., Rehman, W. U., Al-Ghamdi, Y. O., and Ghani, U. (2021). Adsorption, kinetics and thermodynamics studies of methyl orange dye sequestration through chitosan composites films. *International Journal of Biological Macromolecules*, 168, 383-394.

Jabbari, E., Leijten, J., Xu, Q., and Khademhosseini, A. (2016). The matrix reloaded: the evolution of regenerative hydrogels. *Materials Today*, 19(4), 190-196.

Jiang, X., Xiong, D. A., An, Y., Zheng, P., Zhang, W., and Shi, L. (2007). Thermoresponsive hydrogel of poly (glycidyl methacrylate-co-N-isopropylacrylamide) as a nanoreactor of gold nanoparticles. *Journal of Polymer Science Part A: Polymer Chemistry*, 45(13), 2812-2819.

Kaewmanee, T., Phuruangrat, A., Thongtem, T., and Thongtem, S. (2020). Solvothermal synthesis of Mn-Zn Ferrite (core)@SiO_2 (shell)/BiOBr0. 5Cl0. 5 nanocomposites used for adsorption and photocatalysis combination. *Ceramics International*, 46(3), 3655-3662.

Karthikeyan, P., Vigneshwaran, S., Preethi, J., and Meenakshi, S. (2020). Preparation of novel cobalt ferrite coated-porous carbon composite by simple chemical co-precipitation method and their mechanistic performance. *Diamond and Related Materials*, 108, 107922.

Kaur, M., Singh, M., Mukhopadhyay, S. S., Singh, D., and Gupta, M. (2015). Structural, magnetic and adsorptive properties of clay ferrite nanocomposite and its use for effective removal of Cr (VI) from water. *Journal of Alloys and Compounds*, 653, 202-211.

Kerroum, M. A., Essyed, A., Iacovita, C., Baaziz, W., Ihiawakrim, D., Mounkachi, O., Hamedoun, M., Benyoussef, A., Benaissa, M., and Ersen, O. (2019). The effect of basic pH on the elaboration of $ZnFe_2O_4$ nanoparticles by co-precipitation method: structural, magnetic and hyperthermia characterization. *Journal of Magnetism and Magnetic Materials*, 478, 239-246.

Konwar, A., Kalita, S., Kotoky, J., and Chowdhury, D. (2016). Chitosan-iron oxide coated graphene oxide nanocomposite hydrogel: a robust and soft antimicrobial bio-film. *ACS Applied Materials and Interface,* 8(32), 20625-20634.

Kulal, P., and Badalamoole, V. (2020). Magnetite nanoparticle embedded Pectin-graft-poly (N-hydroxyethylacrylamide) hydrogel: evaluation as adsorbent for dyes and heavy metal ions from waste water. *International Journal of Biological Macromolecules*, 156, 1408-1417.

Kumar, R., Singh, R. K., Alaferdov, A. V., and Moshkalev, S. A. (2018). Rapid and controllable synthesis of Fe_3O_4 octahedral nanocrystals embedded-reduced graphene oxide using microwave irradiation for high performance lithium-ion batteries. *Electrochimica Acta*, 281, 78-87.

Kumar, S., Nair, R. R., Pillai, P. B., Gupta, S. N., Iyengar, M. A., and Sood, A. K. (2014). Graphene oxide-$MnFe_2O_4$ magnetic nanohybrids for efficient removal of lead and arsenic from water. *ACS Applied Materials & Interfaces*, 6(20), 17426-17436.

Laftah, W. A., Hashim, S., and Ibrahim, A. N. (2011). Polymer hydrogels: a review. *Polymer-Plastics Technology and Engineering*, 50(14), 1475-1486.

Lasheen, M. R., El-Sherif, I. Y., Sabry, D. Y., El-Wakeel, S. T., and El-Shahat, M. F. (2015). Adsorption of heavy metals from aqueous solution by magnetite nanoparticles and magnetite-kaolinite nanocomposite: equilibrium, isotherm and kinetic study. *Desalination and Water Treatment*, 57(37), 17421-17429.

Lee, W. F., and Tsao, K. T. (2010). Effect of silver nanoparticles content on the various properties of nanocomposite hydrogels by *in situ* polymerization. *Journal of Materials Science*, 45(1), 89-97.

Li, S., Cui, J., Wu, X., Zhang, X., Hu, Q., and Hou, X. (2019). Rapid *in situ* microwave synthesis of Fe_3O_4@MIL-100 (Fe) for aqueous diclofenac sodium removal through integrated adsorption and photodegradation. *Journal of Hazardous Materials*, 373, 408-416.

Li, S., Liu, X., Huang, W., Li, W., Xia, X., Yan, S., and Yu, J. (2011). Magnetically assisted removal and separation of cationic dyes from aqueous solution by magnetic nanocomposite hydrogels. *Polymers for Advanced Technologies*, 22(12), 2439-2447.

Lin, M., Tng, L., Lim, T., Choo, M., Zhang, J., Tan, H. R., and Bai, S. (2014). Hydrothermal synthesis of octadecahedral hematite (α-Fe_2O_3) nanoparticles: an epitaxial growth from goethite (α-FeOOH). *The Journal of Physical Chemistry C*, 118(20), 10903-10910.

Lu, L., Yuan, S., Wang, J., Shen, Y., Deng, S., Xie, L., and Yang, Q. (2018). The formation mechanism of hydrogels. *Current Stem Cell Research & Therapy*, 13(7), 490-496.

Lucas-Granados, B., Sánchez-Tovar, R., Fernández-Domene, R. M., and Garcia-Anton, J. (2018). Influence of electrolyte temperature on the synthesis of iron oxide nanostructures by electrochemical anodization for water splitting. *International Journal of Hydrogen Energy*, 43(16), 7923-7937.

Ma, X., Tao, H., Yang, K., Feng, L., Cheng, L., Shi, X., Li, Y., Guo, L., and Liu, Z. (2012). A functionalized graphene oxide-iron oxide nanocomposite for magnetically targeted drug delivery, photothermal therapy, and magnetic resonance imaging. *Nano Research*, 5(3), 199-212.

Maciel, D. J., da Silva, M. R., and Ferreira, I. L. (2019). Preparation of a superparamagnetic nanocomposite hydrogel for adsorptive performance in the wastewater treatment. *Journal of Applied Polymer Science*, 136(26), 47705.

Madduma-Bandarage, U. S., and Madihally, S. V. (2021). Synthetic hydrogels: synthesis, novel trends, and applications. *Journal of Applied Polymer Science*, 138(19), 50376.

Mahdavinia, G. R., and Etemadi, H. (2014). *In situ* synthesis of magnetic CaraPVA IPN nanocomposite hydrogels and controlled drug release. *Materials Science and Engineering: C*, 45, 250-260.

Mahdavinia, G. R., Massoudi, A., Baghban, A., and Shokri, E. (2014). Study of adsorption of cationic dye on magnetic kappa-carrageenan/PVA nanocomposite hydrogels. *Journal of Environmental Chemical Engineering*, 2(3), 1578-1587.

Mahdavinia, G. R., Rahmani, Z., Mosallanezhad, A., Karami, S., and Shahriari, M. (2016). Effect of magnetic laponite RD on swelling and dye adsorption behaviors of κ-carrageenan-based nanocomposite hydrogels. *Desalination and Water Treatment*, 57(43), 20582-20596.

Mahdavinia, G. R., Soleymani, M., Sabzi, M., Azimi, H., and Atlasi, Z. (2017). Novel magnetic polyvinyl alcohol/laponite RD nanocomposite hydrogels for efficient removal of methylene blue. *Journal of Environmental Chemical Engineering*, 5(3), 2617-2630.

Mahmoud, M. E., Amira, M. F., Zaghloul, A. A., and Ibrahim, G. A. (2016). Microwave-enforced sorption of heavy metals from aqueous solutions on the surface of magnetic iron oxide-functionalized-3-aminopropyltriethoxysilane. *Chemical Engineering Journal*, 293, 200-206.

Malatji, N., Makhado, E., Ramohlola, K. E., Modibane, K. D., Maponya, T. C., Monama, G. R., and Hato, M. J. (2020). Synthesis and characterization of magnetic clay-based carboxymethyl cellulose-acrylic acid hydrogel nanocomposite for methylene blue dye removal from aqueous solution. *Environmental Science and Pollution Research*, 27(35), 44089-44105.

Maneerung, T., Liew, J., Dai, Y., Kawi, S., Chong, C., and Wang, C. H. (2016). Activated carbon derived from carbon residue from biomass gasification and its application for dye adsorption: kinetics, isotherms and thermodynamic studies. *Bioresource Technology*, 200, 350-359.

Marcelo, L. R., de Gois, J. S., da Silva, A. A., and Cesar, D. V. (2021). Synthesis of iron-based magnetic nanocomposites and applications in adsorption processes for water treatment: a review. *Environmental Chemistry Letters*, 19(2), 1229-1274.

Mittal, H., Kumar, V., Saruchi and Ray, S. S. (2016). Adsorption of methyl violet from aqueous solution using gum xanthan/Fe_3O_4 based nanocomposite hydrogel. *International Journal of Biological Macromolecules*, 89, 1-11.

Moghaddam, R. H., Shabani, A. M. H., and Dadfarnia, S. (2019). Synthesis of new hydrogels based on pectin by electron beam irradiation with and without surface modification for methylene blue removal. *Journal of Environmental Chemical Engineering*, 7(1), 102919.

Mohamed, N. A., Al-Harby, N. F., and Almarshed, M. S. (2020). Enhancement of adsorption of Congo red dye onto novel antimicrobial trimellitic anhydride isothiocyanate-cross-linked chitosan hydrogels. *Polymer Bulletin*, 77(12), 6135-6160.

Mohammadi, R., Saboury, A., Javanbakht, S., Foroutan, R., and Shaabani, A. (2021). Carboxymethylcellulose/polyacrylic acid/starch-modified Fe_3O_4 interpenetrating magnetic nano-composite hydrogel beads as pH-sensitive carrier for oral anticancer drug delivery system. *European Polymer Journal*, 153, 110500.

Monerris, M., Broglia, M. F., Yslas, E. I., Barbero, C. A., and Rivarola, C. R. (2019). Highly effective antimicrobial nanocomposites based on hydrogel matrix and silver nanoparticles: long-lasting bactericidal and bacteriostatic effects. *Soft Matter*, 15(40), 8059-8066.

Mufti, N., Atma, T., Fuad, A., and Sutadji, E. (2014). Synthesis and characterization of black, red and yellow nanoparticles pigments from the iron sand. *AIP Conference Proceedings*, 1617(1), 165-169.

Mustaffa, M. S., Azis, R. A. S., Abdullah, N. H., Ismail, I., and Ibrahim, I. R. (2019). An investigation of microstructural, magnetic and microwave absorption properties of multi-walled carbon nanotubes/ $Ni_{0.5}Zn_{0.5}Fe_2O_4$. *Scientific Reports*, 9(1), 1-7.

Nakhjiri, M. T., Bagheri Marandi, G., and Kurdtabar, M. (2019). Adsorption of methylene blue, brilliant green and rhodamine B from aqueous solution using collagen-g-p(AA-co-NVP)/Fe_3O_4@SiO_2 nanocomposite hydrogel. *Journal of Polymers and the Environment*, 27(3), 581-599.

Nakhjiri, M. T., Bagheri Marandi, G., and Kurdtabar, M. (2021). Preparation of magnetic double network nanocomposite hydrogel for adsorption of phenol and p-nitrophenol from aqueous solution. *Journal of Environmental Chemical Engineering*, 9(2), 105039.

Naushad, M., Sharma, G., and Alothman, Z. A. (2019). Photodegradation of toxic dye using Gum Arabic-crosslinked-poly (acrylamide)/$Ni(OH)_2$/FeOOH nanocomposites hydrogel. *Journal of Cleaner Production*, 241, 118263.

Nguyen, T. A., Nguyen, L. T. T., Bui, V. X., Nguyen, D. H., Lieu, H. D., Le, L. M., and Pham, V. (2020). Optical and magnetic properties of $HoFeO_3$ nanocrystals prepared by a simple co-precipitation method using ethanol. *Journal of Alloys and Compounds*, 834, 155098.

Nguyen, V. D., Kynicky, J., Ambrozova, P., and Adam, V. (2017). Microwave-assisted synthesis of goethite nanoparticles used for removal of Cr (VI) from aqueous solution. *Materials*, 10(7), 783.

Nikolic, V. N., Spasojevic, V., Panjan, M., Kopanja, L., Mrakovic, A., and Tadic, M. (2017). Re-formation of metastable ε-Fe_2O_3 in post-annealing of Fe_2O_3/SiO_2 nanostructure: synthesis, computational particle shape analysis in micrographs and magnetic properties. *Ceramics International*, 43(10), 7497-7507.

Nizamuddin, S., Siddiqui, M. T. H., Mubarak, N. M., Baloch, H. A., Abdullah, E. C., Mazari, S. A., Griffin, G. J., Srinivasan, M. P., and Tanksale, A. (2019). Iron oxide nanomaterials for the removal of heavy metals and dyes from wastewater. *Nanoscale Materials in Water Purification,* 447-472.

Odaci, D., Kahveci, M. U., Sahkulubey, E. L., Ozdemir, C., Uyar, T., Timur, S., and Yagci, Y. (2010). *In situ* synthesis of biomolecule encapsulated gold-cross-linked poly (ethylene glycol) nanocomposite as biosensing platform: a model study. *Bioelectrochemistry*, 79(2), 211-217.

Ohemeng-Boahen, G., Sewu, D. D., Tran, H. N., and Woo, S. H. (2021). Enhanced adsorption of Congo red from aqueous solution using chitosan/hematite nanocomposite hydrogel capsule fabricated via anionic surfactant gelation. *Colloids and Surfaces A: Physicochemical and Engineering Aspects*, 625, 126911.

Ohemeng-Boahen, G., Tran, H. N., Sewu, D. D., and Woo, S. H. (2020). Multi-membrane formation in chitosan hydrogel shell by the addition of goethite nanoparticles. *Carbohydrate Polymers*, 229, 115543.

Omidian, H., Rocca, J. G., and Park, K. (2005). Advances in superporous hydrogels. *Journal of Controlled Release*, 102(1), 3-12.

Ozay, O., Ekici, S., Baran, Y., Aktas, N., and Sahiner, N. (2009). Removal of toxic metal ions with magnetic hydrogels. *Water Research*, 43(17), 4403-4411.

Park, H. Y., Schadt, M. J., Wang, L., Lim, I. I. S., Njoki, P. N., Kim, S. H., Jang, M. Y., Luo, J., and Zhong, C. J. (2007). Fabrication of magnetic core@shell Fe oxide@Au nanoparticles for interfacial bioactivity and bio-separation. *Langmuir*, 23(17), 9050-9056.

Peng, N., Hu, D., Zeng, J., Li, Y., Liang, L., and Chang, C. (2016). Superabsorbent cellulose-clay nanocomposite hydrogels for highly efficient removal of dye in water. *ACS Sustainable Chemistry & Engineering*, 4(12), 7217-7224.

Peppas, N. A., and Hoffman, A. S. (2020). Hydrogels. *Biomaterials Science,* 153-166.

Pourjavadi, A., Hosseini, S. H., Seidi, F., and Soleyman, R. (2013). Magnetic removal of crystal violet from aqueous solutions using polysaccharide-based magnetic nanocomposite hydrogels. *Polymer International*, 62(7), 1038-1044.

Pu, S., Ma, H., Zinchenko, A., and Chu, W. (2017). Novel highly porous magnetic hydrogel beads composed of chitosan and sodium citrate: an effective adsorbent for the removal of heavy metals from aqueous solutions. *Environmental Science and Pollution Research*, 24(19), 16520-16530.

Radoń, A., Kubacki, J., Kądziołka-Gaweł, M., Gębara, P., Hawełek, Ł., Topolska, S., and Łukowiec, D. (2020). Structure and magnetic properties of ultrafine superparamagnetic Sn-doped magnetite nanoparticles synthesized by glycol assisted co-precipitation method. *Journal of Physics and Chemistry of Solids*, 145, 109530.

Radwan, M., Rashad, M. A., Sadek, M., and Elazab, H. A. (2019). Synthesis, characterization and selected application of chitosan-coated magnetic iron oxide nanoparticles. *Journal of Chemical Technology & Metallurgy*, 54(2).

Rafieian, S., Mirzadeh, H., Mahdavi, H., and Masoumi, M. E. (2019). A review on nanocomposite hydrogels and their biomedical applications. *Science and Engineering of Composite Materials*, 26(1), 154-174.

Rahmani, S., Zeynizadeh, B., and Karami, S. (2020). Removal of cationic methylene blue dye using magnetic and anionic-cationic modified montmorillonite: kinetic, isotherm and thermodynamic studies. *Applied Clay Science*, 184, 105391.

Ramirez-Muñiz, K., Perez-Rodriguez, F., and Rangel-Mendez, R. (2018). Adsorption of arsenic onto an environmental friendly goethite-polyacrylamide composite. *Journal of Molecular Liquids*, 264, 253-260.

Rebekah, A., Bharath, G., Naushad, M., Viswanathan, C., and Ponpandian, N. (2020). Magnetic graphene/chitosan nanocomposite: a promising nano-adsorbent for the removal of 2-naphthol from aqueous solution and their kinetic studies. *International Journal of Biological Macromolecules*, 159, 530-538.

Rezazadeh, H., Moghadam, P. N., Ehsanimehr, S., and Fareghi, A. R. (2020). Synthesis of a new magnetic nanocomposite hydrogel based on poly (vinyl acetate-co-maleic anhydride)/melamine for efficient dye removal. *Journal of Elastomers & Plastics*, 52(1), 70-89.

Saberi, A., Alipour, E., and Sadeghi, M. (2019). Superabsorbent magnetic Fe_3O_4-based starch-poly (acrylic acid) nanocomposite hydrogel for efficient removal of dyes and heavy metal ions from water. *Journal of Polymer Research*, 26(12), 1-14.

Saber-Samandari, S., Gulcan, H. O., Saber-Samandari, S., and Gazi, M. (2014). Efficient removal of anionic and cationic dyes from an aqueous solution using pullulan-graft-polyacrylamide porous hydrogel. *Water, Air, & Soil Pollution*, 225(11), 1-14.

Sadeghfar, F., Ghaedi, M., Asfaram, A., Jannesar, R., Javadian, H., and Pezeshkpour, V. (2018). Polyvinyl alcohol/Fe_3O_4@carbon nanotubes nanocomposite: electrochemical-assisted synthesis, physicochemical characterization, optical properties, cytotoxicity effects and ultrasound-assisted treatment of aqueous based organic compound. *Journal of Industrial and Engineering Chemistry*, 65, 349-362.

Salama, A. (2018). Preparation of CMC-g-P(SPMA) super adsorbent hydrogels: exploring their capacity for MB removal from waste water. *International Journal of Biological Macromolecules*, 106, 940-946.

Sarkar, N., Sahoo, G., and Swain, S. K. (2020). Graphene quantum dot decorated magnetic graphene oxide filled polyvinyl alcohol hybrid hydrogel for removal of dye pollutants. *Journal of Molecular Liquids*, 302, 112591.

Schwertmann, U., and Cornell, R. M. (2008). Iron oxides in the laboratory: preparation and characterization. John Wiley & Sons.

Sewu, D. D., Lee, D. S., Woo, S. H., and Kalderis, D. (2021). Decolorization of triarylmethane dyes, malachite green, and crystal violet, by sewage sludge biochar: isotherm, kinetics, and adsorption mechanism comparison. *Korean Journal of Chemical Engineering*, 38(3), 531-539.

Shalla, A. H., Bhat, M. A., and Yaseen, Z. (2018). Hydrogels for removal of recalcitrant organic dyes: a conceptual overview. *Journal of Environmental Chemical Engineering*, 6(5), 5938-5949.

Sharma, G., ALOthman, Z. A., Kumar, A., Sharma, S., Ponnusamy, S. K., and Naushad, M. (2017). Fabrication and characterization of a nanocomposite hydrogel for combined photocatalytic degradation of a mixture of malachite green and fast green dye. *Nanotechnology for Environmental Engineering*, 2(1), 1-7.

Sharma, G., Kumar, A., Sharma, S., Ala'a, H., Naushad, M., Ghfar, A. A., Ahamad, T., and Stadler, F. J. (2019). Fabrication and characterization of novel Fe0@Guar gum-crosslinked-soya lecithin nanocomposite hydrogel for photocatalytic degradation of methyl violet dye. *Separation and Purification Technology*, 211, 895-908.

Sherlala, A. I. A., Raman, A. A. A., Bello, M. M., and Buthiyappan, A. (2019). Adsorption of arsenic using chitosan magnetic graphene oxide nanocomposite. *Journal of Environmental Management*, 246, 547-556.

Shi, X., Ruan, W., Hu, J., Fan, M., Cao, R., and Wei, X. (2017). Optimizing the removal of rhodamine B in aqueous solutions by reduced graphene oxide-supported nanoscale zerovalent iron (nZVI/rGO) using an artificial neural network-genetic algorithm (ANN-GA). *Nanomaterials*, 7(6), 134.

Shipley, H. J., Engates, K. E., and Grover, V. A. (2013). Removal of Pb (II), Cd (II), Cu (II), and Zn (II) by hematite nanoparticles: effect of sorbent concentration, pH, temperature, and exhaustion. *Environmental Science and Pollution Research*, 20(3), 1727-1736.

Shirsath, S. R., Hage, A. P., Zhou, M., Sonawane, S. H., and Ashokkumar, M. (2011). Ultrasound assisted preparation of nanoclay Bentonite-FeCo nanocomposite hybrid hydrogel: a potential responsive sorbent for removal of organic pollutant from water. *Desalination*, 281, 429-437.

Singh, R., Munya, V., Are, V. N., Nayak, D., and Chattopadhyay, S. (2021). A biocompatible, pH-Sensitive, and magnetically separable superparamagnetic hydrogel nanocomposite as an efficient platform for the removal of cationic dyes in wastewater treatment. *ACS Omega*, 6(36), 23139-23154.

Sinha, V., and Chakma, S. (2019). Advances in the preparation of hydrogel for wastewater treatment: a concise review. *Journal of Environmental Chemical Engineering*, 7(5), 103295.

Sionkowska, A. (2011). Current research on the blends of natural and synthetic polymers as new biomaterials. *Progress in Polymer Science*, 36(9), 1254-1276.

Song, F., Li, X., Wang, Q., Liao, L., and Zhang, C. (2015). Nanocomposite hydrogels and their applications in drug delivery and tissue engineering. *Journal of Biomedical Nanotechnology*, 11(1), 40-52.

Su, X., and Chen, B. (2018). Tough, resilient and pH-sensitive interpenetrating polyacrylamide/ alginate/montmorillonite nanocomposite hydrogels. *Carbohydrate Polymers*, 197, 497-507.

Sun, X., Yang, L., Li, Q., Zhao, J., Li, X., Wang, X., and Liu, H. (2014). Amino-functionalized magnetic cellulose nanocomposite as adsorbent for removal of Cr(VI): synthesis and adsorption studies. *Chemical Engineering Journal*, 241, 175-183.

Suri, K., Annapoornib, S., Sarkara, A. K., and Tandon, R. P. (2002). Gas and humidity sensors based on iron oxide-polypyrrole nanocomposites. *Sensors and Actuators B: Chemical*, 81(2-3), 277-282.

Tang, P., Wong, C., and Woo, K. (2011). Methods of preparation, characterisation and applications. *Asian Journal of Biological Sciences*, 4(2), 189-195.

Teng, X., and Yang, H. (2005). Synthesis of magnetic nanocomposites and alloys from platinum-iron oxide core-shell nanoparticles. *Nanotechnology*, 16(7), S554-561.

Tesh, S. J., and Scott, T. B. (2014). Nano-composites for water remediation: a review. *Advanced Materials*, 26(35), 6056-6068.

Thommes, M., Kaneko, K., Neimark, A. V., Olivier, J. P., Rodriguez-Reinoso, F., Rouquerol, J., and Sing, K. S. (2015). Physisorption of gases, with special reference to the evaluation of surface area and pore size distribution (IUPAC Technical Report). *Pure and Applied Chemistry*, 87(9-10), 1051-1069.

Torres-Badajoz, S. G., Rodríguez-Núñez, J. R., López-Ramírez, E., Pena-Caballero, V., Villa-Lerma, A. G., Madera-Santana, T. J., Rodríguez-Carrillo, M. G., and Castillo, O. S. (2022). Kinetic and equilibrium studies of Cr (VI) adsorption using glutaraldehyde-crosslinked chitosan beads. *Iranian Journal of Science and Technology, Transactions A: Science*, 46(1), 147-160.

Tran, H. N., You, S. J., Hosseini-Bandegharaei, A., and Chao, H. P. (2017). Mistakes and inconsistencies regarding adsorption of contaminants from aqueous solutions: a critical review. *Water Research*, 120, 88-116.

Ullah, F., Othman, M. B. H., Javed, F., Ahmad, Z., and Akil, H. M. (2015). Classification, processing and application of hydrogels: a review. *Materials Science and Engineering: C*, 57, 414-433.

Vangelista, S., Mantovan, R., Cocco, S., Lamperti, A., Salicio, O., and Fanciulli, M. (2012). Chemical vapor deposition growth of Fe_3O_4 thin films and Fe/Fe_3O_4 bi-layers for their integration in magnetic tunnel junctions. *Thin Solid Films*, 520(14), 4617-4621.

Varaprasad, K., Nunez, D., Yallapu, M. M., Jayaramudu, T., Elgueta, E., and Oyarzun, P. (2018). Nano-hydroxyapatite polymeric hydrogels for dye removal. *RSC Advances*, 8(32), 18118-18127.

Varaprasad, K., Raghavendra, G. M., Jayaramudu, T., Yallapu, M. M., and Sadiku, R. (2017). A mini review on hydrogels classification and recent developments in miscellaneous applications. *Materials Science and Engineering: C*, 79, 958-971.

Vilardi, G., Ochando-Pulido, J. M., Verdone, N., Stoller, M., and Di Palma, L. (2018). On the removal of hexavalent chromium by olive stones coated by iron-based nanoparticles: equilibrium study and chromium recovery. *Journal of Cleaner Production*, 190, 200-210.

Wan, D., Li, W., Wang, G., and Wei, X. (2016). Size-controllable synthesis of Fe_3O_4 nanoparticles through oxidation-precipitation method as heterogeneous Fenton catalyst. *Journal of Materials Research*, 31(17), 2608-2616.

Wang, G., Zeng, Y., Zhou, F., Chen, X., Ma, Y., Zheng, L., Li, M., Sun, Y., Liu, X., and Liu, H. (2020). One-step solvothermal synthesis of porous $MnFe_2O_4$ nanoflakes and their magnetorheological properties. *Journal of Alloys and Compounds*, 819, 153044.

Wang, J., Gao, M., Wang, D., Li, X., Dou, Y., Liu, Y., and Pan, H. (2015). Chemical vapor deposition prepared bi-morphological carbon-coated Fe_3O_4 composites as anode materials for lithium-ion batteries. *Journal of Power Sources*, 282, 257-264.

Wang, J., Li, X., Cheng, Q., Lv, F., Chang, C., and Zhang, L. (2020). Construction of beta-FeOOH@tunicate cellulose nanocomposite hydrogels and their highly efficient photocatalytic properties. *Carbohydrate Polymers*, 229, 115470.

Wang, L. Y., and Wang, M. J. (2016). Removal of heavy metal ions by poly (vinyl alcohol) and carboxymethyl cellulose composite hydrogels prepared by a freeze-thaw method. *ACS Sustainable Chemistry & Engineering*, 4(5), 2830-2837.

Wang, T., Dai, Z., Kang, J., Fu, F., Zhang, T., and Wang, S. (2018). A TiO_2 nanocomposite hydrogel for Hydroponic plants in efficient water improvement. *Materials Chemistry and Physics*, 215, 242-250.

Wang, Y., Gao, Y., Chen, L., and Zhang, H. (2015). Goethite as an efficient heterogeneous Fenton catalyst for the degradation of methyl orange. *Catalysis Today*, 252, 107-112.

Wang, Y., Wu, D., Wei, Q., Wei, D., Yan, T., Yan, L., Hu, L., and Du, B. (2017). Rapid removal of Pb(II) from aqueous solution using branched polyethylenimine enhanced magnetic carboxymethyl chitosan optimized with response surface methodology. *Scientific Reports*, 7(1), 1-11.

Wei, Z., Xing, R., Zhang, X., Liu, S., Yu, H., and Li, P. (2013). Facile template-free fabrication of hollow nestlike α-Fe_2O_3 nanostructures for water treatment. *ACS Applied Materials & Interfaces*, 5(3), 598-604.

Wu, X. L., Wang, L., Chen, C. L., Xu, A. W., and Wang, X. K. (2011). Water-dispersible magnetite-graphene-LDH composites for efficient arsenate removal. *Journal of Materials Chemistry*, 21(43), 17353-17359.

Wu, X., Cao, R., Hu, J., Zhou, C., Fu, L., and Wei, X. (2020). Graphene-supported iron-based composites: a review of applications to wastewater treatment. *Desalination and Water Treatment*, 181, 300-320.

Xu, P., Zeng, G. M., Huang, D. L., Feng, C. L., Hu, S., Zhao, M. H., Lai, C., Wei, Z., Huang, C., and Xie, G. X. (2012). Use of iron oxide nanomaterials in wastewater treatment: a review. *Science of the Total Environment*, 424, 1-10.

Yan, L., Li, S., Yu, H., Shan, R., Du, B., and Liu, T. (2016). Facile solvothermal synthesis of Fe_3O_4/bentonite for efficient removal of heavy metals from aqueous solution. *Powder Technology*, 301, 632-640.

Yang, X., Chen, C., Li, J., Zhao, G., Ren, X., and Wang, X. (2012). Graphene oxide-iron oxide and reduced graphene oxide-iron oxide hybrid materials for the removal of organic and inorganic pollutants. *RSC Advances*, 2(23), 8821-8826.

Yang, X., Liu, Q., Chen, X., Yu, F., and Zhu, Z. (2008). Investigation of PVA/ws-chitosan hydrogels prepared by combined γ-irradiation and freeze-thawing. *Carbohydrate Polymers*, 73(3), 401-408.

Yu, X., Shan, Y., and Chen, K. (2018). Fabrication and characterization of morphology-tuned single-crystal monodisperse Fe_3O_4 nanocrystals. *Applied Surface Science*, 439, 298-304.

Yu, X., Tong, S., Ge, M., Zuo, J., Cao, C., and Song, W. (2013). One-step synthesis of magnetic composites of cellulose@iron oxide nanoparticles for arsenic removal. *Journal of Materials Chemistry A*, 1(3), 959-965.

Zelmanov, G., and Semiat, R. (2013). Selenium removal from water and its recovery using iron (Fe^{3+}) oxide/hydroxide-based nanoparticles sol (NanoFe) as an adsorbent. *Separation and Purification Technology*, 103, 167-172.

Zhang, B., Zhang, T., Zhang, Z., and Xie, M. (2019). Hydrothermal synthesis of a graphene/magnetite/montmorillonite nanocomposite and its ultrasonically assisted methylene blue adsorption. *Journal of Materials Science*, 54(16), 11037-11055.

Zhang, J., Huang, Q., and Du, J. (2016). Recent advances in magnetic hydrogels. *Polymer International*, 65(12), 1365-1372.

Zhang, J., Zhai, S., Li, S., Xiao, Z., Song, Y., An, Q., and Tian, G. (2013). Pb(II) removal of $Fe_3O_4@SiO_2-NH_2$ core-shell nanomaterials prepared via a controllable sol-gel process. *Chemical Engineering Journal*, 215-216, 461-471.

Zhang, T., Zuo, T., Hu, D., and Chang, C. (2017). Dual physically cross-linked nanocomposite hydrogels reinforced by tunicate cellulose nanocrystals with high toughness and good self-recoverability. *ACS Applied Materials & Interfaces*, 9(28), 24230-24237.

Zhao, Q. S., Ji, Q. X., Xing, K., Li, X. Y., Liu, C. S., and Chen, X. G. (2009). Preparation and characteristics of novel porous hydrogel films based on chitosan and glycerophosphate. *Carbohydrate Polymers*, 76(3), 410-416.

Zheng, X., Wu, D., Su, T., Bao, S., Liao, C., and Wang, Q. (2014). Magnetic nanocomposite hydrogel prepared by ZnO-initiated photopolymerization for La (III) adsorption. *ACS Applied Materials & Interfaces*, 6(22), 19840-19849.

Zhou, Q., Li, J., Wang, M., and Zhao, D. (2016). Iron-based magnetic nanomaterials and their environmental applications. *Critical Reviews in Environmental Science and Technology*, 46(8), 783-826.

Zhuang, F., Tan, R., Shen, W., Zhang, X., Xu, W., and Song, W. (2015). Monodisperse magnetic hydroxyapatite/Fe_3O_4 microspheres for removal of lead(II) from aqueous solution. *Journal of Alloys and Compounds*, 637, 531-537.

Chapter 13

Polysaccharide-Based Nanocomposite Hydrogels as Adsorbent Materials for Remediation of Mercuric Ions from Wastewater

Prajwal Kulal[1,2,*]
and Vishalakshi Badalamoole[2]

[1] Karkala Jnana Sudha P.U. College, Ganit Nagar, Kukkundoor, Udupi, Karnataka, India
[2] Department of Post-Graduate Studies and Research in Chemistry,
Mangalore University, Mangalagangothri (D.K.), Karnataka, India

Abstract

The presence of harmful pollutants limits the sustainable source of clean water globally. Mercury and related compounds generally reach the environment by human activities, such as mining, coal burning, petrol refining, and contaminate soil and water. Among several accessible wastewater purification technologies, adsorption is the most extensively practiced method due to its feasible, economic, and supportable characteristics. In recent years, polysaccharides have been widely employed for fabricating adsorption materials owing to their inherent biodegradability, biocompatibility, ease of modification, high adsorptive ability, low cost, and renewability. This chapter is envisioned at providing a report on the systematic development of various polysaccharide-based nanocomposite hydrogel adsorbents for mercuric ion removal from wastewater. The critical factors influencing the adsorption process and mechanism, along with future prospects of these materials are also discussed.

Keywords: polysaccharides, adsorbents, mercuric ion, water purification

13.1. Introduction

Mercury, a silver-white shiny metallic liquid, is one of the oldest metals known and its compounds are used throughout history. It has the features of low boiling point (356.73°C) and melting point (-38.83°C), good electrical conductivity (1.1×10^6 S/m), high specific gravity

* Corresponding Author's Email: prajwal92kulal@gmail.com.

In: Nanocomposite Hydrogels and their Emerging Applications
Editor: Mohammad Sirousazar
ISBN: 979-8-88697-675-5

(13.6), and uniform expansion coefficient (1.8×10^{-4} 1/°C), which make it extensively used in the fields of industry, health, agriculture, medicine, transportation, and national defense (Jeon and Holl, 2003). Archaeologists found mercury in an Egyptian tomb dating to 1500 BC (Pendergast, 1982). The Egyptians and the Chinese have used cinnabar as a red pigment in tattoos for centuries before the birth of Christ (Vane, Beriro, and Turner, 2015). The Carroting process was used from the mid-18th to the mid-19th centuries in the manufacture of felt hats. Animal skins were rinsed in a carroty of mercuric nitrate (Glenn, Cahn, and Siler, 1973). Mercurous chloride has been utilized in traditional medicine as a topical disinfectant, diuretic, and laxative. Mercurochrome is used on wounds as an antibacterial agent (Liu et al., 2008). Earlier, for the treatment of syphilis, mercuric chloride was used (Maprani et al., 2005). "Blue mass" a pill or syrup containing mercury was used as the main ingredient for the treatment of depression, constipation, child-bearing issues, and toothache (Ngim et al., 1992). Mercuric fulminate is an explosive used in tapping caps for weaponries (Garner and Hailes, 1933).

Mercury is usually present in a variety of forms in the environment, and it can be divided into three categories, including elemental mercury and inorganic and organic forms. Mercury and most of its compounds are enormously toxic and are considered as pollutants in the environment which are extremely harmful to living organisms. The chief sources of mercury pollution are chemical manufacturing, mineral industry production, waste incineration, fossil fuel combustion, forest fire, volcanic activity, and atmospheric mercury deposition (Singh and Kumar, 2011). The usage of mercury-contaminated water predominantly disturbs the nervous system, with minor effects on the renal and gastrointestinal systems (Rice et al., 2014). Basically, mercury troubles the neurological and mental functions of the human body by linking with thiol residues present in proteins (Azevedo, Furieri, and Pecanha, 2012). The inactivation of sulfur by intracellular mercury can cause severe disease by inhibiting various hormones, enzymes, and cofactors (Jan et al., 2015). Because mercury tends to bioaccumulate and biomagnify in the food chain, it can even adversely affect human fetuses (Jaishankar et al., 2014). The US Environmental Protection Agency (USEPA) has included Hg(II) ion in the list of priority pollutants with a requisite limitation of 10 mg/L for wastewater. The abatement of mercury content from water is an acute environmental management endeavor.

13.2. Biopolymer-Based Functional Materials

Together with other biomolecules, like proteins and nucleotides, polysaccharides are abundant and constitute a prolific class of macromolecules that almost exist in all living forms. Polysaccharides represent a wide-ranging group of stereoregular carbohydrate polymers containing different monosaccharide units, which are interconnected by glycosidic bonds. Monosaccharide building blocks and glycosidic linkages establish and govern the complexity and diversity of the polysaccharides (Whistler, 1973). Structural variances confer distinct physical and chemical properties to the polysaccharides. The presence of diverse functional groups, the prospective for modification, biocompatibility, biodegradability, significant absorption capacity for water, and so on turn polysaccharides to favorable candidates for numerous applications (Bhattacharya and Misra, 2004). Many polysaccharides are non-toxic and for this reason, there is predominant commercial interest. Based on the originated sources, they can be classified into the plant, animal, algae, microbial, and fungi polysaccharides (Yalpani, 1988) as listed in Table 13.1.

Table 13.1. Classification of polysaccharides according to their origin

Origin	Polysaccharide
Plant	Starch, Cellulose, Hemicellulose, Pectin, Acacia gum, Guar gum, Karaya gum, Tara gum, Locust bean gum, Femigreek gum, Tragacanth gum, Cassia tora gum, Glucomannan, Costus glucans, Arabinogalactan, Xylan, Xyloglucan, Inulin, and β-Glucan
Animal	Glycogen, Hyaluronan, Chitin, Chitosan, Heparin, Chondroitin sulfate, Keratan sulfate, and Dermatan sulfate
Algae	Alginate, Agar, Carrageenan, Fucoidan, Ulvan, Laminarin, Angelan, Porphyran, Spirulan, Agarose, and Rhodymenan
Microbial	Curdlan, Scleroglucan, Pullulan, Dextran, Xanthan, Levan, Gellan, Emulsan, Schizophyllan, Glucuronan, Succinoglycan, and Nigeran
Fungi	Elsinan and Yeast glucans

The chemical modification of the polysaccharides with synthetic monomers allows the combination of the useful features of natural and synthetic polymers and results in the formation of functionalized materials for highly specific applications (Athawale and Lele, 1998). Graft copolymer gels and interpenetrating network systems comprising of polysaccharides and synthetic polymers are an important class of such functionalized materials used for widespread applications (Bagdanov, Schacht, and Blucke, 1997).

Hydrogels can be defined as soft materials composed of polymer chains with either physical or chemical crosslinks that result in three-dimensional networks which render them capable of holding large contents of water. The hydrophilicity and crosslinking between network chains play an important role in controlling the structure and properties of hydrogels (Ahmed, 2015). The capability of hydrogels to absorb water arises from hydrophilic functional groups attached to the polymeric backbone, while their poor solubility in water happens from crosslinks among network chains. Based on the nature of the polysaccharide used as well as on the density of the network linkages, such structures can contain various quantities of water characteristically in the swollen state (Ozmen and Okay, 2005). The word “hydrogel” was first used in literature around 1894 (Bemmelen, 1894). In 1936, DuPont’s research group synthesized poly (2-hydroxyethyl methacrylate) (PHEMA). This polymer was brittle, hard, and glassy and hence did not gain much importance (Dave and Gor, 2018). In 1960, Wichterle and Lim (1960) synthesized crosslinked PHEMA hydrogel which was found to be soft and elastic in nature. They presented its application in soft contact lenses. Thereafter, a great number of hydrogel formulations were made (Coviello et al., 2007). In the last few years, hydrogels have established a superior concern among researchers due to their encouraging applications in a wide range of areas.

13.3. Classification of Hydrogels

The polymer network of hydrogels can be classified as nonporous, microporous, and superporous. In nonporous hydrogels, swelling in water occurs through only the diffusion of water molecules to the surface. Microporous hydrogel contains pores of few microns dimension in its structure and the swelling in these hydrogels is considered to proceed by diffusion as well as leaching of water molecules via pores. Superporous hydrogels have been developed in recent times (Singhal and Gupta, 2016). They contain a vast number of large-sized pores that are helpful in attaining extraordinarily high swelling. The pore size of superporous hydrogels plays a great role in deciding the swelling rates. Extremely porous hydrogel networks have been found to exhibit excessive and faster swelling.

Hydrogels can also be categorized by the nature of crosslinking in their networks, either static or dynamic (Appel et al., 2012). Static hydrogels are crosslinked with inflexible covalent bonds whereas dynamic hydrogels are crosslinked via either physical (non-covalent) or chemical (covalent) bonds. Static hydrogels are always susceptible to permanent collapse, with the lack of the ability to reshuffle their crosslinked networks. They usually break before any noticeable cracks at the failure point. Therefore, such inflexible arrangements result in a breakable hydrogel with limited structural complexity. Dynamic hydrogels have the ability to recover their shapes and restore damages imposed on them. Because of their more durable structures, they are appropriate for use in applications, such as biomedicine, sensors, and adsorbents for pollutant removal from industrial effluents.

13.4. Nanocomposite Hydrogels

Hybrid materials are heterogeneous in nature and are created by the assembly of organic and inorganic components. They possess characteristic shapes, surface appearances, environmental tolerances, and overall durabilities (Thakur and Kessler, 2015). Hybrid hydrogels can be described as water-swollen materials which retain a unique three-dimensional structure and are composed of different polymeric and inorganic materials. Nanotechnology provides an opportunity to develop advanced materials with multifunctionality. The incorporation of nanomaterials in hydrogels is an innovative approach to obtain hydrogel systems with desired properties, like hydrophilicity or hydrophobicity, dispersibility, porosity, and mechanical strength (Kaur et al., 2012).

A composite is defined as a material that is composed of two or more different phases having distinguishable chemical, physical, biological, or structural properties. Nanocomposites are multiphase materials, in which at least one of the phases displays dimensions in the range of 10-1000 Å. Nanocomposites came to existence in 1991 with the discovery of carbon nanotubes which form nanocomposites with extraordinary properties (Iijima, 2002). The amazing associations of properties, like catalytic, electrical, mechanical, magnetic, and biodegradability are accomplished to come together in one substrate as per the application requirements (Asmatulu, Twomey, and Overcash, 2015; Dong et al., 2017). The properties of the polymeric nanocomposites depend upon the synthetic process, type of nanofillers and their distribution, interfacial bonds, the extent of intermixing of fillers and matrix, amount and features of nanofillers, nature of the developed interphase, size and shape of nanomaterials, and morphology of system (Miao and Hamad, 2013). Biopolymer-based nanocomposites or bio-hybrid materials comprise mainly polysaccharides and nanoscopic inorganic substances of diverse structure and morphology. They are different from nanocomposites in preparation method, functionalities, properties, biodegradability, and biocompatibility (Ferreira et al., 2017).

13.5. Classification of Nanocomposite Hydrogels

Depending on several aspects, nanocomposite hydrogels can be classified based on the types of matrix used and the shape of reinforcements.

13.5.1. Based on Matrix Used

According to their matrix materials, the nanocomposite hydrogels are categorized in four different types:

- *Ceramic matrix nanocomposites*: In this group of nanocomposites, a small portion of the volume is occupied by a ceramic material (e.g., borides, nitrides, oxides, and silicides) which is finely dispersed in order to bring about the specific nanoscopic properties.
- *Metal/metal oxide matrix nanocomposites*: Different metal nanoparticles (e.g., Au and Ag) or metal oxides (e.g., iron oxide, titanium oxide, zinc oxide, and copper oxide) with appropriate physico-chemical characteristics have been embedded in polymer networks to synthesize metal/metal oxide nanocomposite hydrogels. The covalent bonds offer strong interactions between nanoparticles and polymer chains, producing substantial variations in the properties of nanocomposite hydrogels.
- *Clay matrix nanocomposites*: Nanoclays are nanoparticles of layered mineral silicates, which provide an anisotropic plate-like structure. The main kinds of clay nanoparticles incorporated into nanocomposite hydrogels contain bentonite, montmorillonite, and other silicate clays. Layered silicates can deliver significant mechanical enrichment due to widespread surface area even in lower concentrations.
- *Carbon matrix nanocomposites*: These are distinct kinds of composites that comprise carbon introduced in a carbonaceous matrix. Carbon-based nanocomposites have been put to a variety of applications, including aeronautical and space applications, because of their lightweight and high-temperature properties.

13.5.2. Based on Shape

Based on the shape of the particle reinforcements, nanocomposite hydrogels can be classified into particulate, elongated particle, and layered structures.

- *Particulate nanocomposite*: In particulate nanocomposites, iso-dimensional particles are typically used as reinforcements. The reinforcing effect is moderate due to the low aspect ratio and the main purpose of exploiting such type of reinforcements is to develop resistance to flammability, diminish permeability, and lower the cost.
- *Elongated particle nanocomposite*: Elongated particles (e.g., carbon nanotubes and cellulose nanofibrils) are used as reinforcement materials in elongated particle nanocomposites. Owing to the higher aspect ratio of the reinforcement, these nanocomposites show greater mechanical strength.

- *Layered particle nanocomposites*: A layered particle nanocomposite is categorized into three subclasses based on the distribution of particles in the matrix. Flocculated or phase-separated nanocomposites are formed when there is no partition between the layers as a result of particle-particle interactions. Intercalated nanocomposites are obtained when polymer chains are intercalated between sheets of layered nanoparticles. Exfoliated nanocomposites are produced when there is a partition of individual layers.

13.6. Fabrication of Polysaccharide-Based Nanocomposite Hydrogels

Various methods have been employed for the synthesis of polysaccharide-based hybrid nanocomposites. The most significant requirement is the uniform distribution of the nanoparticles throughout the matrix. As the nanoparticles have the trend to get agglomerated to create large-sized particles, there is a requirement to regulate these systems inside a matrix in such a manner that no two particles associate to make a macro-sized particle. The approaches that are considered commonly are described below.

13.6.1. Electrospinning

Electrospinning is a physical process of nanocomposite design that consists of the extrusion of a polymeric solution in the presence of dispersed nanoparticles through a syringe needle (spinneret) on a collection plate at a high voltage field (Meli et al., 2010). It includes three ways, wet-dry, wet-wet, and co-axial electrospinning. Wet-dry electrospinning uses a volatile solvent that vaporizes as the nanocomposite is turned into a collection plate. Chitosan has been wet-dry electrospun with collagen from hexafluoro isopropanol/trifluoroacetic acid to get fibers (Chen et al., 2010). Wet-wet electrospinning can make use of a non-volatile solvent, for instance, a room temperature ionic liquid, which is rotated into a second solvent in a collection plate positioned between the collection plate and the spinneret. For example, gelatin can be wet-wet electrospun with the acidic polysaccharide, hyaluronan from a dimethylformamide/water solution (Li et al., 2006). Co-axial electrospinning is a recently adapted technology, which can formulate nanocomposites from two different components concurrently. The resulting nanocomposites have a core-sheath structure. Cellulose can be co-dissolved with other polysaccharides, for instance, the anticoagulant heparin, in room temperature ionic liquid and electrospun together to form composite fibers (Viswanathan et al., 2006).

13.6.2. Film Casting and Dip Coating

The film casting process involves the formation of continuous film by coating the dispersion on a carrier web followed by drying and sintering (Kim et al., 2010). Multiple coatings can be made to obtain the desired thickness. Cellulose dissolved in room temperature ionic liquid and penetrated into the multi-walled carbon nanotube (MWCNT) forms a uniform cellulose-MWCNT cast film with outstanding electrical properties (Pushparaj et al., 2007). Dip coating comprises the deposition of a liquid film by the controlled withdrawal of a substrate from a solution using a dip coater (Stevens et al., 2011). The substrate is dropped into a solution bath until it is fully immersed. A slight delay arises before the substrate is withdrawn. For the period of the withdrawal process, a thin layer of the solution rests on the surface of the substrate. When completely withdrawn, the liquid from the film then initiates to vaporize and leaves behind a dry film. Using this method, cotton fabrics have been coated with ZnO-soluble starch nanocomposites for antibacterial applications (Vigneshwaran et al., 2006).

13.6.3. Layer-by-Layer Assembly

The layer-by-layer process includes the successive deposition of two or more materials that physically bond together (Filip et al., 2011). A charged material is adsorbed on top of a substrate, and after washing, an oppositely charged material is adsorbed above the first layer. These set up a single bilayer with a thickness usually in the order of nanometers, and the deposition procedure can be reiterated until a multilayer film of the desired thickness has been collected. Electrostatically structured polysaccharide films of alginate and chitosan coated with layered silicate are produced by this method and utilized in the inhibition of bacterial growth (Deng et al., 2011).

13.6.4. Colloidal Assembly

The monodisperse nanometer-sized particles in a polymeric suspension can spontaneously develop close-packed colloidal crystal clusters if their free volume is controlled beneath a definite level (Wang et al., 2011). The combination of cyclodextrin and starch nanocrystals has been prepared by this method and developed for the sustained release of drugs (Chang et al., 2010).

13.6.5. Ionotropic Gelation

Ionotropic gelation is recognized as the ability of polysaccharides to crosslink in the presence of counter ions to form nanoparticles (Vongchan et al., 2011). The size and surface charge of nanoparticles can be adjusted by altering the proportion of polysaccharide and stabilizer. Chitosan-hyaluronic acid nanoparticles loaded with heparin has been prepared by this method and utilized for the treatment of asthma (Ampuero et al., 2009).

13.6.6. Grafting Synthetic Polymers by Microwave Irradiation

A popular and flexible method to develop polysaccharide-based nanocomposites is the attaching of synthetic polymers to natural polysaccharides in presence of nanoparticles. Despite of the remarkable physical and chemical properties of polysaccharide-based nanocomposites, the undesired homopolymer formation in the simultaneous competing reaction decreases the copolymer yield and also creates difficulties in the commercialization of the grafting techniques (Da Silva, De Paula, and Feitosa, 2007). Furthermore, the need for an inert atmosphere is an additional drawback for various conventional grafting techniques (Kappe, 2004). To overcome these problems, the usage of microwave irradiation has been encouraged in the production of a variety of modified polysaccharide-containing nanocomposites (Tizzotti et al., 2010).

Microwaves involve electromagnetic radiation in the frequency range of 300 MHz to 300 GHz. They generate electromagnetic energy which can be rapidly transmitted in the bulk of the reaction mixture (Galema, 1997). Microwave radiations cause selective excitation of the polar bonds only, which successively leads to their cleavage. This cleavage of bonds generates large numbers of free radical sites on the polysaccharide backbone. The C-C sequence of the backbone remains unaffected by microwave radiation as it is moderately non-polar (Singh et al., 2006). Microwave irradiation substantially lowers the usage of toxic chemicals as well as the reaction time for the grafting reactions. The *in situ* mode of energy conversion in microwaves confirms high yields, product selectivity, and enhanced purity of the preferred products (Dallinger and Kappe, 2007). Moreover, in many instances, polysaccharide-based nanocomposites produced by microwave assistance exhibited superior properties over conventionally synthesized counterparts. Microwave heating is not only an alternative to conventional thermal heating for polysaccharide modifications, but it also offers an alternative eco-friendly, cleaner, greener methodology ensuring better control and greater reproducibility of percentage grafting in the end product (Singh, Kumar, and Sanghi, 2012).

13.7. Adsorption Process for Water Purification

Intense research and progressive actions are stimulated in the direction of development of efficient methods for the removal of metallic ions from industrial effluents since the early days (Sharma et al., 2009). Diverse techniques, such as biological treatment, ion exchange, chemical oxidation, coagulation, membrane separation, and adsorption have been recognized for effluent treatment (Allegre et al., 2004; Hassan and Hawkyard, 2002; Zaleschi et al., 2014; Tripathi and Shukla, 1991; Visvanathan, Aim, and Parameshwaran, 2000; Gangadhar and Vishalakshi, 2019; Ahn, Chang, and Yoon, 1991). The adsorption process has been established to be superior to other techniques in terms of initial cost, flexibility, simplicity of design, ease of operation, and insensitivity to toxic pollutants. Adsorption also does not produce harmful substances (Mittal, Maity, and Ray, 2015; Prajwal and Vishalakshi, 2020a). Adsorption is a physico-chemical method that usually includes the mass transfer of a constituent (adsorbate) from the gas or liquid phase to the surface and/or boundary of the solid phase (adsorbent) (Yagub et al., 2014). Usually, the adsorption process occurs in four successive steps (Hai et al., 2017):

1. Bulk diffusion.
2. Film diffusion or external mass diffusion: transport from the bulk solution to the adsorbent surface by diffusion through the boundary layer.
3. Pore diffusion or intraparticle diffusion: diffusion from the external surface into the pores of adsorbents.
4. Physical and/or chemical reaction: adsorption due to interaction between adsorbate and active sites of adsorbent.

The adsorbate molecules adsorbed on the surface of the adsorbent can be single or multiple layers. When monolayer adsorption happens, the adsorbates can only cover a single molecular layer on the adsorbent surface, and chemical adsorption is a typical monolayer adsorption. While for multilayer adsorption, the adsorbent surface adsorbs more than one molecular layer of adsorbates. Physical adsorption can be multilayer adsorption, and multiple layers of physical adsorption can be performed onto the chemically adsorbed monolayer.

Adsorption can be categorized into two types, including physical and chemical adsorption (Boparai, Joseph, and Carroll, 2011). Physical adsorption is a weak adsorption process with small energy requirements (5-40 kJ/mol) and the foremost driving forces include electrostatic interaction, van der Waal forces, hydrogen bonding, and π-π interaction, comprising the establishment of outer-sphere surface complexes. Chemical adsorption is described by the creation of chemical bonds or surface coordination compounds (inner-sphere surface complexes) between adsorbate and the adsorbent surface through electron transfer or pairing, such as ion/ligand exchange, reduction/oxidation, Lewis acid-base interaction, demanding larger activation energies (41-800 kJ/mol). The type of adsorption mechanism is governed by the nature of the adsorbate species and the adsorbent. For a specific water decontamination process, multiple interactions may take place instantaneously. The predominant types of mechanisms governing the adsorption of contaminants are discussed below.

13.7.1. Physical Interaction

- *Pore filling or size-selective adsorption*: Pore filling is the usual adsorption mechanism for porous adsorbents. The large mesopores turn out to be saturated during the initial adsorption stage and then it takes an extended time to fill the micropores of adsorbents owing to the stronger diffusion resistance (Wang, Wang, and Ma, 2010a). Furthermore, contaminants with appropriate sizes can be sequestrated by adsorbent with comparable pore size and shape since they are readily available to the pores of the solid (Loganathan et al., 2014). The adsorption behavior of adsorbents varies with their porous arrangements and the sizes of adsorbates. Materials with microporous nature are reflected to be more effective adsorbents for ions (Karau et al., 1997).
- *Electrostatic interaction*: Charged adsorbents can easily interact with oppositely charged pollutants, namely by electrostatic interaction or Coulomb force, which is the typical interactive force when adsorbing charged substances from the aqueous solution. Due to the adsorption of water molecules or by structural reasons, the surfaces of the nanocomposite are covered with a greater number of hydroxyl groups, which also plays a major role in electrostatic interaction towards the adsorbate species (Cho et al., 2015).
- *π-π interaction/stacking*: π-π interaction/stacking is a general mechanism in the adsorption of organic pollutants on adsorbents with C=C bonds or benzene rings, which displays an essential role in the field of emerging high-efficiency adsorbents. Metal oxide nanocomposites normally show high adsorption efficacy for organic pollutants caused by the strong interaction between surface active sites of the adsorbents and organic pollutants (Ouyang et al., 2015).

- *Hydrogen bonding*: Hydrogen bonding arises during the adsorption of organic pollutants on adsorbents with active functional groups. The solid surface often has polar functional groups having hydrogen atoms, such as hydroxyl (-OH), carboxyl (-COOH), amino (-NH_2), and imino (-NH-), which might interact with the lone pair electrons of the electronegative atoms, such as nitrogen, oxygen, fluorine, sulfur, and chlorine in the adsorbate species to form a hydrogen bond. Also, the lone pair electrons of electronegative atoms on the surface of the adsorbent can associate with hydrogen atoms in the adsorbate (Jiang et al., 2015).
- *Hydrophobic interaction*: Normally, the surface that has a weak interaction with water molecules is termed as hydrophobic. Hydrophobic interactions are frequently viewed during the adsorption of organic molecules containing long carbon chains, which are mostly non-polar and water-insoluble molecules (Zhu, Tao, and Pan, 2010).

13.7.2. Chemical Interaction

- *Ion/ligand exchange*: The presence of active functional groups, such as -OH, -NH_2, -COOH on the surface of the adsorbent generates the opportunity for ion/ligand exchange. This means the pollutants interchange the coordinative surface ligands of adsorbents by establishing inner-sphere surface complexes (Zhu et al., 2015).
- *Reduction/oxidation*: Reduction or oxidation between metal oxide-containing nanocomposites and pollutants has been suggested for improving the adsorption affinity (Jiang et al., 2014).
- *Lewis acid-base interaction*: The surface active spots of the nanocomposite are easily protonated in an acidic solution and turn into weak Lewis acids, which behave as electron acceptors. Whereas pollutants become weak Lewis bases and act as electron donors. As stated by Pearson's hard and soft acids and bases (HSAB) principle, the soft Lewis acids have a tendency to form a strong bond with soft Lewis base to produce covalent complexes, while the hard Lewis acids are likely to form a strong bond with hard Lewis base to make ionic complexes (Han et al., 2013).

13.8. Adsorbents Developed by Modification of Polysaccharides by Graft Polymerization and Nanocomposite Formation for Removal of Mercuric Ions from Wastewater

Samaneh et al., (2012) synthesized chitosan-*g*-poly (acrylamide) semi-interpenetrating polymer network superabsorbent hydrogel via ultraviolet (UV) irradiation, and N, N'-methylene-bisacrylamide was used as crosslinker. The mercury sorption capacity under non-buffered conditions was achieved ca. 2001.8 mg/g via the colorimetric method. The kinetic data were fitted to the pseudo-second-order model. Also, the isothermal adsorption equilibrium data were described using the Langmuir model. In addition, the influence of pH, hydrogel dose, and initial mercury concentration on the adsorption capacity of the semi-interpenetrating polymer network (semi-IPN) hydrogel was evaluated. Finally, the hydrogel was regenerated without losing its original activity and stability.

The same research team (2015) has also prepared pullulan-*g*-polyacrylamide semi-IPN hydrogel and the mercury sorption capacity under non-buffered conditions was achieved 8.6 mmol/g in the first 90 min *via the* colorimetric method. The mechanism of adsorption was well presented using a pseudo-second-order kinetic model and the isothermal adsorption equilibrium data were fitted to the Freundlich model. In another study, Zhu et al., (2015) reported a series of granular hydrogels containing carboxyl and thiourea groups, prepared by a facile redox-initiated polymerization in air atmosphere at room temperature. The prepared hydrogels were used as the adsorbents for efficient adsorption of Hg(II) ion from aqueous solutions. The variables affecting the adsorption capacity were systematically investigated including pH, initial concentration, contact time, desorption, and reusability. The results indicated that the as-prepared adsorbent has fast adsorption kinetics and high adsorption capacity, with an equilibrium time of about 40 min and a maximum adsorption capacity of 1106.72 mg/g for Hg(II). Over the pH range studied, this granular hydrogel shows insignificant changes in the adsorption capacity for Hg(II). The adsorption process for Hg(II) can be described using the pseudo-second-order kinetic model and the equilibrium isotherm matches well with the Langmuir model. The reusability demonstrates that after five adsorption-desorption cycles, a high adsorption capacity is still observed for Hg(II), though a decreasing tendency is inevitable.

Wang et al., (2016) chose chitosan-*g*-poly (vinylimidazole-*co*-2-acrylamido-2-methyl-propane sulfonic acid) hydrogel for the selective removal of mercury ions from the aqueous solution. The adsorption results could be well described by the pseudo-second-order kinetic model and the Langmuir isotherm model with a maximum adsorption capacity of 363.55 mg/g. Wang and Wang (2010b) carried out batch adsorption experiments to remove Hg(II) from aqueous solutions using chitosan-*g*-poly (acrylic acid)/attapulgite hydrogel composites as adsorbents. The various factors influencing the adsorption capacity of the composites were investigated. The results indicate that the adsorption equilibrium of the composites can be achieved within about ten minutes and the equilibrium adsorption capacities of the composites with 10, 30, and 50 wt.% of attapulgite content were 785.20, 679.63, and 541.06 mg/g, respectively.

Wang et al., (2014a) synthesized chitosan/polyvinyl alcohol hydrogel adsorbent with a three-dimensional network structure, which showed superior adsorption properties for Hg(II) ions. The pH influence study showed Hg(II) adsorption on the hydrogel adsorbent is strongly pH-dependent. The adsorption capacity at pH=2.0 (200.20 mg/g) was higher than that at pH=11.0 (140.27 mg/g), which indicated that Hg(II) adsorption on the hydrogel adsorbent was not only by electrostatic interactions. The binding force between Hg(II) ions and functional groups was strengthened at high temperatures, which made the adsorption more favorable at high temperatures. The adsorption process fitted well by the Langmuir isotherm model with maximum adsorption capacity increased from 697.70 to 950.62 mg/g with increasing temperature from 20 to 60 °C. Wang et al., (2014b) presented the adsorption of Hg(II) ions onto chitosan-polyvinyl alcohol/bentonite nanocomposite. The bentonite content (10%) had a great influence on the structure of the nanocomposite by providing mesoporous nature, and the maximum adsorption capacity of 455.12 mg/g was observed.

Hashem et al., (2008) prepared the starch hydrogels via the graft polymerization of acrylonitrile onto maize starch using ceric ammonium nitrate as the initiator followed by saponification with sodium hydroxide. This study showed that the prepared hydrogels could be used effectively for the adsorption of Hg(II) ions from aqueous solutions. Factors affecting the adsorption of Hg(II) ions onto the starch hydrogels were studied, including the pH value of the solution, the contact time, and the concentrations of the adsorbent and adsorbate. The adsorption data obtained obeyed the Langmuir and Freundlich isotherms. The adsorption of Hg(II) ions onto the starch hydrogels followed second-order kinetics. The maximum adsorption capacity of the optimized sample towards Hg(II) ions was 1250 mg/g. Thakur et al., (2022) prepared gum tragacanth crosslinked 2-hydroxyethyl methacrylate-*co*-acrylamide hydrogel and gum tragacanth crosslinked 2-hydroxyethyl methacrylate-*co*-acrylamide/zinc oxide hydrogel composite with better Hg(II) adsorption capability. The hydrogel composite exhibited higher Hg(II) adsorption (154.8 mg/g) than the parent hydrogel.

Sonmez and Gazi (2015) reported the synthesis of pullulan-*g*-poly (acrylic acid-*co*-acrylamide) hydrogels by free radical solution polymerization. The novel hydrogel has shown good Hg(II) ion adsorption capacity (8.6 mmol/g in 90 min) from wastewater. Further, 8.3 mmol/g adsorption and 1.9 mmol/g desorption were obtained during the third adsorption-desorption cycle. The adsorption process was best fitted to the pseudo-second-order kinetic model and the Freundlich isotherm model. Naushad et al., (2016) evaluated the ability of starch/SnO_2 nanocomposite to remove Hg(II) ion from an aqueous medium. The adsorption was maximum at pH = 6 and equilibrium was attained within 60 min. The maximum adsorption capacity was found to be 192 mg/g at 25°C which was increased with the temperature, indicating the endothermic adsorption process. The pseudo-second-order equation signified the adsorption kinetics and the Freundlich isotherm model fitted the adsorption data better than the Langmuir. The adsorption capability of the regenerated adsorbents could still be preserved at 94% by the fourth adsorption-desorption cycle. Karmakar et al., (2019) described the preparation of an interpenetrating terpolymer network consisting of pectin-*g*-(methacrylic acid-*co*-3-(*N*-isopropylacrylamido)-2-methylpropanoic acid-*co*-*N*-isopropylacrylamide) which was found to adsorb 1172.97 mg/g of Hg(II) ions from aqueous solution with 0.02 g of adsorbent dosage at 293 K.

Prajwal and Vishalakshi's research group (2020a, 2020b, 2020c, 2021, 2022) synthesized a variety of novel polysaccharide-based hydrogels and their corresponding magnetite nanocomposites. They include pectin-*g*-poly (N-hydroxyethylacrylamide), κappa carrageenan-*g*-poly (N-hydroxyethylacrylamide), gum ghatti-*g*-poly (4-acryloylmorpholine), gum ghatti-*g*-poly (itaconic acid), and gum acacia-*g*-poly (3-chloro-2-hydroxypropylmethacrylate). The applicability of these materials for the treatment of mercuric ion-contaminated wastewater has been studied. The adsorption capacities of the nanocomposites for the adsorbate species follows order: Ggh-*g*-PIA/Fe_3O_4 (401.4 mg/g) > GA-*g*-PCHPMA/Fe_3O_4 (292.8 mg /g) > Pec-*g*-PHEAA/Fe_3O_4 (240.2 mg/g) > κC-*g*-PHEAA/Fe_3O_4 (239.9 mg/g) > Ggh-*g*-PAcM/Fe_3O_4 (235.1 mg/g). The same order is exhibited by the non-magnetic hydrogels as well. The nanocomposites exhibited higher adsorption capacity than the parent hydrogels which is mainly due to the presence of magnetite nanoparticles.

Upon incorporating nanoparticles, the nanocomposite hydrogels exhibited greater surface area along with some additional binding sites which influenced the adsorption capacity to a larger extent. Sirajo and Vishalakshi (2018) prepared pectin-*g*-poly (2-(methacryloyloxyethyl) trimethylammonium chloride-*co*-2-acrylamido-2-methyl-1-propane sulfonic acid) and its composite with montmorillonite. The adsorption capacity of the representative samples for Hg(II) were 12.16 and 19.58 mg/g, respectively. The adsorption isotherm study was best explained by the Langmuir model. The percentage recovery for the above two systems towards Hg(II) ions was found to be 78.53 and 51.92%, respectively. Guo et al., (2014) described novel hydrazine hydrate-modified pectin for the removal of Hg(II). The efficiency of removal by the modified pectin reached 74.8%, which was far greater than the efficiency obtained with unmodified pectin (30.5%). Morsi et al., (2018) synthesized a multifunctional nanocomposite based on magnetite nanoparticles, chitosan, and polythiophene for efficient and selective removal of Hg(II) ions from aqueous solution. The experimental equilibrium data fitted well to the Freundlich isotherm model specifying that the adsorption process is predominantly heterogeneous due to the existence of different functional groups. The kinetic studies indicated that the adsorption process follows the pseudo-second-order model signifying the foremost adsorption mechanism of chemical adsorption.

Sharma's group (2020) used gum tragacanth-*g*-poly (N,N-dimethylacrylamide) nanocomposite hydrogel filled with reduced graphene oxide for the effective elimination of mercuric ions from aqueous solution with a maximum adsorption capacity of 666.6 mg/g. The system exhibited pseudo-second-order kinetics and Langmuir isotherm model fit. Saberi et al., (2020) modified starch with polyethylene glycol-poly (acrylic acid) and silver nanoparticles. The obtained hydrogel and the silver nanocomposite were investigated for the removal of Hg(II) ions from water. The maximum adsorption capacities of the gel and the nanocomposites were found to be 158.21 mg/g and 182.53 mg/g, respectively. Rashidi et al., (2022) designed meso/magnetic nanocomposite hydrogel by the blending of chitosan and cellulose biopolysaccharides in the presence of magnetic laponite RD nanoparticles. The adsorption process of Hg(II) ions best correlated to the Langmuir isotherm and pseudo-second-order models. The maximum uptake of Hg(II) ions was 579.8 mg/g. The thermodynamic data showed that the adsorption process was spontaneous, favorable, and endothermic. Taleb et al., (2021) studied the effect of Ag nanoparticles coated chitosan modified acrylamide/HEMA hydrogel, by examining the prospect of using the prepared hydrogel and nanocomposite for adsorption of Hg(II). The results showed that the adsorption best fits the pseudo-second-order kinetic model. Thermodynamic data revealed a spontaneous exothermic physisorption process. The adsorption capacity increased with an increase in the pH up to 5.0. The desorption studies showed that the 1 M HCl solution was ideal to elute 98% of mercuric ions from the nanocomposite.

According to the work of Singh and Kumar, (2011) tamarind seed kernel polysaccharide-silica nanohybrids fabricated in a base-catalyzed sol-gel reaction were found to be efficient in Hg(II) removal from the aqueous solution. Due to the establishment of nanostructures, a huge number of free sorbent spots were accessible at the hybrid which made the system effectual in Hg(II) binding. Regeneration studies directed that the loaded Hg(II) from the used hybrid can be easily desorbed and effectively reused for eight successive adsorption-desorption cycles. According to the report of Kolya et al., (2014) ionic attractions are mainly responsible for binding Hg (II) ions with the -COOH groups of the starch-*g*-poly-(N-methylacrylamide-*co*-acrylic acid). The maximum adsorption uptake of 11.0 mg/g was found at pH=5.5, contact time of 180 min, and temperature of 45 °C. The sorption data were found to obey the Langmuir isotherm and the pseudo-second-order kinetic model. Ranote et al., (2018) treated Hg(II) ion-contaminated wastewater with *Moringa oleifera* gum modified with the nano-polyacryloyl group. This material presented a maximum adsorption capacity of 840.34 mg/g with rapid adsorption as equilibrium was reached within 20 min. The high desorption tendency was also observed as the reusability study was carried out for 20 repeated cycles with an adsorption capacity of 71.25 mg/g after the 20th cycle and cumulative 1606.25 mg/g of Hg(II) ions were adsorbed with the same dosage of adsorbent (5 mg).

13.9. Future Prospects

Exploiting the structure of polysaccharides to achieve desired properties is a major research need. The rising cost of conventional adsorbents certainly makes polysaccharide-based materials one of the most prominent biosorbents for wastewater treatment. However, the data on the cost of the development of adsorbents is not often reported in the publications, as it is governed by local availability. Also, the performance comparison of diverse adsorbents is challenging because of discrepancies in data presentation, predominantly due to different experimental conditions, such as temperature, pH, ionic strength, presence of competing ions, and particle size. The design of nanocomposite hydrogels with high specific surface area is still the demand of scientific research. Thus, much work is necessary to meet the growing demands of novel materials with high performance at the industrial scale.

Conclusion

In recent years, growing expenses involved in water treatment have encouraged the usage of new low-cost materials derived from renewable resources. Green chemistry methodology emphasizes naturally occurring resources and their use in different industrial applications for economic and environmental concerns. Adsorption is a technique of choice for treating industrial effluents and a beneficial option for protecting the environment. Polysaccharides, in addition to being eco-friendly and non-toxic, possess a variety of functional groups permitting them to be easily modified for the desired applications. The chemical modification of these polysaccharides affords a wide variety of hydrogels with improved properties. The introduction of different functional groups on polysaccharides through grafting with synthetic polymers and incorporation of nanoparticles enhances adsorption properties along with mechanical properties to sustain multiple cycles of adsorption and desorption. Nanocomposite hydrogel technology overcomes traditional hydrogel limitations.

This chapter presented a wide range of nanocomposite hydrogels in the remediation of mercuric ions from wastewater. The compilation of literature reports indicated the development of various polysaccharide-based nanocomposite adsorbents for mercury ion removal with variable adsorption performance making them favorable candidates for practical applications. The polysaccharide-based nanocomposite hydrogels meet the requirements of versatile materials which the emerging world demands. This chapter recommends the use of such sustainable adsorbents with a further search on obtaining the polysaccharides from natural sources.

References

Ahmed, E. M. (2015). Hydrogel: Preparation, characterization and applications: a review. *Journal of Advanced Research*, 6(2), 105-121.

Ahn, D. H., Chang, W. S., and Yoon, T. (1999). Dyestuff wastewater treatment using chemical oxidation, physical adsorption and fixed bed biofilm process. *Process Biochemistry*, 34(5), 429-439.

Allegre, C., Maisseu, M., Charbit, F., and Moulina, P. (2004). Coagulation-flocculation decantation of dye house effluents: Concentrated effluents. *Journal of Hazardous Materials*, 116(1-2), 57-64.

Ampuero, F. A. O., Brea, J., Loza, M. I., Torres, D., and Alonso, M. J. (2009). Chitosan hyaluronic acid nanoparticles loaded with heparin for the treatment of asthma. *International Journal of Pharmaceutics*, 381(2), 122-129.

Appel, E. A., Lasheras, J. D. B., Loh, X. J., and Scherman, O. A. (2012). Supramolecular polymeric hydrogels. *Chemical Society Reviews*, 41(18), 6195-6214.

Asmatulu, E., Twomey, J. M., and Overcash, M. (2015). Recycling of fibre-reinforced composites and direct structural composite recycling concept. *Journal of Composite Materials*, 48(5), 593-608.

Athawale, V. D., and Lele, V. (1998). Graft copolymerization onto starch. II. Grafting of acrylic acid and preparation of its hydrogels. *Carbohydrate Polymers*, 35(1-2), 21-27.

Azevedo, B. F., Furieri, L. B., and Pecanha, F. M. (2012). Toxic effects of mercury on the cardiovascular and central nervous systems. *Journal of Biomedicine and Biotechnology*, 2012, 949048.

Bagdanov, B., Schacht, E., and Blucke, A. V. D. (1997). Thermal and rheological properties of gelatin-dextran hydrogels. *Journal of Thermal Analysis and Calorimetry*, 49(2), 847-856.

Bemmelen, J. M. V. (1894). Der Hydrogel und das kristallinische Hydrat des Kupferoxydes [*The hydrogel and crystalline hydrate of copper oxide*], *Zeitschrift für Chemie und Industrie der Kolloide*, 5, 466-468.

Bhattacharya, A., and Misra, B. N. (2004). Grafting: a versatile means to modify polymers techniques, factors and applications. *Progress in Polymer Science*, 29(8), 767-814.

Boparai, H. K., Joseph, M., and Carroll, D. M. O. (2011). Kinetics and thermodynamics of cadmium ion removal by adsorption onto nano zerovalent iron particles. *Journal of Hazardous Materials*, 186(1), 458-465

Chang, P. R., Li, J., Chen, Y., Wang, D., Yu, J., and Chen, J. (2010). Structure and properties of polysaccharide nanocrystal-doped supramolecular hydrogels based on cyclodextrin inclusion. *Polymer*, 51(19), 4398-4407.

Chen, Z. G., Wang, P. W., Wei, B., Mo, X. M., and Cui, F. Z. (2010). Electrospun collagen-chitosan nanofiber: a biomimetic extracellular matrix for endothelial cell and smooth muscle cell. *Acta Biomaterialia*, 6(2), 372-382.

Cho, D. W., Jeon, B. H., Chon, C. M., Schwartz, F. W., Jeong, Y., and Song, H. (2015) Magnetic chitosan composite for adsorption of cationic and anionic dyes in aqueous solution. *Journal of Industrial and Engineering Chemistry*, 28, 60-66.

Coviello, T., Matricardi, P., Marianecci, C., and Alhaique, F. (2007). Polysaccharide hydrogels for modified release formulations, *Journal of Controlled Release*, 119, 5-24.

Da Silva, D. A., De Paula, R. C. M., and Feitosa, J. P. A. (2007). Graft copolymerisation of acrylamide onto cashew gum. *European Polymer Journal*, 43(6), 2620-2629.

Dallinger, D., and Kappe, C. O. (2007). Microwave-assisted synthesis in water as solvent. *Chemical Reviews*, 2563-2591.

Dave, P. N., and Gor, A. (2018). Natural polysaccharide-based hydrogels and nanomaterials: Recent trends and their applications. *Handbook of Nanomaterials for Industrial Applications*, 36-66.

Deng, H., Wang, X., Liu, P., Ding, B., Du, Y., Li, G., Hu, X., and Yang, J. (2011). Enhanced bacterial inhibition activity of layer-by-layer structured polysaccharide film-coated cellulose nanofibrous mats via addition of layered silicate. *Carbohydrate Polymers*, 83(1), 239-245.

Dong, X., Wei, C., Liang, J., Liu, J. T., Kong, D., and Lv, F. (2017). Thermosensitive hydrogel loaded with chitosan-carbon nanotubes for near infrared light triggered drug delivery. *Colloids and Surfaces B: Biointerfaces*, 154, 253-262.

Ferreira, F. V., Pinheiro, I. F., Gouveia, R. F., Thim, G. P., and Lona, L. M. F. (2017). Functionalized cellulose nanocrystals as reinforcement in biodegradable polymer nanocomposites. *Polymer Composites*, 39, E9-E29.

Filip, J., Sefcovicova, J., Tomcik, P., Gemeiner, P., and Tkac, J. (2011). A hyaluronic acid dispersed carbon nanotube electrode used for a mediatorless NADH sensing and biosensing. *Talanta*, 84(2), 355-361.

Galema, S. A. (1997). Microwave chemistry. *Chemical Society Reviews*, 26(3), 233-238.

Gangadhar, B., and Vishalakshi, B. (2019). Silver nanoparticles embedded pectin-based hydrogel: a novel adsorbent material for separation of cationic dyes. *Polymer Bulletin*, 76(8), 4215-4236.

Garner, W. E., and Hailes, H. R. (1933). Thermal decomposition and detonation of mercury fulminate. *Proceedings of the Royal Society of London*, 139(839), 576-595.

Glenn, C., Cahn, P. H., and Siler, W. (1973). Mercury concentrations in fish, plankton and water from three Western Atlantic estuaries. *Journal of Fish Biology*, 5(6), 641-647.

Gregory, J., and Dhond, R. V. (1972). Wastewater treatment by ion exchange. *Water Research*, 6(6), 681-694.

Guo, J. J., Zhang, J. Y., Yue, Y., and Guo, Y. (2014). Mechanism of mercury removal by a novel hydrazine hydrate-modified pectin. *Bulgarian Chemical Communications*, 46(4), 801-805.

Hai, N. T., You, S. J., Bandegharaei, A. H., and Chao, H. P. (2017). Mistakes and inconsistencies regarding adsorption of contaminants from aqueous solutions: a critical review. *Water Research*, 120, 88-116.

Han, C., Li, H., Pu, H., Yu, H., Deng, L., Huang, S., and Luo, Y. (2013). Synthesis and characterization of mesoporous alumina and their performances for removing arsenic (V). *Chemical Engineering Journal*, 217, 1-9.

Hashem, A., Ahmad, F., and Fahad, R. (2008). Application of some starch hydrogels for the removal of mercury (II) ions from aqueous solutions. *Adsorption Science & Technology*, 26(8), 563-579.

Hassan, M. M., and Hawkyard, C. J. (2002). Ozonation of aqueous dyes and dye house effluent in a bubble-column reactor. *Journal of Environmental Science and Health A*, 37(8), 1563-1579.

Iijima, S. (2002). Carbon nanotubes: past, present, and future. *Physica B*, 323(1-4), 1-5.

Jaishankar, M., Tseten, T., Anbalagan, N., Mathew, B. B., and Beeregowda, K. N. (2014). Toxicity, mechanism and health effects of some heavy metals. *Interdisciplinary Toxicology*, 7(2), 60-72.

Jan, A. T., Azam, M., Siddiqui, K., Ali, A., Choi, I., and Haq, Q. M. R. (2015). Heavy metals and human health: Mechanistic insight into toxicity and counter defense system of antioxidants. *International Journal of Molecular Sciences*, 16(12), 29592-29630.

Jeon, C., and Holl, W. H. (2003). Chemical modification of chitosan and equilibrium study for mercury ion removal. *Water Research*, 37(19), 4770-4780.

Jiang, T., Liang, Y. D., He, Y. J., and Wang, Q. (2015). Activated carbon/$NiFe_2O_4$ magnetic composite: a magnetic adsorbent for the adsorption of methyl orange. *Journal of Environmental Chemical Engineering*, 3(3), 1740-1751.

Jiang, W., Cai, Q., Xu, W., Yang, M., Cai, Y., Dionysiou, D. D., and O'Shea, K. E. (2014). Cr(VI) adsorption and reduction by humic acid coated on magnetite. *Environmental Science & Technology*, 48(14), 8078-8085.

Kappe, C. O. (2004) Controlled microwave heating in modern organic synthesis. *Angewandte Chemie International Edition*, 43(46), 6250-6284.

Karau, A., Benken, C., Thommes, J., and Kula, M. R. (1997). The influence of particle size distribution and operating conditions on the adsorption performance in fluidized beds. *Biotechnology and Bioengineering*, 55(1), 54-64.

Karmakar, M., Mondal, H., Mahapatra, M., Chattopadhyay, K. P., Chatterjee, S., Singha, N. R. (2019). Pectin-grafted terpolymer superadsorbent via N-H activated strategic protrusion of monomer for removals of Cd(II), Hg(II) and Pb(II). *Carbohydrate Polymers*, 206, 778-791.

Kaur, H., Banipal, T. S., Thakur, S., Bakshi, M. S., Kaur, G., and Singh, N. (2012). Novel biodegradable films with extraordinary tensile strength and flexibility provided by nanoparticles. *ACS Sustainable Chemistry and Engineering*, 1(1), 127-136.

Kim, D. H., Park, S. Y., Kim, J., and Park, M. J. (2010). Preparation and properties of the single-walled carbon nanotube/cellulose nanocomposites using N-methylmorpholine-N-oxide monohydrate. *Journal of Applied Polymer Science*, 117(6), 3588-3594.

Kolya, H., Das, S., and Tripathy, T. (2014). Synthesis of starch-*g*-poly-(N-methylacrylamide-*co*-acrylic acid) and its application for the removal of mercury (II) from aqueous solution by adsorption. *European Polymer Journal*, 58, 1-10.

Li, J., He, A., Han, C. C., Fang, D., Hsiao, B. S., and Chu, B. (2006). Electrospinning of hyaluronic acid (HA) and HA/gelatin blends. *Macromolecular Rapid Communications*, 27(2), 114-120.

Liu, J., Shi, J. Z., Yu, L. M., Goyer, R. A., and Waalkes M. P. (2008). Mercury in traditional medicines: Is cinnabar toxicologically similar to common mercurials?. *Experimental Biology and Medicine*, 233(7), 810-817.

Loganathan, P., Vigneswaran, S., Kandasamy, J., and Bolan, N. S. (2014). Removal and recovery of phosphate from water using sorption. *Critical Reviews in Environmental Science and Technology*, 44(8), 847-907.

Maprani, A. C., Al, T. A., MacQuarrie, K. T., Dalziel, J. A., Shaw, S. A., and Yeats, P. A. (2005). Determination of mercury evasion in a contaminated headwater stream. *Environmental Science & Technology*, 39(6), 1679-1687.

Meli, L., Miao, J., Dordick, J. S., and Linhardt, R. J. J. (2010). Electrospinning from room temperature ionic liquids for biopolymer fiber formation. *Green Chemistry*, 12(11), 1883-1892.

Miao, C., and Hamad, W. Y. (2013). Cellulose reinforced polymer composites and nanocomposites: a critical review. *Cellulose,* 20(5), 2221-2262.

Mittal, H., Maity, A., and Ray, S. S. (2015). Effective removal of cationic dyes from aqueous solution using gum ghatti-based biodegradable hydrogel. *International Journal of Biological Macromolecules,* 79, 8-20.

Morsi, R. E., Al-Sabagh, A. M., Moustafa, Y. M., ElKholy, S. G., and Sayed, M. S. (2018). Polythiophene modified chitosan/magnetite nanocomposites for heavy metals and selective mercury removal. *Egyptian Journal of Petroleum,* 27(4), 1077-1085.

Naushad, M., Ahamad, T., Sharma, G., Al-Muhtaseb, A. H., Albadarin, A. B., Alam, M. M., Alothman, Z. A., Alshehri, S. M., and Ghfar, A. A. (2016). Synthesis and characterization of a new starch/SnO_2 nanocomposite for efficient adsorption of toxic Hg^{2+} metal ion. *Chemical Engineering Journal*, 300, 306-316.

Ngim, C. H., Foo, S. C., Boey, K. W., and Keyaratnam, J. (1992). Chronic neurobehavioral effects of elemental mercury in dentists. *Occupational and Environmental Medicine,* 49(11), 782-790.

Ouyang, K., Zhu, C., Zhao, Y., Wang, L., Xie, S., and Wang, Q. (2015). Adsorption mechanism of magnetically separable Fe_3O_4/graphene oxide hybrids. *Applied Surface Science*, 355, 562-569.

Ozmen, M. M., and Okay, O. (2005). Superfast responsive ionic hydrogels with controllable pore size. *Polymer*, 46(19), 8119-8127.

Pendergast, D. M. (1982). Ancient Maya mercury. *Science*, 217(4559), 533-535.

Prajwal, K., Preetha, B. K., and Vishalakshi, B. (2022). Develpoment of gum acacia based magnetic nanocomposite adsorbent for wastewater treatment, *Polymer Bulletin*, 79, 9457-9484.

Prajwal, K., and Vishalakshi, B. (2020a). Magnetite nanoparticle embedded pectin-*graft*-poly(N-hydroxyethylacrylamide) hydrogel: evaluation as adsorbent for dyes and heavy metal ions from waste water, *International Journal of Biological Macromolecules*, 153, 1408-1417.

Prajwal, K., and Vishalakshi, B. (2020b). Efficient removal of dyes and heavy metal ions from wastewater using gum ghatti-*graft*-poly(4-acryloylmorpholine) hydrogel incorporated with magnetite nanoparticles, *Journal of Environmental Chemical Engineering*, 8(5), 104207.

Prajwal, K., and Vishalakshi, B. (2020c). Hybrid nanocomposite of kappa-carrageenan and magnetite as adsorbent material for water purification, *International Journal of Biological Macromolecules*, 165, 542-553.

Prajwal, K., and Vishalakshi, B. (2021). Evaluation of gum ghatti-*g*-poly(itaconic acid) magnetite nanocomposite as an adsorbent material for water purification. *International Journal of Biological Macromolecules*, 193, 2232-2242.

Pushparaj, V. L., Shaijumon, M. M., Kumar, A., Murugesan, S., Ci, L., Vajtai, R., Linhardt, R. J., Nalamasu, O., and Ajayan, P. M. (2007). Flexible energy storage devices based on nanocomposite paper. *Proceedings of the National Academy* of *Sciences*, 104(34), 13574-13577.

Ranote, S., Ram, B., Kumar, D., Chauhan, G. S., and Joshi, V. (2018). Functionalization of Moringa oleifera gum for use as Hg^{2+} ions adsorbent. *Journal of Environmental Chemical Engineering*, 6(2), 1805-1813.

Rashidi, S., Kalhor, E. g., Rashidzadeh, B., and Abolhasani, J. (2022). Comparative adsorption of Cu (II), Hg (II), Co (II), and Ni (II) ions on novel magnetic chitosan/cellulose/laponite RD nanocomposite hydrogel. *International Journal of Environmental Analytical Chemistry*, In press, doi: 10.1080/03067319.2022.2077105.

Rice, K. M., Walker, E. M., Miaozong W., Gillette, C., and Blough E. R. (2014). Environmental Mercury and Its Toxic Effects. *Journal of Preventive Medicine and Public Health*, 47(2), 74-83.

Saberi, A., Sadeghi, M., and Alipour E. (2020). Design of AgNPs-base Starch/PEG-poly (acrylic acid) hydrogel for removal of mercury (II). *Journal of Polymers and the Environment*, 28(3), 906-917.

Samaneh, S. S., and Gazi, M. (2015). Pullulan based porous semi-IPN hydrogel: Synthesis, characterization and its application in the removal of mercury from aqueous solution. *Journal of the Taiwan Institute of Chemical Engineers*, 51, 143-151.

Samaneh, S. S., Gazi, M., and Yilmaz, E. (2012). UV-induced synthesis of chitosan-*g* polyacrylamide semi-IPN superabsorbent hydrogels. *Polymer Bulletin*, 68(6), 1623-1639.

Sharma, B., Thakur, S., Trache, D., Nezhad, H. Y., and Thakur, V. K. (2020). Microwave assisted rapid synthesis of reduced graphene oxide-based gum tragacanth hydrogel nanocomposite for heavy metal ions adsorption. *Nanomaterials*, 10(8), 1616.

Sharma, Y. C., Srivastava, V., Singh, V. K., Kaul, S. N., and Weng, C. H. (2009). Nano-adsorbents for the removal of metallic pollutants from water and wastewater. *Environmental Technology*, 30(6), 583-609.

Singh, V., and Kumar P. (2011). Design of nanostructured tamarind seed kernel polysaccharide-silica hybrids for mercury (II) removal. *Separation Science and Technology*, 46(5), 825-838.

Singh, V., Kumar, P., and Sanghi, R. (2012). Use of microwave irradiation in the grafting modification of the polysaccharides - a review. *Progress in Polymer Science*, 37(2), 340-364.

Singh, V., Tripathi, D. N., Tiwari, A., and Sanghi, R. (2006). Microwave synthesized chitosan-graft-poly(methylmethacrylate): an efficient Zn^{2+} ion binder. *Carbohydrate Polymers*, 65(1), 35-41.

Singhal, R., and Gupta, K. (2016). A review: tailor-made hydrogel structures (classifications and synthesis parameters). *Polymer-Plastics Technology and Engineering*, 55(1), 54-70.

Sirajo, A. Z., and Vishalakshi, B. (2018). Pectin graft copolymer-montmorillonite composite: synthesis, swelling and divalent metal ion adsorption. *Separation Science and Technology*, 53(14), 2170-2185.

Sonmez, S. S., and Gazi, M. (2015). Pullulan based porous semi-IPN hydrogel: synthesis, characterization and its application in the removal of mercury from aqueous solution. *Journal of the Taiwan Institute of Chemical Engineers*, 51, 143-151.

Stevens, K. N. J., Croes, S., Boersma, R. S., Stobberingh, E. E., vander Marel, C., vander Veen, F. H., Knetsch, M. L. W., and Koole, L. H. (2011). Hydrophilic surface coatings with embedded biocidal silver nanoparticles and sodium heparin for central venous catheters. *Biomaterials*, 32(5), 1264-1269.

Taleb, M. F. A., Albalwi, H., and Ismail, F. A. E. (2021). Removal of mercury (II) from aqueous solution using silver nanocomposite: synthesis and adsorption mechanism. *Journal of Inorganic and Organometallic Polymers and Materials*, 31(4), 1825-1835.

Thakur, S., Sharma, B., Thakur, A., and Gupta, V. K. (2022). Synthesis and characterisation of zinc oxide modified biorenewable polysaccharides based sustainable hydrogel nanocomposite for Hg^{2+} ion removal: towards a circular bioeconomy. *Bioresource Technology*, 348, 126708.

Thakur, V. K., and Kessler, M. R. (2015). Self-healing polymer nanocomposite materials: a review. *Polymer*, 69, 369-383.

Tizzotti, M., Charlot, A., Fleury, E., Stenzel, M., and Bernard, L. (2010). Modification of polysaccharides through controlled/living radical polymerization grafting-towards the generation of high performance hybrids. *Macromolecular Rapid Communications*, 31(20), 1751-1772.

Tripathi, B. D., and Shukla, S. C. (1991). Biological treatment of wastewater by selected aquatic plants. *Environmental Pollution*, 69(1), 69-78.

Vane, C. H., Beriro, D. J., and Turner, G. H. (2015). Rise and fall of mercury (Hg) pollution in sediment cores of the Thames Estuary, London, UK. *Earth and Environmental Science Transactions of the Royal Society of Edinburgh*, 105(4), 285-296.

Vigneshwaran, N., Kumar, S., Kathe, A. A., Varadarajan, P. V., and Prasad, V. (2006). Functional finishing of cotton fabrics using zinc oxide soluble starch nanocomposites. *Nanotechnology*, 17(20), 5087-5095.

Visvanathan, C., Aim, R. B., and Parameshwaran, K. (2000). Membrane separation bioreactors for wastewater treatment. *Critical Reviews in Environmental Science and Technology*, 30(1), 1-48.

Viswanathan, G., Murugesan, S., Pushparaj, V., Nalamasu, O., Ajayan, P. M., and Linhardt, R. J. (2006). Preparation of biopolymer fibers by electrospinning from room temperature ionic liquids. *Biomacromolecules*, 7(2), 415-418.

Vongchan P., Wutti-In, Y., Sajomsang, W., Gonil, P., Kothan, S., and Linhardt, R. J. (2011). N, N, N-Trimethyl chitosan nanoparticles for the delivery of monoclonal antibodies against hepatocellular carcinoma cells. *Carbohydrate Polymers*, 85(1), 215-220.

Wang, F. Y., Wang, H., and Ma, J. W. (2010a). Adsorption of cadmium (II) ions from aqueous solution by a new low-cost adsorbent-bamboo charcoal. *Journal of Hazardous Materials*, 177(1-3), 300-306.

Wang, F., Zheng, Y., and Wang A. (2016). Preparation of chitosan-*g*-poly (vinylimidazole-*co*-2-acrylamido-2-methylpropane sulfonic acid) granular hydrogel for selective adsorption of Hg^{2+}. *Water, Air & Soil Pollution*, 227, 110.

Wang, M., Olszewska, A., Walther, A., Malho, J., Schacher, F. H., Ruokolainen, J., Ankerfors, M., Laine, J., Berglund, L. A., Osterberg, M., and Ikkala, O. (2011). Colloidal ionic assembly between anionic native cellulose nanofibrils and cationic block copolymer micelles into biomimetic nanocomposites. *Biomacromolecules*, 12(6), 2074-2081.

Wang, X., and Wang A. (2010b). Adsorption characteristics of chitosan-*g*-poly(acrylic acid)/attapulgite hydrogel composite for Hg(II) ions from aqueous solution. *Separation Science and Technology*, 45(14), 2086-2094.

Wang, X., Su, R., and Wang, C. (2014a). pH dependence and thermodynamics of Hg(II) adsorption onto chitosan-poly(vinyl alcohol) hydrogel adsorbent. *Colloids and Surfaces A Physicochemical and Engineering Aspects*, 441, 51-58.

Wang, X., Yang, L., Zhang, J., Wang, C., and Li, Q. (2014b). Preparation and characterization of chitosan-poly(vinyl alcohol)/bentonite nanocomposites for adsorption of Hg(II) ions. *Chemical Engineering Journal*, 404-412.

Whistler, R. L. (1973). Solubility of polysaccharides and their behavior in solution. *Advances in Chemistry*, 251, 242-255.

Wichterle, O., and Lim, D. (1960). Hydrophilic gels for biological use, *Nature*, 185(4706), 117-118.

Yagub, M. T., Sen, T. K., Afroze, S., and Ang, H. M. (2014). Dye and its removal from aqueous solution by adsorption: a review. *Advances in Colloid and Interface Science*, 209, 172-184.

Yalpani, M. (1988). *Polysaccharides: Synthesis, Modifications and Structure/Property Relations* (Vol. 36 of the series, "Studies in Organic Chemistry"). Elsevier Science Publishers.

Zaleschi, L., Secula, M. S., Teodosiu, C., Stan, C. S., and Cretescu, I. (2014). Removal of rhodamine 6G from aqueous effluents by electrocoagulation in a batch reactor: Assessment of operational parameters and process mechanism. *Water, Air & Soil Pollution*, 225(9), 1-14.

Zhu, J., Lin, X., Wu, P., Zhou, Q., and Luo, X. (2015). Fluoride removal from aqueous solution by Al(III)-Zr(IV) binary oxide adsorbent. *Applied Surface Science*, 357, 91-100.

Zhu, Q., Tao, F., and Pan, Q. (2010). Fast and selective removal of oils from water surface via highly hydrophobic core-shell Fe_2O_3@C nanoparticles under magnetic field. *ACS Applied Materials and Interfaces*, 2(11), 3141-3146.

Zhu, Y., Zheng, Y., Wang, W., and Wang A. (2015). Highly efficient adsorption of Hg(II) and Pb(II) onto chitosan-based granular adsorbent containing thiourea groups. *Journal of Water Process Engineering*, 7, 218-226.

Chapter 14

Nanocomposite Hydrogels for Irrigation Applications

Behnam Rezazadeh[1]
Kazem Adavi[2,*]
and Mehdi Elyasi[1]
[1] Faculty of Chemical Engineering, Sahand University of Technology, Tabriz, Iran
[2] Chemical Engineering Department, Ecole Polytechnique de Montreal, Montreal, Canada

Abstract

In recent years, due to the occurrence of drought periods and water scarcity, agriculture is recognized as the most water-consuming and vulnerable to water shortage among other sectors, e.g., households and industry. In addition, population growth and the need for more agricultural products increased scientists' interest in the reduction of water consumption in the agricultural sector. One of the most promising strategies for conserving water and reducing water consumption in agricultural irrigation is employing hydrogels as superabsorbent materials. Nanotechnology has the potential to revolutionize agricultural irrigating systems and provide solutions for current and future problems. Hydrogels and nanocomposite hydrogels have been the focus of research for irrigation due to their characteristic properties, such as high porosity, permeability, and water absorption capability. The main aim of this chapter is to introduce the principal concepts of agricultural irrigating systems, along with the basic concepts of hydrogels, nanocomposite hydrogels, and superabsorbent polymers. The applications of superabsorbent polymers, hydrogels, and nanocomposite hydrogels in agricultural irrigation focusing on the recent advances in this field are also overviewed.

Keywords: hydrogel, nanocomposite, irrigation, agriculture, superabsorbent

14.1. Introduction

With a growing population, competition for resources, such as water, land, and energy has increased. Climate change creates additional uncertainty about resource availability in the future) Madramootoo and Fyles, 2010). Over the coming decades, the agricultural sector must

---

* Corresponding Author's Email: kazem.adavi@polymtl.ca.

In: Nanocomposite Hydrogels and their Emerging Applications
Editor: Mohammad Sirousazar
ISBN: 979-8-88697-675-5

increase production while using available resources more effectively. Water demand is expected to be 60% higher than today by 2030. The ratio of water withdrawal to the natural water renewal rates may increase by over 60%, resulting in water scarcity. Irrigated areas increase by almost 1% per year and the irrigation water demand will increase to 14% by 2025 (Rosegrant and Cai, 2002). The agricultural ecosystem is susceptible to drought stress and can threaten food security worldwide. The irrigation methods used significantly impact water resource depletion and looking for new ways to reduce water consumption in the agricultural sector is inevitable (Cannazza et al., 2014).

One of the main challenges in agricultural activities is irrigation. Traditional irrigation management is performed by the people themselves and requires the presence and continuous monitoring of irrigation by the farmers in the field area. In fact, the irrigation methods used have a considerable impact in terms of water resource depletion. Thus, under such conditions, we should look for ways to reduce water use in the agricultural sector. One of the most promising strategies for reducing consumption in agricultural irrigation is using superabsorbent polymers (SAPs) (Calcagnile and Sibillano, 2019). SAPs are hydrophilic polymers with the ability to absorb and retain water from tens to thousands of times their dry weight (Zhang et al., 2014). SAP particles are mixed with the soil and used in irrigation systems. They can not only increase soil's ability to retain water due to their high water absorbency, but also improve soil aeration as well as germination and growth of plants when they are mixed with the soil.

Another way to reduce water consumption in the agricultural sector is using subsurface irrigation technology using, e.g., pervaporation irrigation systems (Camp, 1998; Lamm, 2016; Quiñones-Bolaños et al., 2005a). A pervaporation irrigation system provides *in situ* water treatment while supplying it to the plant root zone. More recently, a novel method was invented to reduce water consumption in the agricultural sector based on a molecular irrigating system composed of hydrogel or nanocomposite hydrogel pipes (Rezazadeh et al., 2020; Sirousazar et al., 2019). In these systems, the hydrogel or nanocomposite hydrogel pipe is buried in the ground in the vicinity of the plant's root, while it is connected to a batch water resource. When the surrounding soil is dry, a water chemical potential gradient is created between the pipe and the soil and as a result, water molecules gradually diffuse through into the soil. Due to the very low water consumption rate, hydrogel and nanocomposite hydrogel pipes have been suggested for irrigation in arid regions, e.g., deserts, where freshwater is not accessible. In the following section, the basic concepts of hydrogels and nanocomposite hydrogels are briefly introduced.

14.2. Hydrogel and Nanocomposite Hydrogels

14.2.1. Hydrogel

Hydrogels are polymeric materials having three-dimensional, hydrophilic, and crosslinked networks capable of imbibing a large amount of water or aqueous solutions. The ability of hydrogels to absorb water or aqueous solutions is due to the presence of hydrophilic groups, such as -OH, $-CONH_2$, -COOH, and $-SO_3H$ in their structure (Nascimento et al., 2018). A hydrogel has a state that is neither wholly liquid nor completely solid. These half-liquid-like and half-solid-like properties cause many interesting desirable behaviors that are not found in either a pure solid or a pure liquid state. Hydrogels are usually prepared from crosslinking of hydrophilic monomers or polymers, but hydrophobic moieties can also be used in their synthesis to regulate specific properties required for a given application (Ahmed, 2015). It is possible to produce hydrogels from either natural or synthetic polymers. Most synthetic polymers are hydrophobic and chemically more robust while they present good mechanical properties that result in low degradation rates. Natural polymers can be included in the preparation of hydrogels beside synthetic polymers to produce hybrid hydrogels with desired functions (Shantha and Harding, 2002). The polymeric chains can be connected together to form a hydrogel in two different manners. In the first manner, the chains are physically interconnected without using any chemical crosslinker. The hydrogel prepared by this method is called physical or reversible hydrogel. The second method is based on crosslinking the polymeric chain using crosslinking agents via performing a chemical reaction. The prepared hydrogel by this technique is known as chemical or permanent hydrogel (Hennink and van Nostrum, 2002).

Some hydrogels exhibit drastic volume changes in response to specific external stimuli, such as temperature, pH, light, solvent quality, and electrical and magnetic fields. These hydrogels are known as smart or stimuli-responsive hydrogels which exhibit single or multiple stimuli-responsive characteristics. Smart hydrogels change their structural and volume phase transition as a response to external stimuli resulting in an enormous potential for various advanced biomedical applications (Shojaeiarani et al., 2019). Hydrogels have recently drawn great attention for use in a wide variety of biomedical applications, such as cell therapeutics, wound healing, cartilage/bone regeneration, and the controlled release of drugs. This is due to their biocompatibility and the similarity of their physical properties to natural tissues. Although the main application of hydrogels is in the biomedical field, however, they have great potentials to be used in other fields, such as the food industry, water treatment, cosmetics, and agriculture, as shown in Figure 14.1 (Bao et al., 2019; Dong and Zhang, 2018; Kenawy et al., 1992; Roufegari-Nejhad et al., 2019; Shojaeiarani et al., 2019).

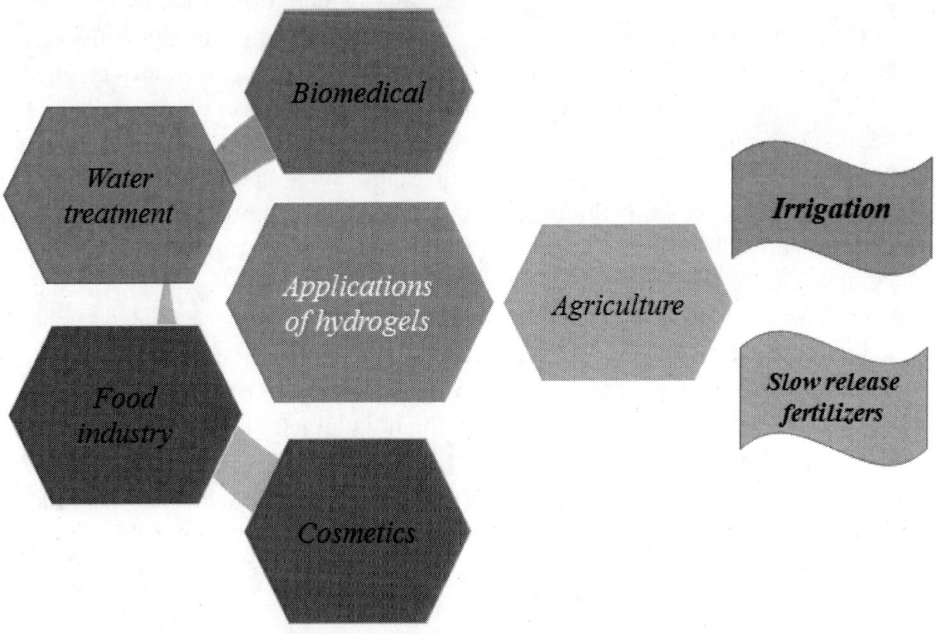

Figure 14.1. Some applications of hydrogels.

14.2.2. Nanocomposite Hydrogels

The combination of nanotechnology with other fields results in advanced sciences and technologies. Multiple procedures to incorporate nanoscale have been met with conventional methods for manufacturing improved materials, such as nanocomposites. Nanocomposites are a novel class of materials consisting of at least a matrix phase (polymer, metal, or ceramic) and a reinforcing phase, where the reinforcing agent, such as nanoparticles, has at least one dimension in the nanoscale range (1-100 nm) (Bao et al., 2019; Sirousazar et al., 2007, 2012). Nanocomposites demonstrate improved mechanical properties, electrical conductivity, thermal stability, gas barrier properties, and fire retardancy compared with conventional composites (Chen et al., 2018). Inorganic nanoparticles, such as silver, carbon nanotubes, and various types of clays, e.g., montmorillonite (MMT), laponite, and kaolinite are widely employed in the production of nanocomposites (Zhang et al., 2014).

As mentioned in the previous section, hydrogels have diverse applications in several fields because of their unique physical, chemical, and biological properties, high biocompatibility, and versatility in fabrication. Despite these features, hydrogels possess some shortcomings, e.g., poor mechanical strength and low strain, which restrict their application in some cases (Jahani-Javanmardi et al., 2016). For improvement of the mentioned limitations, trials have been made by researchers for redesigning new hydrogels, named nanocomposite hydrogels, with improved chemical, physical, and mechanical properties. A nanocomposite hydrogel consists of immiscible stiff or soft components and complex nanoscale structures (Chen et al., 2018).

In general, nanocomposite hydrogels are defined as reinforced hydrophilic three-dimensional networks consisting of at least a crosslinked polymeric matrix and a nanoscale reinforcing agent. They may also be defined as crosslinked polymer networks swollen with water or aqueous solutions in the presence of nanoparticles or nanostructures. Nanocomposite hydrogels mainly utilize nanoscopic inorganic materials which form crosslinking junctions. In these hybrid hydrogels, the ends of the polymer chains may adsorb strongly on the surfaces of the nanoparticles and enough of the chains attach to different particles to provide the bridges that constitute a three-dimensional network structure. They can be synthesized through the polymerization/crosslinking of hydrophilic monomers/polymers in the presence of nanoparticles.

Nanoscale integration of both organic and inorganic components offers controllable mechanical strength and swelling (Sirousazar and Kokabi, 2012). Compared with conventional hydrogels, the nanoparticles play a significant role in enhancing the structural stability of nanocomposite hydrogels due to the multiple interactions between nanoparticles and polymers, such as hydrogen bonding, van der Waals interaction, and electrostatic force. The nanoparticles in the three-dimensional network of nanocomposite hydrogels not only serve as crosslinkers to reinforce the hydrogel structure but also endow them with their characteristic functionalities. In recent years, nanocomposites hydrogels have been created novel applications in different practical areas, such as sensor-actuator systems, oil recovery, biomedical scaffolds, drug delivery devices, wound dressing systems, food industry, and agricultural irrigation systems (Sirousazar and Kokabi, 2012).

14.3. Agricultural Irrigation Methods

Traditional agricultural irrigation methods are surface irrigation (border, furrow, basin, and broad irrigation), general irrigation, and micro-irrigation. Traditional agricultural irrigation methods usually consume a lot of water resources, and they cause many serious problems for humans in terms of water shortage. To solve this problem, researchers have made extensive efforts to provide novel irrigation methods in recent years. These methods are mainly based on the use of SAPs and subsurface irrigation. Subsurface irrigation is often based on the pervaporation irrigation method and the use of permeable hydrogel and nanocomposite hydrogel pipes (Feng et al., 2017; Quiñones-Bolaños et al., 2005b; Rezazadeh et al., 2020; Sirousazar et al., 2019; Zaharia et al., 2018). In the following section, we will focus on subsurface irrigation techniques, including the use of SAPs and subsurface irrigation by hydrogel and nanocomposite hydrogel pipes.

14.4. Irrigation Using SAPs

The lack of water is a severe problem in regions globally, but the worst one is that they compromise the development of agriculture. A tremendous tool to solve these problems could be achieved using materials with high water absorption and retention capacities based on SAPs. The use of SAPs for agricultural applications has shown promising results, such as helping to reduce irrigation water consumption and the death level of plants and improving water retention in the soil (Pucci et al., 2008). SAPs are hydrophilic polymers with the ability to absorb and retain water from tens to thousands of times their dry weight even under pressure and heat. SAPs are mainly very hydrophilic polymers or hydrogels with intense swelling capability. Agricultural SAPs are referred to as water retention materials because they swell too many times compared to their original size when they contact with water. Over the last few years, SAPs have been utilized in agricultural irrigation applications to reduce water wastage. In the following sections, some basic concepts regarding SAPs, including their preparation methods and properties are introduced followed by their recent applications in agricultural irrigation systems.

14.4.1. Preparation Methods of SAPs

Similar to conventional polymers, SAPs can be prepared using different methods, such as bulk, solution, inverse suspension, and radiation polymerization.

14.4.1.1. Bulk Polymerization

In this method, a small quantity of crosslinker is added to the formulation (if needed, e.g., for preparation of superabsorbent hydrogels), and polymerization is chemically initiated. The initiator to be chosen depends on the type of used monomer and solvents. The polymerized SAP can be produced in various forms, such as film, membrane, rod, and particle. It is a straightforward method, which involves only monomer and monomer-soluble initiators. The rate and degree of polymerization are high for the high monomer concentration. With the gradual conversion, the reaction's viscosity increases, producing heat during polymerization that can be kept away by controlling the response at low conversions (Ahmed, 2015).

14.4.1.2. Solution Polymerization

The ionic or neutral monomers are combined with the multifunctional crosslinking agent (if needed) in the solution polymerization method. A redox initiator system or ultraviolet irradiation initiates the polymerization (Kozlovskaya et al., 2006). The advantage of this method over bulk polymerization is that the solvents help control the viscosity and promote proper heat (Peppas and Khare, 1993).

14.4.1.3. Inverse Suspension Polymerization

For the production of spherical SAP beads with particle sizes ranging from 1 μm to 1 mm, the inverse suspension polymerization technique is used. The main feature of this method is that one gets direct powder or beads, reducing the effort of grinding. It is a water-in-oil process different from the most common oil-in-water process. This technique involves two phases, one is the aqueous phase, and another is the organic phase. The organic phase consists of a solvent and a stabilizer, while the aqueous phase consists of the monomer, crosslinker, and initiator. The shape and size of the SAP particle depend on the viscosity of the monomer phase (Bajpai et al., 2007).

14.4.1.4. Radiation Polymerization

High energy radiation (e.g., gamma ray) and electron beam can initiate the polymerization process to produce SAPs of unsaturated compounds (Savich and Graham, 2007). The radiolysis of water molecules leads to hydroxyl radicals, which attack the polymer chains to form macro-radicals. These polymer radicals recombine to form covalent bonds leading to a crosslinked structure. The advantage of this approach over the chemical initiation technique is that the product obtained is comparatively pure as no chemical initiator is involved.

14.4.2. Properties of SAPs

14.4.2.1. Swelling Behavior

The most important property of SAPs and hydrogels for their use as soil moisture absorbers is swelling. The swelling of SAPs depends on their microstructural characteristics and external conditions. The hydrophilic feature of SAPs can be characterized by a thermodynamic interaction parameter of the polymer with water, density of mesh points, proportion of ionic groups, and degree of dissociation. The value of the equilibrium degree of swelling of SAPs is defined by the zero swelling pressure, which in turn consists of the elasticity of mesh and availability of the related ions (Chen et al., 2004). The swelling capacity of SAPs after contact with water or an aqueous solution is the result of a diffusion process, which involves the solvent molecules in the three-dimensional network due to the high hydrophilicity of the material. This feature causes the expansion of the polymeric chains, which increases the average distance between crosslinking points.

For most of the SAP's technical applications, however, polymer swelling is extremely useful for agriculture applications as a device for controlled or sustained release of soil conditioners (Guilherme et al., 2015). Water retention and release over time are the fundamental properties that swollen SAPs must possess for application in agriculture for sustained irrigation purposes (Sharma et al., 2014). It is thus of high importance to study the behavior of the SAP through deswelling, how the SAP performs after several swelling-deswelling cycles and its effect directly on the soil.

14.4.2.2. Water Holding Capacity

One of the leading agricultural applications of SAPs is to enhance the water holding capacity (WHC) of soil. WHC is a measure of the total amount of water that can be absorbed per gram of a SAP, which directly depends on the swelling ability of SAP. This property is based on the direct interaction of SAP with water. Because SAP swelling behavior is influenced by the conditions in which it is in contact with water (pH, temperature, and ionic strength), it is essential to test its effects directly on the targeted substrate. Significant progress has been made in improving the properties of SAPs for their uses in different environments (Bai et al., 2015).

14.4.2.3. Slow Release of Fertilizers

Studies have shown that most fertilizers applied in the field are lost to the environment, which generates environmental pollution. These nutrients are lost because of leaching since plants only need a small amount during their early growth. Thus, employing fertilizers must be taken seriously into consideration for the agricultural sector's economic and social sustainability. The development of materials for the controlled or slow release of fertilizers has been an issue of interest for decades. It is imperative to develop devices, called smart fertilizers, for the slow or controlled release of nutrients in agriculture systems (Calabi-Floody et al., 2018). Fertilizer loading with smart SAP can overcome the current limitations of conventional fertilizers. These systems can deliver nutrients more controlled or slowly for a longer duration to a specified target and at a certain rate (Azeem et al., 2014; Campos et al., 2015; Chagas et al., 2018; Essawy et al., 2016; Olad et al., 2018). The SAP can act as a controlled release system by favoring the uptake of fertilizers and some nutrient elements, holding them tightly, and delaying their dissolution. Consequently, the plant can access some fertilizers, resulting in improved growth and performance rates (Liang et al., 2007).

14.4.3. Irrigation Application

SAPs and superabsorbent hydrogels can potentially affect the permeability, texture, evaporation rate, and water infiltration rate of the soil. Notably, these materials reduce irrigation frequency and increase soil aeration and microbial activity (Bakass et al., 2012; Rehman et al., 2011). SAPs help plants withstand prolonged moisture stress, improve the physical properties of soils and soilless media, enhance seed germination and seedling emergence rate, improve root growth and density, and reduce the nursery establishment period. Several studies have been carried out on the application of SAPs in agriculture, some of them are briefly reviewed here.

Woodhouse and Johnson, (1991) developed SAPs based on starch and polyacrylamide and utilized them for the growth of lettuce and barley. The crops were grown under limited irrigation on a coarse sand substrate. Their experiments showed that the interval between field capacity and the onset of wilting was increased by up to three-fold in the presence of SAPs. It was also found that water use efficiency and dry matter production positively responded to the presence of SAPs. Raafat et al., (2012) prepared a series of SAP based on carboxymethylcellulose and polyvinyl pyrrolidone crosslinked with gamma irradiation for agricultural irrigation applications. It was found that the prepared SAPs had good swelling capability which was extremely sensitive to the ionic strength and cationic kind. The prepared SAPs due to their considerable slow urea release, good water retention capacity, and being economical and environment-friendly were suggested as useful materials for application in agricultural irrigation.

Cannazza et and coworkers (2014) prepared a cellulose-based SAP as an environment-friendly material for the optimization of water consumption in agriculture. The SAP was synthesized and tested for its swelling capability in different aqueous media. The effectiveness of the SAP in agricultural applications was then evaluated by analyzing its performance after several absorption/desorption cycles, over a period of ca. 80 days. The results showed that SAP-amended soil can store a considerable amount of water and release it gradually to the plant roots when needed. Hasija et al., (2018) fabricated a SAP based on acrylic acid grafted over agar and gum arabic for sustainable moisture retention capacity in agricultural land. The biodegradable SAP was synthesized using microwave irradiation under the best conditions to get the maximum percentage of swelling. The synthesized SAP showed great potential as a soil conditioning material to be used for agriculture purposes.

Recently, Kazeminejadfard and Hojjati, (2019) synthesized a novel biocompatible SAP by graft copolymerization of hydroxypropyl distarch phosphate based on tapioca starch with acrylic acid, loaded with clinoptilolite zeolites for reducing water consumption in agricultural applications. Investigation of the effect of SAP composition on the water absorbency exhibited that the optimum SAP had the maximum water absorbencies of 1075 and 67 g/g in distilled water and 0.9 wt.% NaCl solution, respectively. It was also shown that within the first week after irrigation, the soil texture with 100% sand had the maximum water retention, 69.6 g more than its control sample. More recently, Rashad et al., (2021) produced a natural-based environment-friendly SAP based on rice. The prepared SAP was examined for its ability to improve soil water retention, control the release of urea, and biodegradability in soil. They evaluated the influence of rice-based SAP on sweet pepper productivity under severe deficit irrigation conditions. The results showed that the soil water retention and urea release rates improved plant growth in response to the addition of SAP.

14.5. Subsurface Irrigation Using Hydrogel and Nanocomposite Hydrogel Pipes

Another method to reduce water consumption in the agricultural sector is using subsurface irrigation technology, e.g., the pervaporation irrigation systems. Subsurface irrigation is mainly performed using semipermeable hydrophilic pervaporation membranes. These membranes are formed into a tube that water transport occurs through pervaporation (Sule et al., 2013). This method of irrigation provides *in situ* water treatment while supplying it to the plant root zone. The evaporative tube is filled with water and buried in the soil during irrigation. The dryness of the surrounding soil and a chemical potential gradient across the membrane draws water from inside the tube into the soil (Muthu and Brant, 2015; Quiñones-Bolaños and Zhou, 2006; Todman et al., 2018). This feature means the system can interact with the plant by providing an increased irrigation flux when the soil condition is harder, and the plant has more need for water. Conversely, a reduced water flow is donated to the plant when the surrounding soil is wet (Quiñones-Bolaños et al., 2005a).

Recently, a new kind of subsurface irrigation system was introduced based on hydrogel and nanocomposite hydrogel tubes. These novel pipes are soft, flexible, and swellable which create an excellent opportunity for molecular irrigation (Rezazadeh et al., 2020; Sirousazar et al., 2019). For the first time, hydrogel irrigating pipes were introduced by Sirousazar and coworkers in 2019, based on hydrophilic, permeable, and flexible three-dimensional crosslinked polyvinyl alcohol (PVA) hydrogel. These hydrogel pipes can connect to a water reservoir and be buried around the root of the plant and continuously permeate water molecules directly to the root. The PVA hydrogel pipes were prepared via a cyclic freezing-thawing technique. Hydrogel pipes were prepared based on different PVA loading levels (i.e., 9, 12, and 15 wt.%) and pipe thicknesses and their performances in irrigation applications were investigated. The prepared hydrogel pipes were named Px, where x demonstrates the wt.% of PVA in the prepared hydrogel tube. All the prepared pipes were in the form of one-side closed pipes having a length of 150 mm and an outer diameter of 15 mm. The inner diameter of all hydrogel pipes was chosen as 7 and 10 mm, corresponding to the pipe thicknesses of 4 and 2.5 mm, respectively. Some typically prepared PVA hydrogel pipes by Sirousazar et al., (2019) were exhibited in Figure 14.2a.

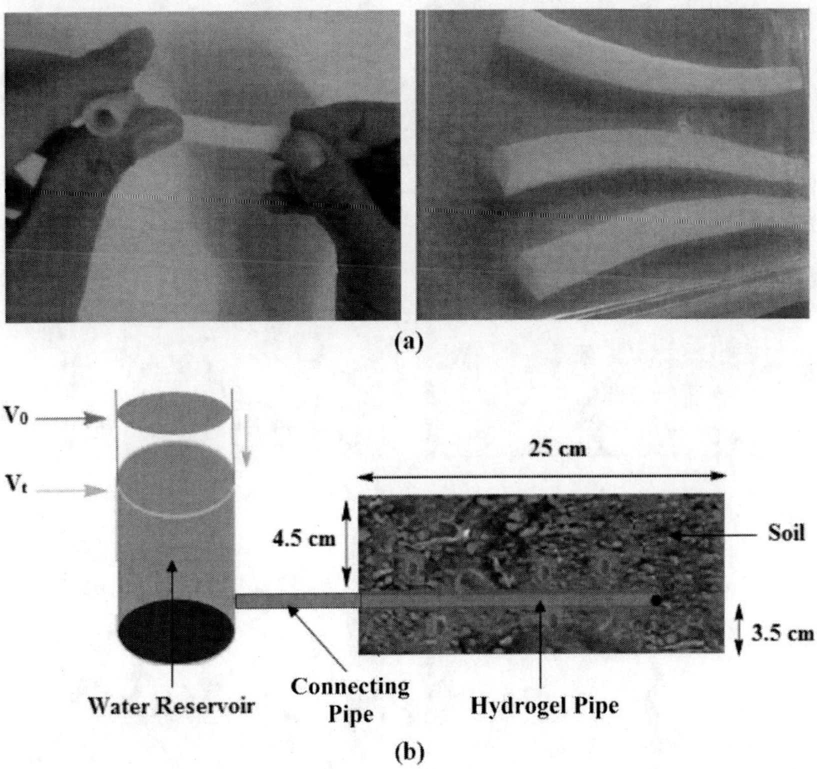

Figure 14.2. (a) Typical PVA hydrogel pipes and (b) experimental setup designed for subsurface irrigation (Sirousazar et al., 2019).

Besides the irrigation experiments, some structural, physical, and mechanical properties of the PVA hydrogel pipes were also studied. Investigating the fracture surfaces of samples showed that all PVA hydrogel pipes had porous structures created due to the crosslinking of PVA chains. The scanning electron microscopy (SEM) images presented in Figure 14.3 show that the P9 sample has a more porous structure with larger pore sizes compared with other samples and the pore size is decreased by increasing the PVA weight percentage and the P15 sample has the most compacted structure among the samples. The effect of the hydrostatic pressure on the equilibrium swelling ratio (q_∞) of the hydrogel pipes was shown in Figure 14.4. As seen, the q_∞ of all hydrogels was decreased by increasing the hydrostatic pressure. The results showed, for example nearly 12% decrease in the q_∞ of the P12 sample by increasing the hydrostatic pressure from 10 to 40 cmH_2O. The decrease in the swelling of hydrogels by increasing the hydrostatic pressure is due to the restricted mobility and relaxation of polymer chains under high external pressure. The results imply that the swelling ability of the PVA hydrogel pipes would be decreased during the practical irrigation if the pipe was buried deep in the ground.

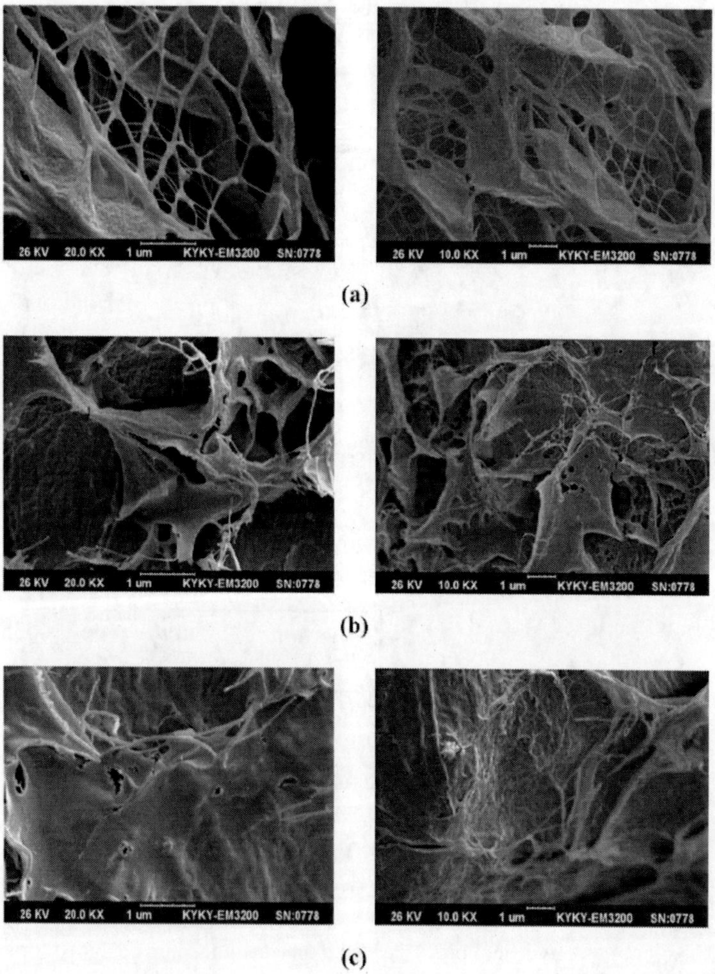

Figure 14.3. SEM images of (a) P9, (b) P12, and (c) P15 hydrogel pipes (Sirousazar et al., 2019).

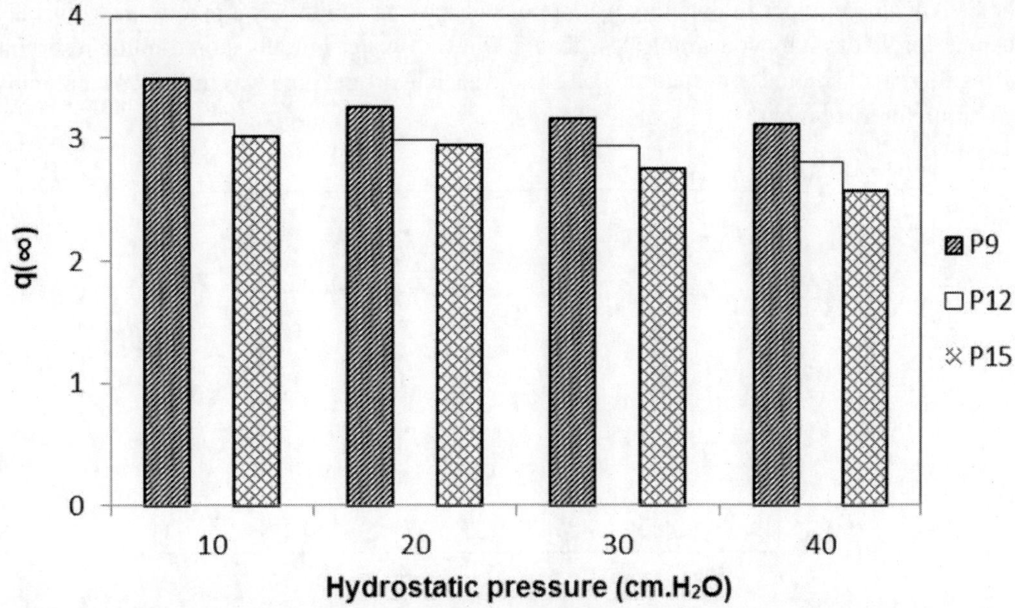

Figure 14.4. The effect of hydrostatic pressure on the equilibrium swelling ratio of PVA hydrogel pipes (Sirousazar et al., 2019).

The performances of the PVA hydrogel pipes in practical irrigation applications were investigated using a designed setup on a laboratory scale. For this purpose, the water flux from each hydrogel pipe to the surrounding soil was experimentally measured. As shown in Figure 14.2b, the utilized setup had a water reservoir, connecting pipe (polyethylene), PVA hydrogel pipe, soil, and a box as the soil container. All water reservoirs were filled initially with 1000 mL fresh water with a height of 20 cm. Since the hydrogel pipes are elastic and soft, therefore a small piece of polyethylene pipe (with a diameter nearly identical to the hydrogel pipe) was used as an interconnecting device between the water reservoir and hydrogel pipe. Then the hydrogel pipes were buried in the dried soil, as schematically shown in Figure 14.2b. As the irrigation test is initiated, the water molecules inside the hydrogel pipe pass through the wall of the hydrogel pipe and diffuse to the soil around the pipe due to the differences between the water chemical potential inside the pipe and in the soil. The irrigation occurs on a molecular scale in this manner and the rate of water diffusion may be decreased as the humidity of the soil around the pipe is increased. Each test was performed for 90 days and the decrease in the volume of water inside the reservoir was measured in some predetermined time intervals. The water flux value at any time (J) was calculated using the Equation (14-1):

$$J = V/At \qquad (14\text{-}1)$$

where V (mL) is the volume of decreased water in the reservoir or the same permeated water from the hydrogel pipe to the soil, A (cm^2) is the effective hydrogel pipe area and t (day) is the permeation time.

Figure 14.5 shows the water permeation flux versus irrigation time for P9, P12, and P15 hydrogel pipes having thicknesses of 2.5 and 4 mm. The results exhibited the unique ability of the PVA hydrogel pipes in donating water molecules to the soil in a continuous and controlled manner for 90 days, by consuming less than 700 mL of water initially stored in the reservoir. Using this small amount of water, the soil around each hydrogel pipe was in a permanent moist condition for three months.

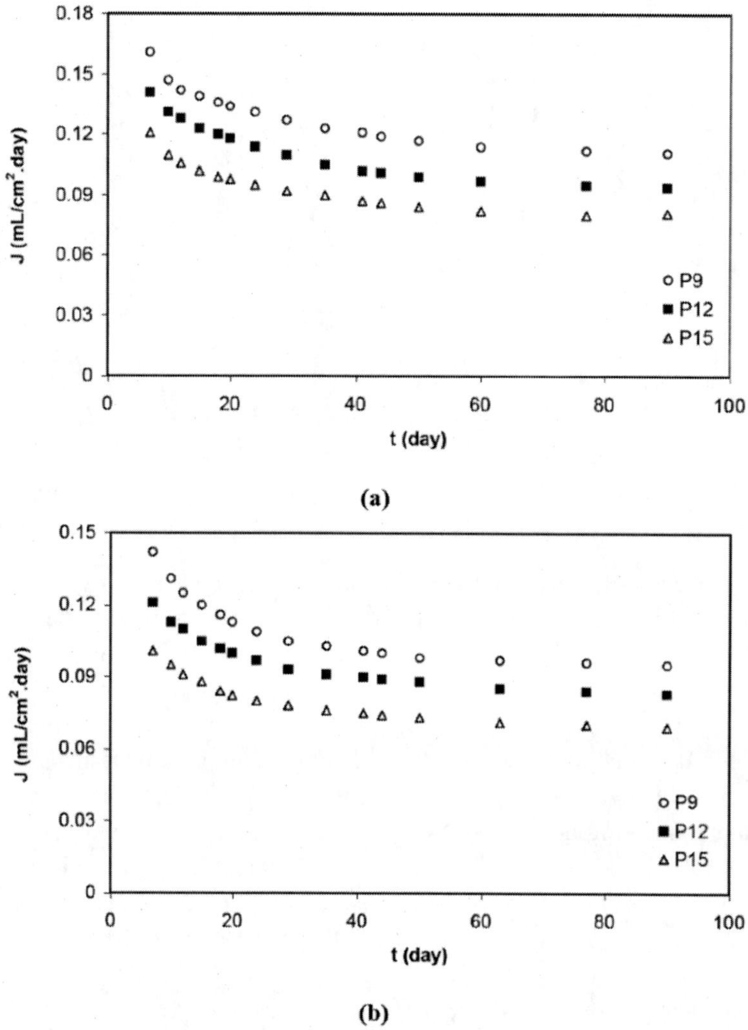

Figure 14.5. Water flux for PVA hydrogel pipes having pipe thicknesses of (a) 2.5 and (b) 4 mm (Sirousazar et al., 2019).

As shown, the water flux rapidly decreased during the initial irrigation days and then the flux approached an almost steady state. This is due to increased soil moisture due to the transfer of water molecules from the hydrogel pipe to the surrounding soil. The moisture of the soil around the pipe is increased over time and the difference between the chemical potential of the water molecules inside and outside the hydrogel pipe, as the driving force of the transfer of water molecules from the pipe to the soil, decreases. The results showed that the water flux would be decreased by increasing the weight percentage of PVA in hydrogel pipes. The transportation and diffusion of water molecules from inside the hydrogel pipe to its environment would be restricted due to the smaller pore sizes, less free volume, and more tortuosity in hydrogel pipes containing a higher content of PVA. The results also showed that the water flux was decreased by increasing the thickness of the hydrogel pipe. The weight percentage of PVA used for the preparation of hydrogel pipes and the thickness of pipes are recognized as two key parameters to design hydrogel pipes with controllable irrigation and one can design a suitable hydrogel pipe with the desired irrigation rate by adjusting these parameters. Generally, it was concluded that the prepared PVA hydrogel pipes are suitable candidates to be used in practical applications for molecular irrigation ends. Molecular irrigation by hydrogel pipes not only drastically reduces water consumption but also creates an ideal condition for better growing the plant due to donating the required and controlled rate of water directly to the root of the plant.

Recently, Rezazadeh et al., (2020) fabricated novel nanocomposite hydrogel pipes based on PVA and MMT clay for subsurface irrigation in the agricultural sector with extraordinarily reduced water consumption. Nanocomposite hydrogel pipes were prepared using PVA, as the matrix phase, and MMT, as the reinforcing phase, via a cyclic freezing-thawing process. All prepared PVA/MMT solutions (before freezing-thawing) had 12 wt.% of PVA in double distilled water (DDW) (based on the total mass of the solution) and 0, 5, and 10 wt.% of MMT (based on the mass of PVA and MMT). The prepared nanocomposite hydrogel pipes were named as NHx, where x shows the weight percentage of MMT in the water-free (completely dried) nanocomposite hydrogel pipe. The prepared pipes were in the form of one-side closed pipes having a length of 200 mm and an outer diameter of 20 mm. Three pipes were prepared using each nanocomposite hydrogel having inner diameters of 7, 10, and 13 mm, corresponding to the pipe thicknesses of 6.5, 5, and 3.5 mm, respectively. The schematic preparation procedure of the PVA/MMT nanocomposite hydrogel pipes was illustrated in Figure 14.6.

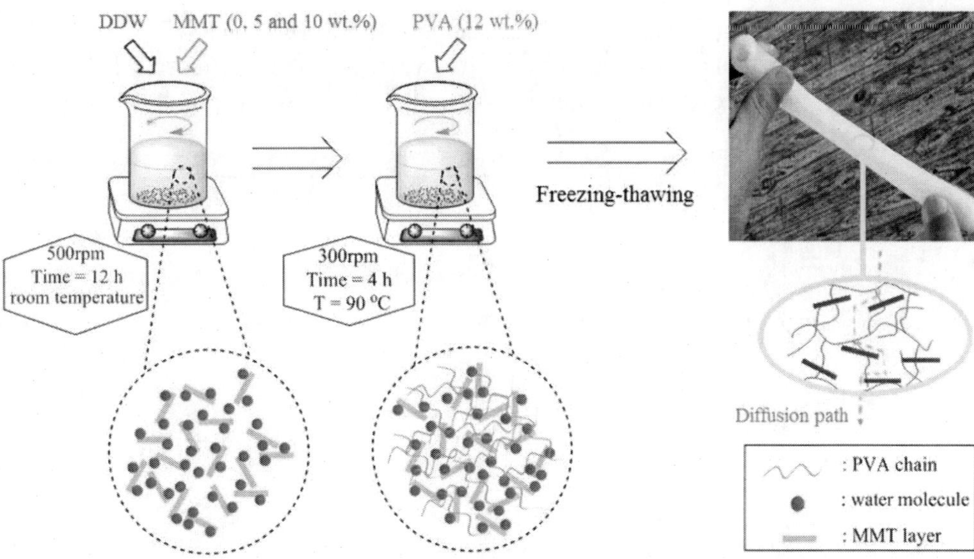

Figure 14.6. A scheme representing the preparation process of the PVA/MMT nanocomposite hydrogel pipes (Rezazadeh et al., 2020).

According to the conducted X-ray diffractometry (XRD), it was found that the nanocomposite hydrogels have an exfoliated morphology which means that the PVA chains have penetrated inside the MMT layers and disrupted their parallel arrangements. The XRD patterns of pristine MMT along with the NH5 and NH10 nanocomposite hydrogels were illustrated in Figure 14.7. As seen, no distinct diffraction peak appeared in the XRD patterns of NH5 and NH10 samples and the diffraction peak related to the MMT (at $2\theta = 7.3°$) disappeared in these samples, which confirms the creation of exfoliated morphology for PVA/MMT nanocomposite hydrogels.

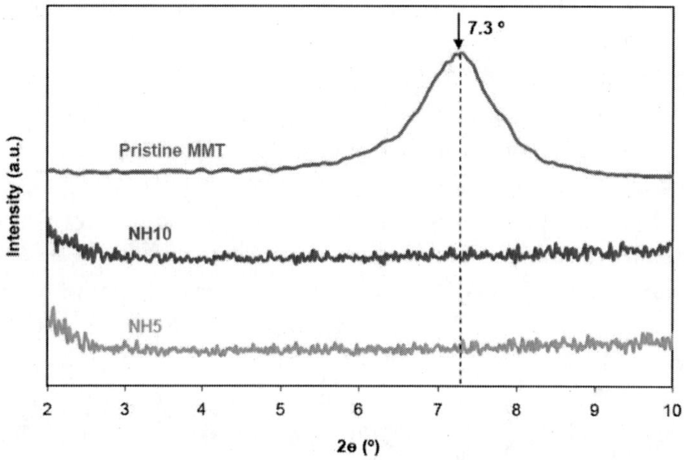

Figure 14.7. XRD patterns of pristine MMT and NH5 and NH10 nanocomposite hydrogels (Rezazadeh et al., 2020).

Figure 14.8 shows the compressive stress-strain curves for NH0, NH5, and NH10 nanocomposite hydrogel pipes. The nanocomposite hydrogel having a higher MMT loading level had higher stress or the same compressive strength at all examined strains.

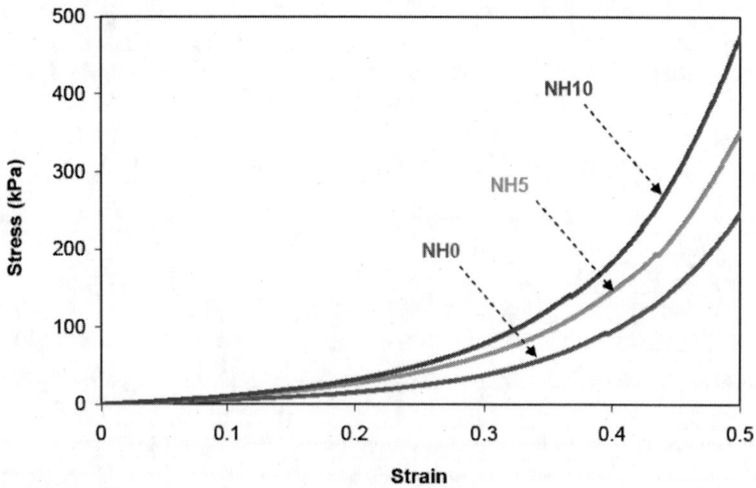

Figure 14.8. Compressive stress-strain curves of PVA/MMT nanocomposite hydrogel pipes (Rezazadeh et al., 2020).

The improved mechanical properties of the nanocomposite hydrogel pipes are attributed to the presence of MMT, as a crosslinking and reinforcing agent, in their networks. The improved compressive strengths of nanocomposite hydrogel pipes could be recognized as an outstanding advantage and may create an ideal condition for pipes in practical irrigation applications. Besides the prolonged longevity, the improved mechanical properties cause the buried nanocomposite hydrogel pipes in the ground to withstand the external pressure applied from the soil around the pipe, as well as the hydrostatic pressure exerted from the level of accumulated water in the water reservoir.

The irrigation study was performed by determining the performances of the prepared nanocomposite hydrogel pipes in practical irrigation applications, by taking into account the effects of the MMT loading level in nanocomposite hydrogels and the thickness of pipes, using a designed setup on a laboratory scale. The irrigation tests were continued for 100 days using water reservoirs filled initially with 2000 mL fresh water, and the decreases in the volume of water inside the reservoirs were measured in some predetermined time intervals.

Figure 14.9 demonstrates the cumulative water consumption volumes against time for NH0, NH5, and NH10 nanocomposite hydrogel pipes having pipe thicknesses of 3.5, 5, and 6.5. All examined pipes demonstrated an extraordinary irrigation performance by keeping a continuous moist environment in the soil surrounding the pipe using a limited amount of water for a long time. The results showed that water consumption could be decreased either by increasing the loading level of MMT used for the preparation of nanocomposite hydrogel pipes or the thickness of pipes. The results indicated a significant decrease in water consumption in nanocomposite hydrogel pipes compared to the pure hydrogel pipe and it was concluded that the PVA/MMT nanocomposite hydrogel pipes are suitable candidates to be used in practical applications for molecular irrigation ends.

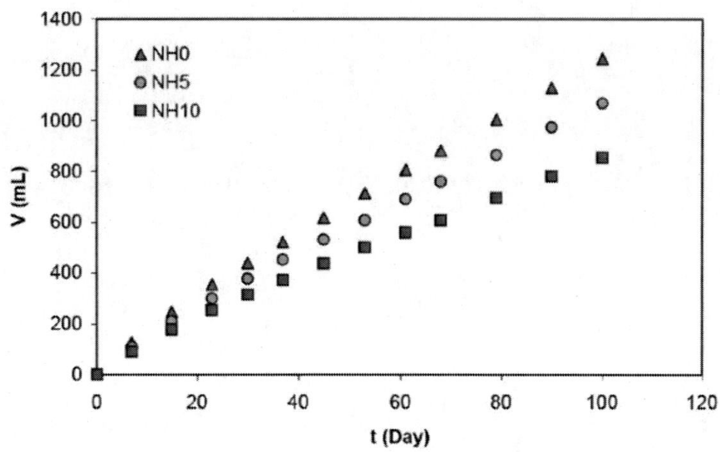

(a)

Figure 14.9. (Continued).

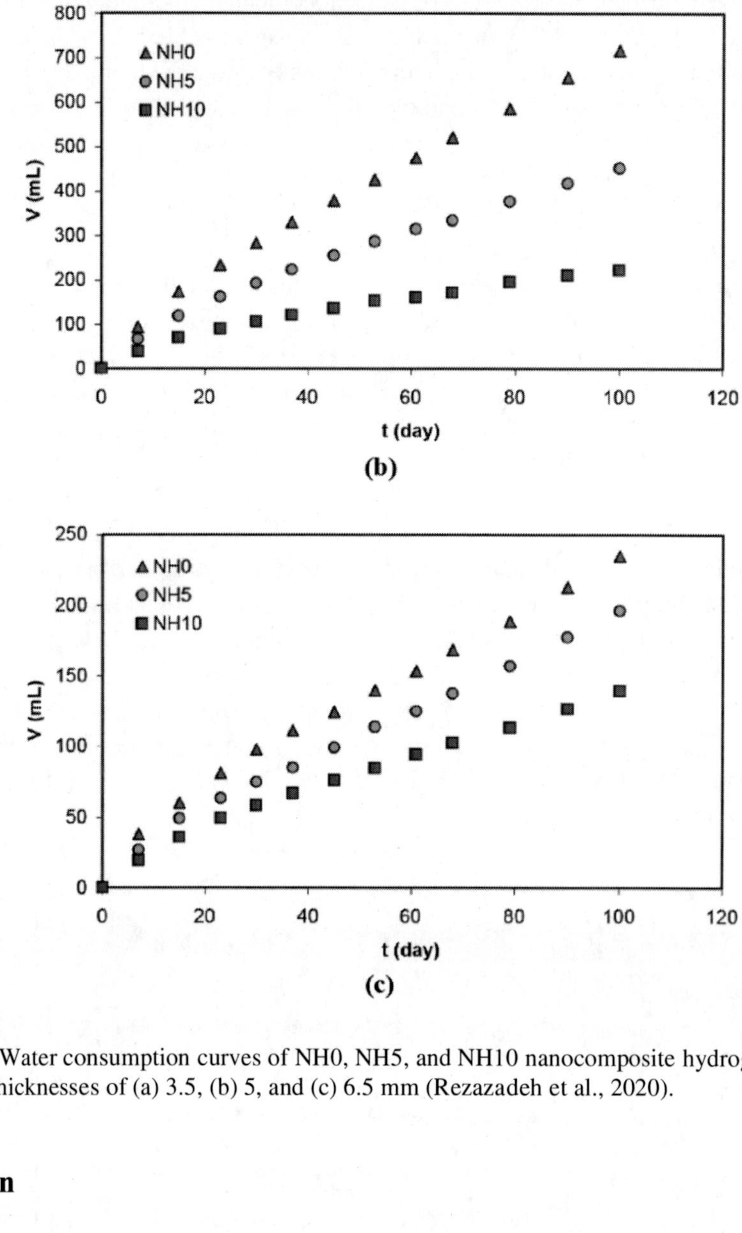

Figure 14.9. Water consumption curves of NH0, NH5, and NH10 nanocomposite hydrogel pipes having pipe thicknesses of (a) 3.5, (b) 5, and (c) 6.5 mm (Rezazadeh et al., 2020).

Conclusion

In this chapter, the recent advances in SAPs and hydrogels and nanocomposite hydrogels for subsurface irrigation were reviewed, along with a summary of the basic topics regarding the fundamental concepts of hydrogels, nanocomposite hydrogels, SAPs, and irrigation systems. An overview of the previous research showed that hydrogel and nanocomposite hydrogel pipes are suitable candidates to be used in practical applications for molecular irrigation in the agricultural sector.

Molecular irrigation by hydrogel and nanocomposite hydrogel pipes not only drastically reduces water consumption but also creates an ideal condition for better growing the plant due to donating the required and controlled rate of water directly to the root of the plant. However, the prepared hydrogel and nanocomposite hydrogel tubes for irrigation are based on a synthetic polymer (PVA). Therefore, the preparation of hydrogel and nanocomposite hydrogel pipes based on natural and environment-friendly materials with improved mechanical properties seems to be the upcoming challenge for researchers in the field of production of hydrogel and nanocomposite hydrogel tubes for subsurface irrigation. An economic study regarding the production of hydrogel and nanocomposite hydrogel pipes on an industrial scale and also feasibility study of the controlled release of fertilizers from these pipes are among the ideas for future research.

References

Ahmed, E. M. (2015). Hydrogel: preparation, characterization, and applications: a review. *Journal of Advanced Research*, 6, 105-121.

Azeem, B., Kushaari, K., Man, Z. B., Basit, A., and Thanh, T. H. (2014). Review materials & methods to produce controlled-release coated urea fertilizer. *Journal of Controlled Release*, 181, 11-21.

Bai, C., Zhang, H. S., Wang, L., and Wang, H. (2015). Starch-based hydrogel loading with carbendazim for controlled-release and water absorption. *Carbohydrate Polymers*, 125, 376-383.

Bajpai, S. K., Bajpai, M., and Sharma, L. (2007). Inverse suspension polymerization of poly (methacrylic Acid-co-partially neutralized acrylic acid) superabsorbent hydrogels: synthesis and water uptake behavior. *Designed Monomers and Polymers,* 10, 181-192.

Bakass, M., Mokhlisse, A., and Lallemant, M. (2002). Absorption and desorption of liquid water by a superabsorbent polymer: effect of polymer in the drying of the soil and the quality of certain plants. *Journal of Applied Polymer Science*, 6, 234-243.

Bao, Z., Xian, C., and Yuan, Q. (2019). Natural polymer-based hydrogels with enhanced mechanical performances: preparation, structure, and property. *Advanced Healthcare Materials*, 8, 258-265.

Calabi-Floody, M., Medina, J., Rumpel, C., Condron, L. M., Hernandez, M., and Dumont, M. (2018). Smart fertilizers as a strategy for sustainable agriculture. *Advances in Agronomy*, 147, 119-157.

Calcagnile, P., and Sibillano, T. (2019). Biodegradable poly (lactic acid)/cellulose-based superabsorbent hydrogel composite material as water and fertilizer reservoir in agricultural applications. *Journal of Applied Polymer Science*, 136, 475-482.

Camp, C. R. (1998). Subsurface drip irrigation: a review. *Transactions of the ASAE*, 41(5), 1353-1367.

Campos, E. V. R., de Oliveira, J. L., Fraceto, L. F., and Singh, B. (2015). Polysaccharides as safer release systems for agrochemicals. *Agronomy for Sustainable Development*, 35(1), 47-66.

Cannazza, G., Cataldo, A., and Sannino, A. (2014). Experimental assessment of the use of a novel superabsorbent polymer (SAP) for the optimization of water consumption in agricultural irrigation process. *Water*, 6(7), 2056-2069.

Chen, S. L., Zommorodi, M., Fritz, E., Wang, S., and Huttermann, A. (2004). Hydrogel modified uptake of salt ions and calcium in Populus euphratica under saline conditions. *Trees - Structure and Function*, 18, 175-183.

Chen, T., Hou, K., Ren, Q., and Chen, G. (2018). Nanoparticle-polymer synergies in nanocomposite hydrogels: from design to application. *Macromolecular Rapid Communications*, 39, 254-259.

Dong, C., and Zhang, J. (2018). Developing stretchable and grapheneoxide-based hydrogel for the removal of organic pollutants and metal ions. *Applied Catalysis B: Environmental*, 22, 146-156.

Essawy, H. A., Ghazy, M. B. M., El-Hai, F. A., and Mohamed, M. F. (2016). Superabsorbent hydrogels via graft polymerization of acrylic acid from chitosan-cellulose hybrid and their potential in controlled release of soil nutrients. *International Journal of Biological Macromolecules*, 89, 144-151.

Feng, D., Bai, B., and Suo, Y. (2017). Novel fabrication of biodegradable superabsorbent microspheres with diffusion barrier through thermo-chemical modification and their potential agriculture applications for water holding and sustained release of fertilizer. *Journal of Agricultural and Food Chemistry,* 65, 5896-5907.

Guilherme, M. R., Aouada, F. A., Fajardo, A. R., Martins, A. F., Paulino, A. T., Davi, M. F. T., Rubira, A. F., and Muniz, E. C. (2015). Superabsorbent hydrogels based on polysaccharides for application in agriculture as soil conditioner and nutrient carrier: a review. *European Polymer Journal,* 72, 365-385.

Hasija, V., Sharma, K., Kumar, V., Sharma, S., and Sharma, V. (2018). Green synthesis of agar/Gum Arabic based superabsorbent as an alternative for irrigation in agriculture. *Vacuum,* 157, 458-464.

Hennink, W. E., and van Nostrum, C. F. (2002). Novel crosslinking methods to design hydrogels. *Advanced Drug Delivery Reviews*, 64, 223-236.

Jahani-Javanmardi, A., Sirousazar, M., Shaabani, Y., and Kheiri, F. (2016). Egg white/poly (vinyl alcohol)/MMT nanocomposite hydrogels for wound dressing. *Journal of Biomaterials Science, Polymer Edition*, 27, 1262-1276.

Kazeminejadfard, F., and Hojjati, M. R. (2019). Preparation of superabsorbent composite based on acrylic acid-hydroxypropyl distarch phosphate and clinoptilolite for agricultural applications. *Journal of Applied Polymer Science*, 136(16), 47365.

Kenawy, E. R., Chemistry, A., and Akelah, A. (1992). Controlled release of agrochemical molecules chemically bound to polymers. *European Polymer Journal*, 28, 841-862.

Kozlovskaya, V., Kharlampieva, E., Mansfield, M. L., and Sukhishvili, S. A. (2006). Poly (methacrylic Acid) hydrogel films and capsules: response to pH and ionic strength, and encapsulation of macromolecules. *Chemistry of Materials,* 18, 328-336.

Lamm, F. R. (2016). Cotton, tomato, corn, and onion production with subsurface drip irrigation: a review. *American Society of Agricultural and Biological Engineers*, 59(1), 263-278.

Liang, M., Zhan, R., Liu, F., and Niu, Z. A. (2007). Preparation of superabsorbent slow release nitrogen fertilizer by inverse suspension polymerization. *Polymer International*, 56, 729-737.

Madramootoo, C. A., and Fyles, H. (2010). Irrigation in the context of today's global food crisis. *Irrigation and Drainage*, 59, 40-52.

Muthu, S., and Brant, J. A. (2015). Interrelationships between flux, membrane properties, and soil water transport in a subsurface pervaporation irrigation system. *Environmental Engineering Science*, 32, 539-550.

Nascimento, D. M., Nunes, Y. L., and Figueirêdo, M. C. (2018). Nanocellulose nanocomposite hydrogels: technological and environmental issues. *Green Chemistry*, 11, 2428-2448.

Olad, A., Zebhi, H., Salari, D., Mirmohseni, A., and Tabar, A. R. (2018). Water retention and slow release studies of a salep-based hydrogel nanocomposite reinforced with montmorillonite clay. *New Journal of Chemistry*, 4, 2758-2766.

Peppas, N. A., and Khare, A. R. (1993). Preparation, structure and diffusional behavior of hydrogels in controlled release. *Advanced Drug Delivery Reviews*, 11(1-2), 1-35.

Quiñones-Bolaños, E., and Zhou, H. (2006). Modeling water movement and flux from membrane pervaporation systems for wastewater microirrigation. *Journal of Environmental Engineering*, 132, 1011-1018.

Quiñones-Bolaños, E., Zhou, H., and Parkin, G. (2005a). Membrane pervaporation for wastewater reuse in microirrigation. *Journal of Environmental Engineering*, 131(12), 1633-1643.

Quiñones-Bolaños, E., Zhou, H., Soundararajan, R., and Otten, L. (2005b). Water and solute transport in pervaporation hydrophilic membranes to reclaim contaminated water for micro-irrigation. *Journal of Membrane Science*, 252, 19-28.

Raafat, A. I., Eid, M., and El-Arnaouty, M. B. (2012). Radiation synthesis of superabsorbent CMC based hydrogels for agriculture applications. *Nuclear Instruments and Methods in Physics Research Section B: Beam Interactions with Materials and Atoms*, 283, 71-76.

Rashad, M., Kenawy, E. R., Hosny, A., Hafez, M., and Elbana, M. (2021). An environmental friendly superabsorbent composite based on rice husk as soil amendment to improve plant growth and water productivity under deficit irrigation conditions. *Journal of Plant Nutrition*, 44(7), 1010-1022.

Rehman, A., Ahmad, R., and Safdar, M. (2011). Effect of hydrogel on the performance of aerobic rice sown under different techniques. *Plant, Soil and Environment*, 57, 321-325.

Rezazadeh, B., Sirousazar, M., Abbasi-Chianeh, V., and Kheiri, F. (2020). Polymer-clay nanocomposite hydrogels for molecular irrigation application. *Journal of Applied Polymer Science*, 137(18), 48631.

Rosegrant, M. W., and Cai, X. (2002). Global water demand and supply projections: part 2. results and prospects to 2025. *Water International*, 27(2), 170-182.

Roufegari-Nejhad, E., Sirousazar, M., Abbasi-Chiyaneh, V., and Kheiri, F. (2019). Removal of methylene blue from aqueous solutions using poly (vinyl alcohol)/montmorillonite nanocomposite hydrogels: Taguchi optimization. *Journal of Polymers and the Environment*, 27, 2239-2249.

Savich, M., and Graham, J. (2007). Biodegradable bag containing superabsorbent polymers. *U.S. Patent Application*, 11/333, 631.

Shantha, K. L., and Harding, D. R. K. (2002). Synthesis and evaluation of sucrose-containing polymeric hydrogels for oral drug delivery. *Journal of Applied Polymer Science*, 84, 2597- 2604.

Sharma, R., Bajpai, J., Bajpai, A. K., Acharya, S., Shrivastava, R. B., and Shukla, S. K. (2014). Designing slow water-releasing alginate nanoreserviors for sustained irrigation in scanty rainfall areas. *Carbohydrate Polymers*, 102, 513-520.

Shojaeiarani, J., Bajwa, D., and Shirzadifar, A. (2019). A review on cellulose nanocrystals as promising biocompounds for the synthesis of nanocomposite hydrogels. *Carbohydrate Polymers*, 216, 247-259.

Sirousazar, M., Ghanizadeh, E., Rezazadeh, B., Abbasi-Chianeh, V., and Kheiri, F. (2019). Polymeric hydrogel pipes for irrigation application. *Journal of Polymers and the Environment*, 27(12), 2842-2852.

Sirousazar, M., and Kokabi, M. (2012). Intelligent nanocomposite hydrogels, *Intelligent Nanomaterials*, 487-531.

Sirousazar, M., Kokabi, M., Hassan Z. M., and Bahramian, A. R. (2012). Nanoporous nanocomposite hydrogels composed of polyvinyl alcohol and Na-montmorillonite. *Journal of Macromolecular Science, Parte B*, 51, 1583-1595.

Sirousazar, M., Yari, M., Achachlouei, B. F., Arsalani, J., and Mansoori, Y. (2007). Polypropylene/ montmorillonite nanocomposites for food packaging. *e-Polymers*, 7(1), 305-313.

Sule, M., Jiang, J., Templeton, M., and Huth, E. (2013). Salt rejection and water flux through a tubular pervaporative polymer membrane designed for irrigation applications. *Environmental Technology*, 34, 1329-1339.

Todman, L. C., Chhang, A., Riordan, H. J., and Brooks, D. (2018). Soil osmotic potential and its effect on vapor flow from a pervaporative irrigation membrane. *Journal of Environmental Engineering*, 144(7), 9357-9372.

Woodhouse, J., and Johnson, M. S. (1991). Effect of superabsorbent polymers on survival and growth of crop seedlings. *Agricultural Water Management*, 20, 63-70.

Zaharia, A., Radu, A. L., and Iancu, S. (2018). Bacterial cellulose-poly (acrylic acid-co-N, N′-methylene-bis-acrylamide) interpenetrated networks for the controlled release of fertilizers. *RSC Advances*, 8, 17635-17644.

Zhang, M., Cheng, Z., and Zhao, T. (2014). Synthesis, characterization, and swelling behaviors of salt-sensitive maize bran-poly (acrylic acid) superabsorbent hydrogel. *Journal of Agricultural and Food Chemistry*, 62, 8867-8874.

Index

C

D

E

F

G

H

Q

R

S

U

V

W

X

Y

Z

Editor's Contact Information

Dr. Mohammad Sirousazar, PhD
Professor of Chemical Engineering,
Faculty of Chemical Engineering,
Urmia University of Technology,
Urmia, Iran
m.sirousazar@uut.ac.ir